Composite and Polymeric Materials for Dentistry: Enhancing Antimicrobial and Mechanical Properties

Composite and Polymeric Materials for Dentistry: Enhancing Antimicrobial and Mechanical Properties

Editor

Grzegorz Chladek

MDPI • Basel • Beijing • Wuhan • Barcelona • Belgrade • Manchester • Tokyo • Cluj • Tianjin

Editor
Grzegorz Chladek
Department of Engineering
Materials and Biomaterials
Silesian University
of Technology
Gliwice
Poland

Editorial Office
MDPI
St. Alban-Anlage 66
4052 Basel, Switzerland

This is a reprint of articles from the Special Issue published online in the open access journal *Materials* (ISSN 1996-1944) (available at: www.mdpi.com/journal/materials/special_issues/composite_polymeric_materials_dentistry_antimicrobial_mechanical_properties).

For citation purposes, cite each article independently as indicated on the article page online and as indicated below:

LastName, A.A.; LastName, B.B.; LastName, C.C. Article Title. *Journal Name* **Year**, *Volume Number*, Page Range.

ISBN 978-3-0365-7183-6 (Hbk)
ISBN 978-3-0365-7182-9 (PDF)

© 2023 by the authors. Articles in this book are Open Access and distributed under the Creative Commons Attribution (CC BY) license, which allows users to download, copy and build upon published articles, as long as the author and publisher are properly credited, which ensures maximum dissemination and a wider impact of our publications.

The book as a whole is distributed by MDPI under the terms and conditions of the Creative Commons license CC BY-NC-ND.

Contents

Grzegorz Chladek
Composite and Polymeric Materials for Dentistry: Enhancing Antimicrobial and Mechanical Properties
Reprinted from: *Materials* **2023**, *16*, 1432, doi:10.3390/ma16041432 1

Izabela Maria Barszczewska-Rybarek
A Guide through the Dental Dimethacrylate Polymer Network Structural Characterization and Interpretation of Physico-Mechanical Properties
Reprinted from: *Materials* **2019**, *12*, 4057, doi:10.3390/ma12244057 7

Katarzyna Kaczmarek, Andrzej Leniart, Barbara Lapinska, Slawomira Skrzypek and Monika Lukomska-Szymanska
Selected Spectroscopic Techniques for Surface Analysis of Dental Materials: A Narrative Review
Reprinted from: *Materials* **2021**, *14*, 2624, doi:10.3390/ma14102624 37

Guillermo Grazioli, Louis Hardan, Rim Bourgi, Leina Nakanishi, Elie Amm and Maciej Zarow et al.
Residual Adhesive Removal Methods for Rebonding of Debonded Orthodontic Metal Brackets: Systematic Review and Meta-Analysis
Reprinted from: *Materials* **2021**, *14*, 6120, doi:10.3390/ma14206120 63

John W. Nicholson, Sharanbir K. Sidhu and Beata Czarnecka
Enhancing the Mechanical Properties of Glass-Ionomer Dental Cements: A Review
Reprinted from: *Materials* **2020**, *13*, 2510, doi:10.3390/ma13112510 75

Lora Mishra, Abdul Samad Khan, Marilia Mattar de Amoedo Campos Velo, Saurav Panda, Angelo Zavattini and Fabio Antonio Piola Rizzante et al.
Effects of Surface Treatments of Glass Fiber-Reinforced Post on Bond Strength to Root Dentine: A Systematic Review
Reprinted from: *Materials* **2020**, *13*, 1967, doi:10.3390/ma13081967 89

Marta Kunert and Monika Lukomska-Szymanska
Bio-Inductive Materials in Direct and Indirect Pulp Capping—A Review Article
Reprinted from: *Materials* **2020**, *13*, 1204, doi:10.3390/ma13051204 101

Marta W. Chrószcz-Porebska, Izabela M. Barszczewska-Rybarek and Grzegorz Chladek
Characterization of the Mechanical Properties, Water Sorption, and Solubility of Antibacterial Copolymers of Quaternary Ammonium Urethane-Dimethacrylates and Triethylene Glycol Dimethacrylate
Reprinted from: *Materials* **2022**, *15*, 5530, doi:10.3390/ma15165530 121

Krzysztof Pałka, Małgorzata Miazga-Karska, Joanna Pawłat, Joanna Kleczewska and Agata Przekora
The Effect of Liquid Rubber Addition on the Physicochemical Properties, Cytotoxicity, and Ability to Inhibit Biofilm Formation of Dental Composites
Reprinted from: *Materials* **2021**, *14*, 1704, doi:10.3390/ma14071704 135

Barbara Lapinska, Aleksandra Szram, Beata Zarzycka, Janina Grzegorczyk, Louis Hardan and Jerzy Sokolowski et al.
An In Vitro Study on the Antimicrobial Properties of Essential Oil Modified Resin Composite against Oral Pathogens
Reprinted from: *Materials* **2020**, *13*, 4383, doi:10.3390/ma13194383 151

Simonetta D'Ercole, Francesco De Angelis, Virginia Biferi, Chiara Noviello, Domenico Tripodi and Silvia Di Lodovico et al.
Antibacterial and Antibiofilm Properties of Three Resin-Based Dental Composites against *Streptococcus mutans*
Reprinted from: *Materials* **2022**, *15*, 1891, doi:10.3390/ma15051891 . . . . **169**

Agata Szczesio-Wlodarczyk, Monika Domarecka, Karolina Kopacz, Jerzy Sokolowski and Kinga Bociong
An Evaluation of the Properties of Urethane Dimethacrylate-Based Dental Resins
Reprinted from: *Materials* **2021**, *14*, 2727, doi:10.3390/ma14112727 . . . . **181**

Małgorzata Fischer, Małgorzata Skucha-Nowak, Bartosz Chmiela and Anna Korytkowska-Wałach
Assessment of the Potential Ability to Penetrate into the Hard Tissues of the Root of an Experimental Preparation with the Characteristics of a Dental Infiltratant, Enriched with an Antimicrobial Component—Preliminary Study
Reprinted from: *Materials* **2021**, *14*, 5654, doi:10.3390/ma14195654 . . . . **197**

Małgorzata Fischer, Anna Mertas, Zenon Paweł Czuba and Małgorzata Skucha-Nowak
Study of Cytotoxic Properties of an Experimental Preparation with Features of a Dental Infiltrant
Reprinted from: *Materials* **2021**, *14*, 2442, doi:10.3390/ma14092442 . . . . **211**

Grzegorz Chladek, Michał Nowak, Wojciech Pakieła and Anna Mertas
Effect of *Candida albicans* Suspension on the Mechanical Properties of Denture Base Acrylic Resin
Reprinted from: *Materials* **2022**, *15*, 3841, doi:10.3390/ma15113841 . . . . **223**

Milica Petrović, Marina Randjelović, Marko Igić, Milica Randjelović, Valentina Arsić Arsenijević and Marijana Mionić Ebersold et al.
Poly(methyl methacrylate) with Oleic Acid as an Efficient *Candida albicans* Biofilm Repellent
Reprinted from: *Materials* **2022**, *15*, 3750, doi:10.3390/ma15113750 . . . . **239**

Grzegorz Chladek, Katarzyna Pakieła, Wojciech Pakieła, Jarosław Żmudzki, Marcin Adamiak and Cezary Krawczyk
Effect of Antibacterial Silver-Releasing Filler on the Physicochemical Properties of Poly(Methyl Methacrylate) Denture Base Material
Reprinted from: *Materials* **2019**, *12*, 4146, doi:10.3390/ma12244146 . . . . **251**

Bozhana Chuchulska, Ilian Hristov, Boyan Dochev and Raycho Raychev
Changes in the Surface Texture of Thermoplastic (Monomer-Free) Dental Materials Due to Some Minor Alterations in the Laboratory Protocol—Preliminary Study
Reprinted from: *Materials* **2022**, *15*, 6633, doi:10.3390/ma15196633 . . . . **275**

Maria Francesca Sfondrini, Pekka Kalevi Vallittu, Lippo Veli Juhana Lassila, Annalisa Viola, Paola Gandini and Andrea Scribante
Glass Fiber Reinforced Composite Orthodontic Retainer: In Vitro Effect of Tooth Brushing on the Surface Wear and Mechanical Properties
Reprinted from: *Materials* **2020**, *13*, 1028, doi:10.3390/ma13051028 . . . . **285**

Rini Behera, Lora Mishra, Darshan Devang Divakar, Abdulaziz A. Al-Kheraif, Naomi Ranjan Singh and Monika Lukomska-Szymanska
The One-Year In Vivo Comparison of Lithium Disilicate and Zirconium Dioxide Inlays
Reprinted from: *Materials* **2021**, *14*, 3102, doi:10.3390/ma14113102 . . . . **299**

Agata Szczesio-Wlodarczyk, Karolina Rams, Karolina Kopacz, Jerzy Sokolowski and Kinga Bociong
The Influence of Aging in Solvents on Dental Cements Hardness and Diametral Tensile Strength
Reprinted from: *Materials* **2019**, *12*, 2464, doi:10.3390/ma12152464 313

Ewa Sobolewska, Piotr Makowiecki, Justyna Drozdowska, Ireneusz Dziuba, Alicja Nowicka and Marzena Wyganowska-Światkowska et al.
Cytotoxic Potential of Denture Adhesives on Human Fibroblasts—In Vitro Study
Reprinted from: *Materials* **2022**, *15*, 1583, doi:10.3390/ma15041583 325

Dawid Łysik, Piotr Deptuła, Sylwia Chmielewska, Robert Bucki and Joanna Mystkowska
Degradation of Polylactide and Polycaprolactone as a Result of Biofilm Formation Assessed under Experimental Conditions Simulating the Oral Cavity Environment
Reprinted from: *Materials* **2022**, *15*, 7061, doi:10.3390/ma15207061 339

M. Mohan Babu, P. Syam Prasad, P. Venkateswara Rao, S. Hima Bindu, A. Prasad and N. Veeraiah et al.
Influence of ZrO_2 Addition on Structural and Biological Activity of Phosphate Glasses for Bone Regeneration
Reprinted from: *Materials* **2020**, *13*, 4058, doi:10.3390/ma13184058 355

Editorial

Composite and Polymeric Materials for Dentistry: Enhancing Antimicrobial and Mechanical Properties

Grzegorz Chladek

Department of Engineering Materials and Biomaterials, Faculty of Mechanical Engineering, Silesian University of Technology, 18a Konarskiego Str., 41-100 Gliwice, Poland; grzegorz.chladek@polsl.pl

Billions of people suffer from dental problems and that number is constantly increasing. Paradoxically, the deteriorating state our teeth is accompanied by the ever-increasing desire to preserve our best facial appearance, which is significantly influenced by dental aesthetics. This favors the dynamic development of dental materials. Special attention has been granted to the possibility of giving materials new or improved properties by the introduction of nano- or submicrometer-size additives, natural oils, new monomers, and other potentially beneficial chemical and structural modifications. Equally important are the new data regarding the colonization of dental materials by pathogenic microbes and their influence on the other properties, as well as the multifactor evaluation of materials recently introduced to the market. Therefore, this Special Issue includes a compilation of different review and research analyses pertaining to the improvement of the antimicrobial and mechanical properties of composite and polymeric materials for dentistry.

The published review articles discuss current knowledge related to selected aspects of the development and function of dental materials. Barszczewska-Rybarek [1] analyzed the relationships between structure and biofunctional properties in the cross-linked dimethacrylate-based matrices used for dental materials. She discussed the influence of the chemical structure, molecular structure represented by the degree of conversion and crosslink density, supramolecular structure related to microgel agglomerate dimensions, and the role of hydrogen bonding in the mechanical properties and water sorption. Kaczmarek et al. [2] presented a review focused on using selected spectroscopic methods for surface analysis of different dental materials, including polymer-based materials. The principles, advantages, limitations, and typical applications of techniques such as Raman spectroscopy, Infrared Spectroscopy, Ultraviolet and Visible Spectroscopy, X-ray spectroscopy, and Mass Spectrometry have been shown. The presented work is of particular interest to dentists conducting research related to materials science. Grazioli et al. [3] summarized current knowledge related to the method of removing adhesives after the debonding of metal orthodontic brackets. Studies on this topic are rare, despite the clinical usefulness. Four different methods of bracket surface preparation were investigated until now: sandblasting, laser, mechanical grinding, and direct flame. All tested methods improved shear bond strength and were clinically acceptable; however, only after using an erbium-doped yttrium aluminum garnet laser were similar values than obtained for new brackets. Nicholson et al. [4] review the experimental methods for the improve mechanical properties of conventional and resin-modified dental cements. Certain fibers (glass, cellulose, basalt) and nanoparticles (TiO_2, Al_2O_3 or ZrO_2) were indicated as the most promising examples. However, in the case of the nanoparticles, the positive effect was influenced by the morphology of the cement matrix; in particular, the increase in porosity. All described modifications were partially tested only in laboratory investigations, and none have yet seen clinical use. Mishra et al. [5] analyzed literature related to the influence of surface preparation of glass fiber posts on the strength of the bond to dentine. Surface treatment with application of phosphoric acid, hydrogen peroxide, and silane enhances post's retentiveness. An analysis of the current stage of knowledge regarding the properties of bioinductive materials in direct and indirect pulp capping procedures was presented by Kunert et al. [6]. The authors indicated that calcium silicate cements are characterized by positive properties confirmed by numerous independent research studies; whereas, for a light-cured calcium silicate-based material and a resin-modified glass-ionomer, evidence is insufficient to support the use of these materials in vital pulp therapy.

A widely represented group of research papers was concerned with dental resin-based composites for fillings. Several works related to the development of new materials were presented. The use of dimethacrylates with introduced quaternary ammonium groups is considered as antibacterial dental composites for filling. Chrószcz-Porębska et al. [7] investigated the mechanical properties, as well as the sorption and solubility of experimental matrices composed of six different types of quaternary ammonium urethane-dimethacrylate triethylene glycol dimethacrylate. Despite very good antimicrobial properties, a significant deterioration of mechanical properties and increases in water absorption and solubility were noted. These materials are not attractive in terms of the considered solution, although they may be an interesting proposition in the case of applications requiring lower mechanical properties. Pałka et al. [8] investigated the influence of the addition of liquid rubber (methacrylate-terminated polybutadiene) on the properties of dental composites. The experimental materials presented enhanced shear bond strength values for enamel and dentine, reduced hydrophilicity, and reduced biofilm activity (*Steptococcus mutans*, *Streptococcus sanguinis*); however, they may show cytotoxicity for some formulations. Lapinska et al. [9] investigated the activity of essential oils (rosemary thyme, anise, clove, geranium, cinnamon, limetta, mint, citronella, lavender) against typical oral pathogenic microorganisms (*Streptococcus mutans*, *Lactobacillus acidophilus*, *Candida albicans*) in the context of their incorporation into resin composites. The cinnamon oil was identified as the most promising and introduced in different concentrations into commercially available composite resin. A strong antimicrobial effect of the experimental materials was obtained; however, the authors point out the need to carry out tests of other properties to confirm the favorable characteristics of the materials. Equally important is the research of the materials available on the market. D'Ercole et al. [10] studied the potential to reduce colonization by *Streptococcus mutans* of three commercially available resin-based composites. Investigations have shown that for surfaces prepared in the same way, the materials presented different adherence of bacteria and biofilm accumulation. The authors concluded that the chemical composition of composites will likely play an important role in the process of bacterial adhesion/proliferation; however, they noted the need for further research to confirm their results. Composite materials are complex systems; therefore, it is important for the scientific community to indicate the initial resin formulations for further, more complex experimental analyses. For this reason, Szczesio-Wlodarczyk et al. [11] studied the mechanical properties of resins intended for dental composite matrices containing urethane dimethacrylate (UDMA) and diversified compilations of other dental monomers such as bisphenol A-glycidyl methacrylate (Bis-GMA), triethylene glycol dimethacrylate (TEGDMA), and ethoxylated bisphenol-A dimethacrylate (Bis-EMA). The authors indicated as the most favorable formulations UDMA/Bis-GMA/TEGDMA in proportions of 70/10/20 wt.% and 40/40/20 wt.% and UDMA/Bis-EMA/TEGDMA in proportions of 40/40/20 wt.%, due to their good compilation of flexural strength, flexural modulus, hardness, diametral tensile strength, and water absorption values.

Other important and new dental materials are dental infiltrants used in the treatment of early carious lesions in line with the idea of microinvasive dentistry. Fisher et al. [12] investigated an experimental infiltrant with a chemical composition similar to a commercially available product, but with the addition of a bacteriostatic component (metronidazole). The pilot results suggest that this formulation is not cytotoxic and may be considered as an alternative to the commercial preparation due to its microbilogical action. The developed infiltrant was also investigated with regard to its ability to penetrate into the root cement [13]. Microscopic investigations have shown that the proposed material may be a potentially beneficial solution for the treatment of early carious lesions of the tooth root, as it exhibited a deep penetration into demineralized tissues.

Denture base polymeric materials have been used in dentistry for decades; however, due to their importance and popularity they are still widely investigated by the scientific community. Colonization and penetration of prosthetic materials by *Candida albicans* is a frequently considered problem. Chladek et al. [14] conducted a 90-day experiment with a different methodology to those previously used to verify whether PMMA is penetrated by *Candida albicans* and to investigate its mechanical properties after exposure to yeast-like fungi. Microscopic observations have not confirmed the penetration of fungi into the material. A decrease in surface hardness was registered, while flexural strength, flexural modulus, tensile strength, impact strength, and ball indentation hardness were at the same level as controlNumerous yeast cells were observed on the surface in crystalized structures and in traces after grinding, suggesting that not penetration, but the deterioration of surface quality may create microareas that are difficult to disinfect in clinical conditions. Other investigations related to the problem of PMMA colonization by *Candida albicans* were presented by Petrovi et al. [15]. They introduced 3 to 12% of oleic acid and registered a decrease in the water contact angle and metabolic activity of yeast cells. Investigations of other biofunctional properties are still needed. An important direction of research into denture base resins is the development of materials with antimicrobial properties. Strong antifungal properties were noted after incorporation of submicrometer inorganic particles of silver sodium hydrogen zirconium phosphate. Modification results in reduced flexural strength, impact strength, and enhanced solubility; however, obtained values were at acceptable levels. On the other hand, favorable changes were also recorded; hardness and flexural modulus increased whereas volume loss during the wear test decreased with the introduction of ceramic particles [16]. Chuchulska et al. [17], in the technological study, tested how the small changes in the laboratory protocol of polyamide prosthetic base materials may influence the surface texture. Investigations have shown that modifying the process by altering the melting temperature by 5 °C and the pressure by 0.5 Bar during injection molding can reduce roughness for some materials.

Other materials, which are extremely important from a practical point of view, have been tested to determine or predict their clinical properties. Sfondrini et al. [18] study the effect of tooth brushing on surface wear and the mechanical properties of a glass fiber-reinforced composite orthodontic retainer during an in vitro experiment. The stainless-steel wires, flowable resin composite covered, and spot-bonded fiber reinforced composites were tested after 26 min and 60 min of tooth brushing. The three-point bending test and SEM investigations showed a significant reduction of flexural strength and signs of wear on both fiber-reinforced materials. The authors suggest that proposed solutions need further tests before routine application in clinical practice. Rini Behera et al. [19] presented a one-year in vivo comparison of lithium disilicate and zirconium dioxide class II inlay restorations. The survival rate was evaluated, and one failure was observed in the zirconium dioxide group (the survival probability was 93%) when no failure was observed in the lithium disilicate group. Both types of inlays presented comparable surface roughness, marginal adaptation, anatomic form, occlusal contact, and proximal contact during the experiment, but the color and translucency match was far worse for the zirconium dioxide restorations. Szczesio-Wlodarczyk et al. [20] investigated the effect of different liquids (ethanol, soda solution, and green tea) on the mechanical properties of dental cements used for cementing restorations. The ethanol solution showed the greatest influence on the hardness of composite cement when soda solution was applied to zinc-polycarboxylate cement. Diametral tensile strength values were unchanged for composite cements, increased after exposure of zinc-polycarboxylate cement to ethanol, and decreased in soda solution for glass-ionomer cement. In vitro experiments conducted by Sobolewska et al. [21] demonstrated a slight-to-moderate toxic effect on human pulp fibroblasts of denture adhesive creams, but the authors suggest the need for in vivo studies to verify these results in patients. Łysik et al. [22] investigated the degradation of polylactide (PLA) and polycaprolactone (PCL) as a consequence of biofilm presence (*Candida krusei* and *Steptococcus mutans*) in the artificial saliva. After 56 days, their experiment has shown that as a consequence of the microorganisms, the surface morphologies were changed and molecular weight and mechanical properties decreased.

Special attention should be paid to the newly developed materials with bioactive properties, which are expected to find application in dentistry, implantology, and dental surgery. Babu et al. [23] conducted an experiment with zirconium-doped calcium phosphate-based bioglasses, which should be considered as a prominent solution. A bioglass system was synthesized by the melt quenching process. The results shown that increases in zirconia concentration increase the glass-forming ability and thermal stability of materials. Spectroscopic investigations confirmed the presence of a thin hydroxyapatite layer on the sample surface after incubation in SBF solution. The process of glass degradation after incubation in SBF increased with time but decreased with the increase of ZrO_2 concentration. The results confirmed the suitability of bioglasses tested for bone-related applications.

The quest editor wishes to give special thanks to all the authors, and to the editorial team of *Materials* for the collaborative peer review and publishing process. I hope that readers will enjoy the Special Issue "Composite and Polymeric Materials for Dentistry: Enhancing Antimicrobial and Mechanical Properties" and will find within its pages new knowledge and ample inspiration for future research.

Conflicts of Interest: The author declares no conflict of interest.

References

1. Barszczewska-Rybarek, I.M. A Guide through the Dental Dimethacrylate Polymer Network Structural Characterization and Interpretation of Physico-Mechanical Properties. *Materials* **2019**, *12*, 4057. [CrossRef] [PubMed]
2. Kaczmarek, K.; Leniart, A.; Lapinska, B.; Skrzypek, S.; Lukomska-Szymanska, M. Selected Spectroscopic Techniques for Surface Analysis of Dental Materials: A Narrative Review. *Materials* **2021**, *14*, 2624. [CrossRef] [PubMed]
3. Grazioli, G.; Hardan, L.; Bourgi, R.; Nakanishi, L.; Amm, E.; Zarow, M.; Jakubowicz, N.; Proc, P.; Cuevas-Suárez, C.E.; Lukomska-Szymanska, M. Residual Adhesive Removal Methods for Rebonding of Debonded Orthodontic Metal Brackets: Systematic Review and Meta-Analysis. *Materials* **2021**, *14*, 6120. [CrossRef] [PubMed]
4. Nicholson, J.W.; Sidhu, S.K.; Czarnecka, B. Enhancing the Mechanical Properties of Glass-Ionomer Dental Cements: A Review. *Materials* **2020**, *13*, 2510. [CrossRef]
5. Mishra, L.; Khan, A.S.; de Velo, M.M.A.C.; Panda, S.; Zavattini, A.; Rizzante, F.A.P.; Arbildo Vega, H.I.; Sauro, S.; Lukomska-Szymanska, M. Effects of Surface Treatments of Glass Fiber-Reinforced Post on Bond Strength to Root Dentine: A Systematic Review. *Materials* **2020**, *13*, 1967. [CrossRef]
6. Kunert, M.; Lukomska-Szymanska, M. Bio-Inductive Materials in Direct and Indirect Pulp Capping—A Review Article. *Materials* **2020**, *13*, 1204. [CrossRef]
7. Chrószcz-Porębska, M.W.; Barszczewska-Rybarek, I.M.; Chladek, G. Characterization of the Mechanical Properties, Water Sorption, and Solubility of Antibacterial Copolymers of Quaternary Ammonium Urethane-Dimethacrylates and Triethylene Glycol Dimethacrylate. *Materials* **2022**, *15*, 5530. [CrossRef]
8. Pałka, K.; Miazga-Karska, M.; Pawłat, J.; Kleczewska, J.; Przekora, A. The Effect of Liquid Rubber Addition on the Physicochemical Properties, Cytotoxicity, and Ability to Inhibit Biofilm Formation of Dental Composites. *Materials* **2021**, *14*, 1704. [CrossRef]
9. Lapinska, B.; Szram, A.; Zarzycka, B.; Grzegorczyk, J.; Hardan, L.; Sokolowski, J.; Lukomska-Szymanska, M. An in Vitro Study on the Antimicrobial Properties of Essential Oil Modified Resin Composite against Oral Pathogens. *Materials* **2020**, *13*, 4383. [CrossRef]
10. D'Ercole, S.; De Angelis, F.; Biferi, V.; Noviello, C.; Tripodi, D.; Di Lodovico, S.; Cellini, L.; D'Arcangelo, C. Antibacterial and Antibiofilm Properties of Three Resin-Based Dental Composites against Streptococcus Mutans. *Materials* **2022**, *15*, 1891. [CrossRef]
11. Szczesio-Wlodarczyk, A.; Domarecka, M.; Kopacz, K.; Sokolowski, J.; Bociong, K. An Evaluation of the Properties of Urethane Dimethacrylate-Based Dental Resins. *Materials* **2021**, *14*, 2727. [CrossRef]
12. Fischer, M.; Mertas, A.; Czuba, Z.P.; Skucha-Nowak, M. Study of Cytotoxic Properties of an Experimental Preparation with Features of a Dental Infiltrant. *Materials* **2021**, *14*, 2442. [CrossRef]
13. Fischer, M.; Skucha-Nowak, M.; Chmiela, B.; Korytkowska-Wałach, A. Assessment of the Potential Ability to Penetrate into the Hard Tissues of the Root of an Experimental Preparation with the Characteristics of a Dental Infiltratant, Enriched with an Antimicrobial Component—Preliminary Study. *Materials* **2021**, *14*, 5654. [CrossRef]
14. Chladek, G.; Nowak, M.; Pakieła, W.; Mertas, A. Effect of Candida Albicans Suspension on the Mechanical Properties of Denture Base Acrylic Resin. *Materials* **2022**, *15*, 3841. [CrossRef]
15. Petrović, M.; Randjelović, M.; Igić, M.; Randjelović, M.; Arsić Arsenijević, V.; Mionić Ebersold, M.; Otašević, S.; Milošević, I. Poly (Methyl Methacrylate) with Oleic Acid as an Efficient Candida Albicans Biofilm Repellent. *Materials* **2022**, *15*, 3750. [CrossRef]

16. Chladek, G.; Pakieła, K.; Pakieła, W.; Żmudzki, J.; Adamiak, M.; Krawczyk, C. Effect of Antibacterial Silver-Releasing Filler on the Physicochemical Properties of Poly (Methyl Methacrylate) Denture Base Material. *Materials* 2019, *12*, 4146. [CrossRef]
17. Chuchulska, B.; Hristov, I.; Dochev, B.; Raychev, R. Changes in the Surface Texture of Thermoplastic (Monomer-Free) Dental Materials Due to Some Minor Alterations in the Laboratory Protocol—Preliminary Study. *Materials* 2022, *15*, 6633. [CrossRef]
18. Sfondrini, M.F.; Vallittu, P.K.; Lassila, L.V.J.; Viola, A.; Gandini, P.; Scribante, A. Glass Fiber Reinforced Composite Orthodontic Retainer: In Vitro Effect of Tooth Brushing on the Surface Wear and Mechanical Properties. *Materials* 2020, *13*, 1028. [CrossRef]
19. Behera, R.; Mishra, L.; Divakar, D.D.; Al-Kheraif, A.A.; Singh, N.R.; Lukomska-Szymanska, M. The One-Year in Vivo Comparison of Lithium Disilicate and Zirconium Dioxide Inlays. *Materials* 2021, *14*, 3102. [CrossRef]
20. Szczesio-Wlodarczyk, A.; Rams, K.; Kopacz, K.; Sokolowski, J.; Bociong, K. The Influence of Aging in Solvents on Dental Cements Hardness and Diametral Tensile Strength. *Materials* 2019, *12*, 2464. [CrossRef]
21. Sobolewska, E.; Makowiecki, P.; Drozdowska, J.; Dziuba, I.; Nowicka, A.; Wyganowska-Świątkowska, M.; Janiszewska-Olszowska, J.; Grocholewicz, K. Cytotoxic Potential of Denture Adhesives on Human Fibroblasts—In Vitro Study. *Materials* 2022, *15*, 1583. [CrossRef] [PubMed]
22. Łysik, D.; Deptuła, P.; Chmielewska, S.; Bucki, R.; Mystkowska, J. Degradation of Polylactide and Polycaprolactone as a Result of Biofilm Formation Assessed under Experimental Conditions Simulating the Oral Cavity Environment. *Materials* 2022, *15*, 7061. [CrossRef] [PubMed]
23. Mohan Babu, M.; Syam Prasad, P.; Venkateswara Rao, P.; Hima Bindu, S.; Prasad, A.; Veeraiah, N.; Özcan, M. Influence of ZrO_2 Addition on Structural and Biological Activity of Phosphate Glasses for Bone Regeneration. *Materials* 2020, *13*, 4058. [CrossRef] [PubMed]

Disclaimer/Publisher's Note: The statements, opinions and data contained in all publications are solely those of the individual author(s) and contributor(s) and not of MDPI and/or the editor(s). MDPI and/or the editor(s) disclaim responsibility for any injury to people or property resulting from any ideas, methods, instructions or products referred to in the content.

Review

A Guide through the Dental Dimethacrylate Polymer Network Structural Characterization and Interpretation of Physico-Mechanical Properties

Izabela Maria Barszczewska-Rybarek

Department of Physical Chemistry and Technology of Polymers, Silesian University of Technology, Strzody 9, 44-100 Gliwice, Poland; Izabela.Barszczewska-Rybarek@polsl.pl

Received: 10 November 2019; Accepted: 28 November 2019; Published: 5 December 2019

Abstract: Material characterization by the determination of relationships between structure and properties at different scales is essential for contemporary material engineering. This review article provides a summary of such studies on dimethacrylate polymer networks. These polymers serve as photocuring organic matrices in the composite dental restorative materials. The polymer network structure was discussed from the perspective of the following three aspects: the chemical structure, molecular structure (characterized by the degree of conversion and crosslink density (chemical as well as physical)), and supramolecular structure (characterized by the microgel agglomerate dimensions). Instrumental techniques and methodologies currently used for the determination of particular structural parameters were summarized. The influence of those parameters as well as the role of hydrogen bonding on basic mechanical properties of dimethacrylate polymer networks were finally demonstrated. Mechanical strength, modulus of elasticity, hardness, and impact resistance were discussed. The issue of the relationship between chemical structure and water sorption was also addressed.

Keywords: dental materials; dimethacrylates; polymer networks; structure; morphology; degree of conversion; crosslink density; physical crosslinking; hydrogen bonds; mechanical properties; water sorption

1. Introduction

Poly(dimethacrylate)s are highly crosslinked polymer networks, irreplaceable in applications requiring fast polymerization processes, such as light-curing dental materials [1–5]. A structural diversity, low manufacturing costs, excellent aesthetics, and easy handling have determined the utilization of poly(dimethacrylate)s mainly as matrices in dental restorative composites [6–9] and provisional restoration materials [10].

Dental composites usually consist of a dimethacrylate resin, inorganic filler (usually silica particles of various sizes—nanofillers, microfillers, and macrofillers or the combination of the latter two, which is called "hybrids" [11]), polymerization initiating system, silane coupling agent, which binds organic matrix with a filler, and pigments [6–9]. The majority of dental materials are single-component photocured systems. Photocuring is usually carried out with the aid of camphorquinone initiator and a tertiary amine reducing agent (usually, *N*,*N*-dimethylaminoethyl methacrylate) [12], under visible light irradiation within the range of high-intensity blue light (470–490 nm) [9]. There are also chemically cured, two-component systems. They contain a peroxide initiator (usually, benzoyl peroxide) and an amine accelerator (usually, *N*,*N*-dimethyl-p-toluidine) [9,12].

The most important dental dimethacrylate monomers include 2,2-bis-[4-(2-hydroxy-3-methacryloxypropoxy)phenyl]propane (Bis-GMA, bisphenol A glycerolate dimethacrylate), bisphenol

A ethoxylate dimethacrylate (Bis-EMA), 1,6-bis-(methacryloyloxy-2-ethoxycarbonylamino)-2,4,4-trimethylhexane, which is called the urethane-dimethacrylate monomer (UDMA), and triethylene glycol dimethacrylate (TEGDMA) (Scheme 1) [6–9]. The dimethacrylate monomers, due to their easy preparation and sufficient working time after light exposure, give an edge to design soft actuable (controlled movable carrier) for oral drug delivery not only for the oropharmacological products but other soft tissues in vitro and in vivo, as mentioned by Singh et al. [13].

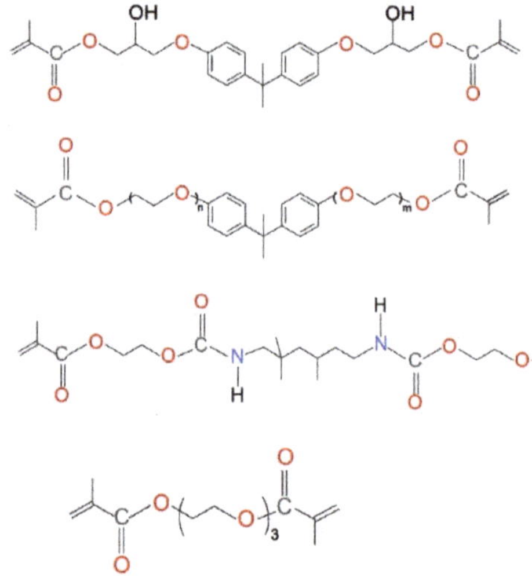

Scheme 1. The chemical structure of popular dental dimethacrylate monomers.

Dimethacrylates are usually used in mixtures of different ratios and then copolymerized [9,14]. There are following exemplary combinations of dimethacrylates used in commercial composites: Bis-GMA and TEGMA (Clearfil ST® (Kuraray), Grandio® (VOCO), Filtek Z100® (3M ESPE)), Bis-GMA, UDMA and TEGDMA (FSB® (3M ESPE), Tetric Ceram® (Ivoclar Vivadent)), Bis-GMA, UDMA and Bis-EMA (Filtek Z250® (3M ESPE)) [14].

Bis-GMA, patented by Bowen in 1962 [15], was the first dental dimethacrylate resin [6–9]. Its large molecular weight and low concentration of double bonds (Table 1) provide low volatility, low polymerization shrinkage, rapid curing, and stiff, durable products of curing [16]. The extremely high viscosity of Bis-GMA limits the degree of conversion and decreases the possibility of filler incorporation. The viscosity of Bis-GMA can be lowered by admixing low molecular weight dimethacrylates. Oligoethylene glycol dimethacrylates may be used for this purpose, of which TEGDMA is the most popular. The lower the viscosity of the mixture, the higher the degree of conversion and the more filler can be incorporated [16]. However, the addition of TEGDMA causes an increase in polymerization shrinkage [16]. In response to Bis-GMA flaws, Bis-EMA and UDMA monomers were developed. These monomers have similar molecular weights to Bis-GMA but are less viscous. Bis-EMA, having no hydroxyl groups is less viscous than UDMA. The mixture of Bis-GMA and Bis-EMA can be used without a reactive diluent, e.g., Renamel® (Cosmedent) [14]. Spectrum TPH® (Dentsply) also does not contain TEGDMA, as it is composed of Bis-GMA, UDMA, and Bis-EMA [14]. The higher viscosity of the UDMA monomer results from the formation of intermolecular hydrogen bonds between urethane species. Due to the good mechanical properties of the UDMA homopolymer, it is the only dimethacrylate that can be used alone in commercial composites, such as Lava Ultimate® (3M ESP) [17]. It can also be combined with Bis-GMA, acting as a viscosity reducer, e.g., Tetric EvoCream® (Vivadent Ivoclar) [14].

Table 1. Properties of popular dimethacrylate monomers and the degree of conversion in the corresponding homopolymers.

Monomer	Molecular Weight (g/mol)	Concentration of Double Bonds (mol/kg)	Viscosity (Pa·s)	Degree of Conversion (%)
Bis-GMA	511	3.90	1200 [1]	39.0 [1]/34.5 [2]
Bis-EMA (n+m = 4)	540	3.70	0.9 [2]	75.5 [2]
UDMA	470	4.25	23.1 [1]	69.6 [1]/72.4 [2]
TEGDMA	286	6.99	0.011 [1]	75.5 [1]/82.5 [2]

[1] Taken from [18]; [2] Taken from [19].

The precise explanation and understanding of the physicomechanical behavior of dimethacrylate polymer networks is a difficult task. Years of research have shown that their properties result from the interaction of the following structural factors: the monomer chemical structure, polymer network molecular structure (crosslink density, which relates to the chemical crosslink density, physical crosslink density, and degree of conversion), as well as morphology.

As far back as the 2000s, the kinetics of dimethacrylate polymerization was precisely described. The theory of the ideal polymer network was developed at that time [20–22]. The real network was characterized by specifying its defects, such as pendant groups and chains, loops, entanglements, and sol fraction (Figure 1) [4,9,21,23]. Attention was paid to stochastic and spatial correlations as well as diffusion control of the reaction [1–4,9,20]. These studies have shown, that the radical polymerization of dimethacrylates is a complex process and involves a series of phenomena, such as auto acceleration (gel effect, occurring when the degree of conversion is only 1–2%), auto deceleration (observed in the further stages of the reaction), the reaction diffusion (the mechanism controlling termination), incomplete conversion of functional groups, the formation of microgel agglomerates (clusters of highly crosslinked polymer of a high degree of cyclization, suspended in the less cross-linked matrix). In effect, such polymer networks are spatially heterogeneous and consist of clusters varying in crosslink density, in which the degree of double bond conversion (DC) is never full (Figure 1) [1–4,20–25].

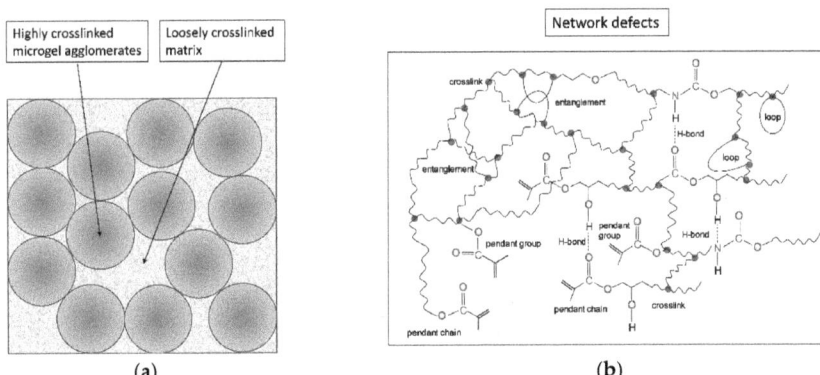

Figure 1. The structural heterogeneity of dimethactylate polymer networks: (a) schematic representation of the morphology; (b) defects in the polymer network microstructure.

With scientific development, the need to clarify the influence of the network's structural features on the material properties has become apparent. The issue of a comprehensive analysis of the structural heterogeneity of dimethacrylate polymer networks, based on a quantitative analysis of each structural aspect arose. A supramolecular structure required quantitative characterization. At the same time, in the literature, it had been suggested, that the presence of microgel agglomerates may adversely affect mechanical properties of poly(dimethacrylate)s, primarily the impact resistance [21,24–28]. In recent years, with advances in characterization techniques, substantial efforts have been made to precisely describe the polymer network spatial heterogeneity, especially the morphology [29–45]. All of this has resulted in many studies on structure-property relationships. Having expertise in this field is in demand for understanding properties of currently used dimethacrylate polymers as well as for designing new dimethacrylate monomers and their compositions in order to achieve the best possible material efficiency.

The purpose of this paper is to review the current state of knowledge on the relationships between structure and properties in the crosslinked matrix of dimethacrylate based dental materials. As their structural characterization is a complex issue, an explanation of the influence of each structural aspect on physicomechanical properties was consistently undertaken. In addition, the article describes major techniques that are currently used to quantitatively analyze the structure of poly(dimethacrylate)s, with reference to the recent developments in this area. Issues presented were based on studies of typical dental dimethacrylate polymer networks as well as on other dimethacrylate polymers, described only in the literature. The analysis of the latter provided support for conclusions on the structure-property relationships, which may be especially useful when defining future trends in the design of new dental dimethacrylate systems.

2. The Chemical Structure

The chemical structure of the dimethacrylate monomer might be recognized as a crucial factor in determining the physicochemical and mechanical performance of the corresponding polymer network. The mechanisms of its action can be divided into direct and indirect effects.

The direct action relates to the influence of chemical groups, constituting the monomer structure, on polymer properties. It also determines the theoretical crosslink density and physical crosslinking.

The indirect action relates to the influence of the monomer chemical structure on the degree of conversion, real crosslink density as well as the supramolecular organization in the corresponding polymer and their consequences in polymer properties.

Speaking of the direct action, it might be said that the dimethacrylate chemical structure determines the overall elasticity of the crosslinked material [22]. Composite dental materials require highly crosslinked polymers that offer high stiffness, hardness, and durability in order to be able to withstand the stresses exerted by orthodontic mechanics and variations in the oral environment [46].

Firstly, the dimethacrylate molecule length determines the theoretical crosslink density of a corresponding polymer network. The shorter the distance between double bonds, the higher the crosslink density. An increase in crosslink density causes a decrease in the possibility of chain reorganization [5,18,31,32,35,47]. Consequently, an increase in modulus is observed [19,23,48]. Physical crosslinks, such as hydrogen bonds, reduce the rotational motion too, causing a stiffening of the network structure [3,49–53].

Secondly, the dimethacrylate monomer molecular elasticity determines the general network elasticity [21,22]. Aliphatic hydrocarbon and oligoether chains give rise to an increase in polymer elasticity, whereas cycloaliphatic and aromatic units cause an increase in polymer stiffness [31,32,48,54–56].

Water sorption (WS) can be also regarded as the chemical structure-dependent physical property of dimethacrylate polymer networks. A certain swelling capacity can be beneficial in dental applications. It was found that polymerization shrinkage deformation can be compensated by hygroscopic expansion. However, the scale of this expansion in dental materials cannot exceed the polymerization shrinkage [57]. For these reasons, the WS of 50 µg/mm^3 was established as an upper limit for dental restorative materials [58]. Like the elastic behavior, water sorption depends on the intrinsic action of the theoretical crosslink density, resulting from both covalent and non-covalent crosslinks, as well as monomer chemical structure, which determines chain elasticity and hydrophilicity. The longer the chains between crosslinks, the lower the crosslink density, the higher the water sorption [48,52,53]. Especially, highly elastic hydrocarbon or oligoether chains give rise to the accommodation of higher water quantity. This effect was shown by several works on WS of dimethacrylate polymers, consisting of oligooxyethylene chains. Park et al. studied a series of Bis-GMA copolymers with mono-, di- and triethyleneglycol dimethacrylates [52]. Ogliari et al. tested the influence of the degree of Bis-EMA ethoxylation on WS [54]. Barszczewska-Rybarek studied WS of homopolymers of six homologous series of urethane-dimethacrylates (Scheme 2) [48]. An increase in the ethylene oxide number in a chain always resulted in an increase of water sorption. Additionally, it was found that if the number of oxyethylene units was greater than three, the WS significantly exceeded the limit of 50 $\mu g/mm^3$.

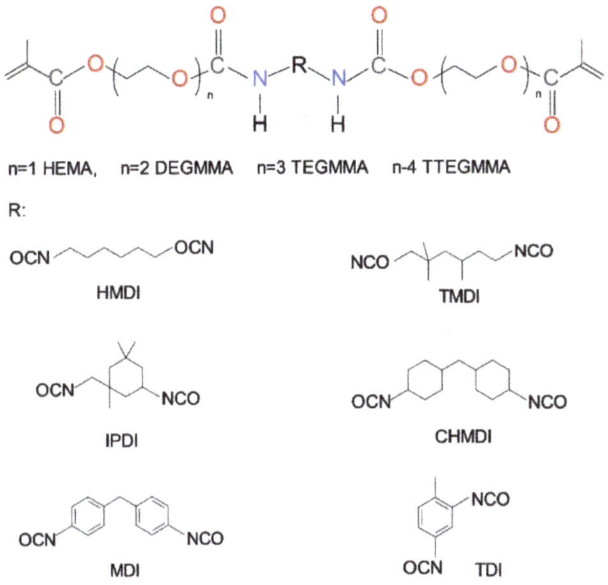

Scheme 2. The chemical structure of a series of urethane-dimethacrylate monomers.

The study on urethane-dimethacrylate polymer networks additionally provided results for the influence of the diisocyanate core on WS. WS increased according to the following order of diisocyanates: Symmetrical cycloaliphatic < symmetrical aromatic < asymmetrical aromatic < substituted aliphatic chain < asymmetrical cycloaliphatic < linear aliphatic chain [48]. This order shows that WS depends on the combination of the diisocyanate chemical character and its symmetry. Generally, fully aliphatic polymers had higher WS than those, having aromatic and alicyclic structures. If compared to the influence of only aliphatic diisocyanates on WS: The linear HMDI and substituted with methylene groups TMDI, the presence of the latter resulted in higher WS.

The influence of cycloaliphatic ring on WS was not defined clearly enough. Both aromatic structures (regardless of the molecular symmetry) and symmetrical cycloaliphatic CHMDI caused a decrease in WS. The presence of asymmetrical cycloaliphatic IPDI resulted in a significant WS increase. This result can be confirmed by findings of Łukaszczyk et al. The comparative analysis of the influence of Bis-GMA and its isosorbide analogue (IS-DMA) (Scheme 3) on WS of their homopolymers and copolymers with TEGDMA 40 wt% was performed in their work [59].

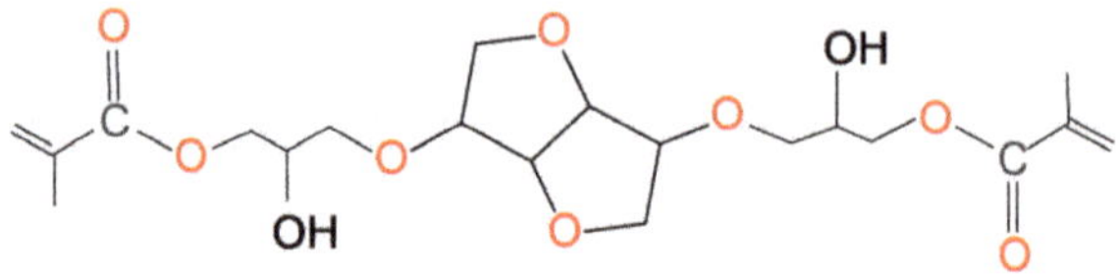

Scheme 3. The chemical structure of IS-DMA.

It was found that WS of IS-DMA homo- and copolymer (respectively, 173 and 100 µg/mm³) was much greater than WS of their Bis-GMA analogues (respectively, 11 and 15 µg/mm³). IS-DMA differs from Bis-GMA by the diol core, deriving from isosorbide (a bicyclic diol, consisting of two fused tetrahydrofuran rings). Polymers of both types were characterized by a similar molecular weight, degree of conversion and mechanical properties [59]. The observed differences in WS were likely due to the reduced strength of intermolecular interactions. It can be explained by the large size of the space the isosorbide occupies and its irregular shape. As a result, larger empty space might be present within polymer network clusters. From these reasons water can easily penetrate through the IS-DMA matrix and accumulate, being caught by hydroxyl, ester and ether groups.

The influence of the presence of hydrogen bond proton donor groups on water sorption exhibits another interesting issue. Obviously, hydroxyl and urethane groups promote water swelling as they can form strong hydrogen bonds involving proton donor as well as proton acceptor atoms [60,61]. That would explain the higher WS of the Bis-GMA homopolymer, having two pendant hydroxyl groups, than the OH-free Bis-EMA homopolymer [53]. On the other hand, the WS of highly crosslinked TEGDMA homopolymer is significantly higher (Table 2). It can be attributed to the lack of proton donors, resulting in an inability to form strong hydrogen bonds and the presence of highly elastic oligoether chains in the TEGDMA molecule [50,53,58]. It leads to the general conclusion that physical crosslinking, resulting from hydrogen bonding, can tighten a polymer network structure and, to some extent, limit water swelling.

Table 2. Water sorption of typical dental dimethacrylate homopolymers.

Monomer	Water Sorption (µg/mm³)
Bis-GMA	32.18 [1], 33.49 [2]
Bis-EMA (n = 4)	20.10 [2]
UDMA	23.85 [1], 29.46 [2]
TEGDMA	66.93 [1], 69.51 [2]

[1] As cited in [48]; [2] As cited in [53].

3. The Chemical Crosslink Density

As mentioned above, the theoretical crosslink density (q_{theor}) of dimethacrylate polymer networks is of great importance for describing the general tightness of the polymer network.

A dimethacrylate polymer network structure differs from the structure of a typical crosslinked polymer obtained by the polymerization of difunctional methacrylate with a certain amount of a tetrafunctional dimethacrylate crosslinker [62,63]. In that latter type of a crosslinked structure, a dimethacrylate molecular weight can be neglected as its body is treated as a volumeless tetrafunctional crosslink, connecting polymethacrylate chains. Some works on dimethacrylate polymer networks lead to the conclusion that, in structural considerations, the repeating unit can be recognized as a primary chain ending at both ends with two volumeless trifunctional crosslinks $-CH_2C(CH_3)-$ [64,65]. It justifies making an approximation that the dimethacrylate monomer molecular weight (MW) corresponds to the network parameter (M_c)–the average molecular weight between cross-links [21]. In this way, the concentration of double bonds (X_{DB}) can be used as a measure of the theoretical crosslink density. The method of calculating q_{theor} from MW in relation to X_{DB} was used in several works [18,31,35,66,67].

Dimethacrylates with higher MW have lower X_{DB} and therefore they are expected to form networks of lower q_{theor}.

The theoretical crosslink density never corresponds to the real crosslink density (q) [21]. The crosslink density in real systems results from q_{theor} and other factors, such as the incomplete conversion and presence of loops [1–5,23–25]. As the degree of conversion (DC) is never full in dimethacrylate polymer networks, the real crosslink density is therefore always lower than its theoretically calculated value. Loops, although they do not cause a decrease in the DC (both chain ends are attached to the same junction point [21]), they do cause a decrease in q (Figure 1).

Several propositions of calculating the real crosslink density (q) are presented in the literature.

In the work of Barszczewska-Rybarek et al., q of a series of urethane-dimethacrylate polymer networks (Scheme 2) was calculated from the degree of conversion (DC). The following formula was developed [68]:

$$q = \frac{2DC - 1}{DC} \tag{1}$$

In the work of Sideridou et al. this equation was further developed by using the mass of unreacted monomer [53]:

$$q = \frac{2DCm_1 - m_3}{DCm_1} \tag{2}$$

where m_1 is the mass of the dried discs obtained through polymerization, m_3 is the mass of the dried discs after water extraction.

Another approach was presented in the work of Barszczewska-Rybarek [35]. The concentration of double bonds in the UDMA monomer (X_{DB}) was reduced by the fraction of unreacted double bonds and the obtained value was treated as a measure of the real crosslink density (q). q was calculated using the following equation:

$$q = X_{DB} \times DC \tag{3}$$

The crosslink density may also be calculated from the network parameter (M_c). M_c is usually determined by utilizing dynamic mechanical analysis [29,65,68–72] and in swelling studies [73], following well known Flory–Rhener equation [74].

The literature provides several equations, which relates M_c to q. The best known and most commonly used is the following equation [75]:

$$M_c = \frac{MW}{q} \tag{4}$$

where MW is a monomer molecular weight.

However, when Equation (4) is applied to highly crosslinked dimethacrylate polymer networks it may give inadequate results [22,29,65,68]. In order to better characterize the structure of poly(dimethacrylate)s the modifications of Equation (4) were developed. Assuming a dimethacrylate molecule to form one tetrafunctional crosslink, the following equation was constructed [65]:

$$M_{c(f=4)} = \frac{MW}{q} - \frac{MW}{2} \tag{5}$$

Assuming a dimethacrylate molecule to form two trifunctional crosslinks the following equation was constructed [65]:

$$M_{c(f=3)} = \frac{2MW}{3q} - \frac{MW}{3} \tag{6}$$

To summarize, it can be said that the network is tighter as q increases and M_c decreases. If M_c is greater than MW and DC is higher than 50%, this procedure can be used for describing the polymer network tightness, otherwise, the value of q does not have a physical meaning [29,68].

4. The Physical Crosslink Density

The physical crosslinking in dental dimethacrylate polymer networks results from hydrogen bonding. A wide range of hydrogen bond options in dental poly(dimethacrylate)s are shown in Figure 2.

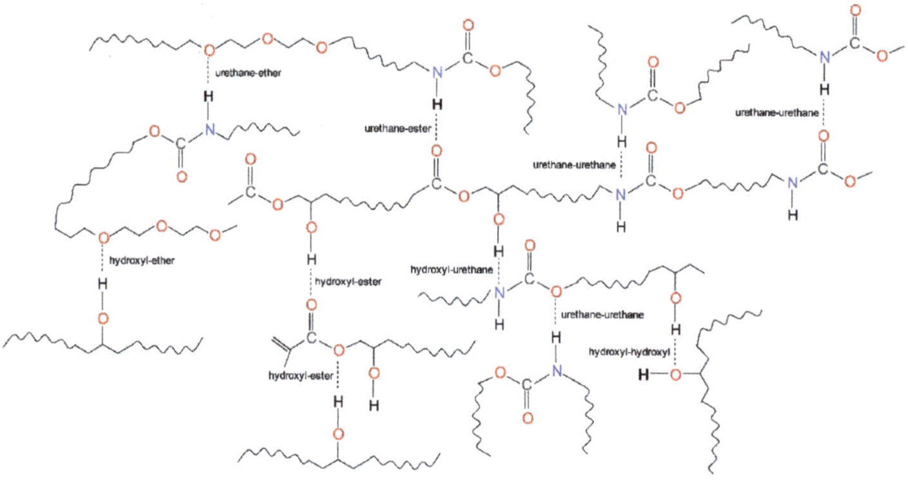

(a)

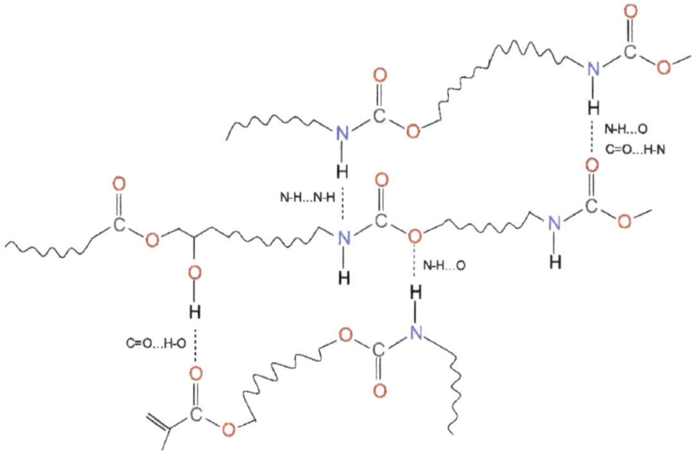

(b)

Figure 2. Hydrogen bonds in dental dimethactylate polymer networks (based on [76]): (**a**) various possibilities for hydrogen bonding; (**b**) hydrogen bonds narrowed to examples presented in Table 3.

Homopolymers and copolymers consisting of Bis-GMA and/or UDMA are therefore crosslinked both chemically and physically. TEGDMA can be involved in physical crosslinking by the acceptor oxygen atom of the ester and ether groups. However, this is only possible if TEGDMA is copolymerized with another dimethacrylate, which can provide a proton donor. This means that the TEGDMA homopolymer cannot be physically crosslinked.

Hydrogen bonding determines the dimethacrylate monomer viscosity. The higher the dissociation energies of hydrogen bonds, which can be formed in the system, the stronger the H-bonds and the higher the monomer viscosity [3,31,45,50,77]. Small molecule monomers, such as TEGDMA, that lacks an H-bond proton donor present very low viscosities (0.05 Pa·s). On the other hand, Bis-GMA exhibits a dramatically high viscosity (from 700 to more than 1300 Pa·s [6,18,19,31]. This is the result of the formation of strong intermolecular hydrogen bonds involving hydroxyl donors.

The influence of hydrogen bonding on a monomer viscosity can be represented by the comparison of Bis-GMA viscosity with its ethoxylated analogue. Bis-GMA is about a thousand times more viscous than Bis-EMA (Table 1) [18,19].

Assuming the analogy in the monomer and polymer chemical structure, this dependence was adopted for the characterization of the strength of physical interactions in a polymer. The overall strength of intermolecular interaction was considered to have followed a trend like viscosity, i.e., the higher the monomer viscosity, the higher the overall strength of physical interactions in a polymer network [3,31,45,66,77,78].

Another approach to quantifying physical crosslinking was based on the analogy between the intermolecular interactions in the solutions of urethane-dimethacrylate monomers (Scheme 2) and in the corresponding polymers. The concentration dependence on 1H-NMR chemical shifts of the NH urethane protons was employed for this purpose [65]. The association constants were then calculated using the Benesi–Hildebrand equation [79]. Their values were discussed from the perspective of the monomer wing length and the core symmetry (only monomers with aromatic MDI and TDI cores were tested). The longer oligooxyethylene chain–the lower the association constant. The presence of asymmetrical TDI resulted in lower association constants if compared to the symmetrical MDI.

Table 3. The influence of hydrogen bonding on the location of absorption bands of the N–H and C=O groups in polyurethane systems, represented in Figure 2b.

Proton Donor Group	H-Bond Type	Wavenumber (cm^{-1})
ν(N–H)	–	3445–3450 [1]
ν(N–H)	N–H...N–H	3315–3340 [1]
ν(N–H)	N–H...O	3260–3290 [1]
Proton Acceptor Group	**H-Bond Type**	**Wavenumber (cm^{-1})**
ν(C=O)	–	1730–1740 [1] 1721–1726 [2]
ν(C=O)	C=O...H–N	1703–1710 [1]
ν(C=O)	C=O...H–O	1712–1719 [2] 1705 [2,3]

[1] As cited in [80]; [2] As cited in [81]; [3] deconvoluted H-bonding.

Pfeifer et al. [5], as well as Lemon et al. [50], estimated the strength of hydrogen bonds by utilizing FTIR spectroscopy in the studies on poly(dimethacrylate)s. The wavenumber and intensity of the -OH absorption peak at approximately 3500 cm^{-1} were monitored. They assumed that, the longer the peak maxima wavelength and the higher the -OH absorbance, the higher the strength of hydrogen bonds. Yilgör et al. provided FTIR data about changes in the ν(N–H) and ν(C=O) band location resulting from hydrogen bonding [80] and Antonucci et al. provided data for the ν(C=O) band location [81]. Those results confirm the correctness of the methodology used in the Pfeifer and Lemon works.

The literature also provides the physical crosslink density calculation procedure, which is analogue to the procedure being utilized to determine the theoretical chemical crosslink density from the concentration of double bonds in a monomer and its molecular weight. The physical crosslink density was calculated from the number of urethane bonds in a series of urethane-dimethacrylate polymer networks (Scheme 2) [35].

5. The Degree of Conversion

The degree of double bond conversion (DC) is the most evident parameter, defining the dimethacrylate polymer network structure. This is also the most often used parameter when the structure-property relationships are being investigated.

The DC in poly(dimethacrylate)s is never full [1–9,18,19,23–25]. Conventional composite dental materials usually reach a DC of about 50–75% [6,8,9,47,82–84]. The limiting conversion in homopolymer networks of spacious dimethacrylates, such as Bis-GMA, can be even less than 50 % [18,19,31,47]. Polymer networks, characterized by the DC lower than 50% are unacceptable in practical applications due to the presence of a sol fraction [1,3,5,68,82,85]. Some authors suggest a DC of 55% as a minimum for clinical success in dentistry [82]. Insufficient curing causes a radical distortion of physicochemical and mechanical behavior as well as decreases material biocompatibility [19,25,85,86]. The leaching of a soluble fraction can cause tissue irritation and have a more serious effect on the organism [87,88]. For these reasons, monomers are usually copolymerized, which provides a satisfactory level of curing and properties.

The DC, in general, depends on the monomer chemical structure [18,19,31,54,77,78,83,89–91]. However, several polymerization parameters, such as: initiation technique [92], curing time [5], sample thickness [84,93], initiator system and its concentration [31,94–98], filler content [99], irradiation time [83], and source [83,97] also play a significant role.

5.1. Methods of the DC Determination

Several instrumental methods are available to allow the DC determination in dimethacrylate polymer networks. The infrared spectroscopy (FTIR, ATR-FTIR, NIR), Raman spectroscopy (RS), differential scanning calorimetry (DSC) and solid-state nuclear magnetic resonance (ssNMR) are particularly readily used.

5.1.1. Fourier Transformed Infrared (FTIR) Spectroscopy

FTIR is the most popular method for the determination of DC in poly(dimethacrylate)s. Polymer samples to be analyzed by FTIR can be prepared in various forms. Usually, they are powdered and then mixed, in small amounts, with a dry potassium bromide (KBr) powder [18,29,45,59,90,92,98–102]. In recent works attenuated total reflection (ATR) sampling technique was effectively used for the DC determination [85,91,103–108]. The main advantage of the ATR-FTIR is simplicity in sample processing. Samples are examined directly in the solid or liquid state without further preparation [109,110]. In the case of the real-time experiments, a monomer drop is sandwiched between NaCl or KBr plates and then polymerized directly in a measuring chamber of the apparatus [5,29,31,54,91,108].

To determine the DC in dimethacrylate polymer networks, the near (NIR, from 4000 to approximately 14,000 cm^{-1}) and mid (MIR, from 400 to 4000 cm^{-1}) spectral infrared regions can be used [83,101]. However, measurements in the MIR region remains fundamental. As Fourier transform spectrometers typically work in the mid-infrared, this technique is abbreviated as FT-IR [111].

Regardless of the spectral region to be used, the determination of the DC in dimethacrylate systems is based on the monitoring of the absorption intensity of double bond vibrations, which decreases due to polymerization. The decreasing absorption intensity of the C=C bonds can be monitored with the use of several bands, present on the spectrum, which refers to the C=C bond various deformations. Their location and intensity depend on the spectrometer working infrared region (NIR and MIR).

The dimethacrylate group provides bands in the FTIR spectra at 1637 cm^{-1}, referring to the C=C stretching vibrations [18,29,54,85,90–92,98–100,103–120], 948 cm^{-1}–referring to the =CH_2 wagging and 816 cm^{-1}–referring to the =CH_2 twisting [112]. Since the peak at 1637 cm^{-1} is stronger [112], it is most commonly used.

In the dimethacrylate NIR spectra, absorption of =C–H bond is located at about 4743 cm^{-1}, while the first overtone is at 6165 cm^{-1}. Because the baseline drops sharply in the 4743 cm^{-1} region, peak area measurements do not provide reliable results. Conversely, the region at 6165 cm^{-1}, is very stable and there is no ambiguity in baseline construction [95]. Therefore, the methacrylate double bond conversion is determined by measuring the overtone band at 6165 cm^{-1} [5,19,31,66,67,81,83,93,95,101,109,121–123].

Since the Beer–Lambert law requires equal sample thickness, the direct ratioing of the C=C absorbance intensity in the polymer and monomer samples is not applicable [101]. For this reason, the relative band ratio method is usually used for the DC determination in poly(dimethacrylate)s [101]. This is, of course, approximation, because, due to the polymerization, the refractive index and the intermolecular interactions in the monomer and polymer are altered [124]. In the internal standard method, the intensity of the double bond vibrations is related to the intensity of the internal standard. This is a band, referring to the bond, whose absorptivity in the monomer and in the polymer is unaffected by a polymerization [101]. The DC is then calculated according to the following general formula [18,29,83,85,90–92,98–101,103,104,107,112,114–119,125]:

$$DC(\%) = \left[1 - \frac{\left(\frac{A_{C=C}}{A_{IS}}\right)_{pol}}{\left(\frac{A_{C=C}}{A_{IS}}\right)_{mon}}\right] \times 100 \qquad (7)$$

where $A_{C=C}$ is the absorbance of the C=C double bonds, A_{IS} is the absorbance of the internal standard, pol is polymer, mon is monomer.

One of the most valuable series of bands utilized as an internal standard derive from in-plane skeletal stretching vibrations of the C=C aromatic ring, occurring in the range of 1620 to 1565 cm^{-1} [101,112]. The band at 1608 cm^{-1} is mostly used for this purpose [18,83,85,90–92,98–100,103,107,115–117]. The second band, at 1582 cm^{-1} can also be used, however, it is usually weaker [82,103]. An aromatic band is present at 4623 cm^{-1} too. It is not recommended to be used as an internal standard, because depending on the dimethacrylate chemical structure, it can undergo a slight variation in its intensity, introducing an error in the DC determination [121].

If the monomer contains no aromatic structures, the band from the C=O stretching vibrations can be used as an internal standard. It is located at around 1715–1720 cm^{-1} [18,29,90,91,101,104,112,114,118,119,125]. It should be noted, that this method produces underestimated DC values [90,91]. It was demonstrated in the studies on a series of Bis-GMA/TEGDMA and Bis-GMA/Bis-EMA copolymers. The average difference between the DC measured with the carbonyl and aromatic internal standards was, respectively of 23% and 17% [91]. Variations were also observed in the DC of urethane-dimethacrylate homopolymers (Scheme 2, TDI and MDI series) [90]. The difference between DC values obtained with the use of the carbonyl and aromatic internal standards depended on the monomer molecule length and decreased from 51% to 0% as the monomer molecule length increased. The reason lies in the loss of conjugation between the C=O and C=C bonds after polymerization. In effect, the C=O bond strength increases, causing a reduction in the C=O peak intensity [18,126]. Due to the increased research activity towards developing new dimethacrylates that have no aromatic bisphenol A, the significance of this method has been growing in importance in recent years.

If a dimethacrylate does not provide a spectrum with a stable, well-resolved internal standard band, a calibration curve can be an option [101,127]. The calibration curve, representing a linear relationship between the absorption intensity and molar concentration of C=C bonds in a standard solution (moles C=C/mL), enables the determination of the C=C molar concentration in a monomer. By multiplying the C=C molarity by the methacrylate unit molecular weight (69.081 g/mol) and dividing the result by a monomer density (g/mL), a weight percentage of the methacrylate groups in monomer was obtained ($WPMG_{mon}$). In order to use the calibration curve for the undiluted monomer, the experiment must be carried out using the same cell as the one used to build the calibration curve.

The weight percentage of the methacrylate groups remaining in a polymer after polymerization ($WPMG_{pol}$) was determined from a polymer IR spectrum. $WPMG_{pol}$ was calculated by dividing the C=C bond absorption intensity by the product of sample thickness and the optical constant (K) [101]. The K value of 0.64 was found by Rueggeberg [127]. Finally, DC was expressed according to the formula [101]:

$$DC(\%) = \left[1 - \frac{(WPMG)_{pol}}{(WPMG)_{mon}}\right] \times 100 \tag{8}$$

Another option for the DC determination in dimethacrylate systems is the real-time infrared spectroscopy (RT FTIR). This method is based on a continuous measurement of the absorbance of the peak deriving from the vibration of C=C bonds during irradiation. Thanks to maintaining constant sample thickness, this method does not require the application of an internal standard or calibration curve. The DC determination is based on the observation of changes in the peak absorbance from the beginning of the experiment to its completion [5,29,31,54,83,91,106,108]:

$$DC(\%) = \left[1 - \frac{A_t}{A_0}\right] \times 100 \tag{9}$$

where A_t is the C=C bond absorbance at time t, A_0 is the C=C bond absorbance before photoactivation ($t = 0$).

Since this method is limited to samples polymerized during the FTIR experiment, it is especially valuable in studies on the photopolymerization kinetics [31,93].

5.1.2. Raman Spectroscopy (RS)

Raman spectroscopy is a complementary vibrational technique for FTIR [112]. It offers the possibility of quantitative characterization of a polymerization extent in dimethacrylate systems, using similar peaks as FTIR. However, sample preparation for RS is much simpler (a sample does not require further processing) [84,112]. The internal standard method is also typically used for the determination of the DC with the use of RS. The DC is typically calculated by monitoring the percentage decrease in the intensity of the methacrylate double bond peak, at around 1640 cm^{-1}, caused by polymerization, in respect to the aromatic peak at 1610 cm^{-1} [47,52,77,83,84,89,94,102,113,120,124,126,128–130]. The C-O-C ether stretching vibration band, occurring in the range of 1294 to 1118 cm^{-1} can be alternatively used [84,112]. Stretching and deformation vibrations of the dipole C–Cl and C–F bonds would be willingly used, occurring in the range of 800 to 200 cm^{-1}, however, they rarely occur in a dimethacrylate structure [63]. NIR RS can also be utilized to monitor DC [128].

5.1.3. Differential Scanning Calorimetry (DSC)

DSC is particularly well suited to studying the dimethacrylate photopolymerization kinetics [2,3,81,131–133]. The DSC apparatus has to be additionally equipped with a UV/VIS radiation source for such measurements. This technique, called differential photocalorimetry, working in an isothermal mode, is a convenient and easy way to determine the polymerization rate and degree of conversion of photoreactive systems.

Antonucci et al. determined DC by measuring the polymerization heat of dental dimetacrylates with the use of DSC working in dynamic mode. Polymerization was thermally activated with benzoyl peroxide [134].

Moszner et al. developed the procedure for the DC determination in real samples (taken from already existing material) using DSC [135]. The exothermal effect of the post-curing reaction was monitored in the dynamic mode. The *DC* was calculated by ratioing the post-polymerization heat and the heat of a total conversion according to the equation [135,136]:

$$DC(\%) = \frac{\Delta H_p}{\Delta H_{100}} \times 100 \qquad (10)$$

where ΔH_{100} (J/mol) is the molar enthalpy of polymerization for the theoretical case of total conversion (*DC* = 100%, ΔH_{100} = 57.8 kJ/mol [137]), ΔH_p (J/mol) is the molar enthalpy determined from the enthalpy value obtained in the DSC measurement ($\Delta H_{p,exp}$ (J/g)), which was calculated according to the following equation:

$$\Delta H_p\left(\frac{J}{mol}\right) = \frac{\Delta H_{p,exp} \times MW}{f} \qquad (11)$$

where *MW* is the monomer molecular weight (g/mol), *f* is the number of functional groups in the monomer (*f* = 2 in dimethacrylates).

5.1.4. Solid State Nuclear Magnetic Resonance (ssNMR)

Solid-state NMR is another valuable technique for characterizing structural features of the polymers in the solid-state. In the works of Pereira et al. [116] and Morgan et al. [138], solid-state NMR has been shown to be a convenient method for measuring the degree of conversion in Bis-GMA containing polymers. ^{13}C signals of carbonyl groups in the monomer and polymer methacrylate groups were observed at about 166 and 177 ppm, respectively. The integral values of these resonances were used to calculate the *DC*.

5.2. The Influence of Chemical Structure on the DC

The compilation of results from many studies leads to the following general conclusions. The *DC* notably depends on the monomer molecular elasticity and distance between methacrylate groups. The more elastic and longer the dimethacrylate molecule, the higher the *DC*. Dimethacrylates of stiff, spacious molecules often do not polymerize to the *DC* sufficiently enough to ensure material compliance with application requirements. Improved curing efficiency can be achieved by the addition of a certain amount of a reactive diluent [3,5,18,19,31,47–50,52,54,55,77,90,91,117,122,129,136]. High strength of intermolecular interactions between monomer molecules (hydrogen bonding) can also negatively affect the *DC* [3,18,31,50,77,78,117].

In the works of Gajewski et al. [19] and Sideridou et al. [18,117], the limiting *DC* in homopolymers of typical dental dimethacrylates was determined and explained. The homopolymers were arranged according to the following order by the increasing *DC*: Bis-GMA < UDMA < Bis-EMA < TEGDMA (Table 1). High molecular stiffness of Bis-GMA, resulting from the presence of aromatic rings and -OH groups, explains the lowest *DC* in the homopolymer. The limitations in rotational movement are so large that they prevent the Bis-GMA homopolymer from achieving the *DC* higher than 50%. Reversibly, the hydroxyl group lacking Bis-EMA was characterized by a higher possibility of rotational conformations, which resulted in a significant increase in *DC*. The combination of high elasticity and hydrogen bonding explains moderated *DC* of the UDMA homopolymer. The fully aliphatic character of TEGDMA and its shortest length was expressed in the highest *DC* [18,19,47,50,117]. In the study of Khatri et al. the Bis-GMA hydroxyl groups were blocked with the n-alkyl pendant substituents [66]. The *DC* in new polymers was higher than the *DC* in the Bis-GMA homopolymer. Additionally, it was found that the longer the substituent, the higher the *DC*. It shows, that due to tangled configuration that can be obtained by the aliphatic linear chains, their presence in the dimethacrylate does not limit curing quality. All these findings revealed a key role of hydrogen bonding in the evolution of a dimethacrylate polymer network. Hydrogen bonds can seriously restrict molecular mobility and diffusion-controlled termination, thus causing a reduction in the crosslinking efficiency.

In research on copolymers, when Bis-GMA was combined with a more flexible comonomer, such as TEGDMA, an increase in DC was observed. The higher the TEGDMA content in the Bis-GMA/TEGMA copolymer, the higher the DC [5,54,55,70,91]. However, in the works of Elliott et al. [23] and Lopez-Suevos et al. [139], the risks associated with the increasing TEGDMA content were shown. The higher the TEGDMA concentration, the greater the tendency to form microgels, due to the greater likelihood of primary and secondary cyclization, which may jeopardize polymer packing density and lead to a looser network structure. It was confirmed in the work of Borges et al., where the influence of the Bis-GMA/TEGDMA ratio on the polymer network DC and crosslink density was determined [70]. The crosslink density increased as the TEGDMA content increased up to a limit of 50 wt% and then decreased, with a further increase of TEGDMA concentration. The DC in all polymers was similar and equaled around 95%. It informs that a certain number of double bonds were engaged in cyclization.

Test results for the DC in the Bis-GMA/Bis-EMA copolymers showed that DC increased as the Bis-EMA content increased [91], which is in agreement with the discussion presented above. The higher the possibility of rotational movement, the higher the DC.

A study of Barszczewska-Rybarek on a series of new urethane-dimethacrylates (Scheme 2) provided knowledge on the influence of the diisocyanate core chemical character and the oligoether chain length on the DC in homopolymers [90]. The DC in fully aliphatic polymers was found to be higher than the DC in polymers having aromatic and cycloaliphatic moieties. The presence of aromatic rings caused a greater decrease in DC than alicyclic structures. Monomers with asymmetrically substituted rings polymerized to higher DC when compared to those having symmetrically substituted rings. Additionally, the DC was closely related to the length of the oligooxyethylene chain, present in the monomer wings. The longer the chain, the higher the DC. However, monomers having tetraoxyethylene chains revealed a slight decrease in DC. The initial increase in DC was explained by the increased flexibility of monomer molecules. An increase in the possibilities of rotational conformations facilitated reaction-diffusion [2]. The final drop in DC was explained by a significant decrease in the concentration of double bonds, causing their remoteness. Park et al. studied DC in a series of copolymers of Bis-GMA with oligoethylene glycols dimethacrylates. The same as above, they found that the higher the number of oxyethylene units in a chain, the higher the DC [52].

Podgórski reported the results for a series of cured new dimethacrylates, synthesized from glycidyl methacrylate with dicarboxylic acid esters. The latter were obtained by the reaction of nadic anhydride with 1,3-propylene, 1,4-butylene, 1,5-penthylene and 1,6-hexylene glycols [119]. Chemical structures of new monomers were like Bis-GMA regarding the wings and differed regarding the cores. A very high DC was achieved by replacing the bisphenol A with the core, being the combination of alicyclic structures and oligomethylene chain. The longer the glycol, the higher the DC. The presence of the -OH groups did not negatively affect DC. The positive effect of the central aliphatic chain on DC was shown to be dominant.

6. The Influence of Chemical Structure and Crosslink Density on Mechanical Properties

The decrease in crosslink density causes the deterioration of physicomechanical properties [5,32,41,47,53,55,56,63,85,92,94]. It comes from the decrease in DC and cyclization. The incomplete conversion results in the presence of pendant groups and chains ending with free methacrylate groups (Figure 1) [1–5,21]. DC does not reveal the presence of loops (Figure 1b). It was found that the longer the dimethacrylate monomer the greater the possibility of forming cycles [5,23,47,139]. For example, when comparing urethane-dimethacrylates with di- and tetraoxyethylene chains (Scheme 2), the latter one has a higher tendency to cyclization [31,48]. On the other hand, when comparing elastic TEGDMA to stiff and spacious Bis-GMA, the latter monomer has a lower tendency to cyclization [23,53,93,139]. Pendant chains and groups, as well as loops, are mechanically ineffective [140,141]. Pfeifer et al. revealed that they can act similarly to a plasticizer–by decreasing the network tightness they give rise to the increase in rotational movement [5].

Entanglements, another type of polymer network defect, work in the opposite way. As they represent additional restraints that magnify the change of entropy with deformation, their presence causes an increase in the physical crosslink density (Figure 1b) [20–22]).

The system composed of homo- and copolymers of Bis-GMA, Bis-EMA, UDMA, and TEGDMA (Scheme 1) was the object of many studies, which led to the general conclusions on the structure-property relationships [18,32,50,52,54–56,77,122,142,143]. The discussion presented below summarizes the results on the influence of chemical structure and degree of crosslinking on selected mechanical properties. The level of crosslinking was characterized by the DC (Table 1), the concentration of double bonds (Table 1) and the strength of hydrogen bonds (Table 4). The values of modulus of elasticity, mechanical strength, hardness and impact resistance of Bis-GMA, Bis-EMA, UDMA, and TEGDMA homopolymers are summarized in Table 5.

Table 4. Types of hydrogen bonds and corresponding energies (adapted from [32]).

The Hydrogen Bond Type	Energy (kJ/mol)
O–H.......N	29
O–H.......O	21
N–H.......N	13
N–H.......O	8

Table 5. Mechanical properties of popular dimethacrylate homopolymers.

Monomer	Young's Modulus (MPa)	Flexural Modulus (MPa)	Flexural Strength (MPa)	Hardness (MPa)	Impact Resistance (kJ/m^2)
Bis-GMA	1427 [1]	1000 [2]	72.4 [2]	73 [3]	6.41 [4]
Bis-EMA (n = 4)	744 [1]	1100 [2]	87.3 [2]	–	–
UDMA	1405 [1]	1800 [2]	133.8 [2]	162 [3]	4.85 [4]
TEGDMA	1134 [1]	1700 [2]	99.1 [2]	129 [3]	8.83 [4]

[1] Taken from [53]; [2] Taken from [19]; [3] Taken from [48]; [4] Taken from [35].

The Bis-GMA homopolymer network is characterized by the lowest crosslink density, which results from the lowest concentration of double bonds in the monomer and the lowest DC in the polymer (Table 1). Simultaneously, the Bis-GMA homopolymer is characterized by a high degree of physical crosslinking, due to the presence of strong OH-type hydrogen bonds. When homopolymers were compared by hardness, the Bis-GMA homopolymer was found to be the softest [32,142]. This indicates that hardness is very sensitive to the DC and chemical crosslink density. Higher values of the remaining mechanical properties suggest that the low level of chemical crosslinking can be compensated by physical crosslinking [32].

In the study of Gajewski et al., the flexural properties of Bis-EMA and Bis-GMA homopolymers were compared [19]. They found that poly(Bis-EMA), which is not physically crosslinked, was characterized by almost 100% higher DC and 20% higher flexural strength. The modulus of both polymers was similar. This indicates that the lack of hydrogen bonding can be partially compensated by the increase in DC. Flexural strength benefited more from the increase in DC than the modulus. In the study of Khatri et al., the influence of n-alkyl Bis-GMA substituent on the DC and flexural strength was shown [66]. By blocking the -OH groups, the flexural strength decreased. It was explained by the decrease in the chemical crosslink density (the DC increased, however, the concentration of double bonds decreased).

The TEGDMA homopolymer is fully aliphatic and has the highest crosslink density, resulting from the smallest monomer molecule and highest DC in the polymer (Table 1). It is not physically crosslinked. It was characterized by a relatively high modulus and the highest impact resistance among homopolymers. However, its flexural strength was the lowest [32]. This means that a high level of chemical crosslinking can ensure high modulus and impact resistance, however, it is insufficient to ensure high mechanical strength.

The UDMA homopolymer is fully aliphatic and chemically as well as physically crosslinked. It was characterized by the highest flexural strength and hardness among homopolymers, which can be explained by high DC and hydrogen bonding. Additionally, this result indicates that the strength of physical crosslinks is not a key factor for these properties, as the comparison of the hydrogen bond strengths demonstrates. The strength of NH-type proton donor H-bond is 40% lower than that of the OH-type proton donor H-bond (Table 4). On the other hand, modulus and impact resistance of poly(UDMA) were the lowest, which leads to the conclusion that these properties of dimethacrylate polymer networks are especially sensitive to the strength of physical crosslinks [32,55].

The influence of chemical and physical crosslinking on tensile and flexural properties was also studied on Bis-GMA, UDMA and TEGDMA copolymers [3,5,18,32,55,56,93,102,122,142,143]. The increase in the TEGDMA content in Bis-GMA/TEGDMA copolymer resulted in an increase in tensile strength [56], but the decrease in flexural strength [56,122]. The decrease in flexural strength was explained by the decrease in physical crosslink density [18,32,93,122]. Tensile strength was found to be less sensitive to physical crosslinking and more sensitive to chemical crosslinking, which increased with the TEGDMA incorporation. The addition of UDMA into copolymers resulted in an increase in both tensile and flexural strength [102,122]. It confirmed that in the case of flexural strength, the presence of physical crosslinks is more important than their strength [32,122].

The results for modulus of Bis-GMA, UDMA, TEGDMA copolymers (filled with silanized glass) were reported in the study of Asmussen et al. [56]. The modulus of Bis-GMA/TEGDMA copolymer increased by adding a certain TEGDMA content, which was 40 wt% (the positive effect of the increasing DC and crosslink density). Higher TEGDMA contents had a negative effect on modulus and caused its reduction. It was explained by the decreasing physical crosslink density and concentration of stiff aromatic structures. The same trend was observed when UDMA was added. A maximum in modulus of elasticity was achieved at 10 wt% of the UDMA content.

In the work of Sideridou et al., the effect of copolymer composition on Young's modulus was determined [53]. It was found, that Bis-GMA/TEGDMA copolymers had significantly higher values for Young's modulus than those predicted by the linear dependence of the values on the copolymer composition. On the contrary, the copolymers of Bis-GMA/UDMA and Bis-GMA/Bis-EMA had Young's modulus values slightly lower than those predicted. It was explained by lower DC in the TEGDMA-free copolymers and the presence of spacious pendant groups, constructed of long, massive Bis-GMA, Bis-EMA, and UDMA molecules, which affected a polymer structure by plasticizing it [5,53].

Hardness is another important mechanical property of poly(dimethacrylate)s. It was found that the hardness of the Bis-GMA homopolymer was lower by around 20% than the Bis-GMA/TEGDMA copolymer [32]. In the other study, copolymers of Bis-GMA with increasing TEGDMA content were characterized. It was found that the hardness of BisGMA/TEGDMA copolymers was unaffected by the TEGDMA content [143]. The results for the DC in the Bis-GMA homo- and copolymers, collected by Collares et al., can help to clear up that discrepancy [91]. The DC in poly(Bis-GMA) can reach the maximum of 40% [18,47,143]. The incorporation of TEGDMA in the amount of 10 wt% caused the DC increase of up to 60%. Further increase in the TEGDMA content resulted in further DC increases. The DC of 76% was found for the Bis-GMA/TEGDMA 10/90 wt% copolymer. It suggests that hardness becomes less sensitive to DC as it exceeds 50%. Copolymerization of Bis-GMA and TEGDMA with UDMA resulted in an additional increase in hardness [32]. This tendency was also noticed when other urethane-dimethacrylate monomers (Scheme 2) were used in copolymerizations [55]. This effect was explained by the synergistic effect of a high degree of double bond conversion and hydrogen bonding.

By comparing the impact resistance of Bis-GMA/TEGDMA and Bis-GMA/TEGDMA/UDMA copolymers it should be noted that, as the UDMA was added, its value decreased, despite the increase in the DC [32]. The moderate strength of hydrogen bonds was pointed at again to explain this result.

In the study of Pfeifer et al., the influence of Bis-EMA on the Bis-GMA/Bis-GMA copolymer impact resistance was determined [102]. It was found that impact resistance decreased as Bis-EMA was added. It was also explained by the decrease in the overall strength of intermolecular interactions. As the -OH group concentration decreased, the number of possible sites of physical crosslinking was reduced [102,115].

The detailed analysis of polymer networks of diacrylate analogous of dental dimethacrylates showed that the higher the crosslink density the higher the impact resistance [45]. As q results from the DC, the limit of 70% of the latter was found, below which intermolecular forces start to play a key role in impact resistance. The higher the strength of hydrogen bonds and the higher their concentration, the higher the impact resistance. The influence of crosslink density on the failure behavior in crosslinked polymers was confirmed by molecular dynamics simulations [72].

The detailed study on a series of Bis-GMA/TEGDMA AgNP nanocomposites [92] and urethane-dimethacrylate homopolymers [35,48] confirmed the conclusion that the incomplete conversion adversely affects mechanical properties. Additionally, results of the studies on AgNP nanocomposites allowed for constructing the following order of properties by increasing sensitivity to the DC: hardness < modulus < bending strength < impact resistance. Increasing AgNP concentration in nanocomposites usually caused decreases in these properties, which coincided with the decreasing DC. The detailed analysis of variations in mechanical properties as a function of the DC of thermally cured samples established that hardness increased throughout the polymer series, regardless of the decrease in DC. Remaining mechanical properties increased at lower AgNP concentrations, and then decreased. Modulus only decreased when DC dropped below 50%. Bending strength decreased when DC was down to 70%. Impact resistance was the most sensitive to DC and decreased when DC was down to 77% [92].

As can be seen, the impact strength of dimethacrylate polymers is a complex phenomenon. It should have high values in order to withstand forces generated during chewing and to prevent accidental fracture [144,145]. However, they are usually lower than expected [26]. Impact strength depends on many physicochemical factors, such as the monomer molecule elasticity, chemical crosslink density, physical crosslinking, degree of conversion and morphology. The higher the chain elasticity [35], crosslink density [32], and the strength of hydrogen bonds [32], the higher the impact resistance. The lower the DC, the lower the impact resistance [32,45,92]. The lower the DC, the greater the role of morphological heterogeneity, resulting from microgel formation [45], which is described in the following section. Additionally, impact strength can be radically reduced due to the presence of air bubbles trapped in the material [92].

7. The Morphology

Generally, poly(dimethacrylate)s are regarded as non-crystalline, amorphous, highly crosslinked polymers. Their supramolecular structure consists of microgel agglomerates [1–5,23–27]. Since it is only a short-range order of supramolecular organization, its characterization is complicated. Many works confirmed the formation of distinct nodular morphology of poly(di(meth)acrylate)s. It was visualized through microscopic observations with the use of atomic force microscopy (AFM) [26,32–34,36,38,71,123,136,146–148] and scanning electron microscopy (SEM) [55,123]. It is only in recent years that methodologies for quantitatively characterizing the morphology of dimethacrylate polymer networks have been developed. They utilize AFM supported with advanced mathematical tools [33,34,38,148], X-ray powder diffractometry (XRPD) [32,35,45,136] as well as techniques of thermal analysis: dynamic-mechanical analysis (DMA) [1,42,52,69–71,149–153], differential scanning calorimetry (DSC) [39] and thermogravimetry (TG) [154–156].

7.1. Qualitative Microscopic Characterization

The AFM images of poly(dimethacrylate) fractures revealed that their morphology consists of globular domains with diameters in the range of 10–100 nm [26,32–34,36,38,71,123,136,146–148]. They were attributed to highly crosslinked microgel agglomerates embedded in a less crosslinked matrix. The qualitative (agglomerate shape) and quantitative (agglomerate dimensions) differences were rarely observed between dimethacrylate polymers. Exceptionally, the Bis-GMA homopolymer revealed the existence of morphology presenting certain specific features. In contrast to the conventional spherical agglomerates, the morphology of poly(Bis-GMA) consisted of elongated "rods", having a diameter of approximately 70 nm and length of up to 500 nm [32]. This phenomenon coincided well with the results from SEM observations. The images of Bis-GMA homopolymer and its copolymer with 20 wt% of TEGDMA revealed a distinctive morphology, which consisted of long, needle-shaped, morphological objects. The remaining dimethacrylate homo- and copolymers showed a more uniform, smooth morphology [55].

Additionally, SEM and AFM observations led to the conclusion that the irradiation direction determines the cross-section morphology. The fracture surfaces perpendicular to the irradiation direction revealed the presence of parallel lines. The anisotropy in the polymerization shrinkage, caused by the high diameter to thickness ratio of mold was identified as the most likely reason for this phenomenon [32,55].

7.2. Quantitative AFM Characterization

7.2.1. Phase Imagining

The AFM morphology was quantified by using the phase imaging AFM of local viscoelasticity of the Bis-GMA and Bis-EMA acrylate analogue polymers [37–39]. The results provided valuable information about the kinetics of diacrylate radical polymerization, which can be transformed into dimethacrylates. The formation of three types of domains, with a varied elastic response, corresponding to varied crosslink density, was detected. However, any quantitative differences in morphologies depending on the chemical composition nor the irradiation type were observed.

7.2.2. Roughness Analysis

The globular AFM topography can be characterized using roughness analysis. There are several parameters available with common image processing software packages. They include the density of summits, mean summit curvature, surface area ratio, root-mean-square roughness. In the work of Munz, these parameters were readily used as descriptors for spatial resolution of diacrylate hydrogel polymer networks, consisting of globular domains with diameters in the range of ~10–100 nm [148]. In dental material studies roughness analysis is typically used to characterize composite surface morphology, depending on the filler type and polishing technique [147,157,158]. In the work of Lungu et al. this procedure was used for the characterization of the UDMA-silsesquioxane nanocomposites [123]. It was found that the addition of silsesquioxane caused a decrease in matrix transparency and it was attributed to the formation of agglomerates.

7.2.3. Fractal Analysis

Fractal analysis of the AFM fracture surfaces of poly(dimethacrylate)s was successfully applied for the quantitative description of their morphology [33]. For that purpose: the fractal dimension (D_F) and its extension–the generalized fractal dimension (D_q) were determined. From the difference between the extreme values of D_q: $D_{-\infty}$ and $D_{+\infty}$, the ΔD parameter was calculated. Its value refers to the degree of surface differentiation and tends towards zero with an increasing degree of homogeneity [159]. Fracture topography was further characterized by analyzing their profile line, which was characterized by the modified fractal dimension (D_β). As a result, several linear relationships between fractal morphology, chemical structure and selected physicomechanical properties of poly(dimethacrylate)s obtained by homo- and copolymerizations of Bis-GMA, UDMA, and TEGDMA were found. The higher the D_F and D_β, the higher the hardness. The higher the D_F, the higher the density. A linear relationship between ΔD and impact resistance was also found. The higher the ΔD, i.e., the greater the polymer heterogeneity, the lower the impact strength. That finding indicated that this mechanical property is governed by the degree of self-similarity.

The detailed analysis of the relationships between the chemical structure and fractal parameters led to the following conclusions [159,160]. The TEGDMA homopolymer morphology was characterized by the highest D_F as well as D_β and the lowest ΔD, indicating the highest structural homogeneity of this polymer network. It was explained by the complex (tangled) conformations, which linear aliphatic chains can adopt [66] and decreased free volume, due to the increased degree of conversion [5], both enabling the highest degree of space fitting. In opposite, the poly(Bis-GMA) fracture was characterized by the lowest D_F and D_β, which corresponds to the lowest degree of space fitting, due to high Bis-GMA rigidity. The results for ΔD lead to an explanation of the lowest impact resistance of the UDMA homopolymer. ΔD of poly(UDMA) was the highest. A lower ΔD of Bis-GMA homopolymer network indicated that its structure can be considered as more homogeneous (tighter), due to a higher strength of physical crosslinks in poly (Bis-GMA) than in poly(UDMA) (Table 4). Copolymerizations resulted in an increase in the fractal dimensions. Thus, the conclusion that TEGDMA plays a crucial role in the development of the dimethacrylate polymer network morphology by reducing its heterogeneity was drawn.

7.2.4. Percolation Theory

Another study on the model system of dental dimethacrylate homo- and copolymers showed that their AFM fracture surfaces have a percolating structure. For this purpose, the percolation probability (*P*) and length of the percolation path (*L*) were determined and their values were related to the selected properties of studied polymer networks [34]. Analysis of the results indicated that both parameters, *P* and *L*, are strongly dependent on the presence of the physical crosslinks, formed by hydrogen bonds. Their strength and the degree of conversion play a further role. However, it was noticed that the higher the crosslink density and hydrogen bond strength, the higher the percolation probability, i.e., the larger clusters are formed during polymerization.

The impact strength of the UDMA homopolymer was a special subject of interest due to its higher than expected brittleness. Studies on the quantitative characterization of the AFM fracture surfaces with the use of the percolation theory tools clearly showed that the impact strength increases with increasing percolation probability. Also, poly(UDMA) was characterized by the lowest *P*.

Additionally, the beneficial role of TEGDMA as a comonomer in the formation of a dimethacrylate polymer network structure was reconfirmed. Its presence allows for the formation of large percolating clusters and long percolation paths, consisting of chemical and physical crosslinks, with the aid of ester and ether proton acceptors.

7.3. Quantitative XRPD Characterization

XRPD can also be recognized as a valuable tool for the quantitative characterization of supramolecular organization in dimethacrylate polymer networks.

In several studies, the average dimensions of microgel agglomerates (*D*) were calculated by using the peak half-width and the Scherrer equation [32,35,45,136]. Their values were of a few nanometers, which confirmed that dimethacrylate polymer networks have a short-range molecular order [161]. Nevertheless, the *D* values were varied and depended on the polymer chemical structure and increased according to the following order: TEGDMA < Bis-GMA/TEGDMA < UDMA < Bis-GMA < Bis-GMA/TEGDMA/UDMA.

These studies revealed that the morphology of dimethacrylate polymer networks is influenced by the monomer molecule dimensions, packing abilities and strength of physical crosslinks. The more spacious the monomer molecule and the higher the hydrogen bond strength, the higher *D*. The more compact packing (higher elasticity, resulting in higher rotational possibilities), the lower *D*.

In the work of Barszczewska-Rybarek, it was found that the presence of massive agglomerates can negatively affect the impact resistance of poly(diacrylate)s. The ability to form the strongest hydrogen bonds in the Bis-GA/TTEGDA/UDA copolymer (the acrylate analogous to Bis-GMA/TEGDMA/UDMA) favored the formation of massive clusters. However, their concentration within the less crosslinked matrix was insufficient to effectively withstand the energy of the impact [45].

7.4. Thermal Analysis

Recently, the quantitative characterization of the structural heterogeneity of dimethacrylate polymer networks was extended to the use of thermal analysis techniques: dynamic-mechanical analysis (DMA) [1,42,52,69–71,149–153], differential scanning calorimetry (DSC) [39] and thermogravimetry (TG) [154–156]. The bimodal character of the peaks present on the thermograms in the glass transition region or the decomposition region, indicates the presence of two phases significantly varying in crosslink density. The temperature range in which the physical change or chemical reaction occurs must be determined. The narrower the temperature range the more homogeneous the network. In the case of DMA, the changes in the dynamic storage modulus and the loss tangent are observed, which occurs during the glass transition. Heterogeneity was quantified by the determination of the peak width at half-height [1,42,52,69–71,149–153]. Additionally, the network heterogeneity can be assessed by the storage modulus decrease in the glass transition region. It was found that, the more heterogeneous the network, the more rapid decrease in the storage modulus in the region of low frequencies. It was also found that the loss modulus in the region of its maximum is very slightly sensitive to the "long-range" network heterogeneity [149]. In the TG studies, the gradual weight loss at the polymer degradation temperature is determined [154–156]. In the DSC studies, the changes in the heat capacity during glass transition as well as the enthalpy of curing are evaluated [39]. However, this information does not provide answers to questions about the size and arrangement of morphological objects.

The literature provides a detailed DMA study on the influence of the Bis-GMA/TEGMA ratio on the polymer network heterogeneity, which was characterized by the crosslink density (calculated from E') and tan delta peak width [70]. The latter decreased as the TEGDMA content increased up to a certain limit and then increased, with a further increase of TEGDMA concentration. The Bis-GMA/TEGDMA 60/40 copolymer was characterized by the lowest tan delta peak width. i.e., it was recognized as the most homogeneous. That result was explained by dense packing and uniform free volume distribution [70]. The Bis-GMA/TEGDMA 20/80 copolymer was characterized by the lowest crosslink density and the broadest tan delta peak. That result, in relation to a very high DC (around 95%), indicated the greatest heterogeneity and the high tendency to cyclization represented by TEGDMA [70,152].

8. Biocompatibility of Dental Dimethacrylate Polymer Networks

The biocompatibility of dental materials is an important issue. It is closely connected with the polymer network structure development. In general, this term describes the material ability to be in contact with a living system without producing adverse effects by not being toxic, injurious, physiologically reactive and not causing immunological rejection [162]. In fact, it is a complex phenomenon, which involves biological interactions, patient risks, clinical experience, and engineering expertise. For that reason, no single test is required to evaluate the biological efficacy of material [163]. There is no specific procedure for characterizing the biocompatibility of dental materials. Several researchers recommend using standardized testing procedures for the evaluation of the biocompatibility of medical devices [163,164]. Those documents include ANSI/ADA standard No. 41 [165] and the ISO 10993 series of standards [166].

Dimethacrylate polymer networks are generally non-toxic [9]. Their toxicity results from leaching of low molecular weight substances, which can have a harmful effect on oral tissues [167–169]. Low molecular substances can come from the incomplete conversion and chemical degradation of dimethacrylate polymer network [168,169]. Several adverse reactions were reported after the application of dimethacrylate dental material. These include migraine, amenorrhea, fatigue, sinusitis, insomnia etc. The reliability of those complaints is difficult to verify, especially since several researchers pointed out that they can have a psychosomatic background [169].

Unreacted dimethacrylate monomer can be eluted by saliva [169]. Water and ethanol are the most popular solvents, occurring in the oral environment [53,58,102]. The local concentrations of leached monomers can be in the millimolar range, which is high enough to induce a variety of adverse biological effects [167]. It was found that the higher the degree of conversion, the lower the possibility of unreacted monomer content in the dimethacrylate matrix [170]. It was also found that the presence of unpolymerized surface layer, resulted from the oxygen inhibition, is a significant source of the unpolymerized monomer and increases composite cytotoxicity [169]. For this reason, mechanical removal of the oxygen inhibited layer could be recommended [171].

Typical dental dimethacrylates were thoroughly examined for cytotoxic properties. Issa et al. [172] and Reichl et al. [173] tested the cytotoxicity of dental dimetacrylates by using the MMT (3-(4,5-dimethylthiazol-2-yl)-2,5-diphenyltetrazolium bromide) and LDH (lactate dehydrogenase) assays. The TC_{50} values (concentrations altering MTT and LDH activity by 50%) were similar in both assays and ranged from around 0.32 to 6.7 mM. Monomers were arranged by increasing toxicity (increasing TC_{50}) according to the following order: HEMA (2-hydroxyethyl methacrylate) < DMAEMA (2-(dimethylamino)ethyl methacrylate, photoinitiation accelerator) < TEGDMA < UDMA < Bis-GMA [172]. Dental dimethacrylate monomers were also tested for acute toxicity. It was found that LD_{50} (the median lethal dose of a toxin, which corresponds to the dose required to kill half the members of a tested population) of Bis-GMA and UDMA was of > 5000 mg/kg body weight (rats), whereas LD_{50} of TEGDMA equaled to 10,837 mg/kg body weight (rats). As all the LD_{50} values were greater than 2000 mg/kg body weight, it means that these monomers cannot be classified as toxic [169]. The ED_{50} (the dose that produces a biological response in 50% of the population that takes it) for dental dimetnacrylates was also determined. According to this parameter, dimethacrylates can be arranged according to the following order: Bis-GMA (0.08–0.14 mM) < TEGDMA (0.12–0.26 mM) < UDMA (0.06–0.47 mM) < Bis-EMA (0.21–0.78 mM) < HEMA (1.77–2.52 mM) [174]. A summary of these results leads to the conclusion that TEGDMA is less toxic than Bis-GMA and UDMA. They are also in line with the general principle that dimetacrylates are more toxic than monometacrylates [169,175].

The second reason of the presence of leachable components in a composite dental material is hydrolysis or enzymatic degradation of a dimethacrylate polymer network [168,169,175]. Several products of Bis-GMA and Bis-EMA biodegradation, such as bisphenol A, bisphenol A diglycidylether and bisphenol A dimethacrylate were found to have estrogenic-like effects [168,169,175–177]. It was also found that Bis-EMA undergoes biodegradation to a lesser degree than Bis-GMA [169]. However, some researchers indicate that use of Bis-GMA based dental materials do not cause a clinically relevant estrogen-like effect on patients. Thus, the postulated estrogen-like effect is no reason to restrict indications for these dimethacrylate-based materials [169].

9. Conclusions

Material characterization and design are essential to better understanding its physicochemical and mechanical properties. In particular, the determination of relationships between structure and properties at different scales can lead to the improvement of material performance. This review is well-suited to the contemporary trends in the detailed characterization of polymeric materials, which are focused on the development of modern research methodologies and advanced interpretation of their properties. It can be expected that the issues presented in this review would be beneficial for the engineering of dimethacrylate dental materials.

It can be concluded that the higher the *DC* and the more homogeneous the morphology of a dimethacrylate polymer network, the better its mechanical performance. Molecular elasticity, resulting from higher possibilities of rotational movement and compact packing, usually positively affects the *DC* and morphology. Morphological heterogeneity arises from the formation of microgel agglomerates. The bigger the monomer molecule and the stronger the hydrogen bonds, the higher the network heterogeneity. On the other hand, hydrogen bonding positively affects many mechanical properties, especially impact resistance. Strong hydrogen bonds present in the Bis-GMA structure, by acting as physical crosslinks, are capable of compensating for the negative effect of insufficient DC on mechanical properties, especially impact resistance. The strength of H-bonds formed by the UDMA imine proton is insufficient to ensure high impact resistance. On the other hand, it is enough to ensure high hardness.

Copolymerization of basic monomers, such as Bis-GMA and UDMA, with reactive diluents, such as TEGDMA, having small and flexible molecules, usually causes an increase in the polymer network homogeneity and mechanical performance. 40 wt% of TEGDMA was found as the optimum content in a copolymer. Its higher amount can cause an increase in the morphological heterogeneity due to cyclization.

Funding: This research was funded by the Polish Budget Founds for Scientific Research in 2019 as a core funding for R&D activities in the Silesian University of Technology, grant number 04/040/BK_19/0101.

Conflicts of Interest: The authors declare no conflict of interest.

References

1. Lovell, L.G.; Berchtold, K.A.; Elliot, J.E.; Lu, H.; Bowman, Ch.N. Understanding the kinetics and network formation of dimethacrylate dental resins. *Polym. Adv. Technol.* **2001**, *12*, 335–345. [CrossRef]
2. Andrzejewska, E. Photopolymerization kinetics of multifunctional monomers. *Prog. Polym. Sci.* **2001**, *26*, 605–665. [CrossRef]
3. Dickens, H.; Stansbury, J.W.; Choi, K.M.; Floyd, C.J.E. Photopolymerization kinetics of methacrylate dental resins. *Macromolecules* **2003**, *36*, 6043–6053. [CrossRef]
4. Anseth, K.S.; Bowman, C.N. Kinetic gelation predictions of aggregation in tetrafunctional monomer polymerizations. *J. Polym. Sci. B Polym. Phys.* **1995**, *33*, 1769–1780. [CrossRef]
5. Pfeifer, C.S.; Shelton, Z.R.; Braga, R.R.; Windmoller, D.; Machado, J.C.; Stansbury, J.W. Characterization of dimethacrylate polymeric networks: A study of the crosslinked structure formed by monomers used in dental composites. *Eur. Polym. J.* **2011**, *47*, 162–170. [CrossRef] [PubMed]
6. Watts, D.C. Dental Restorative Materials. In *Materials Science and Technology: A Comprehensive Treatment*; Cahn, R.W., Haasen, P., Kramer, E.J., Eds.; VCH: New York, NY, USA, 1992; Volume 14, pp. 209–258.
7. Powers, J.M.; Sakaguchi, R.L. Restorative Materials—Composites and Polymers. In *Craig's Restorative Dental Materials*, 13th ed.; Mosby: St. Louis, MI, USA, 2013; ISBN 9780323081085.
8. Vasudeva, G. Monomer systems for dental composites and their future. *J. Calif. Dent. Assoc.* **2009**, *37*, 389–398. [PubMed]
9. Anseth, K.; Newman, S.M.; Bowman, C.N. Polymeric Dental Composites: Properties and Reaction Behavior of Multimethacrylate Dental Restorations. In *Biopolymers II. Advances in Polymer Science*, 1st ed.; Peppas, N.A., Langer, R.S., Eds.; Springer: Berlin/Heidelberg, Germany, 1995; Volume 122, pp. 177–217.
10. Astudillo-Rubio, D.; Delgado-Gaete, A.; Bellot-Arcís, C.; Montiel-Company, J.M.; Pascual-Moscardó, A.; Almerich-Silla, J.M. Mechanical properties of provisional dental materials: A systematic review and meta-analysis. *PLoS ONE* **2018**, *13*, e0193162. [CrossRef]
11. Chan, K.H.S.; Mai, Y.; Kim, H.; Tong, K.C.T.; Ng, D.; Hsiao, J.C.M. Review: Resin composite filling. *Materials* **2010**, *3*, 1228–1243. [CrossRef]
12. Kwon, T.Y.; Bagheri, R.; Kim, Y.K.; Kim, K.H.; Burrow, M.F. Cure mechanisms in materials for use in esthetic dentistry. *J. Investig. Clin. Dent.* **2012**, *3*, 3–16. [CrossRef]
13. Singh, A.V.; Ansari, M.H.D.; Laux, P.; Luch, A. Micro-nanorobots: Important considerations when developing novel drug delivery platforms. *Expert Opin. Drug Deliv.* **2019**, *16*, 1259–1275. [CrossRef]
14. Santulli, C. Nanostructured Composites for Dental Fillings. In *Nanostructured Polymer Composites for Biomedical Applications*; Swain, S.K., Jawaid, M., Eds.; Elsevier: Amsterdam, NL, USA; Oxford, UK; Cambridge, UK, 2019; pp. 277–294.
15. Bowen, R.L. Dental Filling Material Comprising Vinyl Silane Treated Fused Silica and a Binder Consisting of the Reaction Product of Bis Phenol and Glycidyl Acrylate. U.S. Patent 3066112A, 26 June 1962.
16. Peutzfeldt, A. Resin composites in dentistry: The monomer systems. *Eur. J. Oral. Sci.* **1997**, *105*, 97–116. [CrossRef] [PubMed]

17. El-Banna, A.; Sherief, D.; Fawzy, A.S. Resin Based Dental Composites for Tooth Filling. In *Advanced Dental Biomaterials*; Khurshid, Z., Najeeb, S., Zafar, M.S., Sefat, F., Eds.; Elsevier: Duxford, UK; Cambridge, UK; Kidlington, UK, 2019; pp. 127–174.
18. Sideridou, I.; Tserki, V.; Papanastasiou, G. Effect of chemical structure on degree of conversion in light-cured dimethacrylate-based dental resins. *Biomaterials* **2002**, *23*, 1819–1829. [CrossRef]
19. Gajewski, V.E.S.; Pfeifer, C.S.; Fróes-Salgado, N.R.G.; Boaro, L.C.C.; Braga, R.R. Monomers used in resin composites: Degree of conversion, mechanical properties and water sorption/solubility. *Braz. Dent. J.* **2012**, *23*, 508–514. [CrossRef] [PubMed]
20. Dusek, K.; MacKnight, W. Cross-Linking and Structure of Polymer Networks. In *Crosslinked Polymers: Chemistry, Properties, and Applications*; Dickie, R.A., Labana, S.S., Bauer, R.S., Eds.; American Chemical Society: Washington, DC, USA, 1988; Volume 1, pp. 2–27.
21. Hild, G. Model networks based on 'endlinking' processes: Synthesis, structure and properties. *Prog. Polym. Sci.* **1998**, *23*, 1019–1149. [CrossRef]
22. Flory, P.J. Molecular Theory of Rubber Elasticity. *Polymer* **1985**, *17*, 1–12. [CrossRef]
23. Elliott, J.E.; Lovell, L.G.; Bowman, C.N. Primary cyclization in the polymerization of bis-GMA and TEGDMA: A modeling approach to understanding the cure of dental resins. *Dent. Mater.* **2001**, *17*, 221–229. [CrossRef]
24. Roshchupkin, V.; Kumaz, S. Network Structure Formation (Radical polymerization). In *Polymeric Materials Encyclopedia*, 1st ed.; Salamone, J.C., Ed.; CRC Press: Boca Raton, FL, USA, 1996; Volume 6, pp. 4576–4581.
25. Kannurpatti, A.; Anseth, J.; Bowman, C.H.N. A study of the evolution of mechanical properties and structural heterogeneity of polymer networks formed by photo-polymerizations of multifunctional (meth)acrylates. *Polymer* **1998**, *39*, 2507–2513. [CrossRef]
26. Rey, L.; Duchet, J.; Galy, J.; Sautereau, H.; Vouagner, D.; Carrion, L. Structural heterogeneities and mechanical properties of vinyl/dimethacrylate networks synthesized by thermal free radical polymerization. *Polymer* **2002**, *43*, 4375–4384. [CrossRef]
27. Cook, W.D. Fracture and structure of highly crosslinked polymer composites. *J. Appl. Polym. Sci.* **1991**, *42*, 1259–1269. [CrossRef]
28. Simon, I.G.P.; Allen, P.E.M.; Williams, D.R.G. Properties of dimethacrylate copolymers of varying crosslink density. *Polymer* **1991**, *32*, 2577–2587. [CrossRef]
29. Wang, E.; Hasheminasab, A.; Guo, Y.; Soucek, M.D.; Cakmak, M. Structure characterization of UV-curing PEG-b-PPG-b-PEG dimethacrylate cross-linked network. *Polymer* **2018**, *153*, 241–249. [CrossRef]
30. Gu, Y.; Zhao, J.; Johnson, J.A. A (Macro)Molecular-level understanding of polymer network topology. *Trends Chem.* **2019**, *1*, 318–334. [CrossRef]
31. Stansbury, J.W. Dimethacrylate network formation and polymer property evolution as determined by the selection of monomers and curing conditions. *Dent. Mater.* **2012**, *28*, 13–22. [CrossRef] [PubMed]
32. Barszczewska-Rybarek, I. Structure-property relationships in dimethacrylate networks based on Bis-GMA, UDMA and TEGDMA. *Dent. Mater.* **2009**, *25*, 1082–1089. [CrossRef]
33. Barszczewska-Rybarek, I.; Krasowska, M. Fractal analysis of heterogeneous polymer networks formed by photopolymerization of dental dimethacrylates. *Dent. Mater.* **2012**, *28*, 695–702. [CrossRef]
34. Krasowska, M.; Barszczewska-Rybarek, I. The percolation theory in studying the morphology of polymer networks formed by photopolymerization of dental dimethacrylates. *Eur. Polym. J.* **2016**, *76*, 77–78. [CrossRef]
35. Barszczewska-Rybarek, I. The role of molecular structure on impact resistance and bending strength of photocured urethane-dimethacrylate polymer networks. *Polym. Bull.* **2017**, *74*, 4023–4040. [CrossRef]
36. Di Lorenzo, F.; Seiffert, S. Nanostructural heterogeneity in polymer networks and gels. *Polym Chem* **2015**, *6*, 5515–5528. [CrossRef]
37. Krzeminski, M.; Molinari, M.; Defoort, B.; Coqueret, X. Nanoscale heterogeneities in radiation-cured diacrylate networks: Weakness or asset? *Radiat. Phys. Chem.* **2013**, *84*, 79–84. [CrossRef]
38. Krzeminski, M.; Molinari, M.; Troyon, M.; Coqueret, X. Characterization by atomic force microscopy of the nanoheterogeneities produced by the radiation-induced cross-linking polymerization of aromatic diacrylates. *Macromolecules* **2010**, *43*, 8121–8127. [CrossRef]
39. Krzeminski, M.; Molinari, M.; Troyon, M.; Coqueret, X. Calorimetric characterization of the heterogeneities produced by the radiation-induced cross-linking polymerization of aromatic diacrylates. *Macromolecules* **2010**, *43*, 3757–3763. [CrossRef]

40. Panyukov, S.V. Theory of heterogeneities in polymer networks. *Polym. Sci. A* **2016**, *58*, 886–898. [CrossRef]
41. Seiffer, S. Origin of nanostructural inhomogeneity in polymer-network gels. *Polym. Chem.* **2017**, *8*, 4472–4487. [CrossRef]
42. Guo, Z.; Sautereau, H.; Kranbuehl, D. Evidence for spatial heterogeneities observed by frequency dependent dielectric and mechanical measurements in vinyl/dimethacrylate systems. *Polymer* **2010**, *26*, 416–425. [CrossRef]
43. Husar, B.; Commereuc, S.; Chmela, S.; Verney, V. Characterization of networks from photoreactive copolymers: An attempt to correlate chemical composition to network structure. *Polym. Int.* **2010**, *59*, 1563–1570. [CrossRef]
44. Szczepanski, R.C.; Pfeifer, C.S.; Stansbury, J.W. A new approach to network heterogeneity: Polymerization induced phase separation in photo-initiated, free-radical methacrylic systems. *Polymer* **2012**, *53*, 4694–4701. [CrossRef]
45. Barszczewska-Rybarek, I.M. A new approach to morphology studies on diacrylate polymer networks using X-ray powder diffraction. *Macromol. Chem. Phys.* **2013**, *214*, 1019–1026. [CrossRef]
46. Moore, R.J.; Watts, J.T.F.; Hood, J.A.A.; Burritt, D.J. Intra-oral temperature variation over 24 hours. *Eur. J. Orthod.* **1999**, *21*, 249–261. [CrossRef]
47. Leprince, J.G.; Palin, W.M.; Hadis, M.A.; Devaux, J.; Leloup, G. Progress in dimethacrylate-based dental composite technology and curing efficiency. *Dent. Mater.* **2013**, *29*, 139–156. [CrossRef]
48. Barszczewska-Rybarek, I.M. Characterization of urethane-dimethacrylate derivatives as alternative monomers for the restorative composite matrix. *Dent. Mater.* **2014**, *30*, 1336–1344. [CrossRef]
49. Lee, D.W.; Kim, H.N.; Lee, D.S. Introduction of reversible urethane bonds based on vanillyl alcohol for efficient self-healing of polyurethane elastomers. *Molecules* **2019**, *24*, 2201. [CrossRef] [PubMed]
50. Lemon, M.T.; Jones, M.S.; Stansbury, J.W. Hydrogen bonding interactions in methacrylate monomers and polymers. *J. Biomed. Mater. Res. A* **2007**, *83*, 734–746. [CrossRef] [PubMed]
51. Kanlayakan, N.; Kerdpol, K.; Prommin, C.; Salaeh, R.; Chansen, W.; Sattayanon, C.; Kungwan, N. Effects of different proton donor and acceptor groups on excited-state intramolecular proton transfers of amino-type and hydroxy-type hydrogen-bonding molecules: Theoretical insights. *New J. Chem* **2017**, *41*, 8761–8771. [CrossRef]
52. Park, J.; Eslick, J.; Ye, Q.; Misra, A.; Spencer, P. The influence of chemical structure on the properties in methacrylate-based dentin adhesives. *Dent. Mater.* **2011**, *27*, 1086–1093. [CrossRef]
53. Sideridou, I.; Tserki, V.; Papanastasiou, G. Study of water sorption, solubility and modulus of elasticity of light-cured dimethacrylate-based dental resins. *Biomaterials* **2003**, *24*, 655–665. [CrossRef]
54. Ogliari, F.A.; Ely, C.; Zanchi, C.H.; Fortes, C.B.; Samuel, S.M.; Demarco, F.F.; Petzhold, C.L.; Piva, E. Influence of chain extender length of aromatic dimethacrylates on polymer network development. *Dent. Mater.* **2007**, *24*, 165–171. [CrossRef]
55. Barszczewska-Rybarek, I.; Jurczyk, S. Comparative Study of Structure-Property Relationships in Polymer Networks Based on Bis-GMA, TEGDMA and Various Urethane-Dimethacrylates. *Materials* **2015**, *8*, 1230–1248. [CrossRef]
56. Asmussen, E.; Peutzfeldt, A. Influence of UEDMA, BisGMA, and TEGDMA on selected mechanical properties of experimental resin composites. *Dent. Mater.* **1998**, *14*, 51–56. [CrossRef]
57. Bociong, K.; Szczesio, A.; Sokolowski, K.; Domarecka, M.; Sokolowski, J.; Krasowski, M.; Lukomska-Szymanska, M. The influence of water sorption of dental light-cured composites on shrinkage stress. *Materials* **2017**, *10*, 1142. [CrossRef]
58. Sideridou, I.; Achilias, D.S.; Spyroudi, C.; Karabela, M. Water sorption characteristics of light-cured dental resins and composites based on Bis-EMA/PCDMA. *Biomaterials* **2004**, *25*, 367–376. [CrossRef]
59. Łukaszczyk, J.; Janicki, B.; Frick, A. Investigation on synthesis and properties of isosorbide based bis-GMA analogue. *J. Mater. Sci. Mater. Med.* **2012**, *23*, 1149–1155. [CrossRef] [PubMed]
60. Bian, L. Proton donor is more important than proton acceptor in hydrogen bond formation: A universal equation for calculation of hydrogen bond strength. *Phys. Chem. A* **2003**, *107*, 11517–11524. [CrossRef]
61. Jeffrey, G. *An Introduction to Hydrogen Bonding*, 1st ed.; Oxford University Press: New York, NY, USA, 1997; pp. 33–55.
62. Barszczewska-Rybarek, I.M.; Korytkowska-Wałach, A.; Kurcok, M.; Chladek, G.; Kasperski, J. DMA analysis of the structure of crosslinked poly(methyl methacrylate)s. *Acta. Bioeng. Biomech.* **2017**, *19*, 47–53. [CrossRef] [PubMed]

63. Reiche, A.; Sandner, R.; Weinkauf, A.; Sandner, B.; Fleischer, G.; Rittig, F. Gel electrolytes on the basis of oligo(ethylene glycol) (n) dimethacrylates—Thermal, mechanical and electrochemical properties in relationship to the network structure. *Polymer* **2000**, *41*, 3821–3836. [CrossRef]
64. Barszczewska-Rybarek, I. Prediction of physical properties of dimethacrylate polymer networks by a group contribution approach. *Int. J. Polym. Anal. Charact* **2013**, *18*, 93–104. [CrossRef]
65. Barszczewska-Rybarek, I.; Korytkowska, A.; Gibas, M. Investigations on the structure of poly(dimethacrylate)s. *Des. Monomers. Polym.* **2001**, *4*, 301–314. [CrossRef]
66. Khatri, C.A.; Stansbury, J.W.; Schultheisz, C.R.; Antonucci, J.M. Synthesis, characterization and evaluation of urethane derivatives of Bis-GMA. *Dent. Mater.* **2003**, *19*, 584–588. [CrossRef]
67. Ge, J.; Trujillo, M.; Stansbury, J.W. Synthesis and photopolymerization of low shrinkage methacrylate monomers containing bulky substituent groups. *Dent. Mater.* **2005**, *21*, 1163–1169. [CrossRef]
68. Barszczewska-Rybarek, I.; Gibas, M.; Kurcok, M. Evaluation of the network parameter in aliphatic poly(urethane dimethacrylate)s by dynamic thermal analysis. *Polymer* **2000**, *41*, 3129–3135. [CrossRef]
69. Litvinov, V.M.; Dias, A.A. Analysis of network structure of UV-cured acrylates by 1H NMR relaxation, 13C NMR spectroscopy, and dynamic mechanical experiments. *Macromolecules* **2001**, *34*, 4051–4060. [CrossRef]
70. Borges, M.G.; Barcelos, L.M.; Menezes, M.S.; Soares, C.J.; Fugolin, A.P.P.; Navarro, O.; Huynh, V.; Lewis, S.H.; Pfeifer, C.S. Effect of the addition of thiourethane oligomers on the sol–gel composition of BisGMA/TEGDMA polymer networks. *Dent. Mater.* **2019**, *35*, 1523–1531. [CrossRef] [PubMed]
71. Béhin, P.; Stoclet, G.; Ruse, D.; Sadoun, M. Dynamic mechanical analysis of high pressure polymerized urethane dimethacrylate. *Dent. Mater.* **2014**, *30*, 728–734. [CrossRef] [PubMed]
72. Zhao, J.; Yu, P.; Dong, S. The influence of crosslink density on the failure behavior in amorphous polymers by molecular dynamics simulations. *Materials* **2016**, *9*, 234. [CrossRef] [PubMed]
73. Malana, M.A.; Bukhari, J.D.; Zohra, R. Synthesis, swelling behavior, and network parameters of novel chemically crosslinked poly (acrylamide-co-methacrylate-co-acrylic acid) hydrogels. *Des. Monom. Polym* **2014**, *17*, 266–274. [CrossRef]
74. Flory, P.J.; Rehner, J. Statistical mechanics of crosslinked polymers. Rubber like elasticity. *J. Chem. Phys.* **1943**, *11*, 521. [CrossRef]
75. Alger, M.S.M. *Polymer Science Dictionary*, 3rd ed.; Springer: Berlin, Germany, 2017.
76. Redman, R.P. Developments in Polyurethane Elastomers. In *Developments in Polyurethane*; Buist, J.M., Ed.; Applied Science Publishers: London, UK, 1978; pp. 33–176.
77. Gonçalves, F.; Kawano, Y.; Pfeifer, C.S.; Stansbury, J.W.; Braga, R.R. Influence of bis-GMA, TEGDMA, and bis-EMA contents on viscosity, conversion, and flexural strength of experimental resins and composites. *Eur. J. Oral. Sci.* **2009**, *117*, 442–446. [CrossRef]
78. Baroudi, K.; Saleh, A.M.; Silikas, N.; Watts, D.C. Shrinkage behaviour of flowable resin-composites related to conversion and filler-fraction. *J. Dent.* **2007**, *35*, 651–655. [CrossRef]
79. Benesi, H.A.; Hildebrand, J.H. A spectrophotometric investigation of the interaction of iodine with aromatic hydrocarbons. *J. Amer. Chem. Soc.* **1949**, *71*, 2703–2707. [CrossRef]
80. Yilgör, E.; Burgaz, E.; Yurtsever, E.; Yilgör, İ. Comparison of hydrogen bonding in polydimethylsiloxane and polyether based urethane and urea copolymers. *Polymer* **2000**, *41*, 849–857. [CrossRef]
81. Antonucci, J.M.; Fowler, B.O.; Weir, M.D.; Skrtic, D.; Stansbury, J.W. Effect of ethyl-alpha-hydroxymethylacrylate on selected properties of copolymers and ACP resin composites. *J. Mater. Sci. Mater. Med.* **2008**, *19*, 3263–3271. [CrossRef]
82. Alshali, R.Z.; Silikas, N.; Satterthwaite, J.D. Degree of conversion of bulk-fill compared to conventional resin-composites at two time intervals. *Dent. Mater.* **2013**, *29*, e213–e217. [CrossRef] [PubMed]
83. Randolph, L.D.; Palin, W.M.; Bebelman, S.; Devaux, J.; Gallez, B.; Leloup, G.; Leprince, J.G. Ultra-fast light-curing resin composite with increased conversion and reduced monomer elution. *Dent. Mater.* **2014**, *30*, 594–604. [CrossRef] [PubMed]
84. Par, M.; Gamulin, O.; Marovic, D.; Klaric, E.; Tarle, Z. Raman spectroscopic assessment of degree of conversion of bulk-fill resin composites—Changes at 24 hours post cure. *Oper. Dent.* **2015**, *40*, E92–E101. [CrossRef] [PubMed]
85. Moldovan, M.; Balazsi, R.; Soanca, A.; Roman, A.; Sarosi, C.; Prodan, D.; Vlassa, M.; Cojocaru, I.; Saceleanu, V.; Cristescu, I. Evaluation of the degree of conversion, residual monomers and mechanical properties of some light-cured dental resin composites. *Materials* **2019**, *12*, 2109. [CrossRef] [PubMed]

86. Mackert, J.R. Physical Properties and Biocompatibility. In *Dental Materials and Their Selection*, 4th ed.; O'Brien, W.J., Ed.; Quintessence Publishing: Chicago, IL, USA, 2009; pp. 12–24.
87. Roman, A.; Páll, E.; Moldovan, M.; Rusu, D.; Soritau, O.; Festila, D.; Lupse, M. Cytotoxicity of experimental resin composites on Mesenchymal Stem Cells isolated from two oral sources. *Microsc. Microanal.* **2016**, *22*, 1018–1033. [CrossRef] [PubMed]
88. Manhart, J.; Chen, H.; Hamm, G.; Hickel, R. Buonocore Memorial Lecture. Review of the clinical survival of direct and indirect restorations in posterior teeth of the permanent dentition. *Oper. Dent.* **2004**, *29*, 481–508.
89. Lempel, E.; Czibulya, Z.; Kovács, B.; Szalma, J.; Tóth, Á.; Kunsági-Máté, S.; Varga, Z.; Böddi, K. Degree of Conversion and BisGMA, TEGDMA, UDMA Elution from Flowable Bulk Fill Composites. *Int. J. Mol. Sci.* **2016**, *17*, 732. [CrossRef]
90. Barszczewska-Rybarek, I.M. Quantitative determination of degree of conversion in photocured poly(urethane-dimethacrylate)s by FTIR spectroscopy. *J. Appl. Polym. Sci.* **2012**, *123*, 1604–1611. [CrossRef]
91. Collares, F.M.; Portella, F.F.; Leitune, V.C.B.; Samuel, S.M.W. Discrepancies in degree of conversion measurements by FTIR. *Braz. Oral. Res.* **2014**, *28*, 9–15. [CrossRef]
92. Barszczewska-Rybarek, I.; Chladek, G. Studies on the curing efficiency and mechanical properties of Bis-GMA and TEGDMA nanocomposites containing silver nanoparticles. *Int. J. Mol. Sci.* **2018**, *19*, 3937. [CrossRef]
93. Zhang, Y.; Kranbuehl, D.E.; Sautereau, H.; Seytre, G.; Dupuy, J. Modelling and measuring UV cure kinetics of thick dimethacrylate samples. *Int. J. Biol. Macromol.* **2009**, *42*, 203–210. [CrossRef]
94. Guimaraes, T.; Schneider, L.F.; Braga, R.R.; Pfeifer, C.S. Mapping camphorquinone consumption, conversion and mechanical properties in methacrylates with systematically varied CQ/amine compositions. *Dent. Mater.* **2014**, *30*, 1274–1279. [CrossRef] [PubMed]
95. Schroeder, W.F.; Vallo, C. Effect of different photoinitiator systems on conversion profiles of a model unfilled light-cured resin. *Dent. Mater.* **2007**, *23*, 1313–1321. [CrossRef] [PubMed]
96. Emami, N.; Söderholm, K.J. Influence of light-curing procedures and photoinitiator/co-initiator composition on the degree of conversion of light-curing resins. *J. Mater. Sci. Mater. Med.* **2005**, *16*, 47–52. [CrossRef] [PubMed]
97. Denis, A.B.; Diagone, C.A.; Plepis, A.M.G.; Viana, R.B. The effect of the polymerization initiator and light source on the elution of residual BisGMA and TEGDMA monomers: A study using liquid chromatography with UV detection. *Spectr. Acta A Mol. Biomol. Spectr.* **2015**, *151*, 908–915. [CrossRef] [PubMed]
98. Kim, M.; Suh, B.I.; Shin, D.; Kim, K.M. Comparison of the Physical and Mechanical Properties of Resin Matrix with Two Photoinitiator Systems in Dental Adhesives. *Polymers* **2016**, *8*, 250. [CrossRef] [PubMed]
99. Halvorson, R.H.; Erickson, R.L.; Davidson, C.L. The effect of filler and silane content on conversion of resin-based composite. *Dent. Mater.* **2003**, *19*, 327–333. [CrossRef]
100. Sahebalam, R.; Boruziniat, A.; Mohammadzadeh, F.; Rangrazi, A. Effect of the time of salivary contamination during light curing on degree of conversion and microhardness of a restorative composite resin. *Biomimetics* **2018**, *3*, 23. [CrossRef]
101. Moraes, L.G.; Rocha, R.S.; Menegazzo, L.M.; de Araújo, E.B.; Yukimito, K.; Moraes, J.C. Infrared spectroscopy: A tool for determination of the degree of conversion in dental composites. *J. Appl. Oral. Sci.* **2008**, *16*, 145–149. [CrossRef]
102. Pfeifer, C.S.; Silva, L.R.; Kawano, Y.; Braga, R.R. Bis-GMA co-polymerizations: Influence on conversion, flexural properties, fracture toughness and susceptibility to ethanol degradation of experimental composites. *Dent. Mater.* **2009**, *25*, 1136–1141. [CrossRef]
103. Stencel, R.; Pakieła, W.; Barszczewska-rybarek, I.; Żmudzki, J.; Kasperski, J.; Chladek, G. Effects of different inorganic fillers on mechanical properties and degree of conversion of dental resin composites. *Arch. Metall. Mater.* **2018**, *63*, 1361–1369. [CrossRef]
104. Zhang, M.; Puska, M.A.; Botelho, M.G.; Säilynoja, E.S.; Matinlinna, J.P. Degree of conversion and leached monomers of urethane dimethacrylate-hydroxypropyl methacrylate-based dental resin systems. *J. Oral. Sci.* **2016**, *58*, 15–22. [CrossRef] [PubMed]
105. Yuan, S.; Liu, F.; He, J. Preparation and characterization of low polymerization shrinkage and Bis-GMA-free dental resin system. *Adv. Polym. Technol.* **2015**, *34*, 21503. [CrossRef]
106. Scherzer, T.; Tauber, A.; Mehnert, R. UV curing of pressure sensitive adhesives studied by real-time FTIR-ATR spectroscopy. *Vib. Spectrosc.* **2002**, *29*, 125–131. [CrossRef]

107. Al-Odayni, A.B.; Alfotawi, R.; Khan, R.; Saeed, W.S.; Al-Kahtani, A.; Aouak, T.; Alrahlah, A. Synthesis of chemically modified BisGMA analog with low viscosity and potential physical and biological properties for dental resin composite. *Dent. Mater.* **2019**, *35*, 1532–1544. [CrossRef] [PubMed]
108. Scherzer, T. Real-time FTIR-ATR spectroscopy of photopolymerization reactions. *Macromol. Symp.* **2002**, *184*, 79–98. [CrossRef]
109. Fringeli, U.P. ATR and Reflectance IR Spectroscopy, Applications. In *Encyclopedia of Spectroscopy and Spectrometry*, 3rd ed.; Lindon, J.C., Tranter, G.E., Koppenaal, D.W., Eds.; Academic Press: Kidlington, UK, 2017; pp. 115–129. [CrossRef]
110. Ferrer, N. *Forensic Science, Applications of IR Spectroscopy, In Encyclopedia of Spectroscopy and Spectrometry*, 3rd ed.; Lindon, J.C., Tranter, G.E., Koppenaal, D.W., Eds.; Academic Press: Kidlington, UK, 2017; pp. 695–706. [CrossRef]
111. Griffiths, P. FTIR vs. FT-IR vs. Mid-IR. *Appl. Spectrosc.* **2010**, *64*, 40A. [CrossRef]
112. Lin-Vien, D.; Colthup, N.B.; Fateley, W.G.; Grasselli, J.G. *The Handbook of Infrared and Raman Characteristic Frequencies of Organic Molecules*; Academic Press: London, UK, 1991; pp. 137–281.
113. Saen, P.; Atai, M.; Nodehi, A.; Solhi, L. Physical characterization of unfilled and nanofilled dental resins: Static versus dynamic mechanical properties. *Dent. Mater.* **2016**, *32*, e185–e197. [CrossRef]
114. Yu, B.A.; Liu, D.L.; Liu, F.; He, J.W. Preparation and characterization of light-cured dental resins without methacrylate monomers derived from bisphenol A. *Adv. Polym. Technol.* **2016**, *33*, 21417. [CrossRef]
115. da Silva, E.M.; Miragaya, L.; Noronha-Filho, J.D.; Amaral, C.M.; Poskus, L.T.; Guimarães, J.G.A. Characterization of an experimental resin composite organic matrix based on a tri-functional methacrylate monomer. *Dent. Mater. J.* **2016**, *35*, 159–165. [CrossRef]
116. Pereira, S.; Nunes, T.; Kalachandra, S. Low viscosity dimethacrylate comonomer compositions [Bis-GMA and CH3Bis-GMA] for novel dental composites; analysis of the network by stray-field MRI, solid-state NMR and DSC & FTIR. *Biomaterials* **2002**, *23*, 3799–3806. [CrossRef] [PubMed]
117. Sideridou, I.D.; Karabela, M.M. Effect of the amount of 3-methacryloxypropyltrimethoxysilane coupling agent on physical properties of dental resin nanocomposites. *Dent. Mater.* **2009**, *25*, 1315–1324. [CrossRef] [PubMed]
118. Podgórski, M. Structure–property relationship in new photo-cured dimethacrylate based dental resins. *Dent. Mater.* **2012**, *28*, 398–409. [CrossRef] [PubMed]
119. Podgorski, M. Synthesis and characterization of novel dimethacrylates of different chain lengths as possible dental resins. *Dent. Mater.* **2010**, *26*, e188–e194. [CrossRef]
120. Gauthier, M.A.; Stangel, I.; Ellis, T.H.; Zhu, X.X. A new method for quantifying the intensity of the C=C band of dimethacrylate dental monomers in their FTIR and Raman spectra. *Biomaterials* **2005**, *26*, 6440–6448. [CrossRef]
121. Stansbury, J.W.; Dickens, S.H. Determination of double bond conversion in dental resins by near infrared spectroscopy. *Dent. Mater.* **2001**, *17*, 71–79. [CrossRef]
122. Floyd, C.J.; Dickens, S.H. Network structure of Bis-GMA- and UDMA-based resin systems. *Dent. Mater.* **2006**, *22*, 1143–1149. [CrossRef]
123. Lungu, A.; Şulcă, N.M.; Vasile, E.; Badea, N.; Pârvu, C.; Iovu, H. The influence of POSS substituent on synthesis and properties of hybrid materials based on urethane dimethacrylate (UDMA) and various polyhedral oligomeric silsesquioxane (POSS). *J. Appl. Polym. Sci.* **2011**, *121*, 2919–2926. [CrossRef]
124. Johnck, M.; Muller, L.; Neyer, A.; Hofstraat, J.W. Quantitative determination of unsaturation in photocured halogenated acrylates and methacrylates by FT-IR and Raman-spectroscopy and by thermal analysis. *Polymer* **1999**, *40*, 3631–3640. [CrossRef]
125. Luo, S.; Zhu, W.; Liu, F.; He, J. Preparation of a Bis-GMA-free dental resin system with synthesized fluorinated dimethacrylate monomers. *Int. J. Mol. Sci.* **2016**, *17*, 2014. [CrossRef]
126. Kammer, S.; Albinsky, K.; Sandner, B.; Wartewig, S. Polymerization of hydroxyalkyl methacrylates characterized by combination of FT-Raman and step-scan FT-i.r. photoacoustic spectroscopy. *Polymer* **1999**, *40*, 1131–1137. [CrossRef]
127. Rueggeberg, F.A. Determination of resin cure using infrared analysis without an internal standard. *Dent. Mater.* **1994**, *10*, 282–286. [CrossRef]
128. Silva Soares, L.E.; Martin, A.A.; Barbosa Pinheiro, A.L. Degree of conversion of composite resin: A Raman study. *J. Clin. Laser Med. Surg.* **2003**, *21*, 357–362. [CrossRef] [PubMed]

129. BinMahfooz, A.M.; Qutub, O.A.; Marghalani, T.Y.; Ayad, M.F.; Maghrabi, A.A. Degree of conversion of resin cement with varying methacrylate compositions used to cement fiber dowels: A Raman spectroscopy study. *J. Prosthet. Dent.* **2018**, *119*, 1014–1020. [CrossRef]
130. Pianelli, C.; Devaux, J.; Bebelman, S.; Leloup, G. The micro-Raman spectroscopy, a useful tool to determine the degree of conversion of light-activated composite resins. *J. Biomed. Mater. Res.* **1999**, *48*, 675–681. [CrossRef]
131. Habib, E.; Zhu, X.X. Photo-calorimetry method optimization for the study of light-initiated radical polymerization of dental resins. *Polymer* **2018**, *135*, 178–184. [CrossRef]
132. Jakubiak, J.; Sionkowska, A.; Lindén, L.Å.; Rabek, F. Isothermal Photo differential scanning calorimetry. Crosslinking polymerization of multifunctional monomers in presence of visible light photoinitiators. *J. Therm. Anal. Calorim.* **2001**, *65*, 435. [CrossRef]
133. Morancho, J.M.; Cadenato, A.; Fernández-Francos, X.; Salla, J.M.; Ramis, X. Isothermal kinetics of photopolymerization and thermal polymerization of Bis-gma/TEGDMA resins. *J. Therm. Anal. Calorim.* **2008**, *98*, 513–522. [CrossRef]
134. Antonucci, J.M.; Toth, E.E. Extent of polymerization of dental resins by differential scanning calorimetry. *J. Dent. Res.* **1983**, *62*, 121–125. [CrossRef]
135. Moszner, N.; Völkel, T.; Fischer, U.K.; Klester, A.; Rheinberger, V. Synthesis and polymerisation of new multifunctional urethane methacrylates. *Angew Makromol. Chem.* **1999**, *265*, 31–35. [CrossRef]
136. Barszczewska-Rybarek, I. Study on the effects of urethane-dimethacrylates' structures on the morphology and properties of polymers based on them. *Polimery* **2008**, *53*, 190–194. [CrossRef]
137. Sawada, H. Thermodynamics of Polymerization. I. *J. Macromol. Sci. C* **1969**, *3*, 313–338. [CrossRef]
138. Morgan, D.R.; Kalachandra, S.; Shobha, H.K.; Gunduz, N.; Stejskal, E.O. Analysis of a dimethacrylate copolymer (BisGMA and TEGDMA) network by DSC and 13C solution and solid-state NMR spectroscopy. *Biomaterials* **2000**, *21*, 1897–1903. [CrossRef]
139. Lopez-Suevos, F.; Dickens, S.H. Degree of cure and fracture properties of experimental acid-resin modified composites under wet and dry conditions. *Dent. Mater.* **2008**, *24*, 778–785. [CrossRef]
140. Jancar, J.; Wang, W.; DiBenedetto, A.T. On the heterogeneous structure of thermally cured bis-GMA/TEGDMA resins. *J. Mater. Sci. Mater. Med.* **2000**, *11*, 675–682. [CrossRef]
141. Wool, R.P. Properties of Triglyceride-Based Thermosets. In *Bio-Based Polymers and Composites*; Wool, R.P., Sun, X.S., Eds.; Elsevier Academic Press: Boston, MA, USA, 2005; pp. 202–255. [CrossRef]
142. Ferracane, J. Correlation between hardness and degree of conversion during the setting reaction of unfilled dental restorative resins. *Dent. Mater.* **1985**, *1*, 11–14. [CrossRef]
143. Asmussen, E. Restorative resins: Hardness and strength vs. quantity of remaining double bonds. *Scand. J. Dent. Res.* **1982**, *90*, 484–489. [CrossRef]
144. Linden, L.A. Dental Polymers (Overview). In *Polymeric Materials Encyclopedia*, 1st ed.; Salamone, J.C., Ed.; CRC Press: Boca Raton, FL, USA, 1996; Volume 12, p. 1849.
145. Badakar, C.M.; Shashibhushan, K.K.; Naik, N.S.; Reddy, V.V. Fracture resistance of microhybrid composite, nano composite and fibre-reinforced composite used for incisal edge restoration. *Dent. Traumatol.* **2011**, *27*, 225–229. [CrossRef]
146. Suzuki, A.; Yamazaki, M.; Kobiki, Y. Direct observation of polymer gel surfaces by atomic force microscopy. *J. Chem. Phys.* **1996**, *104*, 1751–1757. [CrossRef]
147. Caglayan, M.O. Nanomechanical characterization of flowable dental restorative nanocomposite resins using AFM. *Polym. Plast. Technol. Eng.* **2017**, *56*, 1813–1821. [CrossRef]
148. Munz, M. Microstructure and roughness of photopolymerized poly(ethylene glycol) diacrylate hydrogel as measured by atomic force microscopy in amplitude and frequency modulation mode. *Appl. Surf. Sci.* **2013**, *279*, 300–309. [CrossRef]
149. Gurtovenko, A.A.; Gotlib, Y.; Kilian, H. Viscoelastic dynamic properties of heterogeneous polymer networks with domain structure. *Macromol. Theory Simul.* **2000**, *9*, 388–397. [CrossRef]
150. Gotlib, Y.; Gurtovenko, A.A.; Kilian, H.G. Relaxation modulus of heterogeneous polymer networks with the domain structure. *Polym. Sci. A* **2001**, *43*, 308–314. [CrossRef]
151. Sideridou, I.D.; Karabela, M.M.; Vouvoudi, E.C. Dynamic thermo-mechanical properties and sorption characteristics of two commercial light cured dental resin composites. *Dent. Mater.* **2008**, *24*, 737–743. [CrossRef] [PubMed]

152. Dean, K.M.; Cook, W.D. Small angle neutron scattering and dynamic mechanical thermal analysis of dimethacrylate/epoxy IPNs. *Eur. Polym. J.* **2006**, *42*, 2872–2887. [CrossRef]
153. Abedin, F.; Roughton, B.C.; Spencer, P.; Ye, Q.N.; Camarda, K.V. Computational molecular design of water compatible dentin adhesive system. *Comput. Aided Chem. Eng.* **2015**, *37*, 2081–2086. [CrossRef]
154. Pielichowski, K.; Bogdał, D.; Pielichowski, J.; Boroń, A. Thermal decomposition of the copolymers based on long-chained diol dimethacrylates and BIS-GMA/TEGDMA. *Thermochim. Acta* **1997**, *307*, 155–165. [CrossRef]
155. Achilias, S.; Karabela, M.; Sideridou, I. Thermal degradation of light-cured dimethacrylate resins. Part, I. Isoconversional kinetic analysis. *Thermochim. Acta* **2008**, *472*, 74–83. [CrossRef]
156. Achilias, S.; Karabela, M.; Sideridou, I. Thermal degradation and isoconversional kinetic analysis of light-cured dimethacrylate copolymers. *J. Therm. Anal. Calorim.* **2010**, *99*, 917–923. [CrossRef]
157. Kumari, C.M.; Bhat, K.M.; Bansal, R. Evaluation of surface roughness of different restorative composites after polishing using atomic force microscopy. *J. Conserv. Dent.* **2016**, *19*, 56–62. [CrossRef]
158. Janus, J. Surface roughness and morphology of three nanocomposites after two different polishing treatments by a multitechnique approach. *Dent. Mater.* **2010**, *26*, 416–425. [CrossRef] [PubMed]
159. Grzywna, Z.J.; Krasowska, M.; Ostrowski, L.; Stolarczyk, J. Can generalized dimension (D_q) and f(a) be used in structure—Morphology analysis? *Acta Phys. Pol. B* **2001**, *32*, 1561–1578.
160. Krasowska, M.; Strzelewicz, A.; Dudek, G.; Rybak, A.; Barszczewska-Rybarek, I.; Turczyn, R. Fractal geometry characterization of fracture profiles of polymeric materials. *A Phys. Pol. B* **2014**, *45*, 2011. [CrossRef]
161. Cullity, B.D.; Stock, S.R. *Elements of X-ray Diffraction*, 3rd ed.; Pearson: Essex, UK, 2001; pp. 557–572.
162. Michel, V. Terminology for biorelated polymers and applications (IUPAC Recommendations 2012). *Pure Appl. Chem.* **2012**, *84*, 377–410. [CrossRef]
163. Bhola, R.; Bhola, S.M.; Liang, H.; Mishra, B. Biocompatible denture polymers—A review. *Trends Biomater. Artif. Organs* **2010**, *23*, 129–136.
164. Monsees, T.K. Biocompatibility and anti-microbiological activity characterization of novel coatings for dental implants: A primer for non-biologists. *Front. Mater.* **2016**, *3*, 40. [CrossRef]
165. *ANSI/ADA Standard No. 41. Evaluation of Biocompatibility of Medical Devices Used in Dentistry—ADA41-2005D*; American National Standards Institute/American Dental Association: Chicago, IL, USA, 2005; Available online: https://www.ada.org/en/science-research/dental-standards (accessed on 4 November 2019).
166. *ISO/TR 10993-1:22. Biological Evaluation of Medical Devices. International Standard Organization. Technical Committee: ISO/TC 194 Biological and Clinical Evaluation of Medical Devices*; ICS: Geneva, Switzerland, 2018; Available online: www.iso.org (accessed on 4 November 2019).
167. Pawłowska, E.; Loba, K.; Błasiak, J.; Szczepańska, J. Właściwości i ryzyko stosowania metakrylanu bisfenolu A i dimetakrylanu uretanu—Podstawowych monomerów kompozytów stomatologicznych. *Dent. Med. Probl.* **2009**, *46*, 477–485.
168. Mousavinasab, S.M. Biocompatibility of composite resins. *Dent. Res. J. (Isfahan)* **2011**, *8*, S21–S29.
169. Schmalz, G. Resin-Based Composites. In *Biocompatibility of Dental Materials*; Schmalz, G., Arenholt, D., Eds.; Springer: Berlin/Heidelberg, Germany, 2009; pp. 99–130. [CrossRef]
170. Ferracane, J.L. Elution of leachable components from composites. *J. Oral. Rehabil.* **1994**, *21*, 441–445. [CrossRef]
171. Sandborgh-Englund, G.; Nygren, A.T.; Ekstrand, J.; Elinder, C.G. No evidence of renal toxicity from amalgam fillings. *Am. J. Physiol.* **1996**, *271*, R941–945. [CrossRef]
172. Issa, Y.; Watts, D.C.; Brunton, P.A.; Waters, C.M.; Duxbury, A.J. Resin composite monomers alter MTT and LDH activity of human gingival fibroblasts in vitro. *Dent. Mater.* **2004**, *20*, 12–20. [CrossRef]
173. Reichl, F.X.; Simon, S.; Esters, M.; Seiss, M.; Kehe, K.; Kleinsasser, N.; Hickel, R. Cytotoxicity of dental composite (co)monomers and the amalgam component Hg(2+) in human gingival fibroblasts. *Arch. Toxicol.* **2006**, *80*, 465–472. [CrossRef] [PubMed]
174. Geurtsen, W. Biocompatibility of resin-modified filling materials. *Crit. Rev. Oral. Biol. Med.* **2000**, *11*, 333–355. [CrossRef] [PubMed]
175. Bakopoulou, A.; Papadopoulos, T.; Garefis, P. Molecular toxicology of substances released from resin–based dental restorative materials. *Int. J. Mol. Sci.* **2009**, *10*, 3861–3899. [CrossRef] [PubMed]
176. Moharamzadeh, K.; Brook, I.M.; Van Noort, R. Biocompatibility of resin-based dental materials. *Materials* **2009**, *2*, 514–548. [CrossRef]

177. Pulgar, R.; Olea-Serrano, M.F.; Novillo-Fertrell, A.; Rivas, A.; Pazos, P.; Pedraza, V.; Navajas, J.M.; Olea, N. Determination of bisphenol A and related aromatic compounds released from Bis-GMA based composites and sealants by high performance liquid chromatography. *Environ. Health Perspect.* **2000**, *108*, 21–27. [CrossRef]

 © 2019 by the author. Licensee MDPI, Basel, Switzerland. This article is an open access article distributed under the terms and conditions of the Creative Commons Attribution (CC BY) license (http://creativecommons.org/licenses/by/4.0/).

Review

Selected Spectroscopic Techniques for Surface Analysis of Dental Materials: A Narrative Review

Katarzyna Kaczmarek [1,*], Andrzej Leniart [1], Barbara Lapinska [2], Slawomira Skrzypek [1] and Monika Lukomska-Szymanska [2,*]

1. Department of Inorganic and Analytical Chemistry, Faculty of Chemistry, University of Lodz, 12 Tamka St., 91-403 Lodz, Poland; andrzej.leniart@chemia.uni.lodz.pl (A.L.); slawomira.skrzypek@chemia.uni.lodz.pl (S.S.)
2. Department of General Dentistry, Medical University of Lodz, 251 Pomorska St., 92-213 Lodz, Poland; barbara.lapinska@umed.lodz.pl
* Correspondence: katarzyna.kaczmarek@chemia.uni.lodz.pl (K.K.); monika.lukomska-szymanska@umed.lodz.pl (M.L.-S.); Tel.: +48-42-635-5783 (K.K.); +48-42-675-7461 (M.L.-S.)

Abstract: The presented work focuses on the application of spectroscopic methods, such as Infrared Spectroscopy (IR), Fourier Transform Infrared Spectroscopy (FT-IR), Raman spectroscopy, Ultraviolet and Visible Spectroscopy (UV-Vis), X-ray spectroscopy, and Mass Spectrometry (MS), which are widely employed in the investigation of the surface properties of dental materials. Examples of the research of materials used as tooth fillings, surface preparation in dental prosthetics, cavity preparation methods and fractographic studies of dental implants are also presented. The cited studies show that the above techniques can be valuable tools as they are expanding the research capabilities of materials used in dentistry.

Keywords: dental materials; dental ceramics; spectroscopy; IR; FT-IR; Raman spectroscopy; UV-Vis; X-ray spectroscopy; XRF; XRD; MS

Citation: Kaczmarek, K.; Leniart, A.; Lapinska, B.; Skrzypek, S.; Lukomska-Szymanska, M. Selected Spectroscopic Techniques for Surface Analysis of Dental Materials: A Narrative Review. *Materials* **2021**, *14*, 2624. https://doi.org/10.3390/ma14102624

Academic Editor: Grzegorz Chladek

Received: 17 April 2021
Accepted: 11 May 2021
Published: 17 May 2021

Publisher's Note: MDPI stays neutral with regard to jurisdictional claims in published maps and institutional affiliations.

Copyright: © 2021 by the authors. Licensee MDPI, Basel, Switzerland. This article is an open access article distributed under the terms and conditions of the Creative Commons Attribution (CC BY) license (https://creativecommons.org/licenses/by/4.0/).

1. Introduction

For centuries in dentistry, the search remains for the best methods and materials to restore missing tooth structures or replace missing teeth [1–3]. Modern dentistry, in addition to its primary role to improve the health of masticatory system, increasingly focuses on the aesthetics of the preformed reconstructions [4–7]. The rapid development of this field is possible thanks to the constant introduction of new materials and research techniques [8–14]. Due to wide range of (ceramic, metallic, synthetic or composite) dental materials available on the market, it is crucial that dental technicians and/or dentists choose the appropriate method by taking into account its limitations and features [13,15–20].

The application of the material in the oral cavity, i.e., in direct contact with tissues, places special demands on this group of biomaterials regarding their physicochemical and biological properties, i.e., biocompatibility, local and general harmlessness to the organism [18,21,22], resistance to the effects of physicochemical factors in the oral cavity [21] and biophysical indifference [18,19]. In addition, the accuracy in mimicking natural tooth shapes [19], stability of mechanical properties [19,22], ease processing [21,23], appropriate aesthetics [19,24], and finally a moderate price [19,23] are further requirements that need to be met. The first phenomenon that occurs after the introduction of the biomaterial into the oral cavity is the formation of a biofilm on its surface [25–27]. In order to prevent the micro-leakage and the biofilm formation, it is important to know the material structural and chemical components [7,28,29]. Surface structure, roughness, chemical composition and reactivity are just a few of the main characteristics that should be assessed before qualifying particular material for dental applications [18,19,27,30,31]. In addition, surface characteristics are related to other properties of the material, such as mechanical or chemical features, enabling detailed understanding how the material behaves under oral conditions [32,33].

The advanced development of microscopic and spectroscopic techniques enables a detail study of the chemical structure and surface phenomena [20,34] of studied materials. Microscopy enables observation of surfaces, and structural and intra-material changes at high magnification [35–37]. It includes, inter alia, Optical Microscopy [38–40], Electron Microscopy [39–42], X-ray Microscopy [39,40], and Scanning Probe Microscopy [39,41].

These techniques allow for advanced experimental research, which in combination with powerful acquisition procedures and data processing, enable detailed analysis of the studied system. Besides morphological and structural characterizations, microscopic techniques provide additionally information on the quantitative and qualitative chemical composition of the analysed sample either at a selected point, or along a given line, thus providing elemental maps (Energy Dispersive X-ray Spectroscopy—EDS), crystal orientation (Electron Backscatter Diffraction—EBSD), and many other characteristics of dental biomaterials [40,43–45]. The use of microscopic methods for material testing is widely described in the literature; therefore, in this paper we present the application of spectroscopic techniques, basic principles of operation, their advantages and limitations, as well as their application for studying dental materials.

Spectroscopy covers a lot of techniques based on the interaction of the electromagnetic radiation with the matter; sometimes, the term spectroscopy refers to an analytical technique involving generation and interpretation of spectra. The use of spectroscopy, especially in combination with microscopic techniques, provides valuable information on the chemical structure of dental materials. The major advantages of spectroscopic techniques are seen in the fact that they are non-destructive and using small amount of sample weights [34,46–52]. Spectroscopic techniques including: Infrared spectroscopy (IR), Fourier Transform Infrared Spectroscopy (FT-IR), Raman spectroscopy, Ultraviolet and Visible spectroscopy (UV-Vis), X-ray spectroscopy, and Mass Spectrometry (MS) are very useful in the dental material studies [53].

The aim of the current review was to provide the reader with a "broad view" of spectroscopic techniques and their application in a dental material research.

2. Search Strategy

An electronic search was carried out in SCOPUS and PubMed digital databases using the keywords related to the topic search and combining the keywords using "AND" and "OR". The search strategy employed was as follows: (Dental material spectroscopic methods) OR Dental material spectroscopy) OR Dental ceramic spectroscopic method) OR Dental ceramics spectroscopy) OR Dental implant spectroscopy) AND (Dental material IR) OR Dental ceramic IR) AND (Dental material FTIR) OR Dental material FT-IR) OR Dental ceramic FTIR) AND ((Dental material Raman spectroscopy) OR Dental ceramic Raman spectroscopy) AND (Dental material UV-Vis spectroscopy) OR Dental material UV-Vis) Dental ceramic UV-Vis spectroscopy) AND (Dental material X-ray spectroscopy) OR Dental ceramic X-ray spectroscopy) AND (Dental material Mass Spectrometry) OR Dental ceramic Mass Spectrometry).

The review of potentially eligible data articles on a given spectroscopic technique was narrowed down to books, book chapters and overviews in Subject Area-Dentistry. If no information could be found in this regard, it was extended to research articles on chemical and material subject fields. The review is mainly based on articles on spectroscopic methods excluding microscopic methods.

Initially 12,904 articles were found. After removal of duplicates and restriction to the years 2008–2021, 1534 articles were identified from digital databases and a manual search. Full texts of papers were obtained from the journals. The inclusion and exclusion criteria for articles are presented in Table 1.

Table 1. The inclusion and exclusion criteria for articles.

Inclusion Criteria	Exclusion Criteria
Research on only dental biomaterials used for restorations.	Literature on dental materials and fluids, equipment used as instruments and equipment for a dental office.
Research including ceramics, calcium phosphates, glasses, polymers, adhesives, composites, glass ionomers, silver amalgam, alloys and titanium implants.	All papers in other than the English language, where the full text was not available.
Dental material research published no later than 5 years ago.	Same data that was published at different times.

3. Fundamentals and Division of Spectroscopy

For the purpose of a comprehensive overview of spectroscopic techniques, it is necessary to apply various criteria for classification of the techniques. Most often, the following classification criteria are adopted: the range of electromagnetic radiation, properties of the studied systems, and the method of collecting a spectrum, referring to the way of energy exchange between the radiation and the matter. The division of spectroscopy according to the radiation range is actually related to various experimental techniques as listed in Table 2. Depending on the radiation range, different radiation sources, detectors and dispersion devices are used. In optical spectroscopy, prisms and diffraction gratings are most frequently used as dispersion devices [54–57].

Table 2. Spectroscopic techniques in different ranges of electromagnetic spectrum radiation.

Region of Electromagnetic Spectrum	Wavelength Range λ (m)	Spectroscopic Technique
Microwave	1–10^{-3}	Microwave spectroscopy
Infrared	10^{-3}–10^{-6}	Infrared spectroscopy Raman spectroscopy
Ultraviolet and visible	10^{-6}–10^{-8}	UV-Visible spectroscopy Atomic absorption spectroscopy Fluorescence spectroscopy Phosphorescence spectroscopy
X-ray	10^{-9}–10^{-12}	X-ray diffraction X-ray fluorescence X-ray photoelectron spectrometry Mass spectrometry
γ-ray	10^{-12}–10^{-14}	Mossbauer spectroscopy

Spectroscopy can be analysed based on the intrinsic aspects of the studied process. In this regard, the following types of spectroscopy can be differentiated: nuclear spectroscopy, atomic spectroscopy, and molecular spectroscopy with particular emphasis on the spectroscopy of condensed systems. Such division is related to the specific energy levels taking part in the energetic transition of the studied system. Each type of spectroscopy is associated with specific motion of the constituents of the system at a microscopic level, and thus differs in the magnitude of the energy between transition energetic states [34,53,58,59].

Depending on the radiation measurement process, one distinguishes three types of spectra: absorption, emission and Raman spectroscopy, respectively. Absorption spectra can be defined as a set of all transitions from lower to higher levels so, they correspond to an increase of the system energy (photon uptake). The simplest type of absorption spectrum arises when the lowest energy level, i.e., the basic level, is occupied. The occupation of energy levels is related to the thermodynamic equilibrium, which is determined by the temperature of the system. It is assumed that only the basic level is filled at room temperature. Emission spectra can be defined as a set of all transitions from higher to lower energy levels. Transitions in emission spectra correspond to the reduction of the energy, i.e., the radiation of photons. The characteristic feature of Raman spectra is the change in the frequency of the scattered radiation (νr) in relation to the frequency of the incident radiation (νp) [60,61].

4. Infrared Spectroscopy (IR) and Fourier Transform Infrared Spectroscopy (FT-IR)

4.1. Principle of the Technique

The area of the electromagnetic spectrum with the wavenumber (reciprocal of the wavelength) from approx. 14,000 to 200 cm^{-1}, i.e., between the visible and the microwave region, is called the infrared (IR). The absorption of such energy amount within this region is large enough to cause the chemical bonds to oscillate, but not enough to cause their breakage. Molecules rotate around their symmetry axes and at the same time their atoms oscillate between equilibrium positions. The IR absorption spectra are obtained by measuring the relative intensity of the transmitted or absorbed radiation as a function of the wavenumber, and they are presented in form of a plot of the transmittance or absorbance of the radiation versus the wavenumber (cm^{-1}).

There are many types of IR spectrophotometers, e.g., filter photometers, double-beam spectrophotometer, Fourier transform spectrometer, attenuated total reflectance (ATR) FT-IR instrument [55]. In modern Fourier transform devices, which consists of illuminating simultaneously the sample with a beam of radiation from the entire tested IR range. After this beam has passed through the sample, the beam from the same source has not been interfered with, and the spectrum is obtained using the Fourier transform of the recorded interference spectrum. This requires the use of equipment with a software that performs this mathematical operation and provides information about vibrations in the form of an interferogram. FT-IR spectroscopy allows to determine the characteristic vibrations of atomic groups of the studied compound.

Due to the fact that the FT-IR spectrum is characteristic for a given substance, infrared spectrophotometric analysis is most often used for qualitative analysis. In addition to characterizing a pure substance by means of FT-IR analysis, the presence of additional substances in a studied mixture, as well as interactions between atomic groups of different compounds in the mixture can be examined. Changes in the spectrum (presence of additional peaks) indicate the presence of other functional groups (presence of another compound), while the shifting of the peaks relative to the spectrum of the original pure compound indicates interactions of a given atomic group with other groups of the studied mixture [61,62]. Besides, the following infrared spectroscopic techniques can be currently differentiated: transmission spectroscopy (TS), internal reflection spectroscopy (IRS), which is also referred to as attenuated total reflection (ATR), mirror reflection (ERS—external reflection spectroscopy), diffuse reflection spectroscopy (DRS), emission spectroscopy (ES) and photoacoustic spectroscopy (PAS) [63,64].

4.2. Type of Tested Samples

In infrared spectroscopy substances can be examined in a gaseous, liquid and solid state. Gas-sample spectra can be obtained by inserting the gaseous sample into a purged cuvette. Liquids can be tested in a pure form, or in a solution form. Liquid samples are placed between the sodium chloride plates; approx. One to ten milligrams of a liquid substance is needed. For liquids that dissolve sodium chloride, silver chloride plates can be used instead. Solid samples are typically tested in a form of suspension, pellet, or deposited glassy film (a solid film on a glass substrate). The spectrum of a solid sample most frequently is obtained by dispersing the sample within an alkali halide pellet. The spectra of solids can also be obtained by preparing a film of the solid sample, following evaporation of the solvent from a droplet deposited on a sodium chloride plate of a solution containing the studied compound [55,65].

4.3. Sample Characteristics

FT-IR spectroscopy allows to determine the characteristic vibrations for groups present in a compound. Due to the fact that the FT-IR spectrum is a spectrum characteristic for a given substance, infrared spectrophotometric analysis is most often used for qualitative research. In addition to distinguishing between pure substances by means of FT-IR analysis, it is possible to investigate the presence of additional substances as well as their interactions with individual groups of the original compound. Changes in the spectrum (presence of additional peaks) indicate the presence of other functional groups (presence of another compound), while shifts in the directions of other wavelengths than for the spectrum of the original sample indicate the interaction of a given group with a different group of the added substance [61,62,66].

For example, the FT-IR method was used to identify the structure of a dental material, i.e., fibroin thanks to application of FT-IR, a NH functional group in fibroin was recognised as an amino acid structure with a peak of 3309.80 cm^{-1} [67]. Measurement of absorbance using FT-IR led to a spectra of the various bioceramic powders and finally their identification [68].

4.4. Advantages and Limitations

Currently, infrared spectroscopy is widely used technique to identify functional groups and chemical compounds (both organic and inorganic), as well as to assess the purity of a compound. It is inexpensive, instruments are easy to be operated, and IR spectra are obtained quickly [55]. Infrared spectroscopy is a promising alternative to other techniques, as it is not time consuming with respect to sample preparation, measurement and interpretation of the results; it is also non-invasive and relatively simple measuring technique. As all frequencies are measured simultaneously in FT-IR spectroscopy, the spectrum is obtained for a few seconds, being advantageous in comparison with conventional infrared spectroscopy measurements lasting [34,69–71]. The undoubted advantage of infrared spectroscopy techniques is the ability to test small amounts of the material [62], and in ATR technique, the ability to penetrate of the light beam an into a sample depth of about 0.5–3 µm [72].

The limitation of IR is the fact that in classic spectrometers, the IR spectra are obtained by examining the absorption of a specific beam of monochromatic radiation (one wavelength), and then sweeping the sample by gradually changing the wavelength during measurement with a dispersion element (prism, diffraction grating). Before the measurement, it is often necessary to properly prepare the sample, e.g., grinding the test material, thoroughly mixing with the potassium bromide (KBr) matrix and compressing under pressure to form a tablet. The sample should not contain water, because water strongly absorbs radiation with the wavenumber from approximately 3700 cm^{-1} and 1630 cm^{-1} (this absorption may obscure the bands of the tested substance) [55].

In the ATR sampling system, the main source of errors during the measurement is the imperfect contact between the sample and the diamond crystal [65]. The next drawback of FT-IR instruments is the fact that they are equipped only with a single beam, whereas dispersive instruments generally have a double beam [69].

4.5. Applications

IR spectroscopy was used to track the polymerization kinetics of dental resins [73,74] and adhesives [75–78] to improve the mechanical features of the dental material. The IR technique was also used to determine the role of intermolecular collagen cross-linking in the mechanical behaviour of dentin [79], to evaluate the structure of heterocyclic compounds as candidates for pulp regeneration [80] and to analyse new generation biomimetic materials replicating the mineral organic dentin and enamel complex [81].

The FT-IR technique was used for the structural characterization of dental materials such as: implant materials [82], biopolymers [83], ceramics [84], resin nanocomposites [85], implant coatings [86–88], bioceramics [89], resins [90–94], cements [95], bioglass [96], and self-curing materials [97].

4.6. Spectrum Example

Figure 1 shows the FT-IR spectrum of self-curing polymethyl methacrylate (PMMA)-based dental materials (UNIFAST III, GC Corporation, Tokyo, Japan) control samples prepared by mixing UNIFAST III resin powder (GC Corporation, Tokyo, Japan) with UNIFAST liquid monomer (GC Corporation, Tokyo, Japan). The powder spectrum of GC UNIFAST III is depicted in the inset of Figure 1. The band at around 1132 cm^{-1} is the characteristic absorption vibration of PMMA. The bands at about 1218 cm^{-1}, 1361 cm^{-1}, 1735 cm^{-1} and 2927 cm^{-1} are assigned to υ(C–O) stretching vibration, wagging vibration of C–H, C=O stretching, and C–H stretching, respectively [97].

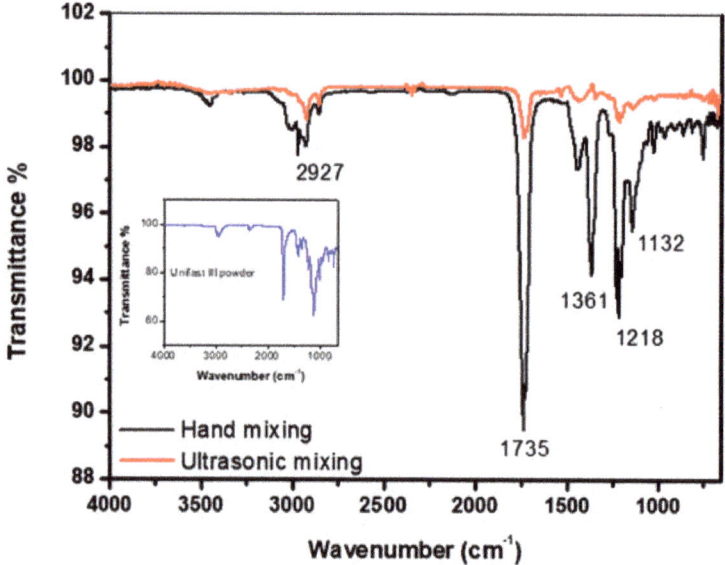

Figure 1. FT-IR spectrum of control specimen made by hand and ultrasonic mixing methods as well as the spectrum corresponding to UNIFAST III powder [97].

Figure 2 shows the FT-IR data of samples prepared by mixing UNIFAST III (GC Corporation, Tokyo, Japan) resin powder with ultrasonic-mixed UNIFAST (GC Corporation, Tokyo, Japan) liquid monomer, which includes starting materials and reinforced, nanosized hexagonal boron nitride h-BN (US Research Nanomaterials Co., Ltd., Houston, TX, USA) at various concentrations. Most researchers agree that there are two distinct IR absorption bands in boron nitride films. These are the band around 1380 cm^{-1} (in plane) and the band around 780 cm^{-1} (out of plane), which is due to B–N stretching and B–N–B bending, respectively [97].

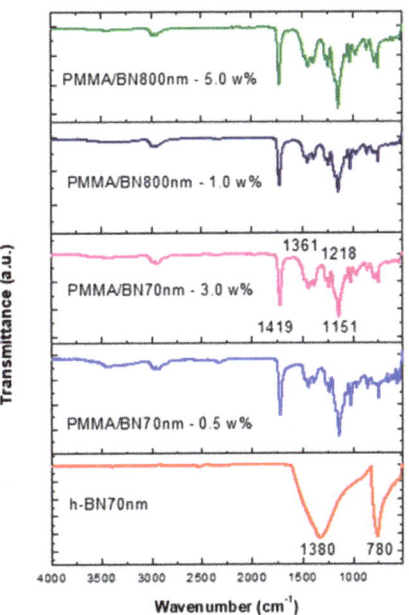

Figure 2. FT-IR data of specimens made by ultrasonic mixing for nano-sized h-BN reinforcement with different concentrations [97].

5. Raman Spectroscopy

5.1. Principle of Technique

Raman spectroscopy is a technique used to study vibrational-rotational spectra of molecules. In the light scattered by the examined medium, apart from the component with the same frequency as the incident light (Rayleigh scattering), there are components with a changed frequency. The Raman spectrum is represented as a function of the intensity of the scattered light versus the frequency, (wavenumber), which is calculated as a difference between the frequencies of the incident and scattered radiation (Raman shift).

Two main techniques are used to obtain the Raman spectrum: dispersion and Fourier transform (FT). They differ in the wavelength of the incident radiation and the method of detection. Typically, lasers of different wavelengths result in different penetration depths in the sample surface. The shorter the laser wavelength, the information closer to the surface it provides. On the other hand, when the selected laser wavelength is too long, information about the material near the surface may remain hidden (or omitted). Therefore, the selection of the correct laser wavelength is very important to obtain an accurate Raman characterization result, especially in the case of multi-layer substrates or surface sensitive samples. Currently, many new modifications are used in Raman spectroscopy, which primarily increase the sensitivity of the technique, such as: Resonance Raman Spectroscopy (RRS), Surface Enhanced Raman Spectroscopy (SERS), Confocal Raman Microscopy and Coherent Antistokes Raman Spectroscopy (CARS). Raman spectroscopy provides a lot of important information about the geometric structure of molecules and the nature of chemical bonds. In the case of crystalline samples, among others, information on the structure, crystal lattice arrangement, elastic properties, stresses and the nature of phase transitions can be obtained. Hence, Raman spectroscopy can be used in qualitative analysis, because each compound is associated with a characteristic Raman spectrum, which is a "dactyloscopic trace", as well as it provides quantitative information based on the dependence of the intensity of signals on the concentration of substances in the analysed sample [56,98–101].

5.2. Type of Tested Samples

In the abovementioned techniques, samples can be tested in a wide variety of states; for example, in the form of solids, liquids or vapours, hot or cold, in bulk, as microscopic particles, or as surface layers. Typical accessories for Raman spectroscopy are powder sample holders, cuvette holders, small liquid sample holders (nuclear magnetic resonance sample tubes), and irregularly shaped object clamps [56].

5.3. Sample Characterisctics

In the Raman spectrum, the intensity of the radiation coming from the sample is a linear function of the molar concentration and the thickness of the layer. The Raman spectrum is an oscillating-rotational spectrum that carries information, e.g., about the structure of the test substance and its chemical composition. Raman spectroscopy is now an eminent technique for the characterisation of 2D materials and phonon modes in crystals. Properties, such as number of monolayers, inter-layer breathing and shear modes, in-plane anisotropy, doping, disorder, thermal conductivity, strain and phonon modes can be extracted using Raman spectroscopy [102,103].

For example, Raman mappings confirmed that the graphene nanocoating covering the Biomedical-Grade Ti-6Al-4V alloys showed high structural stability and resistance to mechanical stress and chemical degradation, maintaining >99% coverage after corrosion [104]. Raman spectroscopies have been used to study dental materials were mostly find the degree of conversion (DC) of dental composites, adhesives and setting reaction of cements [103].

5.4. Advantages and Limitations

Similar to other spectroscopic techniques in Raman spectroscopy the time of measurement is short and sample preparation is simple. Additionally, the technique is not invasive and the obtained results can be easily and rapidly analysed [56].

For all spectrometer systems using visible incident light, the main disadvantage is the phenomenon of fluorescence. This is bigger problem in the visible region than in UV or near-infrared region. Since Raman scattering is a weak effect a strong excitation source is required to provide a high-power density in the sample. Hence, besides the fluorescence caused by the tested sample, any tiny impurities that are fluorescent can give large interfering fluorescence signals. Since fluorescence occurs at energies below the excitation value, it can be quite intense in the energy region covered by Raman Stokes scattering [56].

5.5. Applications

Raman spectroscopy was used to determine the rate of polymerization of bioglass ceramic [105], identification and testing of the presence of various compounds in implant materials [106], ceramics [107–109], bioglass [110]; analysis of the phase composition of bioglass [111], ceramics [112,113]; and evaluation of the degree of conversion of resins [114] and ceramics [115].

5.6. Spectrum Example

Figure 3a shows seven Raman spectra of bovine (B-Raw) and human (H-Raw) bone powders, calcinated bovine bone powder (B-560) and commercial samples based on β-tricalcium phosphate (RTR, Septodont, Saint-Maur-des-Fossés, France), calcium phosphosilicate combined with polyethylene glycol and glycerine (Novabone®, Osteogenics Biomedical, Lubbock, TX, USA) and hydroxyapatite (Nukbone®, Biocriss, Mexico City, Mexico and Biograft®, Biograft, Mexico City, Mexico). All Raman bands are presented in detail in the article [106]. The bands A, H, O, V, and W correspond to the amide, amine, and DNA atomic groups, while the C and F bands are assigned to C–H groups from proteins and fats of the bone. On the other hand, the bands B, E, G, I, S, U, A′, and G′ (groups C–H) in the Novabone are the organic compounds of PEG and glycerol. Bands corresponding to the carbonate group (CO$_3$) are designated with D, J, and E′. Inorganic phase bands of the samples are designated with K, L, N, O, P, Q, R, Y, Z, A′, B′, C′, D′, and F′, corresponding to the phosphate group (PO$_4$). Finally, the silicate group (SiO$_3$) is represented by the bands M and T. Figure 3b refers to the hydroxyapatite band located at 960 cm^{-1}. Figure 3c shows the full width at the half-maximum (FWHM) of the bands presented in Figure 3b. It is observed that the crystal size affects the width and size of the peak, which may be an indication of the crystalline quality too [106]. Furthermore, Castillo-Paz et al. [116] observed changes of raw hydroxyapatite obtained from porcine bones during heat treatment with heating rates of 2.5 and 5.0 °C/min between 600 °C and 1100 °C. They found that crystal growth and the transition from nano to a micro-size occurred above 700 °C. The result was a narrowing of the analysed Raman band of the characteristic Raman band at 960 cm^{-1} for ν1 PO$_4^{3-}$ and a decrease of full width at half the maximum (FWHM), which was confirmed by HR-SEM images and analysis thermogravimetric (TG). A similar effect was observed by Ramirez-Gutierrez et al. [117], who studied the influence of the temperature and sintering time on structural, morphological, thermal, and vibrational properties of hydroxyapatite obtained from pig bone (BHA). They showed that the higher intensity of Raman peaks is associated with an increase in the size of the crystallites. Based on the FWHM, they investigated changes in crystal quality. They found that the FWHM value decreased with increasing crystal quality for samples sintered at 600 °C. Under these conditions, during a longer sintering time, removal of the organic phase takes place without any structural transformation. This treatment preserves the polycrystalline properties of BHA with improved crystal quality. This was confirmed by the results of XRD and SEM.

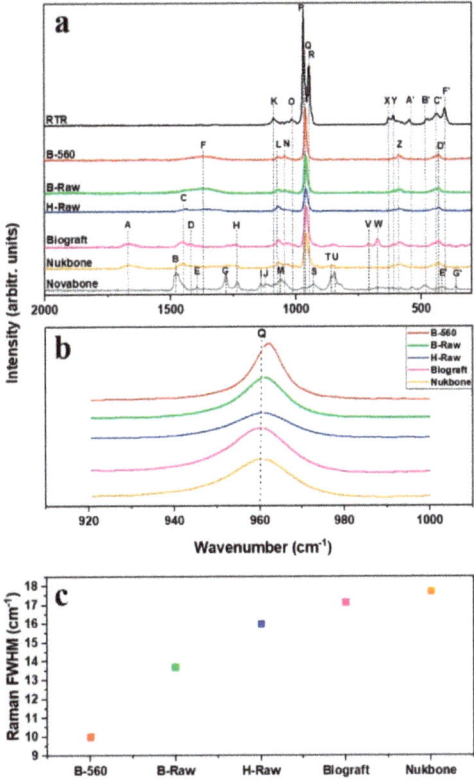

Figure 3. Raman spectra of (**a**) B-560, B-Raw, H-Raw, and commercial bone grafts; (**b**) bands at 960 cm^{-1} of the hydroxyapatites samples, and (**c**) FWHM values of (**b**) bands [106].

6. Ultraviolet and Visible Spectroscopy (UV-Vis)

6.1. Principle of Technique

UV-Vis spectroscopy is one of the oldest instrumental techniques. The analysis is based on the measurement of electron spectra generated during transitions between the energy levels of valence electrons in the spectral range of 200–750 nm. Energy absorption in the UV-Vis range can cause electron transitions from the lowest energy baseline (π and n orbitals) to higher energy levels (anti-bonding π^* orbitals). The method provides information on the structure of the studied molecule, the presence of conjugated double bond systems and aromatic systems. The instrument used in UV-Vis spectroscopy is the UV-Vis spectrophotometer [53,58,118].

6.2. Type of Tested Samples

Spectroscopic analysis is commonly carried out in solutions, but solids and gases may also be studied. The liquids can be contained in a vessel made of a transparent material such as silica, glass, or plastic, known as a cell or cuvette. Gases can be in similar cells that are sealed or plugged to make them gas-tight. In general, measurements are performed for compounds are demonstrated absorption in the UV-Vis range, which contain chromophores and auxochromes in their structure; within them, there is an electronic transition from the ground to the excited state. Absorption in the UV-Vis range is demonstrated by compounds that contain chromophores and auxochromes in their structure; within them, there is an electronic transition from the ground state to the excited state [53,119–121].

6.3. Sample Characteristics

According to Beer Lambert's law, the measurement of the absorbance of a certain substance in a solution is directly proportional to its concentration; hence absorption spectroscopy can be used for quantitative analysis. Therefore, the technique is an excellent analytical tool for the characterization and evaluation of many materials, including biomaterials and dental composites [100].

6.4. Advantages and Limitations

UV-Vis spectroscopy is used for both qualitative and quantitative analysis. Its main advantages are its high sensitivity [119,122], high marking precision [119] and selectivity of determinations [58]. The objective measure for the sensitivity of spectrophotometric method is the molar absorptivity ε, corresponding to the λ_{max} of the tested solution. The ε values for the sensitive methods are above 10,000 $dm^3\ mol^{-1}\ cm^{-1}$, and the ε coefficient with values below 1000 $dm^3\ mol^{-1}\ cm^{-1}$ corresponds to the less sensitive methods [119,122]. The accuracy of the determination depends on the concentration range and the class of devices used [119]. It is conditioned by the efficiency of absorption and the selectivity of reagents causing a coloured product with the substance to be analysed [58].

Organic compounds with chromophore groups, in which an electronic transition from the ground state to the excited state can occur, are capable of absorbing radiation in the UV-Vis range. Most frequently aromatic systems and groups with multiple bonds (e.g., >C=C<, >C=O, –N=N–, –NO$_2$) are chromophores [119].

6.5. Applications

UV-Vis spectroscopy has been applied to study fillers for dental adhesive resins [123], to assess the risk of chemical irritation, allergic reactions and oral hypersensitivity due to the elution of polymers [124], to evaluate new co-initiators useful in the radical photopolymerization of dental polymers [94,125]. This technique was also used to characterize the colour and spectral reflectance and thickness oxide layer [126] and resin [127], to evaluate the transparency of ceramics [128,129], and to assesses the degradation rate of CH-based scaffolds which play a key role in endodontic regeneration periodontal regeneration material [130].

6.6. Spectrum Example

Figure 4 shows the UV-Vis spectra from the camphorquinone (CQ) and the coinitiators ethyl-4-dimethylaminobenzoate (EDAB), monomers with tertiary amines and four methacrylic (MBTTM) or acrylic (MBTTA) useful in dentistry. The CQ exhibits a strong absorption band in the UV region due to $\pi \rightarrow \pi^*$ transition and an absorption band in the visible region at 445 nm (transition n $\rightarrow \pi^*$), with a low extinction coefficient. This absorption band at 445 nm gives to CQ the ability to be used as photoinitiator in the visible region, allowing to form an excited state and the subsequent hydrogen abstraction from the coinitiator. Finally, in the case of EDAB, MBTTA and MBTTM used as hydrogen donors, they show only absorption bands in the ultraviolet region, with high extinction coefficient values [94].

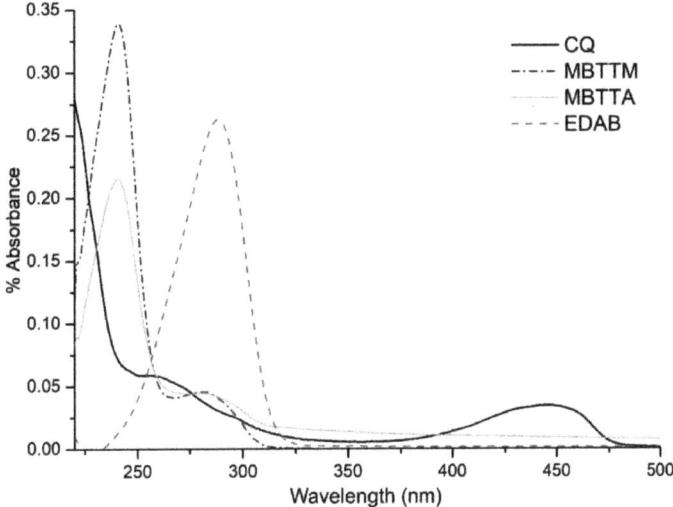

Figure 4. UV-Vis spectra of the CQ initiator and EDAB, MBTTM and MBTTA coinitiators [94].

7. X-ray Spectroscopy

7.1. Principle of Technique

X-rays constitute a part of the spectrum of electromagnetic waves covering the wavelength range from 10^{-9}–10^{-12} m. They are generated in the X-ray tube during the bombardment of the anode (anti-cathode) with a beam of electrons emitted by the cathode and accelerated in the electric field to a high speed. The X-ray spectrum of the anode material consists of lines associated with the corresponding electron transitions characteristic for a given element. X-ray analysis is possible after splitting it into separate lines. X-rays can be detected with an energy dispersed spectrometer (EDS) or a wavelength dispersed spectrometer (WDS). The analysis can be performed point by point or can be mapped. Elemental analyses are quantified by comparison with standard reference materials.

There are other methods of X-ray spectroscopy that are used in the dental materials testing, such as X-ray absorption spectroscopy (XAS), X-ray emission spectroscopy (XES), X-ray fluorescence spectroscopy (XRF), X-ray Photoelectron Spectroscopy (XPS), and X-ray diffraction (XRD). In the XAS method the relationship between the X-ray intensities before and after passing through the sample of defined thickness is investigated. The XES is the method in which the relaxation products of the interaction of photon or electron beams with the sample is studied. This method is related to the spontaneous emission of X-ray photons in the dipole transition between two electronic states. The XRF is an analytical method that measures the emission of the characteristic secondary or fluorescent X-rays from an excited sample bombarded with a high-energy radiation. Another, relatively new technique is the total-reflection X-ray fluorescence (TXRF), where the total reflection effect is obtained by irradiating the sample, placed on a suitable support (quartz), with the primary radiation from the X-ray tube at an angle smaller than the boundary angle [57,60,131–135].

The XPS method measures the kinetic energy of the emitted electrons from the upper layers of the analysed material (1–10 nm), as a result of the exciting a samples surface with X-rays. XPS allows you to determine what elements are in the ultra-thin layers and thin surface layers of the material, the chemical state and the amount of the element detected. The XRD method is based on measuring the intensity and scattering angles of X-rays, leaving the material with the same energy as the incident radiation. There are many databases of XRD spectra, so that there is rapid phase identification for many different crystal samples [136].

7.2. Type of Tested Samples

For testing dental materials, XRF and XRD are mainly used X-ray techniques. The XRF method is used to study metals, ceramics and glass, and the test samples can be of almost any shape and in a form of a powder, paste, solid or liquid. The TXRF method enables the analysis of solutions, thin layers, solids, and different surfaces [137]. Mostly solids are used in XRD testing. The preparation should have a flat and smooth surface. The technique of shaping the preparation consists in filling the window in the aluminium holder with a powder sample and gently pressing the glass plate on the preparation surface [138,139].

7.3. Sample Characteristics

Methods using X-rays allow for the analysis of the elemental composition in a small area closely adjacent to or being the surface of the test sample. For example, the effects of high temperature are observed on different restorative dental materials by detecting changes in their microstructural and elemental composition and using X-ray spectroscopy to determine the content of trace elements. The ability to distinguish dental materials by elemental analyses has had an important impact on the identification process [140,141].

7.4. Advantages and Limitations

The XRF method is used in both qualitative and quantitative analysis. The advantage of this method is the good selectivity and low quantification limit. The XRF method is of particular importance in the case of surface analysis. In the trace analysis 10^{-9} g of the substance can be determined in a sample weighing about 0.1 g. The technique is non-destructive, which is an additional advantage. The TXRF method can determine trace amounts of elements in the range of 10^{-7}–10^{-12} g in 1 g of a sample; small samples (approx. $5 \cdot 10^{-2}$ cm^3) can be used [57,60,131–135].

In the XRF method, it is necessary to use special crystals instead of diffraction gratings; it is only suitable for the analysis of elements with low excitation potential and the presence of matrix effects [57,60,131–135]. XRD is a technique that characterizes only crystalline materials [136].

7.5. Applications

The XRF is used to test the composition of materials, the coating thickness, and to identify elements in dental materials: implant alloy [142] and ceramics [143–152]. XRD is used to obtain information of crystalline structure of solids such as dental ceramics [143–150].

7.6. Spectrum Example

Figure 5 shows that the crystalline phases resulting from the heat treatment of the glass sample were determined using XRD. Diffraction images were obtained in the range of 2θ from 15° to 90° continuously at 0.6°/min. XRD results proved that spherulite crystals detected in the early phase of crystallization were enstatite [143].

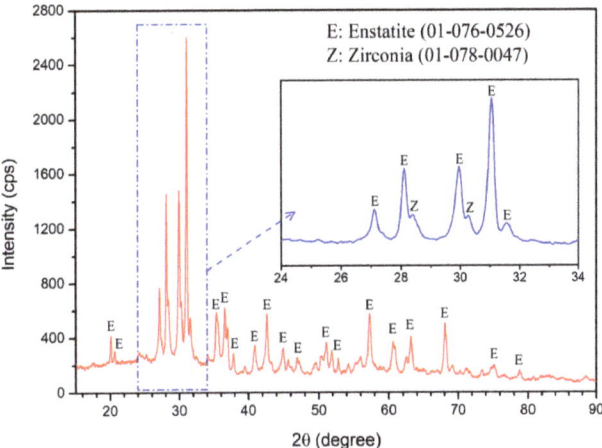

Figure 5. XRD pattern of the glass-ceramic heat-treated at 700 °C for 12 h and then at 1090 °C for 3 min [143].

8. Mass Spectrometry (MS)

8.1. Principle of Operation

Mass spectrometry is an analytical technique used to test or confirm the structure of organic compounds as well as to conduct qualitative and quantitative determination of specific compounds present in a mixture. It enables detecting substances in complex chemical mixtures even in minimal amounts (femtograms). The mode of action of each spectrometer, regardless of its construction, is based on the molecule ionization of the tested substance, allowing for their acceleration in an electric field under vacuum. The heterogeneous flux of positive or negative ions is split into a number of components depending on the mass to charge ratio (m/z). From the weight to charge ratio of an ion, it is usually possible to deduce the molecular weight of the studied compound or the molecular weight of its fragments. The ionization methods in some mass spectrometers are chosen such that the charge (z) is 1 for most ions, so when interpreting the spectrum, it can be assumed that m/z corresponds simply to the molecular weight of the ion. Sample ionization in mass spectroscopy can be performed using one of the following methods: Electron Ionisation, (EI) Electrospray (ESI) Fast-Atom Bombardment (FAB), Secondary Ion Mass Spectrometry (SIMS), Laser Desorption (LD), Matrix Assisted Laser Desorption Ionisation (MALDI), and Inductively Coupled Plasma (ICP). Research indicates that secondary ion mass spectrometry (SIMS) is a sensitive technique that characterizes biomaterials and biomineralized bone and dental tissues [153–157].

8.2. Type of Tested Samples

Samples such as metals, ceramics, organic and biological materials, polymers, biomaterials, and composites can be analysed. For SIMS testing, samples must be compatible with high vacuum, have a flat surface with minimal topography, and must not contain any "loose parts" that could enter the mass analyser and damage it [156].

8.3. Sample Characteristics

Mass spectrometry is an analytical technique based on measuring the mass to electric charge ratio of a given ion. Interpretation of the obtained data allows for the identification of chemical compounds. It enables the determination of their source of origin, precise determination of the composition of complex mixtures of compounds with high molar masses, research on dental materials and polymer chemistry [158,159]. For example, this information can be used to explain the organic composition and eluates of three resin-based pulp coatings in relation to their indications and safety data sheets [159], or to determine the molecular toxicology of substances released by dental materials [160].

8.4. Advantages and Limitations

The Time of Flight Secondary Ion Mass Spectrometry (ToF-SIMS) is a qualitative technique used to analyse surface characteristics that provides information related to molecular compounds, typically fragments of much larger organic macromolecules from the outermost surface of the sample [153]. This surface technique is considered highly sensitive.

The SIMS technique combines high mass resolution and the possibility of quantifying elements with the limitation that only a fixed number of predefined ions can be analysed simultaneously (up to seven ions). Due to the high sensitivity of this technique, only high purity solvents should be used [156]. This equipment has limited optical capabilities; experiencing difficulties in collecting positive or negative ion data and, depending on the type of sample, analysis may last from 30 min to 5 h. After testing, the surface of tested sample is left damaged. The technique provides complementary information to XPS technique [153].

8.5. Applications

The SIMS technique is used to analyse the composition and chemical interactions of dental resins [161,162], ceramics [161–165] and composite materials [166].

8.6. Spectrum Example

Appropriately prepared zirconium oxide samples together with silane-containing primer (MPS), and 10-Methacryloyloxydecyl dihydrogen phosphate primer (MDP) were subjected to the ToF-SIMS test. Figure 6 shows ToF-SIMS spectra of the studied system. The peaks of Zr^+ and ZrO^+ were used as references peaks in the cumulative positive ion spectra (Figure 6A). All groups showed the presence of $C_2H_3^+$ and $C_3H_5^+$ derived from organic contamination. All groups show the characteristic ions ZrO_2^- (121.9) and ZrO^- (138.9), while many differences are revealed among the five treatments in the negative ion spectra from m/z of 0 to 200 (Figure 6B). OH^- and C_2H^- appear in all spectra, which might be caused by the water adsorption and organic contamination after exposure on air. The increased intensity at m/z 122 was caused by the accretion of $SiO_3C_3H_9^-$ (121) on ZrO_2^- (121.9) [161].

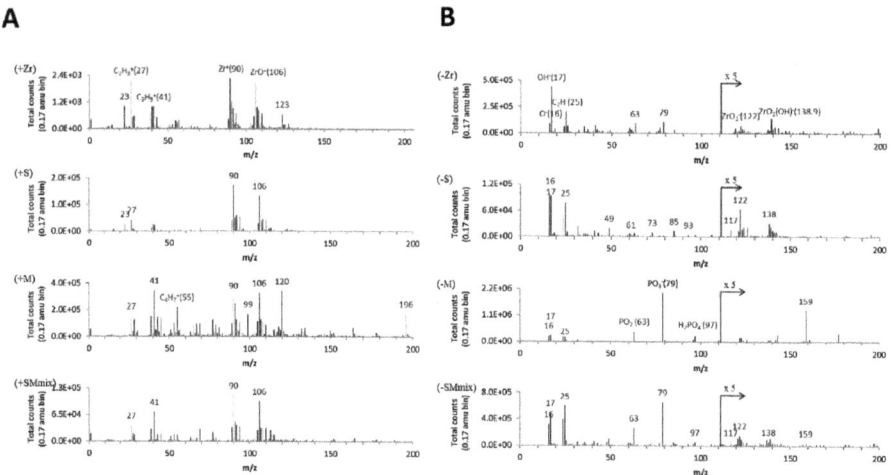

Figure 6. *Cont.*

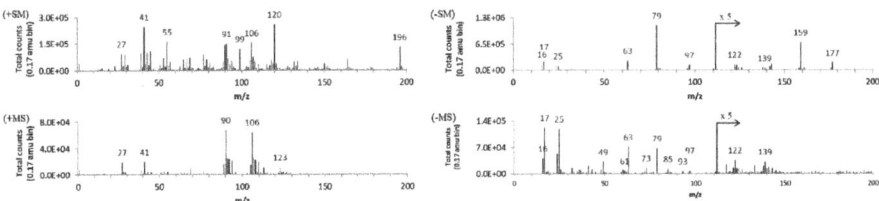

Figure 6. ToF-SIMS spectra (m/z = 0–200 amu) for the experimental groups. (**A**) The positive ion spectra. The characteristic ion peaks of zirconia are Zr^+ (m/z 90), ZrO^+ (m/z 106). The ion peaks under m/z 55 are mainly from organic components. Peaks at m/z 99 and 120 could be the fragments of MDP monomer. (**B**) In the negative ion spectra, the signals after m/z 111 were amplified by 5× to reveal the characteristic negative ion peaks of ZrO_2^- and $ZrO_2(OH)^-$ at m/z 121.9 and 138.9. The characteristic ion $SiO_3C_3H_9^-$ (121) in silane is overlapped with ZrO_2^- and thus the peak at m/z 122 increases [161].

9. Summary and Future Aspect

The article presents a review of spectroscopic methods for characterizing surfaces and application in the study of various state-of-the-art dental materials, i.e., ceramics, steel, glasses, cements, composites and resins. The review describes application of different type of spectroscopy such as infrared (FT-IR, and Raman), ultraviolet and visible: (UV-Vis), X-ray (XAS, XES, XRF, XRD) and mass spectrometry (ToF-SIMS).

Table 3 shows the use of spectroscopic methods for studying dental materials, together with the depth of the analysed layer, the nature of information provided (element, depth profile, surface mapping, chemical groups and bonds, etc.) and sample damage (or lack thereof). According to the last criterion, most of the techniques are non-destructive, and some of them—using X-rays for excitation—are almost non-invasive.

As expected, it turns out that not all methods can be generally applied to different types of biomaterial; only some techniques proved to be excellent in certain types of applications. Each method has its own specificity; therefore, the optimization of the approach depends on the correct selection of the method. This is possible due to the wide range of possibilities for characterizing studied materials.

The purpose of this overview is to facilitate the application of these techniques in the field of dental materials orientation in the drawn image. Of course, a single method cannot provide all of required information; in terms of dental materials, no method can provide complete sample information at all levels. Usually, several complementary methods of mutual confirmation and support were used in the study of a given material. However, the complementarity of the provided information were at different levels of elemental composition, chemistry and structure. The research in the field of dental materials has fully benefited from the multi-method approach, which is particularly suitable as no suitable reference materials are yet available for most of the cases. Therefore, the problem of standards has been carefully considered as it provides the most accurate solution to the correction of matrix effects. The latter are inherent especially in the case of thicker preparations, which is usually the most convenient form of preparation of dental materials. We hope this review helps dentists choose the appropriate method and sample preparation, and help the chemist to optimize the experiment setup according to the specimen characteristics.

Table 3. Comparison of spectroscopic methods for examining the surface of dental materials.

Method	Type of Sample	Analytical Depth	Sample Degradation	Type of Information	Application Examples in Dental Biomaterials and Related Research
Fourier Transform Infrared Spectroscopy (FT-IR)	Gas, liquid, solid	The penetration depth is about 0.5–3 µm [72].	Non-destructive [69]	Quantitative analysis of complex mixtures; the investigation of surface and interfacial phenomena [69]	Implant materials (e.g., to characterize the functional groups of the synthesized apatite particles [82], to study the vibrational states of commercial bone grafts, B-Raw, H-Raw, and B-560 to determine the presence of other functional groups in the samples that do not belong to hydroxyapatite [106]); biopolymers (e.g., characterization of the functional groups in samples of peptide modified demineralized dentin matrix [83]; ceramics (e.g., to complement XRD results, and to determine dental zirconia superficial molecular compositions [84]; to identify functional groups of HAp nanostructures in resin nanocomposites [85]; recording chemical constituents of implant coatings (e.g., metronidazole decorated Ti interfaces [86], to detect chemical groups of the modified PEEK films with covalently grafted osteogenic growth peptide [87]); to analyse bioceramics (e.g., to analyse phase stable β-tricalcium phosphate (β-TCP) in powder samples [89], to determine bulk composition of calcium phosphates [165]); dental resins (e.g., to investigate double bond conversion of dental resin matrix [90] and to calculate the degree of double bond conversion and polymerization rate of photopolymerizable co-initiators in dental monomers [94], to analyse microstructural and surface properties of tricalcium silicate-based pulp capping materials [91], to confirm the final structures of the functional nanoparticles (triazole functional silica) as well as nanocomposites incorporating the functional nanoparticles [92], to analyse powders of monomers: TAT, nt-TiO$_2$, and nt-TiO$_2$:TAT to evaluate a possible chemical interaction between TAT and nt-TiO$_2$ [123]; cements (e.g., to identify the degree of conversion of chemically cured resin modified glass-ionomer cements (RMGICs) testing unset liquids and set materials [93], to provide insight of the setting reactions of a hydraulic calcium silicate cement by taking the FTIR spectra of components before and during the setting reaction [95]); bioglass (e.g., to indicated the integration of the *Calcarea phosphorica* with nano-bioglass ceramic particles [96]); self-curing materials e.g., to compare the structure of boron nitride reinforced PMMA for dental restorations after hand and ultrasonic mixing [97]
Raman Spectroscopy	Gas, liquid, solid (in bulk, as microscopic particles, or as surface layers)	The penetration depth is about 0.01–2300 µm [101].	Non-invasive [56]	Qualitative and quantitative: Investigation of rotational and oscillating spectra of molecules; identification of chemicals component [99]	structure assessment of anti-corrosion coatings e.g., to confirm the growth of graphene and its transfer onto Ti-6Al-4V discs [104]; bioglass (e.g., to investigate the mineral and organic composition of dentin surfaces; demineralized dentin and dentin remineralized with bioglass [110], to analyse the modification of the Ti-Zr-45S5 bioglass alloy surface after oxidation [111]); implant materials (e.g., bovine and human bio hydroxyapatites [106]); ceramics (e.g., chemical analysis of the surface by micro-Raman spectroscopy to establish the presence of MDP monomer on the surface of the zirconia after bonding procedures using MDP containing silane or adhesive [107], to determine the resistance of the titanium substrate to oxidation during the firing of subsequent porcelain layers [108], to assess the chemical composition of the fracture surface in the region of the lithium disilicate ceramic, in the ceramic/staining interface and in the staining applied on the ceramic [109], the complementarily (to XRD) use of micro-Raman to characterize the phase composition of different positions at occlusal loaded area of fixed dental prostheses fabricated from three zirconia grades with varying yttria content [112], to determine phase transformation of the surface of monolithic zirconia submitted to different surface treatments [113], to investigate structural aspects of the glass-ceramic i.e., differently formed crystals, the vitreous area around the crystals, the interface between the TZ3Y substrate and the glass-ceramic, as well as the outer surface of the glass-ceramic [164]; dental resin composites and cements e.g., to evaluate degree of conversion and maximum rate of polymerization [105,114,115]; to analyse powders of monomers: TAT, nt-TiO$_2$, and nt-TiO$_2$:TAT to evaluate a possible chemical interaction between TAT and nt-TiO$_2$ [123]
UV-Vis Spectroscopy	Liquid, solid, gas.	The penetration depth is about 0.02–5 µm [167].	Allows sample recovery [168]	Quantitative: Identification of chemical compounds containing chromophores [168]	resins (e.g., to analyse powders of monomers: TAT, nt-TiO$_2$, and nt-TiO$_2$:TAT to evaluate a possible chemical interaction between TAT and nt-TiO$_2$ [123], to collect optical properties data to calculate colour measurements of dental resin composites containing different opacifiers [127], to investigate the optical properties of Ca$_{10}$(PO$_4$)$_6$(OH)$_2$/Li-BioMOFs structures of dental resin nanocomposites [85]), polymers (e.g., to determine the maximum absorption of conventional polymethyl methacrylate and the absorption of residual conventional polymethyl methacrylate of specimen eluted in the storage liquid [124]), characterization of co-initiators in photopolymerization of polymers [94,125], oxide layers [126]; ceramics e.g., to analyse the translucency of color-gradient multilayered zirconia, whereas quantitative measurements of translucency can be implemented by analysing the definite transmission of light through each specimen [128,129]

Table 3. *Cont.*

Method	Type of Sample	Analytical Depth	Sample Degradation	Type of Information	Application Examples in Dental Biomaterials and Related Research
X-ray Spectroscopy	Powder, paste, solid or liquid	The penetration depth: of XRD is about 50–200 mm [169], XPS 1–10 nm, and XRF 0.5–3 μm [170].	Non-destructive and non-invasive [34,171]	Quantitative: Analysis of crystal structure and phase composition [60]	XPS: biopolymers (e.g., chemical composition of peptide-modified demineralized dentin matrix [83]); anti-corrosion coatings (e.g., to confirm that the graphene film was free of copper residues after ammonium persulfate etching [104]); silver-alloy reinforced GI, zirconia reinforced GI, and conventional GI using X-ray analysis for obtaining elemental compositions before and after the incineration [141]; bioceramics (e.g., to determine the elemental compositions of the outer layers of calcium phosphates [165]); implant material coatings (e.g., to detect the surface chemical constituents and to confirm the presence of osteogenic growth peptide on PEEK surfaces [87]); XRF: implant alloys e.g., to evaluate the fixture and abutment surface of internal hexagonal connection systems [142] and to evaluate chemical composition of dental ceramics [143–152]; XRD: implant materials (e.g., to characterize the structure of strontium apatite particles [82]), to measure the crystallinity of hydroxyapatite particles [88], to characterize phase stable β-tricalcium phosphate (β-TCP) [89] or to distinguishing products with the same gross chemical composition but different crystal structures (e.g., different crystal structures of calcium phosphate) [165], to obtain information on the degree of crystallinity of the tricalcium silicate-based pulp capping materials [91]; ceramics (e.g., determination of the crystalline phases in dental zirconia [84,144,146,164], to evaluate phase transformations on the outer surface of fixed dental prostheses fabricated from three zirconia grades with varying yttria content [112], to determine the crystalline phases resulting from the heat treatments of the glass sample in development of strong glass-ceramics based on the crystallization of micron-sized enstatite and nano-sized zirconia and Ti-containing crystals by controlled crystallization of a $51SiO_2$–$35MgO$–$6Na_2O$–$4ZrO_2$–$4TiO_2$ (mol%) glass) [143]; bioceramics (e.g., to confirm the crystalline nature of nano-bioglass ceramic particles doped with *Calcarea phosphorica* [96], to identify the crystalline and amorphous phases of partially crystallized lithium disilicate ceramics in lithium metasilicate phase [109,143–150], to analyse phase composition of the Ti-Zr-45S5 bioglass alloy [111]; to characterise the phases, the crystallography and the examination of the crystallite size of the $Ca_{10}(PO_4)_6(OH)_2$/Li-BiOMOFs [85]; bone grafts (e.g., to identify the crystalline phases and changes in full width at the half maximum (FWHM) of commercial bone grafts, bovine and human bones as well as their BIO-HAps obtained by calcination [106]); self-curing materials (e.g., to observe patterns of boron nitride reinforced PMMA for dental restorations after hand and ultrasonic mixing [97]; to precisely determine the composition of complex mixtures of compounds e.g., to elucidate the organic composition and eluates of three resin-based pulp-capping materials [159];
Mass Spectrometry	solid	Surface nano-layer [34].	Non-destructive [34]	Qualitative: Composition analysis of solid surfaces and thin films [34]	resins [161,162]; ceramics e.g., to analyse the compositions and chemical interactions of the 3-methacryloyloxypropyltrimethoxysilane (MPS)- and 10-methacryloyloxydecyl-dihydrogen-phosphate (MDP)-base primers, in their single or sequential applications, to zirconia [161,162], chemical analysis of saliva contaminated glass ceramic surface and after different cleaning regimens [163], to investigate ion diffusion between the veneer ceramic and the ZrO_2-based substrate [164], to analyse chemical composition of calcium phosphates [165]; composite materials e.g., the release of BPA from two conventional Bis-GMA-containing and two "BPA-free" restorative resin-based composites, which are commonly used as tooth-coloured filling materials, was examined using liquid chromatography—tandem mass spectrometry [166]

The current representative literature review is by no means comprehensive (which would probably grow to around a thousand titles). Figure 7 presents the number of publications on the use of spectroscopic methods in the study of dental materials and related topics based on 4173 titles, indicating that this field is progressively growing.

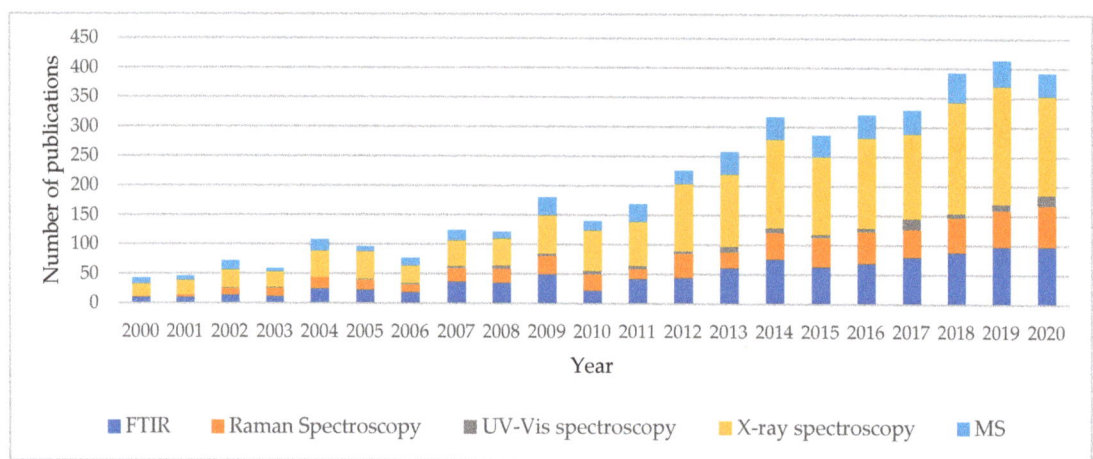

Figure 7. Histogram of the cited publications published in the years 2000–2020 according to Scopus on the application of spectroscopic methods in the analysis of surface phenomena in the study of dental materials and related issues.

Within the period 2000–2020, an increase in the number of publications can be noticed. The data suggest that the use of spectroscopic methods to characterize surfaces in the study of dental materials has reached maturity; though it is a highly specialized, the filed shows a high and sustained dynamics. X-ray spectroscopy and FT-IR techniques are the most popular among researchers. Due to the characteristics of these methods, it can be expected their will continue in the upcoming years, especially as dental materials—despite enormous advances made in recent decades—still have many imperfections and the number of experimental objects will increase worldwide.

Author Contributions: Conceptualization, K.K. and M.L.-S.; methodology, M.L.-S.; software, A.L.; validation, A.L., B.L. and S.S.; formal analysis, K.K.; investigation, K.K.; resources, B.L.; data curation, A.L.; writing—original draft preparation, K.K.; writing—review and editing, B.L. and M.L.-S.; visualization, K.K. and A.L.; supervision, S.S.; project administration, S.S. and M.L.-S.; funding acquisition, M.L.-S. All authors have read and agreed to the published version of the manuscript.

Funding: This research was funded by the National Centre for Research and Development (Warsaw, Poland) within the grant InterChemMed (WND-POWR.03.02.00-00-I029/16-01).

Institutional Review Board Statement: Not applicable.

Informed Consent Statement: Not applicable.

Data Availability Statement: Data sharing is not applicable to this article.

Acknowledgments: The authors thank Valentin Mircerski (University of Lodz, Poland) for fruitful discussion on the problems presented in the work.

Conflicts of Interest: All authors declare no conflict of interest in this paper.

References

1. Alzahrani, K.M. Implant Bio-mechanics for Successful Implant Therapy: A Systematic Review. *J. Int. Soc. Prev. Commun. Dent.* **2020**, *10*, 700–714. [CrossRef]
2. Yu, S.H.; Hao, J.; Fretwurst, T.; Liu, M.; Kostenuik, P.; Giannobile, W.V.; Jin, Q. Sclerostin-Neutralizing Antibody Enhances Bone Regeneration Around Oral Implants. *Tissue Eng. Part A* **2018**, *24*, 1672–1679. [CrossRef] [PubMed]
3. Beltramini, G.; Russillo, A.; Baserga, C.; Pellati, A.; Piva, A.; Candotto, V.; Bolzoni, A.; Beltramini, G.; Rossi, D.; Bolzoni, A.; et al. Collagenated heterologous cortico-cancelleus bone mix stimulated dental pulp derived stem cells. *J. Biol. Regul. Homeost. Agents* **2020**, *34*, 1–5. [PubMed]
4. Jiang, X.; Cao, Z.; Yao, Y.; Zhao, Z.; Liao, W. Aesthetic evaluation of the labiolingual position of maxillary lateral incisors by orthodontists and laypersons. *BMC Oral Health* **2021**, *21*, 42. [CrossRef]
5. Gomez-Meda, R.; Esquivel, J.; Blatz, M.B. The esthetic biological contour concept for implant restoration emergence profile design. *J. Esthet. Restor. Dent.* **2021**, *33*, 173–184. [CrossRef]
6. Furukawa, M.; Wang, J.; Kurosawa, M.; Ogiso, N.; Shikama, Y.; Kanekura, T.; Matsushita, K. Effect of green propolis extracts on experimental aged gingival irritation in vivo and in vitro. *J. Oral Biosci.* **2021**, *63*, 58–65. [CrossRef]
7. Koutouzis, T. Implant-abutment connection as contributing factor to peri-implant diseases. *Periodontology 2000* **2019**, *81*, 152–166. [CrossRef]
8. Birch, S.; Ahern, S.; Brocklehurst, P.; Chikte, U.; Gallagher, J.; Listl, S.; Lalloo, R.; O'Malley, L.; Rigby, J.; Tickle, M.; et al. Planning the oral health workforce: Time for innovation. *Commun. Dent. Oral Epidemiol.* **2021**, *49*, 17–22. [CrossRef]
9. Joda, T.; Yeung, A.W.K.; Hung, K.; Zitzmann, N.U.; Bornstein, M.M. Disruptive Innovation in Dentistry: What It Is and What Could Be Next. *J. Dent. Res.* **2020**, *100*, 448–453. [CrossRef]
10. Joda, T.; Bornstein, M.M.; Jung, R.E.; Ferrari, M.; Waltimo, T.; Zitzmann, N.U. Recent trends and future direction of dental research in the digital era. *Int. J. Environ. Res. Public Health* **2020**, *17*, 1987. [CrossRef]
11. Bastos, N.A.; Bitencourt, S.B.; Martins, E.A.; De Souza, G.M. Review of nano-technology applications in resin-based restorative materials. *J. Esthet. Restor. Dent.* **2020**. [CrossRef]
12. Glied, A.; Mundiya, J. Implant Material Sciences. *Dent. Clin. N. Am.* **2021**, *65*, 81–88. [CrossRef]
13. Saeidi Pour, R.; Freitas, R.C.; Engler, M.L.P.D.; Edelhoff, D.; Klaus, G.; Prandtner, O.; Berthold, M.; Liebermann, A. Historical development of root analogue implants: A review of published papers. *Br. J. Oral Maxillofac. Surg.* **2019**, *57*, 496–504. [CrossRef]
14. Wojda, S.M. Comparative Analysis of Two Methods of Assessment Wear of Dental Materials. *Acta Mech. Autom.* **2015**, *9*, 105–109. [CrossRef]
15. Duhn, C.; Thalji, G.; Al-Tarwaneh, S.; Cooper, L.F. A digital approach to robust and esthetic implant overdenture construction. *J. Esthet. Restor. Dent.* **2021**, *33*, 118–126. [CrossRef]
16. Deb, S.; Chana, S. Biomaterials in Relation to Dentistry. *Front. Oral Biol.* **2015**, *17*, 1–12. [CrossRef]
17. Orsini, G.; Pagella, P.; Mitsiadis, T.A. Modern Trends in Dental Medicine: An Update for Internists. *Am. J. Med.* **2018**, *131*, 1425–1430. [CrossRef]
18. Guglielmotti, M.B.; Olmedo, D.G.; Cabrini, R.L. Research on implants and osseointegration. *Periodontology 2000* **2019**, *79*, 178–189. [CrossRef]
19. Zhang, Y.; Kelly, J.R. Dental Ceramics for Restoration and Metal Veneering. *Dent. Clin. N. Am.* **2017**, *61*, 797–819. [CrossRef]
20. Bhatavadekar, N.B.; Gharpure, A.S.; Balasubramanium, N.; Scheyer, E.T. In Vitro Surface Testing Methods for Dental Implants-Interpretation and Clinical Relevance: A Review. *Compend. Contin. Educ. Dent.* **2020**, *41*, e1–e9.
21. Meschi, N.; Patel, B.; Ruparel, N.B. Material Pulp Cells and Tissue Interactions. *J. Endod.* **2020**, *46*, S150–S160. [CrossRef]
22. Cionca, N.; Hashim, D.; Mombelli, A. Zirconia dental implants: Where are we now, and where are we heading? *Periodontology 2000* **2017**, *73*, 241–258. [CrossRef]
23. Della Bona, A.; Cantelli, V.; Britto, V.T.; Collares, K.F.; Stansbury, J.W. 3D printing restorative materials using a stereolithographic technique: A systematic review. *Dent. Mater.* **2021**, *37*, 336–350. [CrossRef]
24. Wang, Y.; Bäumer, D.; Ozga, A.K.; Körner, G.; Bäumer, A. Patient satisfaction and oral health-related quality of life 10 years after implant placement. *BMC Oral Health* **2021**, *21*, 30. [CrossRef]
25. Eick, S. Biofilms. *Monogr. Oral Sci.* **2020**, *29*, 1–11.
26. Díaz-Garrido, N.; Lozano, C.P.; Kreth, J.; Giacaman, R.A. Competition and Caries on Enamel of a Dual-Species Biofilm Model with Streptococcus mutans and Streptococcus sanguinis. *Appl. Environ. Microbiol.* **2020**, *86*. [CrossRef]
27. Sterzenbach, T.; Helbig, R.; Hannig, C.; Hannig, M. Bioadhesion in the oral cavity and approaches for biofilm management by surface modifications. *Clin. Oral Investig.* **2020**, *24*, 4237–4260. [CrossRef] [PubMed]
28. Bhadila, G.; Menon, D.; Wang, X.; Vila, T.; Melo, M.A.S.; Montaner, S.; Arola, D.D.; Weir, M.D.; Sun, J.; Xu, H.H.K. Long-term antibacterial activity and cytocompatibility of novel low-shrinkage-stress, remineralizing composites. *J. Biomater. Sci. Polym. Ed.* **2021**, 1–16. [CrossRef]
29. Tettamanti, L.; Cura, F.; Andrisani, C.; Bassi, M.A.; Silvestre-Rangil, J.; Tagliabue, A. A new implant-abutment connection for bacterial microleakage prevention: An in vitro study. *ORAL Implantol.* **2017**, *10*, 172–180. [CrossRef] [PubMed]
30. Revilla-León, M.; Morillo, J.A.; Att, W.; Özcan, M. Chemical Composition, Knoop Hardness, Surface Roughness, and Adhesion Aspects of Additively Manufactured Dental Interim Materials. *J. Prosthodont.* **2020**. [CrossRef] [PubMed]
31. Ercoli, C.; Caton, J.G. Dental prostheses and tooth-related factors. *J. Periodontol.* **2018**, *89*, S223–S236. [CrossRef]

32. Revilla-León, M.; Husain, N.A.H.; Methani, M.M.; Özcan, M. Chemical composition, surface roughness, and ceramic bond strength of additively manufactured cobalt-chromium dental alloys. *J. Prosthet. Dent.* **2020**. [CrossRef]
33. Revilla-León, M.; Meyers, M.J.; Zandinejad, A.; Özcan, M. A review on chemical composition, mechanical properties, and manufacturing work flow of additively manufactured current polymers for interim dental restorations. *J. Esthet. Restor. Dent.* **2019**, *31*, 51–57. [CrossRef]
34. Preoteasa, E.A.; Preoteasa, E.S.; Suciu, I.; Bartok, R.N. Atomic and nuclear surface analysis methods for dental materials: A review. *AIMS Mater. Sci.* **2018**, *5*, 781–844. [CrossRef]
35. Karova, E. Application of Atomic Force Microscopy in Dental Investigations. *Int. J. Sci. Res.* **2020**, *9*, 1319–1326. [CrossRef]
36. Roa, J.J.; Oncins, G.; Díaz, J.; Capdevila, X.G.; Sanz, F.; Segarra, M. Study of the friction, adhesion and mechanical properties of single crystals, ceramics and ceramic coatings by AFM. *J. Eur. Ceram. Soc.* **2011**, *31*, 429–449. [CrossRef]
37. Vilá, J.F.; García, J.C.; Guestrin, A. SEM Applied to the Development of Bioactive Surface of Dental Implants. *Microsc. Microanal.* **2020**, *26*, 147–148. [CrossRef]
38. Banaszek, K.; Sawicki, J.; Wołowiec-Korecka, E.; Gorzędowski, J.; Danowska-Klonowska, D.; Sokołowski, J. Use of optical microscopy for evaluation of tooth structure. *J. Achiev. Mater. Manuf. Eng.* **2016**, *79*, 31–40. [CrossRef]
39. Chander, N.G. Characterization of dental materials. *J. Indian Prosthodont. Soc.* **2018**, *18*, 289–290. [CrossRef]
40. Zhou, X.; Thompson, G.E. Electron and Photon Based Spatially Resolved Techniques. In *Reference Module in Materials Science and Materials Engineering*; Elsevier: New York, NY, USA, 2017; pp. 1–30.
41. Yanikoglu, N.D.; Sakarya, R.E. Test methods used in the evaluation of the structure features of the restorative materials: A literature review. *J. Mater. Res. Technol.* **2020**, *9*, 9720–9734. [CrossRef]
42. Naves, L.Z.; Gerdolle, D.A.; de Andrade, O.S.; Markus Maria Gresnigt, M. Seeing is believing? When scanning electron microscopy (SEM) meets clinical dentistry: The replica technique. *Microsc. Res. Tech.* **2020**, *83*, 1118–1123. [CrossRef] [PubMed]
43. Siekaniec, D.; Kopyciński, D. Analysis Phases and Crystallographic Orientation of the grain of High Chromium Cast Iron Using EBSD Technique. *J. Cast. Mater. Eng.* **2017**, *1*, 15. [CrossRef]
44. Koblischka-Veneva, A.; Koblischka, M.R.; Schmauch, J.; Hannig, M. Comparison of human and bovine dental enamel by TEM and t-EBSD investigations. *IOP Conf. Ser. Mater. Sci. Eng.* **2019**, *625*, 012006. [CrossRef]
45. Tomota, Y. Crystallographic characterization of steel microstructure using neutron diffraction. *Sci. Technol. Adv. Mater.* **2019**, *20*, 1189–1206. [CrossRef]
46. Shukla, A.K. *Electron Spin Resonance Spectroscopy in Medicine*; Shukla, A.K., Ed.; Springer: Singapore, 2018; ISBN 9789811322303.
47. Pandoleon, P.; Kontonasaki, E.; Kantiranis, N.; Pliatsikas, N.; Patsalas, P.; Papadopoulou, L.; Zorba, T.; Paraskevopoulos, K.M.; Koidis, P. Aging of 3Y-TZP dental zirconia and yttrium depletion. *Dent. Mater.* **2017**, *33*, e385–e392. [CrossRef]
48. Lopes, C.d.C.A.; Limirio, P.H.J.O.; Novais, V.R.; Dechichi, P. Fourier transform infrared spectroscopy (FTIR) application chemical characterization of enamel, dentin and bone. *Appl. Spectrosc. Rev.* **2018**, *53*, 747–769. [CrossRef]
49. Lach, S.; Jurczak, P.; Karska, N.; Kubiś, A.; Szymańska, A.; Rodziewicz-Motowidło, S. Spectroscopic methods used in implant material studies. *Molecules* **2020**, *25*, 579. [CrossRef]
50. Kreve, S.; Cândido dos Reis, A. Influence of the electrostatic condition of the titanium surface on bacterial adhesion: A systematic review. *J. Prosthet. Dent.* **2021**, *125*, 416–420. [CrossRef]
51. Kim, I.H.; Son, J.S.; Min, B.K.; Kim, Y.K.; Kim, K.H.; Kwon, T.Y. A simple, sensitive and non-destructive technique for characterizing bovine dental enamel erosion: Attenuated total reflection Fourier transform infrared spectroscopy. *Int. J. Oral Sci.* **2016**, *8*, 54–60. [CrossRef]
52. Furuhashi, K.; Uo, M.; Kitagawa, Y.; Watari, F. Rapid and non-destructive analysis of metallic dental restorations using X-ray fluorescence spectra and light-element sampling tools. *Appl. Surf. Sci.* **2012**, *262*, 13–18. [CrossRef]
53. Malik, A.K.; Kumar, R. Heena Spectroscopy: Types. In *Encyclopedia of Food and Health*; Elsevier Ltd.: New York, NY, USA, 2015; pp. 64–72. ISBN 9780123849533.
54. Pignataro, M.F.; Herrera, M.G.; Dodero, V.I. Evaluation of peptide/protein self-assembly and aggregation by spectroscopic methods. *Molecules* **2020**, *25*, 4854. [CrossRef]
55. Kafle, B.P. Infrared (IR) spectroscopy. In *Chemical Analysis and Material Characterization by Spectrophotometry*; Elsevier: New York, NY, USA, 2020; pp. 199–243.
56. Smith, E.; Dent, G. Modern Raman Spectroscopy—A Practical Approach. John Wiley and Sons Ltd.: Chichester, UK, 2005; p. 210. ISBN 0471496685. [CrossRef]
57. Dolenko, G.N.; Poleshchuk, O.K.; Latosińska, J.N. X-ray emission spectroscopy, methods. In *Encyclopedia of Spectroscopy and Spectrometry*; Elsevier: New York, NY, USA, 2016; pp. 691–694. ISBN 9780128032244.
58. Sharma, R.K. Various Spectroscopic Techniques. In *Environmental Pollution: Monitoring Modelling and Control*; Studium Press, LLC: Houston, TX, USA, 2017; pp. 181–206. ISBN ISBN:1626991022/978-1626991026.
59. Patonay, G.; Beckford, G.; Hänninen, P. UV-Vis and NIR Fluorescence Spectroscopy. In *Handbook of Spectroscopy*, 2nd Enlarged ed.; Wiley-VCH Verlag GmbH & Co. KGaA: Weinheim, Germany, 2014; Volume 3–4, pp. 999–1036. ISBN 9783527654703.
60. Adams, F.C. X-ray absorption and Diffraction | Overview. In *Encyclopedia of Analytical Science*; Elsevier: New York, NY, USA, 2019; pp. 391–403. ISBN 9780081019832.
61. Brader, M.L. UV-absorbance, fluorescence and FT-IR spectroscopy in biopharmaceutical development. In *Biophysical Characterization of Proteins in Developing Biopharmaceuticals*; Elsevier: New York, NY, USA, 2020; pp. 97–121. ISBN 9780444641731.

62. Akhtar, S.; Ali, S. Characterization of nanomaterials: Techniques and tools. In *Applications of Nanomaterials in Human Health*; Springer: Singapore, 2020; pp. 23–43. ISBN 9789811548024.
63. Fa, K.; Jiang, T.; Nalaskowski, J.; Miller, J.D. Optical and spectroscopic characteristics of oleate adsorption as revealed by FTIR analysis. *Langmuir* **2004**, *20*, 5311–5321. [CrossRef]
64. Le Pevelen, D.D.; Tranter, G.E. FT-IR and raman spectroscopies, polymorphism applications. In *Encyclopedia of Spectroscopy and Spectrometry*; Elsevier: New York, NY, USA, 2016; pp. 750–761. ISBN 9780128032244.
65. Bell, S.E.J.; Xu, Y. Infrared spectroscopy | Industrial applications. In *Encyclopedia of Analytical Science*; Elsevier: New York, NY, USA, 2019; pp. 124–133. ISBN 9780081019832.
66. Kowalczuk, D.; Pitucha, M. Application of FTIR Method for the Assessment of Immobilization of Active Substances in the Matrix of Biomedical Materials. *Materials* **2019**, *12*, 2972. [CrossRef]
67. Puspita, S.; Sunarintyas, S.; Mulyawati, E.; Anwar, C.; Sukirno; Soesatyo, M.H.N.E. Molecular weight determination and structure identification of Bombyx mori L. Fibroin as material in dentistry. In *AIP Conference Proceedings*; American Institute of Physics Inc.: College, MA, USA, 2020; Volume 2260, p. 40018.
68. Rafeek, A.D.; Choi, G.; Evans, L.A. Morphological, spectroscopic and crystallographic studies of calcium phosphate bioceramic powders. *J. Aust. Ceram. Soc.* **2018**, *54*, 161–168. [CrossRef]
69. Dutta, A. Fourier Transform Infrared Spectroscopy. In *Spectroscopic Methods for Nanomaterials Characterization*; Elsevier: New York, NY, USA, 2017; Volume 2, pp. 73–93. ISBN 9780323461467.
70. Rosi, F.; Cartechini, L.; Sali, D.; Miliani, C. Recent trends in the application of fourier transform infrared (FT-IR) spectroscopy in Heritage Science: From micro: From non-invasive FT-IR. *Phys. Sci. Rev.* **2019**, *4*. [CrossRef]
71. Margariti, C. The application of FTIR microspectroscopy in a non-invasive and non-destructive way to the study and conservation of mineralised excavated textiles. *Herit. Sci.* **2019**, *7*, 1–14. [CrossRef]
72. Munajad, A.; Subroto, C. Suwarno Fourier transform infrared (FTIR) spectroscopy analysis of transformer paper in mineral oil-paper composite insulation under accelerated thermal aging. *Energies* **2018**, *11*, 364. [CrossRef]
73. Puppin-Rontani, J.; Fugolin, A.P.P.; Costa, A.R.; Correr-Sobrinho, L.; Pfeifer, C.S. In vitro performance of 2-step, total etch adhesives modified by thiourethane additives. *Int. J. Adhes. Adhes.* **2020**, *103*, 102688. [CrossRef]
74. Shim, J.S.; Lee, S.Y.; Song, S.-Y.; Jha, N.; Ryu, J.J. Polymerization efficiency of dental dual-cured resin cement light-cured at various times after the initiation of chemical activation. *Int. J. Polym. Mater. Polym. Biomater.* **2020**, *69*, 622–628. [CrossRef]
75. Fugolin, A.P.; Lewis, S.; Logan, M.G.; Ferracane, J.L.; Pfeifer, C.S. Methacrylamide–methacrylate hybrid monomers for dental applications. *Dent. Mater.* **2020**, *36*, 1028–1037. [CrossRef]
76. Fugolin, A.P.; Dobson, A.; Ferracane, J.L.; Pfeifer, C.S. Effect of residual solvent on performance of acrylamide-containing dental materials. *Dent. Mater.* **2019**, *35*, 1378–1387. [CrossRef] [PubMed]
77. Fugolin, A.P.P.; Navarro, O.; Logan, M.G.; Huynh, V.; França, C.M.; Ferracane, J.L.; Pfeifer, C.S. Synthesis of di- and triacrylamides with tertiary amine cores and their evaluation as monomers in dental adhesive interfaces. *Acta Biomater.* **2020**, *115*, 148–159. [CrossRef] [PubMed]
78. Fugolin, A.P.; Dobson, A.; Huynh, V.; Mbiya, W.; Navarro, O.; Franca, C.M.; Logan, M.; Merritt, J.L.; Ferracane, J.L.; Pfeifer, C.S. Antibacterial, ester-free monomers: Polymerization kinetics, mechanical properties, biocompatibility and anti-biofilm activity. *Acta Biomater.* **2019**, *100*, 132–141. [CrossRef] [PubMed]
79. Alania, Y.; dos Reis, M.C.; Nam, J.-W.; Phansalkar, R.S.; McAlpine, J.; Chen, S.-N.; Pauli, G.F.; Bedran-Russo, A.K. A dynamic mechanical method to assess bulk viscoelastic behavior of the dentin extracellular matrix. *Dent. Mater.* **2020**, *36*, 1536–1543. [CrossRef]
80. Zhang, P.; Zhao, X.M. Synthesis, crystal structure and bioactivity evaluation of a heterocyclic compound. *Jiegou Huaxue* **2020**, *39*, 1892–1897. [CrossRef]
81. Seredin, P.V.; Uspenskaya, O.A.; Goloshchapov, D.L.; Ippolitov, I.Y.; Vongsvivut, J.; Ippolitov, Y.A. Organic-mineral interaction between biomimetic materials and hard dental tissues. *Sovrem. Tehnol. V Med.* **2020**, *12*, 43–51. [CrossRef]
82. Gurgenc, T. Structural characterization and dielectrical properties of Ag-doped nano-strontium apatite particles produced by hydrothermal method. *J. Mol. Struct.* **2021**, *1223*, 128990. [CrossRef]
83. Jing, X.; Xie, B.; Li, X.; Dai, Y.; Nie, L.; Li, C. Peptide decorated demineralized dentin matrix with enhanced bioactivity, osteogenic differentiation via carboxymethyl chitosan. *Dent. Mater.* **2021**, *37*, 19–29. [CrossRef]
84. Ramos, N.C.; Alves, L.M.M.; Ricco, P.; Santos, G.M.A.S.; Bottino, M.A.; Campos, T.M.B.; Melo, R.M. Strength and bondability of a dental Y-TZP after silica sol-gel infiltrations. *Ceram. Int.* **2020**, *46*, 17018–17024. [CrossRef]
85. Asadi, F.; Forootanfar, H.; Ranjbar, M. A facile one-step preparation of Ca10(PO4)6(OH)2/Li-BioMOFs resin nanocomposites with Glycyrrhiza glabra (licorice) root juice as green capping agent and mechanical properties study. *Artif. Cells Nanomed. Biotechnol.* **2020**, *48*, 1331–1339. [CrossRef]
86. Fu, D.; Lu, Y.; Gao, S.; Peng, Y.; Duan, H. Chemical Property and Antibacterial Activity of Metronidazole-decorated Ti through Adhesive Dopamine. *J. Wuhan Univ. Technol. Mater. Sci. Ed.* **2019**, *34*, 968–972. [CrossRef]
87. Yakufu, M.; Wang, Z.; Wang, Y.; Jiao, Z.; Guo, M.; Liu, J.; Zhang, P. Covalently functionalized poly(etheretherketone) implants with osteogenic growth peptide (OGP) to improve osteogenesis activity. *RSC Adv.* **2020**, *10*, 9777–9785. [CrossRef]
88. Ding, Y.; Zhang, H.; Wang, X.; Zu, H.; Wang, C.; Dong, D.; Lyu, M.; Wang, S. Immobilization of Dextranase on Nano-Hydroxyapatite as a Recyclable Catalyst. *Materials* **2020**, *14*, 130. [CrossRef]

89. Roopavath, U.K.; Sah, M.K.; Panigrahi, B.B.; Rath, S.N. Mechanochemically synthesized phase stable and biocompatible β-tricalcium phosphate from avian eggshell for the development of tissue ingrowth system. *Ceram. Int.* **2019**, *45*, 12910–12919. [CrossRef]
90. Zeng, W.; Liu, F.; He, J. Physicochemical Properties of Bis-GMA/TEGDMA Dental Resin Reinforced with Silanized Multi-Walled Carbon Nanotubes. *Silicon* **2019**, *11*, 1345–1353. [CrossRef]
91. Voicu, G.; Didilescu, A.C.; Stoian, A.B.; Dumitriu, C.; Greabu, M.; Andrei, M. Mineralogical and microstructural characteristics of two dental pulp capping materials. *Materials* **2019**, *12*, 1772. [CrossRef]
92. Yushau, U.S.; Almofeez, L.; Bozkurt, A. Novel Polymer Nanocomposites Comprising Triazole Functional Silica for Dental Application. *Silicon* **2020**, *12*, 109–116. [CrossRef]
93. Agha, A.; Parker, S.; Patel, M. Polymerization shrinkage kinetics and degree of conversion of commercial and experimental resin modified glass ionomer luting cements (RMGICs). *Dent. Mater.* **2020**, *36*, 893–904. [CrossRef]
94. Pérez-Mondragón, A.A.; Cuevas-Suárez, C.E.; González-López, J.A.; Trejo-Carbajal, N.; Herrera-González, A.M. Evaluation of new coinitiators of camphorquinone useful in the radical photopolymerization of dental monomers. *J. Photochem. Photobiol. A Chem.* **2020**, *403*, 112844. [CrossRef]
95. Alotaibi, J.; Saji, S.; Swain, M.V. FTIR characterization of the setting reaction of biodentine™. *Dent. Mater.* **2018**, *34*, 1645–1651. [CrossRef]
96. Dinesh Kumar, S.; Mohamed Abudhahir, K.; Selvamurugan, N.; Vimalraj, S.; Murugesan, R.; Srinivasan, N.; Moorthi, A. Formulation and biological actions of nano-bioglass ceramic particles doped with Calcarea phosphorica for bone tissue engineering. *Mater. Sci. Eng. C* **2018**, *83*, 202–209. [CrossRef]
97. Alqahtani, M. Effect of hexagonal boron nitride nanopowder reinforcement and mixing methods on physical and mechanical properties of self-cured PMMA for dental applications. *Materials* **2020**, *13*, 2323. [CrossRef]
98. Kafle, B.P. Raman spectroscopy. In *Chemical Analysis and Material Characterization by Spectrophotometry*; Elsevier: New York, NY, USA, 2020; pp. 245–268.
99. Marcott, C.; Padalkar, M.; Pleshko, N. 3.23 Infrared and raman microscopy and imaging of biomaterials at the micro and nano scale. In *Comprehensive Biomaterials II*; Elsevier: New York, NY, USA, 2017; pp. 498–518. ISBN 9780081006924.
100. Omidi, M.; Fatehinya, A.; Farahani, M.; Akbari, Z.; Shahmoradi, S.; Yazdian, F.; Tahriri, M.; Moharamzadeh, K.; Tayebi, L.; Vashaee, D. Characterization of biomaterials. In *Biomaterials for Oral and Dental Tissue Engineering*; Elsevier: New York, NY, USA, 2017; pp. 97–115. ISBN 9780081009673.
101. Xu, Z.; He, Z.; Song, Y.; Fu, X.; Rommel, M.; Luo, X.; Hartmaier, A.; Zhang, J.; Fang, F. Topic review: Application of raman spectroscopy characterization in micro/nano-machining. *Micromachines* **2018**, *9*, 361. [CrossRef]
102. Jones, R.R.; Hooper, D.C.; Zhang, L.; Wolverson, D.; Valev, V.K. Raman Techniques: Fundamentals and Frontiers. *Nanoscale Res. Lett.* **2019**, *14*, 1–34. [CrossRef]
103. Khan, A.S.; Khalid, H.; Sarfraz, Z.; Khan, M.; Iqbal, J.; Muhammad, N.; Fareed, M.A.; Rehman, I.U. Vibrational spectroscopy of selective dental restorative materials. *Appl. Spectrosc. Rev.* **2017**, *52*, 507–540. [CrossRef]
104. Malhotra, R.; Han, Y.M.; Morin, J.L.P.; Luong-Van, E.K.; Chew, R.J.J.; Castro Neto, A.H.; Nijhuis, C.A.; Rosa, V. Inhibiting Corrosion of Biomedical-Grade Ti-6Al-4V Alloys with Graphene Nanocoating. *J. Dent. Res.* **2020**, *99*, 285–292. [CrossRef]
105. Gomes de Araújo-Neto, V.; Sebold, M.; Fernandes de Castro, E.; Feitosa, V.P.; Giannini, M. Evaluation of physico-mechanical properties and filler particles characterization of conventional, bulk-fill, and bioactive resin-based composites. *J. Mech. Behav. Biomed. Mater.* **2021**, *115*, 104288. [CrossRef] [PubMed]
106. Zubieta-Otero, L.F.; Londoño-Restrepo, S.M.; Lopez-Chavez, G.; Hernandez-Becerra, E.; Rodriguez-Garcia, M.E. Comparative study of physicochemical properties of bio-hydroxyapatite with commercial samples. *Mater. Chem. Phys.* **2021**, *259*, 124201. [CrossRef]
107. Gutierrez, M.F.; Perdigao, J.; Malaquias, P.; Cardenas, A.M.; Siqueira, F.; Hass, V.; Reis, A.; Loguercio, A.D. Effect of methacryloyloxydecyl dihydrogen phosphate-containing silane and adhesive used alone or in combination on the bond strength and chemical interaction with zirconia ceramics under thermal aging. *Oper. Dent.* **2020**, *45*, 516–527. [CrossRef] [PubMed]
108. Lubas, M. Au interface effect on Ti-dental porcelain bond strength investigated by spectroscopic methods and mechanical tests. *J. Mol. Struct.* **2020**, *1208*, 127870. [CrossRef]
109. Miranda, J.S.; Barcellos, A.S.d.P.; Campos, T.M.B.; Cesar, P.F.; Amaral, M.; Kimpara, E.T. Effect of repeated firings and staining on the mechanical behavior and composition of lithium disilicate. *Dent. Mater.* **2020**, *36*, e149–e157. [CrossRef]
110. Ubaldini, A.L.M.; Pascotto, R.C.; Sato, F.; Soares, V.O.; Baesso, M.L. Mechanical and Chemical Changes in the Adhesive-Dentin Interface after Remineralization. *J. Adhes. Dent.* **2020**, *22*, 297–309. [CrossRef]
111. Lubas, M.; Przerada, I.; Zawada, A.; Jasinski, J.J.; Jelen, P. Spectroscopic and microstructural investigation of novel Ti–10Zr–45S5 bioglass composite for dental applications. *J. Mol. Struct.* **2020**, *1221*, 128545. [CrossRef]
112. Spies, B.C.; Zhang, F.; Wesemann, C.; Li, M.; Rosentritt, M. Reliability and aging behavior of three different zirconia grades used for monolithic four-unit fixed dental prostheses. *Dent. Mater.* **2020**, *36*, e329–e339. [CrossRef]
113. Tzanakakis, E.; Kontonasaki, E.; Voyiatzis, G.; Andrikopoulos, K.; Tzoutzas, I. Surface characterization of monolithic zirconia submitted to different surface treatments applying optical interferometry and raman spectrometry. *Dent. Mater. J.* **2020**, *39*, 111–117. [CrossRef]

114. Par, M.; Spanovic, N.; Mohn, D.; Attin, T.; Tauböck, T.T.; Tarle, Z. Curing potential of experimental resin composites filled with bioactive glass: A comparison between Bis-EMA and UDMA based resin systems. *Dent. Mater.* **2020**, *36*, 711–723. [CrossRef]
115. Gugelmin, B.P.; Miguel, L.C.M.; Filho, F.B.; da Cunha, L.F.; Correr, G.M.; Gonzaga, C.C. Colorstability of ceramic veneers luted with resin cements and pre-heated composites: 12 months follow-up. *Braz. Dent. J.* **2020**, *31*, 69–77. [CrossRef]
116. Castillo-Paz, A.M.; Londoño-Restrepo, S.M.; Tirado-Mejía, L.; Mondragón, M.A.; Rodríguez-García, M.E. Nano to micro size transition of hydroxyapatite in porcine bone during heat treatment with low heating rates. *Prog. Nat. Sci. Mater. Int.* **2020**, *30*, 494–501. [CrossRef]
117. Ramirez-Gutierrez, C.F.; Londoño-Restrepo, S.M.; del Real, A.; Mondragón, M.A.; Rodriguez-García, M.E. Effect of the temperature and sintering time on the thermal, structural, morphological, and vibrational properties of hydroxyapatite derived from pig bone. *Ceram. Int.* **2017**, *43*, 7552–7559. [CrossRef]
118. Yu, J.; Wang, H.; Zhan, J.; Huang, W. Review of recent UV-Vis and infrared spectroscopy researches on wine detection and discrimination. *Appl. Spectrosc. Rev.* **2018**, *53*, 65–86. [CrossRef]
119. Kafle, B.P. Application of UV–VIS spectrophotometry for chemical analysis. In *Chemical Analysis and Material Characterization by Spectrophotometry*; Elsevier: New York, NY, USA, 2020; pp. 79–145.
120. Kafle, B.P. Theory and instrumentation of absorption spectroscopy. In *Chemical Analysis and Material Characterization by Spectrophotometry*; Elsevier: New York, NY, USA, 2020; pp. 17–38.
121. Kafle, B.P. Sample preparation methods and choices of reagents. In *Chemical Analysis and Material Characterization by Spectrophotometry*; Elsevier: New York, NY, USA, 2020; pp. 39–50.
122. Kafle, B.P. Spectrophotometry and its application in chemical analysis. In *Chemical Analysis and Material Characterization by Spectrophotometry*; Elsevier: New York, NY, USA, 2020; pp. 1–16.
123. Stürmer, M.; Garcia, I.M.; Souza, V.S.; Visioli, F.; Scholten, J.D.; Samuel, S.M.W.; Leitune, V.C.B.; Collares, F.M. Titanium dioxide nanotubes with triazine-methacrylate monomer to improve physicochemical and biological properties of adhesives. *Dent. Mater.* **2021**, *37*, 223–235. [CrossRef] [PubMed]
124. Keul, C.; Seidl, J.; Güth, J.F.; Liebermann, A. Impact of fabrication procedures on residual monomer elution of conventional polymethyl methacrylate (PMMA)—a measurement approach by UV/Vis spectrophotometry. *Clin. Oral Investig.* **2020**, *24*, 4519–4530. [CrossRef]
125. Castellanos, M.; Delgado, A.J.; Sinhoreti, M.A.C.; de Oliveira, D.C.R.S.; Abdulhameed, N.; Geraldeli, S.; Roulet, J.-F. Effect of Thickness of Ceramic Veneers on Color Stability and Bond Strength of Resin Luting Cements Containing Alternative Photoinitiators. *J. Adhes. Dent.* **2019**, *21*, 67–76. [CrossRef] [PubMed]
126. Mallaiah, M.; Gupta, R.K. Surface Engineering of Titanium Using Anodization and Plasma Treatment. *IOP Conf. Ser. Mater. Sci. Eng.* **2020**, *943*, 12016.
127. Haas, K.; Azhar, G.; Wood, D.J.; Moharamzadeh, K.; van Noort, R. The effects of different opacifiers on the translucency of experimental dental composite resins. *Dent. Mater.* **2017**, *33*, e310–e316. [CrossRef]
128. Michailova, M.; Elsayed, A.; Fabel, G.; Edelhoff, D.; Zylla, I.M.; Stawarczyk, B. Comparison between novel strength-gradient and color-gradient multilayered zirconia using conventional and high-speed sintering. *J. Mech. Behav. Biomed. Mater.* **2020**, *111*, 103977. [CrossRef] [PubMed]
129. Xiao, Z.; Yu, S.; Li, Y.; Ruan, S.; Kong, L.B.; Huang, Q.; Huang, Z.; Zhou, K.; Su, H.; Yao, Z.; et al. Materials development and potential applications of transparent ceramics: A review. *Mater. Sci. Eng. R Rep.* **2020**, *139*, 100518. [CrossRef]
130. Qasim, S.B.; Najeeb, S.; Delaine-Smith, R.M.; Rawlinson, A.; Ur Rehman, I. Potential of electrospun chitosan fibers as a surface layer in functionally graded GTR membrane for periodontal regeneration. *Dent. Mater.* **2017**, *33*, 71–83. [CrossRef]
131. Bostedt, C.; Gorkhover, T.; Rupp, D.; Möller, T. Clusters and Nanocrystals. In *Synchrotron Light Sources and Free-Electron Lasers*; Springer International Publishing: Cham, Switzerland, 2020; pp. 1525–1573. ISBN 9783030232016.
132. Fracchia, M.; Ghigna, P.; Vertova, A.; Rondinini, S.; Minguzzi, A. Time-Resolved X-ray Absorption Spectroscopy in (Photo)Electrochemistry. *Surfaces* **2018**, *1*, 11. [CrossRef]
133. Taylor, A. Atomic spectroscopy, biomedical applications. In *Encyclopedia of Spectroscopy and Spectrometry*; Elsevier: New York, NY, USA, 2016; pp. 76–80. ISBN 9780128032244.
134. Hirai, N. Surface analysis. In *Corrosion Control and Surface Finishing: Environmentally Friendly Approaches*; Springer: Tokyo, Japan, 2016; pp. 47–439. ISBN 9784431559573.
135. Janssens, K. X-ray Based Methods of Analysis. In *Modern Methods for Analysing Archaeological and Historical Glass*; John Wiley & Sons Ltd.: Oxford, UK, 2013; Volume 1, pp. 79–128. ISBN 9780470516140.
136. Rajiv, K.; Mittal, K.L. Methods for Assessing Surface Cleanliness. In *Developments in Surface Contamination and Cleaning*; Elsevier: New York, NY, USA, 2019; Volume 12, pp. 23–105.
137. Watts, J.F. Use of surface analysis methods to probe the interfacial chemistry of adhesion. In *Handbook of Adhesion Technology*, 2nd ed.; Springer International Publishing: Cham, Switzerland, 2018; Volume 1–2, pp. 227–255. ISBN 9783319554112.
138. Bunaciu, A.A.; Udriştioiu, E.G.; Aboul-Enein, H.Y. X-ray Diffraction: Instrumentation and Applications. *Crit. Rev. Anal. Chem.* **2015**, *45*, 289–299. [CrossRef]
139. Khan, H.; Yerramilli, A.S.; D'Oliveira, A.; Alford, T.L.; Boffito, D.C.; Patience, G.S. Experimental methods in chemical engineering: X-ray diffraction spectroscopy—XRD. *Can. J. Chem. Eng.* **2020**, *98*, 1255–1266. [CrossRef]
140. Luo, Q. Electron Microscopy and Spectroscopy in the Analysis of Friction and Wear Mechanisms. *Lubricants* **2018**, *6*, 58. [CrossRef]

141. Çarıkçıoğlu, B.; Misilli, T.; Deniz, Y.; Aktaş, Ç. Effects of high temperature on dental restorative materials for forensic purposes. *Forensic Sci. Med. Pathol.* **2021**, *17*, 78–86. [CrossRef]
142. Fiorillo, L.; D'Amico, C.; Campagna, P.; Terranova, A.; Militi, A. Dental materials implant alloys: A X-ray fluorescence analysis on FDS76®. *Minerva Stomatol.* **2020**, *69*, 370–376. [CrossRef]
143. Montazerian, M.; Zanotto, E.D. Tough, strong, hard, and chemically durable enstatite-zirconia glass-ceramic. *J. Am. Ceram. Soc.* **2020**, *103*, 5036–5049. [CrossRef]
144. Cokic, S.M.; Vleugels, J.; Van Meerbeek, B.; Camargo, B.; Willems, E.; Li, M.; Zhang, F. Mechanical properties, aging stability and translucency of speed-sintered zirconia for chairside restorations. *Dent. Mater.* **2020**, *36*, 959–972. [CrossRef]
145. Gunawan, J.; Taufik, D.; Takarini, V.; Hasratinigsih, Z. Self-synthesize and flexural strength test porcelain from Indonesian natural sand. *IOP Conf. Ser. Mater. Sci. Eng.* **2019**, *550*, 012030.
146. Borges, M.A.P.; Alves, M.R.; dos Santos, H.E.S.; dos Anjos, M.J.; Elias, C.N. Oral degradation of Y-TZP ceramics. *Ceram. Int.* **2019**, *45*, 9955–9961. [CrossRef]
147. Belli, R.; Lohbauer, U.; Goetz-Neunhoeffer, F.; Hurle, K. Crack-healing during two-stage crystallization of biomedical lithium (di)silicate glass-ceramics. *Dent. Mater.* **2019**, *35*, 1130–1145. [CrossRef]
148. Kolakarnprasert, N.; Kaizer, M.R.; Kim, D.K.; Zhang, Y. New multi-layered zirconias: Composition, microstructure and translucency. *Dent. Mater.* **2019**, *35*, 797–806. [CrossRef]
149. Hurle, K.; Belli, R.; Götz-Neunhoeffer, F.; Lohbauer, U. Phase characterization of lithium silicate biomedical glass-ceramics produced by two-stage crystallization. *J. Non. Cryst. Solids* **2019**, *510*, 42–50. [CrossRef]
150. Salimkhani, H.; Asghari Fesaghandis, E.; Salimkhani, S.; Abdolalipour, B.; Motei Dizaji, A.; Joodi, T.; Bordbar-Khiabani, A. In situ synthesis of leucite-based feldspathic dental porcelain with minor kalsilite and Fe_2O_3 impurities. *Int. J. Appl. Ceram. Technol.* **2019**, *16*, 552–561. [CrossRef]
151. Nurdin, D.; Primathena, I.; Farah, R.A.; Cahyanto, A. Comparison of Chemical Composition between Indonesian White Portland Cement and MTA as Dental Pulp Capping Material. *Key Eng. Mater.* **2019**, *829*, 34–39. [CrossRef]
152. Bilandžić, M.D.; Wollgarten, S.; Stollenwerk, J.; Poprawe, R.; Esteves-Oliveira, M.; Fischer, H. Glass-ceramic coating material for the CO2 laser based sintering of thin films as caries and erosion protection. *Dent. Mater.* **2017**, *33*, 995–1003. [CrossRef]
153. Yahia, L.H.; Mireles, L.K. X-ray photoelectron spectroscopy (XPS) and time-of-flight secondary ion mass spectrometry (ToF SIMS). In *Characterization of Polymeric Biomaterials*; Elsevier: New York, NY, USA, 2017; pp. 83–97. ISBN 9780081007372.
154. Gosetti, F.; Marengo, E. Mass spectrometry | Selected ion monitoring. In *Encyclopedia of Analytical Science*; Elsevier: New York, NY, USA, 2019; pp. 500–510. ISBN 9780081019832.
155. Lermyte, F. Modern Mass Spectrometry and Advanced Fragmentation Methods. In *New Developments in Mass Spectrometry*; Royal Society of Chemistry: Cambridge, UK, 2021; pp. 1–14.
156. Schaepe, K.; Jungnickel, H.; Heinrich, T.; Tentschert, J.; Luch, A.; Unger, W.E.S. Secondary ion mass spectrometry. In *Characterization of Nanoparticles*; Elsevier: New York, NY, USA, 2020; pp. 481–509. ISBN 9780128141830.
157. Walker, A.V. Secondary ion mass spectrometry. In *Encyclopedia of Spectroscopy and Spectrometry*; Elsevier: New York, NY, USA, 2016; pp. 44–49. ISBN 9780128032244.
158. Paital, B. Mass Spectrophotometry: An Advanced Technique in Biomedical Sciences. *Adv. Tech. Biol. Med.* **2015**, *4*, 1–8. [CrossRef]
159. Nilsen, B.W.; Jensen, E.; Örtengren, U.; Michelsen, V.B. Analysis of organic components in resin-modified pulp capping materials: Critical considerations. *Eur. J. Oral Sci.* **2017**, *125*, 183–194. [CrossRef] [PubMed]
160. Bakopoulou, A.; Papadopoulos, T.; Garefis, P. Molecular Toxicology of Substances Released from Resin–Based Dental Restorative Materials. *Int. J. Mol. Sci.* **2009**, *10*, 3861–3899. [CrossRef] [PubMed]
161. Chuang, S.F.; Kang, L.L.; Liu, Y.C.; Lin, J.C.; Wang, C.C.; Chen, H.M.; Tai, C.K. Effects of silane- and MDP-based primers application orders on zirconia–resin adhesion—A ToF-SIMS study. *Dent. Mater.* **2017**, *33*, 923–933. [CrossRef] [PubMed]
162. Lima, R.B.W.; Barreto, S.C.; Alfrisany, N.M.; Porto, T.S.; De Souza, G.M.; De Goes, M.F. Effect of silane and MDP-based primers on physico-chemical properties of zirconia and its bond strength to resin cement. *Dent. Mater.* **2019**, *35*, 1557–1567. [CrossRef] [PubMed]
163. Lapinska, B.; Rogowski, J.; Nowak, J.; Nissan, J.; Sokolowski, J.; Lukomska-Szymanska, M. Effect of Surface Cleaning Regimen on Glass Ceramic Bond Strength. *Molecules* **2019**, *24*, 389. [CrossRef]
164. Möncke, D.; Ehrt, R.; Palles, D.; Efthimiopoulos, I.; Kamitsos, E.I.; Johannes, M. A multi technique study of a new lithium disilicate glass-ceramic spray-coated on ZrO2 substrate for dental restoration. *Biomed. Glas.* **2017**, *3*, 41–55. [CrossRef]
165. França, R.; Samani, T.D.; Bayade, G.; Yahia, L.; Sacher, E. Nanoscale surface characterization of biphasic calcium phosphate, with comparisons to calcium hydroxyapatite and β-tricalcium phosphate bioceramics. *J. Colloid Interface Sci.* **2014**, *420*, 182–188. [CrossRef]
166. Šimková, M.; Tichý, A.; Dušková, M.; Bradna, P. Dental Composites-a Low-Dose Source of Bisphenol A? *Physiol. Res.* **2020**, *69*, S295–S304. [CrossRef] [PubMed]
167. Cheung, K.H.; Pabbruwe, M.B.; Chen, W.F.; Koshy, P.; Sorrell, C.C. Thermodynamic and microstructural analyses of photocatalytic TiO2 from the anodization of biomedical-grade Ti6Al4V in phosphoric acid or sulfuric acid. *Ceram. Int.* **2021**, *47*, 1609–1624. [CrossRef]
168. NicDaéid, N. Forensic sciences | Systematic drug identification. In *Encyclopedia of Analytical Science*; Elsevier: New York, NY, USA, 2019; pp. 75–80. ISBN 9780081019832.

169. Liu, J.; Saw, R.E.; Kiang, Y.H. Calculation of effective penetration depth in X-ray diffraction for pharmaceutical solids. *J. Pharm. Sci.* **2010**, *99*, 3807–3814. [CrossRef] [PubMed]
170. Bauer, L.J.; Mustafa, H.A.; Zaslansky, P.; Mantouvalou, I. Chemical mapping of teeth in 2D and 3D: X-ray fluorescence reveals hidden details in dentine surrounding fillings. *Acta Biomater.* **2020**, *109*, 142–152. [CrossRef]
171. Pate, M.L.; Aguilar-Caballos, M.P.; Beltrán-Aroca, C.M.; Pérez-Vicente, C.; Lozano-Molina, M.; Girela-López, E. Use of XRD and SEM/EDX to predict age and sex from fire-affected dental remains. *Forensic Sci. Med. Pathol.* **2018**, *14*, 432–441. [CrossRef]

Review

Residual Adhesive Removal Methods for Rebonding of Debonded Orthodontic Metal Brackets: Systematic Review and Meta-Analysis

Guillermo Grazioli [1,†], Louis Hardan [2,†], Rim Bourgi [2], Leina Nakanishi [3], Elie Amm [4], Maciej Zarow [5], Natalia Jakubowicz [5], Patrycja Proc [6], Carlos Enrique Cuevas-Suárez [7,*] and Monika Lukomska-Szymanska [8,*]

1. Department of Dental Materials, School of Dentistry, Universidad de la República. Av. General Las Heras 1925, Montevideo 11300, Uruguay; ggrazioli@odon.edu.uy
2. Department of Restorative Dentistry, School of Dentistry, Saint-Joseph University, Beirut 1107 2180, Lebanon; louis.hardan@usj.edu.lb (L.H.); rim.bourgi@net.usj.edu.lb (R.B.)
3. Graduate Program in Dentistry, School of Dentistry, Federal University of Pelotas, Rua Gonçalves Chaves, 457, Pelotas 96015560, Brazil; leinaa_@hotmail.com
4. Department of Orthodontics, School of Dental Medicine, Saint Joseph University, Beirut 1107 2180, Lebanon; elie.el-amm@usj.edu.lb
5. "NZOZ SPS Dentist" Dental Clinic and Postgraduate Course Centre—pl. Inwalidow 7/5, 30-033 Cracow, Poland; dentist@dentist.com.pl (M.Z.); nljakubowicz@gmail.com (N.J.)
6. Department of Pediatric Dentistry, Medical University of Lodz, Pomorska 251, 92-213 Lodz, Poland; patrycja.proc@umed.lodz.pl
7. Dental Materials Laboratory, Academic Area of Dentistry, Autonomous University of Hidalgo State, Circuito Ex Hacienda La Concepción S/N, San Agustín Tlaxiaca 42160, Mexico
8. Department of General Dentistry, Medical University of Lodz, 251 Pomorska St., 92-213 Lodz, Poland
* Correspondence: cecuevas@uaeh.edu.mx (C.E.C.-S.); monika.lukomska-szymanska@umed.lodz.pl (M.L.-S.); Tel.: +48-42-675-74-61 (M.L.-S.)
† These authors contributed equally to this work.

Abstract: Debonding of orthodontic brackets is a common occurrence during orthodontic treatment. Therefore, the best option for treating debonded brackets should be indicated. This study aimed to evaluate the bond strength of rebonded brackets after different residual adhesive removal methods. This systematic review and meta-analysis was conducted according to the Preferred Reporting Items for Systematic Reviews and Meta-Analyses (PRISMA) statement. PubMed, Web of Science, The Cochrane Library, SciELO, Scopus, LILACS, IBECS, and BVS databases were screened up to December 2020. Bond strength comparisons were made considering the method used for removing the residual adhesive on the bracket base. A total of 12 studies were included for the meta-analysis. Four different adhesive removal methods were identified: sandblasting, laser, mechanical grinding, and direct flame. When compared with new orthodontic metallic brackets, bond strength of debonded brackets after air abrasion ($p = 0.006$), mechanical grinding ($p = 0.007$), and direct flame ($p < 0.001$) was significantly lower. The use of an erbium-doped yttrium aluminum garnet (Er:YAG) laser showed similar shear bond strength (SBS) values when compared with those of new orthodontic brackets ($p = 0.71$). The Er:YAG laser could be considered an optimal method for promoting the bond of debonded orthodontic brackets. Direct flame, mechanical grinding, or sandblasting are also suitable, obtaining clinically acceptable bond strength values.

Keywords: adhesive; bonding; bracket

1. Introduction

The effectiveness of fixed orthodontic treatment requires an adequate bonding between brackets and enamel surfaces [1]. Orthodontic brackets are fixed appliances that are bonded to the tooth and should remain in place until the end of treatment [2], to achieve this, the bond strength between bracket base and enamel surfaces should be strong enough to resist orthodontic forces and masticatory loads [3]. In this sense, many factors can lead to bracket–enamel bond failure, including the type of enamel conditioner, composition of adhesive, bracket base design, bracket material, as well as clinician skills [4].

Debonding of orthodontics from teeth is a common occurrence during orthodontic treatment, varying between 1.8% [5] and 20.1% [6]. Debonding of brackets during treatment is an unpleasant occurrence for the clinician and the patient resulting in increased treatment costs and duration [7]. During orthodontic treatment, the clinician may decide to debond one bracket intentionally and to rebond it on the tooth in a better position [8]. Therefore, clinicians have often to deal with what is the best option for treating with unintentional/intentional debonded brackets, and regardless of the cause of debonding, the orthodontist must decide whether to rebond the same bracket or to bond a new one [9].

One solution is to recycle or re-condition these brackets to reuse them for the same patient during the same visit. The re-condition process consists of removing bonding agent remnants from the bracket base, thus allowing the brackets to be rebonded [10]. Once a bracket is rebonded for its use again, it should exhibit sufficient bond strength. Thus, the main challenge in rebonding brackets is restoring the bracket base to a retentive pattern without damaging the bracket itself [11].

Adhesive remnants of the dislodged brackets had been conventionally removed in-office by using green stones [12], direct flame [13], tungsten-carbide bur [14], sandblasting [15], silica coating [16], or laser application [17]. Even though these methods can be easily performed out in the dental office with minimal cost, there is no consensus as to which is the best method to remove adhesive remnants from the bracket base. Accordingly, this systematic review and meta-analysis aims to evaluate the bond strength of rebonded brackets after different residual adhesive removal methods. The hypothesis to be tested is that different residual adhesive removal methods would provide similar bond strength of recycled/reused brackets when compared to new orthodontic brackets.

2. Materials and Methods

This systematic review and meta-analysis was reported by following the guidelines of the PRISMA statement [18]. The following PICOS framework was used: population, debonded orthodontic brackets; intervention, methods for residual adhesive removal; control, new orthodontic brackets; outcomes, bond strength; and study design, in vitro studies. The research question was: is there an optimal method to remove the residual adhesive of debonded orthodontic brackets?

2.1. Literature Search

The literature search was performed by two independent reviewers until 15 December 2020. The following five electronic databases were screened: PubMed (MedLine), ISI Web of Science, Cochrane Library, SciELO, and Scopus. The search strategy used is listed in Table 1. The reviewers also hand-searched the reference lists of included articles for the identification of additional manuscripts. After the initial screening, all studies were imported into Mendeley Desktop 1.17.11 software to remove duplicates.

Table 1. Keywords used in the search strategy.

	Search Strategy
# 1	Orthodontic bracket OR bracket OR braces OR stainless steel bracket OR recycled bracket.
# 2	Rebonded OR rebonding OR reconditioning OR recycling OR recycling methods OR recycled brackets OR rebonded brackets OR electropolishing OR sandblasting OR ultrasonic scaling OR heating OR Er:YAG laser OR CO2 laser
# 3	#1 and #2

2.2. Study Selection

Two reviewers independently assessed the titles and abstracts of all the manuscripts. Manuscripts for full-text review were selected according to the following eligibility criteria: (1) evaluated the bond strength of new orthodontic metallic brackets; (2) evaluated the bond strength of debonded orthodontic metallic brackets after using a method to remove the adhesive of the orthodontic metallic bracket base; (3) evaluated the bond strength of debonded orthodontic metallic bracket on new intact enamel; (4) included mean and standard deviation data in MPa; (5) published in the English language. Case reports, case series, pilot studies, and reviews were excluded. Full copies of all the potentially relevant studies were analyzed. Those that appeared to meet the inclusion criteria or had insufficient data in the title and abstract to make a clear decision were selected for full analysis. The full-text papers were independently assessed by two authors. Any disagreement regarding the eligibility of the included studies was resolved through discussion and consensus by a third reviewer.

2.3. Data Extraction

Data of interest from the manuscripts included was extracted using Microsoft Office Excel 2019 sheets (Microsoft Corporation, Redmond, WA, USA). These data included the year of publication, country, type of bracket, type of tooth, orthodontic adhesive used, the method for adhesive removal, the mean and standard deviation of the bond strength, and storage conditions.

2.4. Quality Assessment

The methodological quality of each study was assessed by two reviewers, according to the parameters of a previous systematic review of in vitro studies [19]. The risk of bias in each article was evaluated according to the description of the following parameters: specimen randomization, single-operator protocol implementation, blinding of the operator, the presence of a control group, standardization of the sample preparation, adhesive remnant index evaluation (ARI), use of all materials according to the manufacturer's instructions, and description of the sample size calculation. If the authors reported the parameter, the study received a "YES" for that specific parameter. In case of missing information, the parameter received a "NO." The risk of bias was classified according to the sum of "YES" answers received: 1 to 3 indicated a high bias, 4 to 6 medium, and 7 to 8 indicated a low risk of bias.

2.5. Statistical Analysis

Meta-analyses were carried out by using a software program (Review Manager Software version 5.4, The Cochrane Collaboration, Copenhagen, Denmark). The analyses were carried out using a random-effect model, and pooled-effect estimates were obtained by comparing the mean difference between bond strength values obtained using new orthodontic brackets or after removing the adhesive resin. Bond strength comparisons were made considering the method used for removing the residual adhesive on the bracket base. A *p*-value < 0.05 was considered statistically significant. Statistical heterogeneity of the treatment effect among studies was assessed using the Cochran Q test and the inconsistency I^2 test.

3. Results

A total of 3748 publications were collected from all databases (Figure 1).

After duplicates were removed, the literature review yielded 3337 manuscripts for initial examination. From these studies, 3300 studies were excluded after reviewing their titles and summaries. In total, 37 studies were examined by full-text reading. Of these studies, 23 were not included in the qualitative analysis: 2 studies evaluated the bond strength to other substrates different than enamel [20,21], 1 study combined several methods for adhesive removal in the same group [22], and 20 studies performed the rebonding process in the same tooth where the initial bonding process was performed [9,10,12,23–39], of the remaining 14 studies, 2 were excluded from the quantitative analysis because the mean and standard deviation was not available [40,41], totalizing 12 studies for the quantitative analysis.

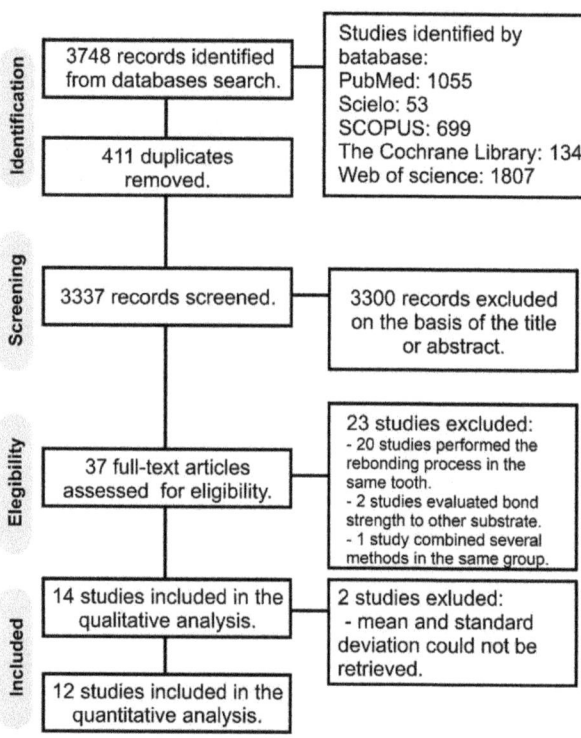

Figure 1. Prisma flow diagram of the study.

Four different adhesive removal methods were identified in this review. These included air abrasion [13–15,17,42–47], laser [17,42], mechanical grinding [14,43,44], and direct flame [12,41,43]. The characteristics of these studies are summarized in Table 2.

Table 2. Demographic data of included studies.

Study	Bracket Used	Tooth Used	Orthodontic Adhesive Used	Storing Conditions	Residual Adhesive Removal Method Used	Secondary Outcome
Achio, 2015	Stainless-steel premolar bracket (Unitek™ Gemini Bracket, 3M Unitek, Monorovia, CA, USA)	Human premolar	Transbond™ Plus Self Etching Primer (3M Unitek)/Transbond™ XT Light Cure Composite (3M Unitek)	Thermocycling (500 cycles between 5 °C and 55 °C)	Sandblasting (Al_2O_3; 50 μm, 90 psi, 10 mm, 10–15 s)	Adhesive remnant index
Bahnasi, 2013	Stainless steel upper premolar bracket (Unitek™ Gemini Bracket (3M Unitek)	Human premolar	Light Cure Orthodontic Adhesive Primer (3M Unitek)/Transbond™ XT Light Cure Composite (3M Unitek)	Thermocycling (500 cycles between 5 °C and 55 °C)	Sandblasting (Al_2O_3; 50 μm, 90 psi, 10 mm, 20–30 s)	Adhesive remnant index

Table 2. Cont.

Study	Bracket Used	Tooth Used	Orthodontic Adhesive Used	Storing Conditions	Residual Adhesive Removal Method Used	Secondary Outcome
Bahnasi, 2013 (b)	Stainless steel upper premolar bracket (Unitek™ Gemini Bracket, 3M Unitek)	Human premolar	Light Cure Orthodontic Adhesive Primer (3M Unitek)/Transbond™ XT Light Cure Composite (3M Unitek)	Thermocycling (500 cycles between 5 °C and 55 °C)	Sandblasting (Al$_2$O$_3$; 50 μm, 90 psi, 10 mm, 20–30 s). Mechanical grinding with a carbide bur with high-speed hand piece. Direct flame with a gas torch flame for 5 s.	Qualitative analysis of the distortion of the base with SEM
Egan, 1996	Stainless steel upper premolar brackets (GAC International Inc., New York, NY, USA)	Human premolar	Rely a Bond (Reliance Orthodontic Products Inc., Itasca, IL, USA) and Phase II paste-paste (Reliance Orthodontic Products Inc.)	Distilled water at 37 °C for 1 week	Mechanical grinding with a green stone	Failure mode
Harini, 2011	Stainless steel premolar brackets *	Human premolar	All Bon-2 (Bisco Inc., Schaumburg, IL, USA.	Distilled water for 24 h	Direct flame with a soldered torch for 5 s.	Adhesive remnant index
Heravi, 2006	Standard Edgewise metal brackets (Dentarum Corp., Ispringen, Germany)	Human upper premolar	No-mix composite (Dentarum Corp., Germany)	Distilled water at 37 °C for 48 h	Mechanical grinding with a tungsten carbide bur with high-speed hand piece	Adhesive remnant index
Ishida, 2011	Metal premolar bracket (Unitek™ Victory series, 3M Unitek)	Human premolar	Transbond™ Plus Self Etching Primer (3M Unitek)/Transbond™ XT Light Cure Composite (3M Unitek)	Artificial saliva at 37 °C for 24 h	Er,Cr:YSGG laser (Power output of 3.75 W, wavelength of 2.78 μm, a pulse duration of 140 μs, a frequency of 20 Hz, and air and water levels, each 50%)	Adhesive remnant index
Kachoei, 2016	Maxillary central incisors (Ortho-Organizer, Carlsbad, CA, USA)	Bovine upper central incisors	Unite Bonding System (3M Unitek, USA)	Distilled water at 37 °C for 1 week	Sandblasting (Al$_2$O$_3$; 50 um, 5 mm). CO$_2$ laser (wavelength of 10,600 nm and a 3 W output power, for 15 s)	Adhesive remnant index
Kamissety, 2015	Stainless steel lower premolar brackets (Gemini, 3M Unitek)	Lower human premolar	Transbond XT adhesive (3M Unitek).	Artificial saliva for 24 h at 37 °C	Mechanical grinding with a green stone with low-speed hand piece. Sandblasting (Al$_2$O$_3$, 50 μm, 10 mm, 90 PSI) Direct flaming with a micro torch Direct flaming with a Bunsen flame	UV/Vis transmittance analysis
Maaitah, 2013	Premolar brackets (Omni 0.022″ Roth, GAC International Inc, New York, NY, USA)	Human premolar teeth	Transbond™ XT Adhesive (3M Unitek)	Thermocycling (500 cycles between 5 °C and 55 °C)	Mechanical grinding with slow speed round tungsten carbide bur. Sandblasting (CoJet™ System Set; 3M Espe)	Adhesive remnant index
Montero, 2015	Upper central incisor brackets (Unitek™ Victory series, 3M Unitek)	Bovine upper central incisors	Transbond Plus Self Etching Primer (3M Unitek)/Transbond XT (3M Unitek)	Distilled water at 37 °C for 24 h	Sandblasting (Al$_2$O$_3$; 25 μm, 50 μm, or 110 μm at 5 mm)	SEM observation
Shahamfar, 2014	Premolar bracket (Equilibrium, Dentaaurum Inc., Ispringen, Germany)	Human premolar teeth	Light Bond™ (Reliance Orthodontic products, IL, USA)	Distilled water at 37 °C for 24 h	Mechanical grinding with slow speed multi blade tungsten carbide bur.	Adhesive remnant index
Sonis, 1996	Lower premolar brackets (GAC International, Inc., Central Islip, Long Island, NY, USA)	Lower human premolar	Rely-a-bond (Reliance, Inc., Itasca, IL, USA)	Thermocycling (1000 cycles between 10 °C and 50 °C)	Sandblasting (90 μm; 90 PSI, 15 to 30 s)	Scanning electron micrograph of base surface
Wheeler, 1983	Stainless steel premolar brackets	Human premolar	Dyna Bond II Series B (Unitek Corporation, Monrovia, CA, USA)	Non-specified	Heating in an oven for 50 min at 454 °C	

A meta-analysis was performed with 12 in vitro studies. Separate analyses for each adhesive removal method were performed (Figure 2). As the control for each study, the SBS value of new orthodontic brackets was considered. Direct flame methods for removing the residual adhesive were evaluated (Figure 2A). The meta-analysis demonstrated that these methods achieved significantly lower bond strength values of rebonded brackets when compared with the new bonded brackets ($p < 0.001$). With regards to the use of mechanical grinding methods to remove the residual adhesive from the base of orthodontic brackets, significantly lower SBS values were also observed (Figure 2B; $p = 0.007$). SBS of rebonded orthodontic brackets after adhesive removal with sandblasting was analyzed (Figure 2C). The meta-analysis performed demonstrated that bond strength values after adhesive removal through sandblasting were significantly lower than the bond strength of new orthodontic brackets ($p = 0.006$). With regards to the use of laser, two different types of laser were identified (Figure 2D). When a CO_2 laser was used for adhesive removal, the SBS of rebonded brackets was lower than the bond strength of new orthodontic brackets ($p < 0.001$). On the other hand, the use of an Er:YAG laser for adhesive removal showed similar SBS values when compared with those of new orthodontic brackets ($p = 0.71$).

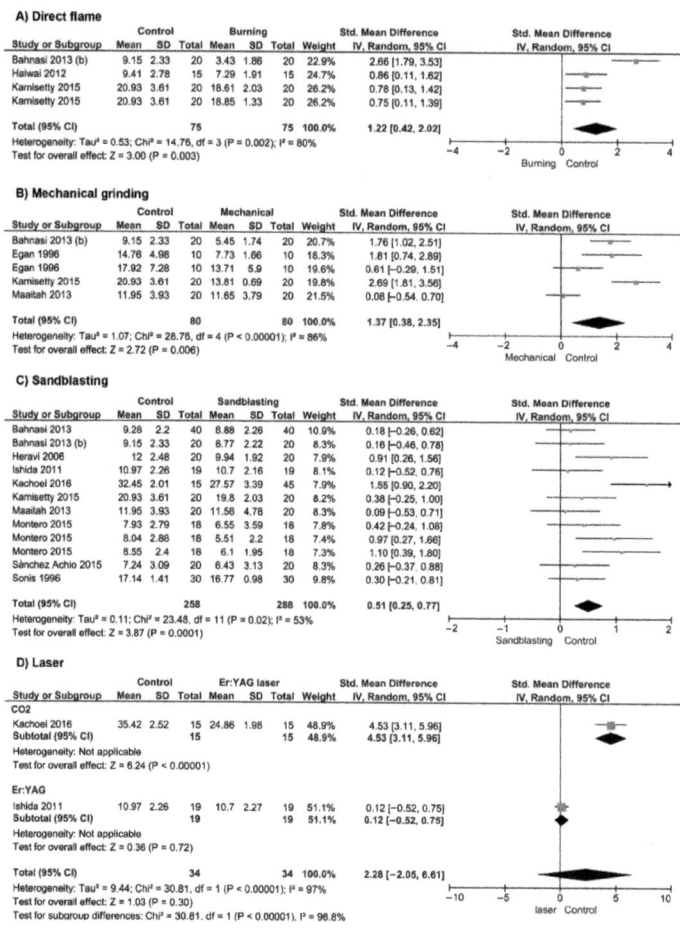

Figure 2. Results of the meta-analysis of bond strength of debonded orthodontics brackets after residual adhesive removal using; (**A**) Direct flame; (**B**) Mechanical grinding; (**C**) Sandblasting; and (**D**) Laser.

According to the parameters considered in the risk of bias assessment, the majority of studies were classified with a medium risk of bias (Table 3). Several of the studies failed to report single-operator, operator-blinded, and sample size calculation parameters.

Table 3. Qualitative synthesis (risk of bias assessment).

Study	Specimen Randomization	Single Operator	Operator Blinded	Control Group	Standardized Specimens	ARI	Manufacturer's Instructions	Sample Size Calculation	Risk of Bias
Achio, 2015	Yes	No	No	Yes	Yes	Yes	Yes	No	Medium
Bahnasi, 2013	No	No	No	Yes	Yes	Yes	Yes	No	Medium
Bahnasi (b), 2013	Yes	No	No	Yes	Yes	No	Yes	No	Medium
Egan, 1996	No	No	No	Yes	Yes	Yes	Yes	No	Medium
Harini, 2011	No	No	No	Yes	Yes	Yes	Yes	No	Medium
Heravi, 2006	No	No	No	Yes	Yes	Yes	Yes	No	Medium
Ishida, 2011	Yes	Yes	No	Yes	Yes	Yes	Yes	No	Medium
Kachoei, 2016	Yes	Yes	Yes	Yes	Yes	Yes	Yes	Yes	Low
Kamissety, 2015	No	No	No	Yes	Yes	No	Yes	No	High
Maaitah, 2013	Yes	Yes	No	Yes	Yes	Yes	Yes	No	Medium
Montero, 2015	Yes	No	No	Yes	Yes	No	Yes	Yes	Medium
Shahamfar, 2014	No	No	No	Yes	Yes	Yes	Yes	No	Medium
Sonis, 1996	Yes	No	No	Yes	Yes	No	Yes	No	Medium
Wheeler, 1983	Yes	Yes	No	Yes	Yes	No	Yes	No	Medium

4. Discussion

This systematic review and meta-analysis aimed to evaluate the bond strength of debonded brackets after different residual adhesive removal methods. Direct flame, mechanical grinding, sandblasting, and laser were the methods found in the literature used for this purpose. Except for the Er:YAG laser, none of the methods evaluated managed to restore SBS values of new orthodontic brackets values, thus our hypothesis was partially rejected.

One of the methods proposed to remove the adhesive remnant after bracket debonding is direct flaming of the bracket base. Under the use of this method, removal of the bonding agent is the most critical part of the recycling process and requires long exposure to heat [44]. The results of the present meta-analysis helped to demonstrate that this method was unable to recover the original values achieved by new orthodontic brackets. Several explanations may be suggested to explain this behavior. First, direct flaming increases the temperature of the bracket base to a temperature in the range of 600–800 °C, which can lead to the disintegration of the metal alloy, and consequently weakens its structure, making it more vulnerable to damage [30]. Also, as most of the metallic orthodontic brackets are made of austenitic stainless steel, application of heat leads to the formation of chrome-carbide compounds, which can render them more susceptible to tarnish and corrosion, and this, in turn, could be responsible for its failure in the mouth [48]. Finally, it has been found that the heat treatment could lead to a decrease in the diameter of the support mesh, which is caused by the presence of large amounts of adhesive residues on the base [30].

When observing the data about mechanical grinding, four studies reported this method. For this purpose, a green stone [43,48], or a carbide bur [14,41] at slow speed were used to grind the bracket surface. The meta-analysis revealed that, when this method was used for the removing of adhesive residual, significantly lower values in the bond strength of rebonded brackets were achieved. During the adhesive removal from the bracket, the preservation of the integrity of the bracket mesh is crucial to ensure an adequate bond strength to the enamel. By grinding the bracket base using a green stone or a carbide bur, there is a high risk of damage or grinding off the mesh base itself, resulting in a decrease in bond strength. [12] Also, grinding the bracket mesh has been proved to leave a considerable amount of the adhesive, obliterating the mesh and decreasing the contact area, thus eliminating virtually any mechanical retention [10,12,14,44].

Sandblasting has been described as a viable procedure for rebonding accidentally lost brackets. This method was the most used in the studies included in the present systematic review. The findings obtained by the meta-analysis suggest that the bond strength observed by debonded, cleaned brackets with sandblasting is significantly lower when compared with new brackets. Previous research has demonstrated that sandblasting of the bracket base could provoke distortion of the mesh [33]. In this sense, the air abrasion procedure causes macro and microscopic alterations in the structure of the bonding surface, consequently affecting the bond strength outcomes [12]. Also, it has been described that after sandblasting, some abrasive particles adhere to the blasted surface, and it is possible that bond strength between any luting material and the abrasive particle remnants might exceed the bond strength of the abrasive particles and the bonding surface, causing premature debonding [49]. On the other hand, the sandblasting process is not able to remove all the resin attached to the bonding mesh [17], directly affecting the bond strength outcomes.

When observing the data about lasers, two different methods were analyzed separately [17,43]. This technology selectively ablates composite by high pulse repetition rates [49]. When analyzing the CO_2 laser, it was found that it is not a suitable method for recycling brackets because considerable amounts of adhesive remnants were left on the base of CO_2 laser-irradiated brackets [49]. As explained before, the remaining adhesive on the bracket base lessens the contact area between the meshwork and adhesive and leads to a decrease in bond strength values. On the other hand, the analysis of the results from the Er:YAG laser method demonstrated that this method is efficient for removing the residual adhesive, being that the values obtained were similar to those of new orthodontic brackets. This result could be explained due to the complete removal of the residual adhesive from the bracket bases without altering the micro and macrostructure of the mesh, resembling the appearance and bonding performance of new brackets [17]. Nevertheless, it should be advised that the use of the Er:YAG laser could melt the meshwork of the bracket base due to heat, and some precautions should be taken when using this method [17].

Regarding the limitations of this systematic review, it is important to highlight that all analyses performed showed high heterogeneity values, which could be attributed to the lack of standardization of the methods for evaluation of the SBS; actually, none of the included studies indicated the following of the international standards for bond strength tests to dental tissues. Future research with more standardized methods is desired to reduce the heterogeneity between the studies focused on this topic and also to establish the optimal protocol for the adhesive removal for rebonding of debonded orthodontic brackets. Also, it is important to encourage researchers for designing and conducting clinical trials evaluating this outcome.

On the other hand, it should be pointed out that despite the meta-analysis showing statistical differences between the SBS of debonded and new brackets, such differences are not clinically relevant, this is because the mean bond strength values of the methods evaluated succeeded to achieve at least 6 MPa, which is the minimum bond strength required for successful orthodontic treatment [17,23]. This could also lead to the perspective that rebonding of debonded orthodontic brackets in the same patient is a reliable treatment option, as long as the adhesive residual within the orthodontic base is completely removed using the above-mentioned procedures.

5. Conclusions

Within the limitations of this systematic review, it could be concluded that the Er:YAG laser could be considered as an optimal method for promoting the bond of debonded orthodontic brackets, this conclusion is based on the fact that the bond strength of rebonded orthodontic brackets was the same as that of the new brackets. Nevertheless, the data suggest that the use of direct flame, mechanical grinding, or sandblasting are suitable options for the removal of residual adhesive from the orthodontic bracket base, obtaining clinically acceptable bond strength values.

Author Contributions: Conceptualization, G.G. and L.N.; methodology, G.G., L.N. and C.E.C.-S.; software, L.H., R.B. and M.L.-S.; validation, R.B., G.G., P.P. and E.A.; formal analysis, C.E.C.-S. and R.B.; investigation, L.H. and P.P.; resources, M.Z., N.J., L.H. and R.B.; data curation, C.E.C.-S.; writing—original draft preparation G.G., L.N., R.B., M.Z., E.A. and L.H.; writing—review and editing,

M.L.-S., L.H. and C.E.C.-S.; visualization, E.A., R.B., M.L.-S., N.J., L.H. and M.Z.; supervision, M.L.-S., L.H. and C.E.C.-S.; project administration, L.H. and C.E.C.-S. All authors have read and agreed to the published version of the manuscript.

Funding: This research received no external funding.

Institutional Review Board Statement: Not applicable.

Informed Consent Statement: Not applicable.

Data Availability Statement: The data that support the findings of this study are available from the corresponding author upon reasonable request.

Conflicts of Interest: The authors declare no conflict of interest.

References

1. Ahmed, T.; Rahman, N.A.; Alam, M.K. Assessment of in Vivo Bond Strength Studies of the Orthodontic Bracket-Adhesive System: A Systematic Review. *Eur. J. Dent.* **2018**, *12*, 602–609. [CrossRef] [PubMed]
2. Tsichlaki, A.; Chin, S.Y.; Pandis, N.; Fleming, P.S. How Long Does Treatment with Fixed Orthodontic Appliances Last? A Systematic Review. *Am. J. Orthod. Dentofac. Orthop.* **2016**, *149*, 308–318. [CrossRef]
3. Knox, J.; Hubsch, P.; Jones, M.L.; Middleton, J. The Influence of Bracket Base Design on the Strength of the Bracket–Cement Interface. *J. Orthod.* **2000**, *27*, 249–254. [CrossRef] [PubMed]
4. Bakhadher, W.; Halawany, H.; Talic, N.; Abraham, N.; Jacob, V. Factors Affecting the Shear Bond Strength of Orthodontic Brackets—A Review of In Vitro Studies. *Acta Med.* **2015**, *58*, 43–48. [CrossRef]
5. Roelofs, T.; Merkens, N.; Roelofs, J.; Bronkhorst, E.; Breuning, H. A Retrospective Survey of the Causes of Bracket- and Tube-Bonding Failures. *Angle Orthod.* **2017**, *87*, 111–117. [CrossRef] [PubMed]
6. Barbosa, I.V.; de Miranda Ladewig, V.; Almeida-Pedrin, R.R.; Cardoso, M.A.; Santiago Junior, J.F.; de Castro Ferreira Conti, A.C. The Association between Patient's Compliance and Age with the Bonding Failure of Orthodontic Brackets: A Cross-Sectional Study. *Prog. Orthod.* **2018**, *19*, 11. [CrossRef]
7. Sukhia, H.R.; Sukhia, R.H.; Mahar, A. Bracket De-Bonding and Breakage Prevalence in Orthodontic Patients. *Pak. Oral. Dent. J.* **2011**, *31*, 1–5.
8. Koo, B.C.; Chung, C.-H.; Vanarsdall, R.L. Comparison of the Accuracy of Bracket Placement between Direct and Indirect Bonding Techniques. *Am. J. Orthod. Dentofac. Orthop.* **1999**, *116*, 346–351. [CrossRef]
9. Mui, B.; Rossouw, P.E.; Kulkarni, G.V. Optimization of a Procedure for Rebonding Dislodged Orthodontic Brackets. *Angle Orthod.* **1999**, *69*, 276–281. [CrossRef]
10. Yassaei, S.; Aghili, H.; KhanPayeh, E.; Goldani Moghadam, M. Comparison of Shear Bond Strength of Rebonded Brackets with Four Methods of Adhesive Removal. *Lasers Med. Sci.* **2014**, *29*, 1563–1568. [CrossRef]
11. Sohrabi, A.; Jafari, S.; Kimyai, S.; Rikhtehgaran, S. Er,Cr:YSGG Laser as a Novel Method for Rebonding Failed Ceramic Brackets. *Photomed. Laser Surg.* **2016**, *34*, 483–486. [CrossRef]
12. Halwai, H.K.; Kamble, R.H.; Hazarey, P.V.; Gautam, V. Evaluation and Comparison of the Shear Bond Strength of Rebonded Orthodontic Brackets with Air Abrasion, Flaming, and Grinding Techniques: An in Vitro Study. *Orthodontics* **2012**, *13*, e1–e9. [PubMed]
13. Bahnasi, F.I.; Abd-Rahman, A.N.; Abu-Hassan, M.I. Effects of Recycling and Bonding Agent Application on Bond Strength of Stainless Steel Orthodontic Brackets. *J. Clin. Exp. Dent.* **2013**, *5*, e197–e202. [CrossRef] [PubMed]
14. Al Maaitah, E.F.; Alomari, S.; Abu Alhaija, E.S.; Safi, A.A.M. The Effect of Different Bracket Base Cleaning Method on Shear Bond Strength of Rebonded Brackets. *J. Contemp. Dent. Pract.* **2013**, *14*, 866–870. [CrossRef] [PubMed]
15. Sanchez Achio, T. A Comparative Study of Shear Debonding Strengh between New Brackets, Air-Abrasion and Recycled Brackets: An In Vitro Analysis. *Odovtos-Int. J. Dent. Sci.* **2015**, *17*, 59–69.
16. Guarita, M.K.; Moresca, A.H.K.; Losso, E.M.; Moro, A.; Moresca, R.C.; Correr, G.M. Effect of Different Surface Treatments for Ceramic Bracket Base on Bond Strength of Rebonded Brackets. *Braz. Dent. J.* **2015**, *26*, 61–65. [CrossRef]
17. Ishida, K.; Endo, T.; Shinkai, K.; Katoh, Y. Shear Bond Strength of Rebonded Brackets after Removal of Adhesives with Er,Cr:YSGG Laser. *Odontology* **2011**, *99*, 129–134. [CrossRef]
18. Elshafay, A.; Omran, E.S.; Abdelkhalek, M.; El-Badry, M.O.; Eisa, H.G.; Fala, S.Y.; Dang, T.; Ghanem, M.A.T.; Elbadawy, M.; Elhady, M.T.; et al. Reporting Quality in Systematic Reviews of in Vitro Studies: A Systematic Review. *Curr. Med. Res. Opin.* **2019**, *35*, 1631–1641. [CrossRef]
19. Bourgi, R.; Hardan, L.; Rivera-Gonzaga, A.; Cuevas-Suárez, C.E. Effect of Warm-Air Stream for Solvent Evaporation on Bond Strength of Adhesive Systems: A Systematic Review and Meta-Analysis of in Vitro Studies. *Int. J. Adhes. Adhes.* **2021**, *105*, 102794. [CrossRef]

20. Regan, D.; van Noort, R.; O'Keeffe, C. The Effects of Recycling on the Tensile Bond Strength of New and Clinically Used Stainless Steel Orthodontic Brackets: An In Vitro Study. *Br. J. Orthod.* **1990**, *17*, 137–145. [CrossRef]
21. Basudan, A.M.; Al-Emran, S.E. The Effects of In-Office Reconditioning on the Morphology of Slots and Bases of Stainless Steel Brackets and on the Shear/Peel Bond Strength. *J. Orthod.* **2001**, *28*, 231–236. [CrossRef]
22. Wendl, B.; Muchitsch, P.; Pichelmayer, M.; Droschl, H.; Kern, W. Comparative Bond Strength of New and Reconditioned Brackets and Assessment of Residual Adhesive by Light and Electron Microscopy. *Eur. J. Orthod.* **2011**, *33*, 288–292. [CrossRef] [PubMed]
23. Abe, R.; Endo, T.; Shimooka, S. Effects of Tooth Bleaching on Shear Bond Strength of Brackets Rebonded with a Self-Etching Adhesive System. *Odontology* **2011**, *99*, 83–87. [CrossRef] [PubMed]
24. Atsu, S.; Catalbas, B.; Gelgor, I.E.; Atsü, S.; Çatalbaş, B.; Gelgör, I.E. Effects of Silica Coating and Silane Surface Conditioning on the Bond Strength of Rebonded Metal and Ceramic Brackets. *J. Appl. Oral. Sci.* **2011**, *19*, 233–239. [CrossRef] [PubMed]
25. Bansal, N.; Valiathan, A.; Bansal, K. The Effects of Various In-Office Reconditioning Methods on Shear Bond Strength, Morphology of Slots and Bases of Stainless Brackets: An in Vitro Study. *J. Indian Orthod. Soc.* **2011**, *45*, 175–182. [CrossRef]
26. Bishara, S.E.; VonWald, L.; Laffoon, J.F.; Warren, J.J. The Effect of Repeated Bonding on the Shear Bond Strength of a Composite Resin Orthodontic Adhesive. *Angle Orthod.* **2000**, *70*, 435–441. [CrossRef] [PubMed]
27. Chacko, P.K.; Kodoth, J.; John, J.; Kumar, K. Recycling Stainless Steel Orthodontic Brackets with Er:YAG Laser—An Environmental Scanning Electron Microscope and Shear Bond Strength Study. *J. Orthod. Sci.* **2013**, *2*, 87–94. [CrossRef]
28. Chung, C.H.; Fadem, B.W.; Levitt, H.L.; Mante, F.K. Effects of Two Adhesion Boosters on the Shear Bond Strength of New and Rebonded Orthodontic Brackets. *Am. J. Orthod. Dentofac. Orthop.* **2000**, *118*, 295–299. [CrossRef] [PubMed]
29. Grabouski, J.K.; Staley, R.N.; Jakobsen, J.R. The Effect of Microetching on the Bond Strength of Metal Brackets When Bonded to Previously Bonded Teeth: An in Vitro Study. *Am. J. Orthod. Dentofac. Orthop.* **1998**, *114*, 452–460. [CrossRef]
30. Gupta, N.; Kumar, D.; Palla, A. Evaluation of the Effect of Three Innovative Recyling Methods on the Shear Bond Strength of Stainless Steel Brackets-an in Vitro Study. *J. Clin. Exp. Dent.* **2017**, *9*, e550–e555. [CrossRef]
31. Kumar, M.; Maheshwari, A.; Lall, R.; Navit, P.; Singh, R.; Navit, S. Comparative Evaluation of Shear Bond Strength of Recycled Brackets Using Different Methods: An In Vitro Study. *Int. J. Oral Health Dent.* **2014**, *6*, 5–11.
32. Lew, K.K.; Chew, C.L.; Lee, K.W. A Comparison of Shear Bond Strengths between New and Recycled Ceramic Brackets. *Eur. J. Orthod.* **1991**, *13*, 306–310. [CrossRef]
33. Lunardi, N.; Gameiro, G.H.; de Araújo Magnani, M.B.; Nouer, D.F.; de Siqueira, V.C.; Consani, S.; Pereira-Neto, J.S. The Effect of Repeated Bracket Recycling on the Shear Bond Strength of Different Orthodontic Adhesives. *Braz. J. Oral Sci.* **2008**, *7*, 1648–1652.
34. Montasser, M.A.; Drummond, J.L.; Roth, J.R.; Al-Turki, L.; Evans, C.A. Rebonding of Orthodontic Brackets. *Angle Orthod.* **2008**, *78*, 537–544. [CrossRef]
35. Jimenez, E.E.O.; Hilgenberg, S.P.; Rastelli, M.C.; Pilatti, G.L.; Orellana, B.; Coelho, U. Rebonding of Unused Brackets with Different Orthodontic Adhesives. *Dent. Press J. Orthod.* **2012**, *17*, 69–76. [CrossRef]
36. Quick, A.N.; Harris, A.M.P.; Joseph, V.P. Office Reconditioning of Stainless Steel Orthodontic Attachments. *Eur. J. Orthod.* **2005**, *27*, 231–236. [CrossRef] [PubMed]
37. Regan, D.; LeMasney, B.; van Noort, R. The Tensile Bond Strength of New and Rebonded Stainless Steel Orthodontic Brackets. *Eur. J. Orthod.* **1993**, *15*, 125–135. [CrossRef]
38. Tavares, S.W.; Consani, S.; Nouer, D.F.; de Araújo Magnani, M.B.B.; Nouer, P.R.A.; Martins, L.M. Shear Bond Strength of New and Recycled Brackets to Enamel. *Braz. Dent. J.* **2006**, *17*, 44–48. [CrossRef] [PubMed]
39. Harini, T.; Sreedhar, R. Effect of an Adhesion Booster on the Bond Strength of New and Recycled Brackets. *Ann. Dent.* **2011**, *III*, 20–22. [CrossRef]
40. Wheeler, J.J.; Ackerman, R.J. Bond Strength of Thermally Recycled Metal Brackets. *Am. J. Orthod. Dentofac. Orthop.* **1983**, *83*, 181–186. [CrossRef]
41. Bahnasi, F.I.; Rahman, A.N.A.A.; Abu-Hassan, M.I. The Impact of Recycling and Repeated Recycling on Shear Bond Strength of Stainless Steel Orthodontic Brackets. *Orthod. Waves* **2013**, *72*, 16–22. [CrossRef]
42. Kachoei, M.; Mohammadi, A.; Esmaili Moghaddam, M.; Rikhtegaran, S.; Pourghaznein, M.; Shirazi, S. Comparison of Multiple Rebond Shear Strengths of Debonded Brackets after Preparation with Sandblasting and CO2 Laser. *J. Dent. Res. Dent. Clin. Dent. Prospects* **2016**, *10*, 148–154. [CrossRef]
43. Kamisetty, S.K.; Verma, J.K.; Arun, S.; Chandrasekhar, S.; Kumar, A. SBS vs Inhouse Recycling Methods-An Invitro Evaluation. *J. Clin. Diagn. Res.* **2015**, *9*, ZC04-8. [CrossRef]
44. Haro Montero, M.M.; Vicente, A.; Alfonso-Hernández, N.; Jiménez-López, M.; Bravo-González, L.A. Comparison of Shear Bond Strength of Brackets Recycled Using Micro Sandblasting and Industrial Methods. *Angle Orthod.* **2015**, *85*, 461–467. [CrossRef] [PubMed]
45. Heravi, F.; Naseh, R.A. Comparative Study between Bond Strength of Rebonded and Recycled Orthodontic Brackets. *Dent. Res. J.* **2006**, *2*, 1–6.
46. Sonis, A.L. Air Abrasion of Failed Bonded Metal Brackets: A Study of Shear Bond Strength and Surface Characteristics as Determined by Scanning Electron Microscopy. *Am. J. Orthod. Dentofac. Orthop.* **1996**, *110*, 96–98. [CrossRef]
47. Huang, T.-H.; Yen, C.-C.; Kao, C.-T. Comparison of Ion Release from New and Recycled Orthodontic Brackets. *Am. J. Orthod. Dentofac. Orthop.* **2001**, *120*, 68–75. [CrossRef]

48. Egan, F.R.; Alexander, S.A.; Cartwright, G.E. Bond Strength of Rebonded Orthodontic Brackets. *Am. J. Orthod. Dentofac. Orthop.* **1996**, *109*, 64–70. [CrossRef]
49. Yassaei, S.; Aghili, H.; Firouzabadi, A.H.; Meshkani, H. Effect of Er: YAG Laser and Sandblasting in Recycling of Ceramic Brackets. *J. Lasers Med. Sci.* **2017**, *8*, 17–21. [CrossRef]

Review

Enhancing the Mechanical Properties of Glass-Ionomer Dental Cements: A Review

John W. Nicholson [1,2,*], **Sharanbir K. Sidhu** [3] **and Beata Czarnecka** [4]

[1] Dental Materials Unit, Bart's and the London Institute of Dentistry, Queen Mary University of London, Mile End Road, London E1 4NS, UK
[2] Bluefield Centre for Biomaterials, 67-68 Hatton Garden, London EC1N 8JY, UK
[3] Centre for Oral Bioengineering, Institute of Dentistry, Bart's & The London School of Medicine and Dentistry, Queen Mary University of London, Turner Street, London E1 2AD, UK; s.k.sidhu@qmul.ac.uk
[4] Department of Biomaterials and Experimental Dentistry, Poznań University of Medical Sciences, ul. Bukowska 70, 60-812 Poznań, Poland; czarnecka@ump.edu.pl
* Correspondence: john.nicholson@bluefieldcentre.co.uk

Received: 11 April 2020; Accepted: 26 May 2020; Published: 31 May 2020

Abstract: This paper reviews the strategies that have been reported in the literature to attempt to reinforce glass-ionomer dental cements, both conventional and resin-modified. These cements are widely used in current clinical practice, but their use is limited to regions where loading is not high. Reinforcement might extend these applications, particularly to the posterior dentition. A variety of strategies have been identified, including the use of fibres, nanoparticles, and larger particle additives. One problem revealed by the literature survey is the limited extent to which researchers have used International Standard test methods. This makes comparison of results very difficult. However, it does seem possible to draw conclusions from this substantial body of work and these are (1) that powders with conventional particle sizes do not reinforce glass-ionomer cements, (2) certain fibres and certain nanoparticles give distinct improvements in strength, and (3) in the case of the nanoparticles these improvements are associated with differences in the morphology of the cement matrix, in particular, a reduction in the porosity. Despite these improvements, none of the developments has yet been translated into clinical use.

Keywords: glass-ionomer cements; resin-modified; fibre; reinforcement; nanoparticles; testing; strength

1. Introduction

Glass-ionomer cements (GICs) are widely used in clinical dentistry with uses including full restorations, liners and bases, luting agents, fissure sealants, and adhesives for orthodontic brackets [1]. Their properties are generally attractive for these applications and include biocompatibility towards tooth tissue [2], the ability to release fluoride [3,4], and adhesion to the tooth surface [5,6]. They also match the natural tooth tissue in terms of coefficient of thermal expansion [7].

Glass-ionomers are made from special basic glass powders that are either calcium or strontium alumino-fluoro-silicate with additions of phosphate, typically $AlPO_4$, and also sodium salts [8]. These are able to react with a solution of polymeric water-soluble acid to form salts that effectively crosslink the polymer chain and cause the material to harden [1,9–11]. The most widely used polymer is polyacrylic acid but cements can also be prepared with acrylic/maleic or acrylic/itaconic copolymer or the copolymer of 2-methylene butanedioic acid with propenoic acid [11]. Commercial formulations often include some of the polymeric acid mixed as a dry powder with the glass. This effectively increases the concentration of the acidic polymer in the final cement without making the liquid to be too viscous to mix. Large amounts of polymeric acid make the resulting cement strong [10], a feature which is necessary for clinical durability.

As well as the formation of carboxylate crosslinks, there is also a secondary setting reaction of certain inorganic species released from the glass [12]. The most important of these components appears to be the phosphate, since without phosphate, ionomer glasses do not give insoluble cements with simple monomeric acids [13]. In glass-ionomer cements, the gradual formation of a modified network from these inorganic species has been proposed as being partly responsible for the changes that take place as the glass-ionomer matures. These changes include decreased plasticity, greater compressive strength, and improved translucency [14].

Glass-ionomer cements of this type were as originally reported in the early 1970s [15] and are now known as conventional glass-ionomers. This is to distinguish them from the resin-modified glass-ionomers (RMGICs), materials first described about 20 years later [16,17]. RMGICs contain the same components as conventional glass-ionomers, namely basic glass powder, water, and polyacid. They also contain a monomer component and associated initiator system. The monomer is usually 2- hydroxyethyl methacrylate, HEMA, and the initiator is the light-sensitive substance camphorquinone [10]. Resin-modified glass-ionomers set by a combination of neutralization (acid-base reaction) and addition polymerization. The set material has a complicated structure based on the blended polymer and polysalt components that arise from these two reactions [18]. Competition between these two network-forming systems results in a delicate balance between them [19], meaning that delay in applying light to initiate the photochemical polymerization reaction can alter the properties of the resulting material [19,20]. Hence, it is important that the manufacturer's recommendations on the duration and timing of the light-cure step are followed closely so that the set material has optimal properties [19].

RMGICs can release the monomer HEMA in varying amounts early in the life of the restoration [21]. The extent of light-curing is an important factor in controlling the amount released, and release also varies with depth of the restoration, because lower layers receive less light intensity and so polymerize to a smaller extent. This, in turn, leaves more unreacted HEMA monomer in the structure, and this free HEMA is able to leach from the cement into the adjacent tooth tissue. HEMA can diffuse through human dentine [22] and is cytotoxic to the cells of the pulp [23,24].

HEMA from resin-modified glass-ionomers can also be the cause of problems for dental personnel. It is a contact allergen and is volatile, so it can be inhaled [24], where it can cause respiratory problems. Dentists should work in a well-ventilated area and make sure that no vapor is inhaled [25]. They are also advised to light-cure left-over material before disposing of it.

Resin-modified glass-ionomers have mainly the same clinical applications as conventional glass-ionomers [26,27], though they are not recommended for the atraumatic restorative treatment (ART) technique. They can be used in Class I, Class II, and Class III restorations, all mainly in the primary dentition, Class V restorations, and also as liners and bases [28]. Their other uses include as fissure sealants [28] and as bonding agents for orthodontic brackets [29].

Despite the many years of development of these materials and the corresponding amount of experience in using them clinically, these materials still have limitations [1]. These limitations arise from the mechanical properties of the set cement. In particular, conventional glass-ionomers are brittle materials with compressive strengths in the range 150–220 MPa and resin-modified glass-ionomers, while tougher and with better flexural strength, have comparable compressive properties. As a result, both types of material have similar limitations for clinical use [1,10]. There have been numerous attempts to reinforce these materials, some of which have been used in clinical materials and others of which still remain experimental. This paper reviews the various approaches to reinforcement that have been reported in the literature for these materials, covering both conventional and resin-modified glass-ionomers.

2. Mechanical Properties of Conventional and Resin-Modified Glass-Ionomer Cements

Before considering the topic of reinforcement, it is appropriate to consider the mechanical properties of glass-ionomer cements of both types. This is important not only to establish the baseline from which reinforcement has to be achieved, but also because some authors refer to the latter group of glass-ionomers as "resin-reinforced". This is not the preferred term, and the correct term is resin-modified, as first proposed in the mid-1990s [30], and widely used throughout the literature [31].

It is difficult to compare these classes of material, because the relevant international standards specify different strengths [32]. Conventional glass-ionomer cements are tested for compressive strength and have a minimum requirement of 100 MPa for restorative use in patients. On the other hand, resin-modified glass-ionomers are specified to be tested for flexural strength and should have a minimum strength of 20 MPa for clinical use [33].

A few published studies have reported compressive strength values for resin-modified glass-ionomers and flexural strength values for conventional glass-ionomers in attempts to compare the materials. Unfortunately, the reported values vary widely, and make it impossible to draw any reliable conclusions, despite the claims of the individual studies. For example, the conventional glass-ionomer Fuji IX (GC, Tokyo, Japan) in its hand-mixed form has been reported to have a compressive strength of 83.6 MPa at 24 h in one study [34] and 350.87 MPa in another [35]. The former study then concludes that resin-modified glass-ionomers are stronger in all test modes, but with such a low reported compressive strength for the conventional cement, how can the data be trusted? This is not to question the ability or skill of the researchers: These are difficult materials to mix [36] and there are many factors to be controlled when preparing specimens for testing. Doing so reliably is extremely challenging.

Other low values of compressive strength have been reported for conventional glass-ionomers. For example, the material Ionofil Molar (VOCO, Cuxhaven, Germany) was reported to have a compressive strength of 78.78 MPa, well below that of the resin-modified materials Vitremer (3M-ESPE, Seefeld, Germany), which had a compressive strength of 169.50 MPa [37]. While it is interesting to observe how high the compressive strength of a resin-modified glass-ionomer can be, the value for the conventional glass-ionomer must be questioned. It is lower than that recommended by the appropriate international standard [32], an improbable outcome for a material from a reputable manufacturer.

Despite these variations in reported values in individual papers, there does seem to be an overall consensus that RMGICs are superior to conventional materials in their diametral tensile strength [38,39] and in their flexural strength [39]. In a recent study, the superiority of the toughness and flexural properties was shown to be consistent across a range of brands [40]. For example, the fracture toughness at 24 h of a number of commercial conventional glass-ionomers ranged from 0.18 to 0.30 MPa m$^{1/2}$ compared with 0.49–0.67 MPa m$^{1/2}$ for a similar set of commercial RMGICs. For the same materials, flexural strength at 24 h ranged from 18 to 34 MPa for the conventional glass-ionomers compared with 49–76 MPa for the RMGICs [40]. These differences are attributable to the presence of the polymerized resin component, which toughens the cement and improves its ability to withstand loading in flexure.

3. Comparison with Tooth Materials

Glass-ionomers of both types are used to repair teeth that have been damaged, mainly by caries. In the context of considering how to improve the mechanical properties of glass-ionomers, it is appropriate to consider the mechanical properties of the natural tooth.

Teeth consist of two main types of material, the dentine and the enamel, and they have very different mechanical properties. Both are anisotropic [41], but whereas dentine is relatively tough, enamel is brittle. As a consequence of their geometry, the main properties of these tissues that have been evaluated are those that can be determined by nano-indentation [42,43], namely hardness, modulus, and fracture toughness. Values of these properties are shown Table 1.

Table 1. Mechanical properties of human tooth enamel and dentine.

Substance	Property	Value	Reference
Enamel	Hardness	2.0–3.5 GPa	[42]
	Young's Modulus	80–120 GPa	[43–45]
	Fracture Toughness	0.67–3.93 MPa m$^{1/2}$	[46]
Dentine	Hardness	0.3–0.7 GPa	[47]
	Young's Modulus	10–40 GPa	[43,48,49]
	Fracture Toughness	1.1–2.3 MPa m$^{1/2}$	[50,51]

In the case of enamel, mechanical properties vary widely across the enamel layer. For example, external enamel has a fracture toughness of around 0.67 ± 0.12 MPa m$^{1/2}$, whereas internal enamel has values of up to 3.93 MPa m$^{1/2}$. This means that the internal enamel has a much greater crack resistance [48] and this arises because the detailed composition and arrangement of the enamel rods differ in these varied locations.

The dentine has more consistent properties throughout its structure, though even this material varies close to the dentino-enamel junction, DEJ. Dentine is a viscoelastic material [52], which means it is much tougher than enamel, and more resistant to fracture.

The DEJ is the region in which the dentine and the enamel are joined [42]. Although enamel and dentine have very different properties, there is a transition in properties of both in the regions close to the DEJ [53]. Although the DEJ is itself brittle, as are the two structures adjacent to it, the overall effect of its presence is that cracks do not propagate through it. This has the effect of making the tooth mechanically strong.

The exact value of strength is difficult to measure, due to the difficulty of preparing uniform test specimens, or even of determining cross-sectional areas of loaded teeth. One paper dealt with the problem by simplifying the experimental data and recording the load at failure of teeth as they underwent fracture [54]. Teeth were extracted and tested in compression in a universal testing machine, either as obtained, with a cut cavity, or with a cavity filled with composite resin (10 samples per condition). Results are shown in Table 2, and they demonstrate that the mere act of cutting a cavity weakens the tooth significantly. Conversely, repairing the cavity goes a long way to restoring the strength of the tooth, even though the final value was still below that of the natural tooth.

Table 2. Load at failure of human teeth ($n = 10$) (standard deviations in parentheses).

Tooth Condition	Load at Failure/kg
Sound, uncut	104.65 (13.59)
Cavity prepared	48.88 (6.25)
Restored with composite layered obliquely	84.05 (14.03)

There is evidence that these values change throughout the lifetime of individuals. The teeth of the elderly are more susceptible to fracture than those of the young [55], though whether this is because they become more brittle with age, suffer fatigue, or simply become weaker, is not clear.

Given the differences in mechanical properties of enamel and dentine, and the variations in the mechanical properties of these tissues with precise location within the tooth and also with the age of the patient, the suggestion by some researchers that glass-ionomers should match the properties of natural teeth is clearly not practicable. Instead, improving the properties of glass-ionomers to enhance their durability should be the goal. Probably the key requirement is to improve the flexural strength and Young's modulus, because these properties correlate with wear and, hence, clinical durability [37,56], though other properties, such as hardness, are also important in influencing durability. Steps that have been studied as possible means of improving the mechanical properties of glass-ionomers are considered in the following sections of this review.

4. Metal Reinforcement

Metal reinforcement has been applied to conventional glass-ionomers used in the dental clinic for many years [57]. The original concept was to develop a glass-ionomer cement to replace amalgams. Two approaches have been used, namely the incorporation of metal powders, mainly silver-tin alloy [58], and the incorporation of silver-tin alloy fused with the ionomer glass powder, a so-called cermet [59].

The earliest report of metal powders for reinforcement of glass-ionomer cements was in a patent by Wilson and Sced granted in 1980 [60]. They described the addition of powders of aluminum, chromium, nickel-aluminum alloy, and silver-tin alloy, all of which were of relatively large particle size, i.e., not nanoparticles. All were claimed to improve the flexural strength [60,61]. Of these, the silver-tin alloy gave the greatest increase, raising the reported value to 40 MPa from a value of 10 MPa for the parent cement.

Silver-tin alloy of the type typically used in amalgams incorporated into a glass-ionomer was marketed as a material known as "Miracle Mix" by the GC company from 1983 [61]. It is still on the market as of the start of 2020, and is a useful material for certain niche applications [62]. Results with it in terms of reinforcement have been mixed, with some authors claiming it is stronger than the typical unmodified glass-ionomer cement Fuji IX (GC, Tokyo, Japan) [63], and others claiming it is weaker [64]. The overall conclusion is that there is no clear improvement in strength [57], but this does not seem to matter, because its continued use does not seem to be because of its extra strength, but because it is radio-opaque.

The alternative approach, namely of the silver alloy being fused to the glass to form a cermet (from the words _cer_amic and _met_al), involves a change at the manufacturing stage. Like the inclusion of metal powder, the result is a cement with compromised aesthetics but improved radio-opacity. The cermet system was commercialized in 1986, when the material Chelon Silver was launched by the ESPE GmbH company [61]. As with Miracle Mix, this material is still on the market at the time of writing (April 2020).

Early studies on strength were carried out, with the cermet cement being compared with the conventional glass-ionomer cement Chelon Fil (ESPE, Seefeld, Germany) Williams et al. reported that the cermet cement had lower strengths than the conventional glass-ionomer [65]. For example, diametral tensile strength was found to be 13 ± 2 MPa compared with 19 ± 3 MPa for Chelon Fil. Walls et al. [66] confirmed these findings for flexural strength, using a three-point bending test, which showed Chelon Silver to have a strength of 29 ± 7 MPa and Chelon Fil to have a strength of 45 ± 5 MPa. The overall conclusion from these and other studies is that cermet-based glass-ionomers are not reinforced at all and are actually weaker than cements made with conventional glass powders [57].

Another metal filler that has been considered for inclusion in glass-ionomers is stainless steel powder [67]. When compared with Miracle Mix, a cement containing finely divided stainless steel was shown to be stronger. For example, the diametral tensile strength was 23 ± 2 MPa compared with 11 ± 2 for Miracle Mix. However, despite this superiority, it is Miracle Mix that has been on the market since 1983, rather than a formulation containing stainless steel powder.

5. Fibre Reinforcement

The possibility of reinforcing glass-ionomer cements with fibres was also mentioned in the 1980 patent of Sced and Wilson [58]. They used both carbon and alumina fibres in a now-obsolete commercial glass-ionomer called Chembond (Dentsply deTrey, Konstanz, Germany), and examined the effect of fibres on flexural strength. Fibres employed were less than 1000 μm long and ranged in diameter from 10 to 200 μm. They were incorporated at 25 volume %, and increased the flexural strength substantially. The control cement had a flexural strength of 10 MPa, but alumina fibres raised this to 44 MPa and carbon fibres raised it to 53 MPa. Similar improvements in flexural strength have been reported by other authors for these types of fibre in different glass-ionomer formulations [64]. Carbon fibres have been claimed to be superior because, unlike alumina fibres, they do not cause the cement to increase in brittleness [68]. On the other hand, they alter the appearance and compromise the aesthetics.

Reactive glass fibres based on ionomer glass formulations have also been used to reinforce glass-ionomer cements. These glasses have been based on either the SiO_2-Al_2O_3-CaO-P_2O_5 [69,70] or the SiO_2-Al_2O_3-CaF_2-Na_3AlF_6 [71] systems. The concept behind the use of these glasses is that, in order for fibres to properly reinforce a material, there must be good interfacial coupling of the matrix to the fibre [57]. Deploying fibres made of basic ionomer glasses ensures that such good coupling occurs and that the fibres are genuinely able to reinforce the cement. In principle, optimal results are obtained when the fibres are aligned with the direction of the primary load. In practice, though, short fibres become dispersed in random directions during mixing, so that improvements in strength may be only slight.

Typical results for the use of reactive glass fibres are shown in Table 3, and these are based on the findings of Lohbauer et al. [71]. The critical fibre length had previously been calculated to be 546 µm, so the fibres used to reinforce the cements, which had a diameter of 26 µm, were prepared to be 600 µm in length. The cement itself was based on components prepared specially for the study, rather than a commercial material. When fibres were incorporated, the highest flexural strength was achieved with 20 volume % loading, as shown in Table 3.

Table 3. Effect of 20 volume % reactive glass fibres in the strength of glass-ionomer cement.

Cement	Compressive Strength/MPa	Flexural Strength/MPa
Unreinforced	64	8.9
Fibre reinforced	170	15.6

Following fracture, specimens were examined by fractography, and it was shown that specimens failed with fibre pull-out. This showed that the presence of the fibres increased strength by increasing the work of fracture, causing a corresponding increase in the fracture toughness.

Other studies have confirmed the reinforcing effect of glass fibres, though not with reactive glasses. Instead, nonbasic glass fibres have been used and incorporated into glass-ionomer cement formulations. In one study, relatively low amounts (3 and 5 weight %) were added to a commercial cement (Medifill, Promedica, Germany) and found to raise the diametral tensile strength from 7.49 ($\pm$ 1.5) MPa to 9.15 ($\pm$ 1.35) MPa at 3 weight % loading and 11.86 ($\pm$ 2.27) MPa at 5 weight % loading [72].

Resin-modified glass-ionomers, too, have been reinforced by adding glass fibres. In this case, too, the fibres were prepared from nonreactive glasses [73], with both hollow and solid short length fibres being added at two different loadings, namely 5 and 10 weight %. A conventional and a resin-modified glass-ionomer were used as parent cements in the study. Results for the hollow-fibre modifications are shown in Table 4 and from these it can be seen that compressive strength was generally unaffected by the inclusion of these fibres, but that flexural strength and fracture toughness increased significantly. Both the conventional and the resin-modified glass-ionomer materials were substantially toughened by adding these fibres, and would be expected to show superior wear and durability if used clinically in patients.

Table 4. Results of incorporating hollow glass fibres into glass-ionomer cements (standard deviations in parentheses).

Material	Type	Compressive Strength/MPa	Flexural Strength/MPa	Diametral Tensile Strength/MPa
Fuji IX	Conventional	98.0 (12.0)	26.3 (9.0)	7.8 (1.7)
FIX/5% fibre		82.8 (13.8)	29.8 (6.0)	14.2 (2.6)
FIX/10% fibre		96.5 (9.0)	44.6 (4.0)	17.1 (3.5)
Fuji II LC	Resin-modified	123.0 (17.0)	55.2 (9.3)	17.7 (2.6)
FII/5% fibre		118.0 (8.3)	58.0 (8.4)	18.6 (4.2)
FII/10% fibre		154.2 (13.0)	75.2 (13.0)	23.0 (3.9)

Other fibres have also been used to reinforce glass-ionomers with some success. One material used has been basalt in fibre form [74]. Basalt is a mixture of the silicate minerals plagioclase, pyroxene, and olivine, and it can be readily formed into fibres. Technical uses for these fibres include the aerospace and automotive industries, where it is used a fireproof textile [75]. It has also been used to reinforce organic polymers, such as epoxy systems [76].

Basalt fibres were included in the conventional glass-ionomer Fuji IX in a study, with levels up to 7 weight % being used [74]. Early strength was found to increase significantly, but there was evidence that the interaction between the fibres and the matrix was weak, which suggested that the properties were likely to decline over the long term. Experimental studies confirmed this suggestion, as both flexural strength and flexural modulus of the fibre-reinforced cements declined over one month and became the same as the control material. No further reports have appeared on this material as a reinforcing fibre in glass-ionomers.

Cellulose microfibres have been studied as possible reinforcing agents for conventional glass-ionomers, along with cellulose nanocrystals [77,78]. The latter are discussed in more detail in the following section and, for the moment, results for the cellulose fibres only will be considered. In fact, these fibres were found to have only slight effects on the mechanical properties of the glass-ionomers (compressive strength, Young's modulus, and diametral tensile strength) though, by contrast, biocompatibility was enhanced [79]. Fibres were added to the commercial conventional glass-ionomer Vidrion R (S.S. White, Brazil), a material with a reported compressive strength of only 49.15 MPa [78], which is too low for clinical use according to the relevant ISO standard [32]. Addition of cellulose fibres improved this and also enhanced both Young's modulus and diametral tensile strength [78].

6. Nanoparticle Reinforcement

The influence of nanoparticles on the properties of glass-ionomers (mainly conventional) has been reported in some recent papers. For example, nanoparticle titanium dioxide has been incorporated into the commercial material Kavitan Plus (SpofaDental, Czech Republic) [80]. Following 24 h of storage in water, the flexural strength determined by three-point bending rose from 14 ± 3 MPa to 23 ± 3 MPa with 3 weight % addition of TiO_2 [75]. However, using the same type of additive in the commercial material Ionofil Molar (VOCO, Germany) failed to produce any observable strengthening effect [81].

Other studies, however, have confirmed the reinforcing effect of nanoparticles [80,82,83]. A variety of nanoparticles have been used, such as titanium dioxide nanotubes [83] and nanoparticles [81,82], alumina nanoparticles [83], and zirconia nanoparticles [84]. Glass-ionomer cements with added nanoparticles have been found to be easier to mix than those without nanoparticles [85] and to contain fewer air voids or internal micro-cracks [86]. All of these features contribute to the observed increase in compressive strength.

One recent study used additions of TiO_2, Al_2O_3 or ZrO_2 nanoparticles to try to improve the strength of conventional glass-ionomers [85]. In addition to determining strength, the authors studied ion release by the cements. The cements used were two modern high-viscosity materials, Equia Fil (GC, Japan) and ChemFil Rock (Dentsply, Germany), with nanoparticles added at levels of 2, 5, and 10 weight %.

Results showed that the nanoparticles had varying effects, depending on type, amount, and storage time. Alumina significantly weakened Equia Fil at 24 h at all three loadings, whereas after 1 week, it weakened only the 5 weight % samples. In ChemFil Rock, for both the 1 day and 1 week storage times, it generally had no effect on strength, though at 5 weight % loading it caused a significant weakening. Findings were similarly mixed for zirconia nanoparticles, whereas titania nanoparticles generally increased strength, particularly at 10 weight % loading [85].

Studying the fracture surfaces by scanning electron microscopy confirmed that the addition of nanoparticles reduced the porosity, with no differences between the materials [86]. Ion release data showed that incorporating nanoparticles typically increased the levels of certain species released from the cement, mainly silica/silicate and phosphate. However, it did not cause release of the metal ions associated with the nanoparticles, i.e., Al, Ti, or Zr [86]. In all cases, no release occurred from the nanoparticle additives and, overall, the authors concluded that their addition to these high-viscosity cements was beneficial.

Other inorganic nanoparticles examined as additives for glass-ionomer cements have been hydroxyapatite [87] and fosterite [88]. The first of these was found to show enhanced antimicrobial activity against *Streptococcus mutans* bacteria, though no information was given on any changes in mechanical properties [87]. The latter was studied entirely for its effects on mechanical properties when added at levels in the range 1–4 weight %. Depending on the loading, this type of nanoparticle gave significant increases in compressive, flexural, and diametral tensile strength [88].

Organic nanoparticles have also been found to have positive effects on the properties of glass-ionomers, as shown by the cellulose nanocrystals previously mentioned [78,89]. These were found to have significant effects on the mechanical properties of a variety of commercial glass-ionomers (Maxxion, Vidrion R, Vitro Molar, Ketac Molar Easy Mix and Fuji Gold Label 9), raising the compressive strength by 10%, Young's modulus by 61%, and the diametral tensile strength by 53% [78]. Fluoride release was also improved [88]. This was despite the fact that there appeared to be no chemical reaction between the cellulose nanocrystals and the cement matrix, as shown by the absence of any changes in the FTIR spectra [89]. Like the addition of cellulose microfibres, the nanocrystals enhanced the biocompatibility of the cements [79] and, overall, showed promise as additives for these cements.

Nanoparticulate silver has also been employed in experimental studies of conventional glass-ionomers [90,91]. The main reason for this has been that silver nanoparticles are capable of releasing Ag^{2+} ions for sustained periods of time, and these are extremely effective antibacterial species [92]. Silver ions have been found to show broad-spectrum antibacterial and antiviral properties at low concentrations as a result of their ability to disrupt bacterial cell walls and reduce the ability of the bacterial DNA to replicate [92]. Studies have confirmed that including silver nanoparticles in glass-ionomer cements leads to substantial improvements in their antibacterial properties [89]. As with nanoparticulate inorganic powders, compressive strength was improved when silver nanoparticles were present [90,91]. So, too, were micro-hardness and micro-shear strength to dentine [92]. Levels of addition have been low, between 0.1 and 0.5%, at which levels there is minimal discoloration of the cement.

7. Other Inorganic Powders

Inorganic powders with particle sizes bigger than the nanoscale have been added to glass-ionomers in attempts at reinforcement, but they have been uniformly unsuccessful [61]. An obvious candidate material is hydroxyapatite, which more or less corresponds to the mineral phase of the tooth and which is able to interact chemically with the polyacid component of the glass-ionomer cement [93]. An early study [94], using hydroxyapatite powder of unknown particle size at levels of 2.5, 5, 10, 20, and 25 weight % showed progressive reductions in compressive strength with increasing loading of additive in two experimental glass-ionomer cements.

In another study, the cement Fuji IX GP was used, and hydroxyapatite was added by substituting the glass powder at levels of 4, 12, and 28 volume % [95]. The mean particle size of the hydroxyapatite powder was 17 μm. Low amounts added made no significant difference to the strength, either compressive or diametral tensile, but higher levels resulted in substantial weakening. This confirmed the earlier findings.

Since these initial studies, there have been several other reports of the use of hydroxyapatite powders to reinforce various commercial glass-ionomer cements [96–98]. Fluorapatite has also been used [94], but in all cases, the addition of hydroxyapatite did not lead to reinforcement.

Other inorganic powders of reasonable particle size have been studied as possible reinforcing fillers. These include bioactive glass [99], montmorillonite clay [100], and borax [101]. In all cases, relatively low loadings were used, typically up to 10 weight % and often lower than that. Findings were consistent, with these additives causing increases in working/setting times but, unfortunately, reductions, often quite substantial, occur in compressive strength [99–101]. None of these powders has proven to be able to provide any reinforcement.

The question of reinforcement with larger particle inorganic powders has been complicated by glass-ionomer brands that the manufacturers claim are reinforced by the addition of ceramic powders, either of unspecified composition or zirconia [102,103]. These materials include "Amalgomer" (Advanced Healthcare Ltd., Tonbridge, UK) and "Zirconomer" (Shofu, Kyoto, Japan). Claims for them are considerable, with one group of authors describing them as "white amalgams", a term which is completely unacceptable. Amalgams are, by definition, alloys of mercury [104], and this sort of inaccurate nomenclature should not be allowed to infiltrate the field of dental materials' science.

The main issue with these materials is that there are no data to support the claim that they are reinforced. Some papers have appeared describing their properties [63,105–107], though most are in low-ranking journals of questionable provenance, and never reported with comparative data for control cements. One study, which is in a reputable journal, aimed to compare the properties of amalgam and "Amalgomer" using Hertzian indentation [102]. It showed that there were distinct differences between the materials, with "Amalgomer" being much weaker [103]. The authors, Wang and Darvell, described the "Amalgomer" material as an ordinary acid-base reaction GIC "but claimed to be reinforced with a ceramic powder" [107], showing that they had at least some skepticism about the claim.

This type of material is certainly strong, for example, "Amalgomer" has been reported to have a compressive strength of 323 MPa [63]. However, in the absence of information on the strength of the parent cement, there is nothing to support the claim of reinforcement. These materials are held to be suitable for use in the posterior dentition [103,105,106] but until better scientific data appear in more prestigious scientific journals, the case for this is unproven. Certainly, what information exists is not sufficiently compelling to alter the general conclusion that inorganic additives do not reinforce conventional glass-ionomer cements.

8. Conclusions

There have been many attempts to improve the strength of glass-ionomer cements over the years since they were first introduced to the dental profession in the mid-1970s. These have mainly been aimed at conventional glass-ionomers, though there are some reports involving resin-modified glass-ionomers. So far, very few of these approaches have proven to be successful. Good results have been obtained with certain fibres, notably hollow glass fibres, and with particular nanoparticles at specific loadings. In the latter case, the cement morphology is altered, with notable reductions in porosity. There are also commercial materials that claim to be reinforced with ceramics such as zirconia and, though these cements appear to be very strong, it is not clear that they are genuinely reinforced.

A major difficulty with work on this topic is that the literature contains a wide variety of mixing and testing conditions [61], despite the existence of relevant International Standards [32,33]. These standards fully specify test conditions in terms of specimen geometry, storage time, and conditions. In this way, they aim to eliminate wide variations in test regimes and the associated uncertainty in the results obtained; researchers really should use these defined test methods. Without this, their results are neither completely valid nor useful to the wider research community.

So far, none of the improvements in strength with certain additives have been translated into clinical materials, and existing commercial glass-ionomers should be used in posterior dentition with caution, paying careful attention to individual clinical conditions.

Author Contributions: Initial planning: J.W.N., S.K.S., and B.C.; literature search and initial drafting, J.W.N.; text revision: S.K.S. and B.C. All authors have read and agreed to the published version of the manuscript.

Funding: This research received no external funding.

Conflicts of Interest: The authors declare no conflict of interest.

References

1. Mount, G.J. *Color Atlas of Glass Ionomer Cement*, 2nd ed.; Martin Dunitz: London, UK, 2002.
2. Sidhu, S.K.; Schmalz, G. The biocompatibility of glass-ionomer materials: A status report for the American Journal of Dentistry. *Am. J. Dent.* **2001**, *14*, 387–396. [PubMed]
3. Wiegand, A.; Buchalla, W.; Attin, T. Review on fluoride-releasing restorative materials-fluoride release and uptake characteristics, antibacterial activity and influence on caries formation. *Dent. Mater.* **2007**, *23*, 343–362. [CrossRef] [PubMed]
4. Forsten, L. Fluoride release and uptake by glass-ionomers and related materials and its clinical effect. *Biomaterials* **1998**, *19*, 503–508. [CrossRef]
5. Nicholson, J.W. Adhesion of glass-ionomer cements to teeth: A review. *Int. J. Adhes. Adhes.* **2016**, *69*, 33–38. [CrossRef]
6. Yamakami, S.A.; Ubaldini, A.L.M.; Sato, F.; Medina Neto, A.; Pascotto, R.C.; Baesso, M.L. Study of the chemical interaction between a high-viscosity glass ionomer cement and dentin. *J. Appl. Oral Sci.* **2018**, *26*, e20170384. [CrossRef]
7. Collado-González, M.; Pecci-Lloret, M.R.; Tomás-Catalá, C.J.; García-Bernal, D.; Oñate-Sánchez, R.E.; Llena, C.; Forner, L.; Rosa, V.; Rodriguez-Lozano, F.J. Thermo-setting glass ionomer cements promote variable biological responses of human dental pulp stem cells. *Dent. Mater.* **2018**, *34*, 932–943. [CrossRef]
8. Hill, R.G.; Wilson, A.D. Some structural aspects of glasses used in ionomer cements. *Glass Technol.* **1988**, *29*, 150–158.
9. Nicholson, J.W.; Czarnecka, B. *Materials for the Direct Restoration of Teeth, Ch 6: Conventional Glass-Ionomer Cements*; Woodhead Publishing: Cambridge, UK, 2016.
10. Nicholson, J.W.; Sidhu, S.K. A review of glass-ionomer cements for clinical dentistry. *J. Funct. Biomater.* **2016**, *7*, 16. [CrossRef]
11. Crisp, S.; Lewis, B.G.; Wilson, A.D. Characterization of glass-ionomer cements. 3. Effect of polyacid concentration on the physical properties. *J. Dent.* **1977**, *5*, 51–56. [CrossRef]
12. Wasson, E.A.; Nicholson, J.W. New aspects of the setting of glass-ionomer cements. *J. Dent. Res.* **1993**, *72*, 481–483. [CrossRef]
13. Shahid, S.; Billington, R.W.; Pearson, G.J. The role of glass composition in the glass acetic acid and glass lactic acid cements. *J. Mater. Sci. Mater. Med.* **2008**, *19*, 541–545. [CrossRef] [PubMed]
14. Nicholson, J.W. Maturation in glass-ionomer dental cements. *Acta Biomater. Odontol. Scand.* **2018**, *4*, 63–71. [CrossRef] [PubMed]
15. Wilson, A.D.; Kent, B. The glass-ionomer cement, a new translucent dental filling material. *J. Appl. Chem. Biotechnol.* **1971**, *21*, 313. [CrossRef]
16. Mitra, S.B. Adhesion to dentin and physical properties of a light-cured glass-ionomer liner/base. *J. Dent. Res.* **1991**, *70*, 72–74. [CrossRef] [PubMed]
17. Mitra, S.B. In vitro fluoride release from a light-cured glass-ionomer liner/base. *J. Dent.* **1991**, *70*, 75–78. [CrossRef]
18. Berzins, D.W.; Abey, S.; Costache, M.C.; Wilkie, C.A.; Roberts, H.W. Resin-modified glass-ionomer setting reaction competition. *J. Dent. Res.* **2010**, *89*, 82–86. [CrossRef]
19. Yelamanchili, A.; Darvell, B.W. Network competition in a resin-modified glass-ionomer cement. *Dent. Mater.* **2008**, *24*, 1065–1069. [CrossRef]
20. Thomas, J.T.; Roberts, H.W.; Diaz, L.; Bradley, T.G.; Berzins, D.W. Effect of light-cure initiation time on polymerization efficiency and orthodontic bond strength with a resin-modified glass-ionomer. *Orthod. Craniofac. Res.* **2012**, *15*, 124–134. [CrossRef]
21. Palmer, G.; Anstice, H.M.; Pearson, G.J. The effect of curing regime on the release of hydroxethyl methacylate (HEMA) from resin-modified glass-ionomer cements. *J. Dent.* **1999**, *27*, 303–311. [CrossRef]
22. Hamid, A.; Hume, W.R. Diffusion of resin monomers through human carious dentin in vitro. *Endod. Dent. Traumatol.* **1997**, *13*, 1–5. [CrossRef]
23. Kan, K.C.; Messer, L.B.; Messer, H.H. Variability in cytotoxicity and fluoride release of resin-modified glass-ionomer cements. *J. Dent. Res.* **1997**, *76*, 1502–1507. [CrossRef] [PubMed]

24. Kanerva, L.; Jolanki, R.; Leino, T.; Estlander, T. Occupational allergic contact dermatitis from 2-hydroxethyl methacrylate and ethylene glycol dimethacrylate in a modified acrylic structural adhesive. *Contact Dermat.* **1995**, *33*, 84–89. [CrossRef] [PubMed]
25. Nicholson, J.W.; Czarnecka, B. The biocompatibility of resin-modified glass-ionomer cements for dentistry. *Dent. Mater.* **2008**, *24*, 1702–1708. [CrossRef] [PubMed]
26. Sidhu, S.K. Clinical evaluations of resin-modified glass-ionomer restorations. *Dent. Mater.* **2010**, *26*, 7–12. [CrossRef]
27. Al Shaibani, D.; Bamusa, R.; Bajafar, S.; Al Eidan, S.; Almuhaidib, D.; Alhakeem, F.; Bakhadher, W. Modifications of glass ionomer restorative material: A review of literature. *EC Dent. Sci.* **2019**, *18*, 1001–1006.
28. Smales, R.J.; Wong, K.C. Two-year clinical performance of a resin-modified glass ionomer sealant. *Am. J. Dent.* **1999**, *12*, 62–64.
29. Pameijer, C.H. Crown retention with three resin-modified glass ionomer luting cements. *J. Am. Dent. Assoc.* **2012**, *143*, 1218–1222. [CrossRef]
30. McLean, J.W.; Nicholson, J.W.; Wilson, A.D. Proposed nomenclature for glass-ionomer dental cements and related materials. Guest Editor. *Quintessence Int.* **1994**, *25*, 587–589.
31. Mount, G.J.; Tyas, M.J.; Ferracane, J.L.; Nicholson, J.W.; Berg, J.H.; Simonsen, R.J.; Ngo, H.C. A revised classification for direct tooth-colored restorative materials. *Quintessence Int.* **2009**, *40*, 691–697.
32. International Organization for Standardization. *ISO 9917-1: Dentistry-Water Based Cements, Part 1: Powder/Liquid acid-Base Cements*; ISO: Geneva, Switzerland, 2007.
33. International Organization for Standardization. *ISO 9917-2: Dentistry-Water Based Cements, Part 2: Resin-Modified Cements*; ISO: Geneva, Switzerland, 2010.
34. Ilie, N.; Hickel, R. Mechanical behaviour of glass ionomer cements as a function of loading and mixing procedures. *Dent. Mater. J.* **2007**, *26*, 526–533. [CrossRef]
35. Busanello, L.; Telles, M.; Miranda Junior, W.G.; Imparato, J.C.; Jacques, L.B.; Mallman, A. Compressive strength of glass-ionomer cements used for atraumatic restorative treatment. *Rev. Odont. Cienc.* **2009**, *24*, 295–298.
36. Wasson, E.A.; Nicholson, J.W. Effect of operator skill in determining the physical properties of glass-ionomer cements. *Clin. Mater.* **1994**, *15*, 169–173. [CrossRef]
37. Rodrigues, D.S.; Buciumeanu, M.; Martinelli, A.E.; Nascimento, R.M.; Henriques, B.; Silva, F.S.; Souza, J.C.M. Mechanical strength and wear of dental glass-ionomer and resin composites affected by porosity and chemical composition. *J. Bio Tribo-Corr.* **2015**, *1*, 24. [CrossRef]
38. Uno, S.; Finger, W.J.; Fritz, U. Long-term mechanical characteristics of resin-modified glass ionomer restorative material. *Dent. Mater.* **1996**, *12*, 64–69. [CrossRef]
39. Farret, M.M.; de Lima, E.M.; Mota, E.G.; Oshima, H.M.S.; Maguilnik, G.; Scheld, P.A. Assessment of the mechanical properties of glass ionomer cements for orthodontic cementation. *Dent. Press J. Orthodont.* **2012**, *17*, 154–159. [CrossRef]
40. Moberg, M.; Brewster, J.; Nicholson, J.W.; Roberts, H. Physical property investigation of contemporary glass ionomer and resin modified glass ionomer restorative materials. *Clin. Oral Investig.* **2019**, *23*, 1295–1308. [CrossRef]
41. Zhang, Y.-R.; Du, W.; Zhou, X.-D.; Yu, H.Y. Review of research on the mechanical properties of the human tooth. *Int. J. Oral Sci.* **2014**, *6*, 61–69. [CrossRef]
42. Cuy, J.L.; Mann, A.B.; Livi, K.J.; Teaford, M.F.; Weihs, T.P. Nanoindentation mapping of the mechanical properties of human molar tooth enamel. *Arch. Oral Biol.* **2002**, *47*, 281–291. [CrossRef]
43. Biswas, N.; Dey, A.; Kundu, S.; Chakraborty, H.; Mukhopadhyay, A.K. Mechanical properties of enamel nanocomposite. *ISRN Biomater.* **2013**, 253761. [CrossRef]
44. He, L.H.; Fujisawa, N.; Swain, M.V. Elastic modulus and stress-starin response of human enamel by nano-identation. *Biomaterials* **2006**, *27*, 4388–4398. [CrossRef]
45. Jeng, Y.R.; Lin, T.T.; Hsu, H.M.; Chang, H.J.; Shieh, D.B. Human enamel rod presents anisotropic nanotribological properties. *J. Mech. Behav. Biomed. Mater.* **2011**, *4*, 515–522. [CrossRef] [PubMed]
46. Bajaj, D.; Arola, D.D. On the R-curve behavior of human tooth enamel. *Acta Biomater.* **2009**, *30*, 4037–4046. [CrossRef] [PubMed]
47. Kinney, J.H.; Marshall, S.J.; Marshall, G.W. The mechanical properties of human dentin: A critical review and re-evaluation of the dental literature. *Crit. Rev. Oral Biol. Med.* **2003**, *14*, 13–29. [CrossRef] [PubMed]

48. Mahoney, W.; Holt, A.; Swain, M.; Kilpatrick, N. The hardness and modulus of elasticity of primary molar teeth: An ultra-micro-indentation study. *J. Dent.* **2000**, *28*, 589–594. [CrossRef]
49. Angker, L.; Swain, M.V.; Kilpatrick, N. Characterising the micro-mechanical behavior of carious dentine primary teeth using nano-indentation. *J. Biomech.* **2005**, *38*, 1535–1542. [CrossRef]
50. Yan, J.; Taskonal, B.; Mecholsky, J.J., Jr. Fractography and fracture toughness of human dentin. *J. Mech. Behav. Biomed. Mater.* **2009**, 478–484. [CrossRef]
51. Iwamoto, N.; Ruse, N.D. Fracture toughness of human dentin. *J. Biomed. Mater. Res. A* **2003**, *5*, 3045–3046. [CrossRef]
52. Chuang, S.-F.; Lin, S.-Y.; Wei, P.-J.; Han, C.-F.; Lin, J.-F.; Chang, H.C. Characterization of the elastic and viscoelastic properties of dentin by a nanoindentation creep test. *J. Biomech.* **2015**, *48*, 2155–2161. [CrossRef]
53. Zaytsev, D.; Panfilov, P. The strength properties of human dentinoenamel junction. *Mater. Lett.* **2016**, *178*, 107–110. [CrossRef]
54. Narasimha Bharadwaj, T.P.; Solomon, P.; Parameswaran, A. Tooth restored with composite resins–A comparative analysis. *Trends Biomater. Artif. Organs* **2002**, *15*, 57–60.
55. de Norouha, F.; Scelza, M.F.Z.; de Silva, L.E.; de Carvalho, W.R. Evaluation of compressive strength in the first premolars between young and elderly people: Ex vivo study. *Gerodontology* **2012**, *29*, e898–e901. [CrossRef] [PubMed]
56. Lohbauer, U. Dental glass ionomer cements as permanent filling materials?–Properties, limitations future trends. *Materials* **2010**, *3*, 76. [CrossRef]
57. Wasson, E.A. Reinforced glass-ionomer cements–a review of properties and clinical use. *Clin. Mater.* **1993**, *12*, 181–190. [CrossRef]
58. Simmons, J.J. Silver-alloy powder and glass ionomer cement. *J. Amer. Dent. Assoc.* **1990**, *120*, 49–52. [CrossRef]
59. McLean, J.W. Cermet cements. *J. Am. Dent. Assoc.* **1990**, *120*, 43–47. [CrossRef]
60. Sced, I.; Wilson, A.D. Polycarboxylic Acid Hardenable Compositions. British Patent 2,028,855A, 1980.
61. Baig, M.S.; Fleming, G.J.P. Conventional glass-ionomer materials: A review of the developments in glass powder, polyacid liquid and the strategies for reinforcement. *J. Dent.* **2015**, *43*, 897–912. [CrossRef]
62. Moshaverinia, A.; Roopour, N.; Chee, W.W.L.; Schricker, S. A review of powder modifications in conventional glass-ionomer dental cements. *J. Mater. Chem.* **2011**, *21*, 1319–1328. [CrossRef]
63. Bhatia, H.P.; Singh, S.; Sood, S.; Sharma, N. A comparative evaluation of sorption, solubility and compressive strength of three different glass ionomer cements in artificial saliva: An in vitro study. *Int. J. Clin. Pediatr. Dent.* **2017**, *10*, 49–54. [CrossRef]
64. Nakajima, H.; Watkins, J.H.; Arita, K.; Hanaoka, K.; Okabe, T. Mechanical properties of glass ionomers under static and dynamic loading. *Dent. Mater.* **1996**, *12*, 30–37. [CrossRef]
65. Williams, J.A.; Billington, R.W.; Pearson, G.J. The comparative strengths of commercial glass-ionomer cements with and without metal additions. *Br. Dent. J.* **1992**, *172*, 279–282. [CrossRef]
66. Walls, A.W.G.; Adamson, J.; McCabe, J.F.; Murray, J.J. The properties of glass polyalkenoate (ionomer) cement incorporating sintered metallic particles. *Dent. Mater.* **1987**, *3*, 113–116. [CrossRef]
67. Kerby, R.E.; Bleiholder, R.F. Physical properties of stainless steel and silver-reinforced glass-ionomer cements. *J. Dent. Res.* **1991**, *70*, 1358–1361. [CrossRef] [PubMed]
68. Oldfield, C.W.B.; Ellis, B. Fibrous reinforcement of glass-ionomer cements. *Clin. Mater.* **1991**, *7*, 313–323. [CrossRef]
69. Kobayashi, M.; Kon, M.; Miyai, K.; Asaoka, K. Strengthening of glass-ionomer cement by compounding short fibres with $CaO-P_2O_5-SiO_2-Al_2O_3$ glass. *Biomaterials* **2000**, *21*, 2051–2058. [CrossRef]
70. Kawano, F.; Kon, M.; Kobayashi, M.; Miyai, K. Reinforcement effect of short glass fibres with $CaO-P_2O_5-SiO_2-Al_2O_3$ glass on strength of glass ionomer cements. *J. Dent.* **2001**, *29*, 377–380. [CrossRef]
71. Lohbauer, U.; Walker, J.; Nikolaenko, S.; Werner, J.; Clare, A.; Petschelt, A.; Griel, P. Reactive fibre reinforced glass ionomer cements. *Biomaterials* **2003**, *17*, 2901–2907. [CrossRef]
72. Hamouda, I.M. Reinforcement of conventional glass-ionomer restorative material with short glass fibres. *J. Mech. Behav. Biomed. Mater.* **2009**, *2*, 73–81. [CrossRef]
73. Garoushi, S.; Vallittu, P.; Lassila, L. Hollow glass fibres in reinforcing glass ionomer cements. *Dent. Mater.* **2017**, *33*, e86–e93. [CrossRef]

74. Bao, X.; Garoushi, S.K.; Lui, F.; Lassila, L.L.J.; Vallittu, P.K.; He, J. Enhancing mechanical properties of glass ionomer cements with basalt fibres. *Silicon* **2019**. [CrossRef]
75. Ross, A. Basalt fibres: Alternative to glass? *Compos. Technol.* **2006**, *12*, 44–48.
76. Lee, S.O.; Rhee, K.Y.; Park, S.T. Influence of chemical surface treatment of basalt fibres on interlaminar shear strength and fracture toughness of epoxy-based composites. *J. Ind. Eng. Chem.* **2015**, *32*, 153–156. [CrossRef]
77. Silva, R.M.; Santos, P.H.N.; Souza, L.B.; Dumont, V.C.; Soares, J.A.; Santos, M.H. Effects of cellulose fibres on the physical and chemical properties of glass ionomer dental restorative materials. *Mater. Res. Bull.* **2013**, *48*, 118–126. [CrossRef]
78. Silva, R.M.; Pereira, F.V.; Mota, F.A.P.; Watanabe, E.; Soares, S.M.C.S.; Santos, M.H. Dental glass ionomer cement reinforced by cellulose microfibres and cellulose nanocrystals. *Mater. Sci. Eng. C* **2016**, *58*, 389–395. [CrossRef] [PubMed]
79. Silva, R.M.; Pereira, F.V.; Santos, M.H.; Soares, J.A.; Miranda, J.L. Biocompatibility of a new dental glass ionomer cement with cellulose microfibres and cellulose nanocrystals. *Braz. Dent. J.* **2017**, *28*, 172–178. [CrossRef] [PubMed]
80. Elsaka, S.E.; Hamouda, I.M.; Swain, M.V. Titanium dioxide nanoparticles addition to a conventional glass-ionomer restorative: Influence on physical and antibacterial properties. *J. Dent.* **2011**, *39*, 589–598. [CrossRef] [PubMed]
81. Dowling, A.H.; Schmitt, W.S.; Fleming, G.J.P. Modification of titanium dioxide particles to reinforce glass-ionomer restoratives. *Dent. Mater.* **2014**, *30S*, e159–e160. [CrossRef]
82. Khademolhosseini, M.R.; Barounian, M.H.; Eskandari, A.; Aminzare, M.; Zahedi, A.M.; Ghahremani, D. Development of new Al$_2$O$_3$/TiO$_2$ reinforced glass-ionomer cements (GICs) nanocomposites. *J. Basic Appl. Sci. Res.* **2012**, *2*, 7526–7529.
83. Semyari, H.; Sattari, M.; Atai, M.; Pournasir, M. The effect of nanozirconia mixed with glass-ionomer on proliferation of epithelial cells and adhesive molecules. *J. Periodontol. Implant Dent.* **2011**, *3*, 63–68. [CrossRef]
84. Cibim, D.D.; Saito, M.T.; Giovani, P.A.; Borges, A.F.S.; Pecorari, V.G.A.; Gomes, O.P.; Lisboa-Filho, P.N.; Niciti-Junior, F.H.; Puppin-Rontani, R.M.; Kantovitz, K.R. Novel nanotechnology of TiO$_2$ improves physical-chemical and biological properties of glass ionomer cement. *Int. J. Biomater.* **2017**, *2017*, 7123919. [CrossRef] [PubMed]
85. Gjorgievska, E.; Nicholson, J.W.; Grabić, D.; Guclu, Z.A.; Melitić, I.; Coleman, N.J. Assessment of the impact of the addition of nanoparticles on the properties of glass-ionomer cements. *Materials* **2020**, *13*, 276. [CrossRef] [PubMed]
86. Gjorgievska, E.; Van Tendeloo, G.; Nicholson, J.W.; Coleman, N.J.; Slipper, I.J.; Booth, S. The incorporation of nanoparticles into conventional glass-ionomer dental restorative cements. *Microsc. Microanal.* **2015**, *21*, 392–406. [CrossRef] [PubMed]
87. Alatawi, R.A.S.; Elsayed, N.H.; Mohammed, W.S. Influence of hydroxyapatite nanoparticles on the properties of glass ionomer cement. *J. Mater. Res. Technol.* **2019**, *8*, 344–349. [CrossRef]
88. Sayyedan, F.S.; Fathi, M.H.; Edris, H.; Doostmohammed, A.; Mortazari, V.; Hanifi, A. Effect of forsterite nanoparticles on mechanical properties of glass ionomer cements. *Ceram. Int.* **2014**, *40*, 10743–10748. [CrossRef]
89. Menezes-Silva, R.; de Oliveira, B.M.B.; Fernandes, P.H.M.; Shimohara, L.Y.; Pereira, F.V.; Borges, A.F.S.; Buzalaf, M.A.R.; Pascotto, R.C.; Sidhu, S.K.; de Lima Navarro, M.F. Effects of the reinforced cellulose nanocrystals on glass-ionomer cements. *Dent. Mater.* **2019**, *35*, 564–573. [CrossRef]
90. Paiva, L.; Fidalgo, T.K.S.; da Costa, L.P.; Maia, L.C.; Balan, L.; Anselme, K.; Ploux, L.; Thiré, R.M.S. Antibacterials properties and compressive strength of new one-step preparation silver nanoparticles in glass-ionomer cements (NanoAg-GIC). *J. Dent.* **2018**, *69*, 102–109. [CrossRef]
91. Jowkar, Z.; Jowkar, M.; Shafiei, F. Mechanical and dentin bond strength properties of the nanosilver enriched glass ionomer cement. *J. Clin. Exp. Dent.* **2019**. [CrossRef]
92. Rai, M.; Yadav, A.; Gade, A. Silver nanoparticles as a new generation of antimicrobials. *Biotechnol. Adv.* **2009**, *27*, 76–83. [CrossRef]
93. Yoshida, Y.; Van Meerbeck, B.; Nakayama, Y.; Yoshioka, M.; Snauwaert, J.; Abe, Y.; Lambrechts, P.; Vanherle, G.; Okazaki, O. Adhesion to and decalcification of hydroxyapatite by carboxylic acids. *J. Dent. Res.* **2001**, *80*, 1565–1569. [CrossRef]

94. Nicholson, J.W.; Hawkins, S.J.; Smith, J.E. The incorporation of hydroxyapatite into glass-polyalkenoate ("glass-ionomer") cements: A preliminary study. *J. Mater. Sci. Mater. Med.* **1993**, *4*, 418–421. [CrossRef]
95. Yap, A.U.P.; Pek, Y.S.; Kumar, R.A.; Cheang, P.; Khor, K.A. Experimental studies on a new bioactive material: HA ionomer cements. *Biomaterials* **2002**, *23*, 955–962. [CrossRef]
96. Gu, Y.; Yap, A.U.P.; Cheang, P.; Khor, K.A. Effects of incorporation of HA/ZrO2 into glass-ionomer cement (GIC). *Biomaterials* **2005**, *26*, 713–720. [CrossRef] [PubMed]
97. Lucas, M.E.; Arita, K.; Nishino, M. Toughness, bonding and fluoride-release properties of hydroxyapatite-added glass ionomer cement. *Biomaterials* **2003**, *24*, 3787–3794. [CrossRef]
98. Moshaverinia, A.; Ansari, S.; Moshaverinia, M.; Roohpour, N.; Darr, J.; Rehman, A. Effects of incorporation of hydroxyapatite and fluorapatite nanobioceramics into conventional glass-ionomer cements (GIC). *Acta Biomater.* **2008**, *4*, 432–440. [CrossRef] [PubMed]
99. Yli-Urpo, H.; Lassila, L.V.J.; Narhi, T.; Vallittu, P.K. Compressive strength and surface characterisation of glass-ionomer cements modified by particles of bioactive glass. *Dent. Mater.* **2005**, *21*, 201–209. [CrossRef]
100. Dowling, A.H.; Stamboulis, A.; Fleming, G.J.P. The influence of montmorillonite clay reinforcement on the performance of a glass ionomer restorative. *Dent. Mater.* **2006**, *34*, 802–810. [CrossRef]
101. Bansal, R.K.; Tewari, U.S.; Singh, P.; Murthy, D.V.S. Modified polyalkenoate (glass ionomer) cement–a study. *J. Oral Rehabil.* **1995**, *22*, 533–537. [CrossRef]
102. Bhattacharya, A.; Vaidya, S.; Tomer, A.K.; Raina, A. GIC at its best–A review on ceramic reinforced GIC. *Int. J. Appl. Dent. Sci.* **2017**, *3*, 405–408.
103. Albeshti, R.; Shahid, S. Evaluation of microleakage of Zirconomer^R, a zirconia reinforced glass ionomer cement. *Acta Stomatol. Croat.* **2018**, *52*, 97–101. [CrossRef]
104. Callister, W.D. Chapter 9 in *"Materials Science and Engineering: An Introduction"*, 7th ed.; John Wiley and Sons, Inc.: New York, NY, USA, 2007.
105. Abdulsamee, N.; Elkhadem, A.H. Zirconomer and Zirconomer Improved (white amalgams): Restorative materials for the future. *Rev. EC Dent. Sci.* **2017**, *15*, 134–150.
106. Kamath, U.; Salam, A. Fracture resistance of maxillary premolars with MOD cavities restored with Zirconomer: An in vitro comparative study. *Int. J. Appl. Dent. Sci.* **2016**, *2*, 77–80.
107. Wang, Y.; Darvell, B.W. Hertzian load-bearing capacity of a ceramic-reinforced glass-ionomer stored wet and dry. *Dent. Mater.* **2011**, *25*, 952–955. [CrossRef] [PubMed]

© 2020 by the authors. Licensee MDPI, Basel, Switzerland. This article is an open access article distributed under the terms and conditions of the Creative Commons Attribution (CC BY) license (http://creativecommons.org/licenses/by/4.0/).

Review

Effects of Surface Treatments of Glass Fiber-Reinforced Post on Bond Strength to Root Dentine: A Systematic Review

Lora Mishra [1], Abdul Samad Khan [2], Marilia Mattar de Amoedo Campos Velo [3], Saurav Panda [4,5], Angelo Zavattini [6], Fabio Antonio Piola Rizzante [7], Heber Isac Arbildo Vega [8,9], Salvatore Sauro [10,11] and Monika Lukomska-Szymanska [12,*]

1. Department of Conservative Dentistry and Endodontics, Institute of Dental Sciences, Siksha 'O' Anusandhan Univeristy, Bhubaneswar 751003, India; loramishra@soa.ac.in
2. Department of Restorative Dental Sciences, College of Dentistry, Imam Abdulrahman Bin Faisal University, Dammam 31441, Saudi Arabia; akhan@iau.edu.sa
3. Department of Operative Dentistry, Endodontics and Dental Materials, Bauru School of Dentistry, University of São Paulo, São Paulo 17012-901, Brazil; mariliavelo@usp.br
4. Department of Periodontics and Oral Implantology, Siksha 'O' Anusandhan Univeristy, Bhubaneswar 751003, India; sauravpanda@soa.ac.in
5. Department of Biomedical, Surgical and Dental Sciences, Universita Degli Studi di Milano, 20122 Milano, Italy
6. Department of Endodontics, King's College London Dental Institute, Guy's Hospital, London SE1 9RT, UK; angelo.zavattini@kcl.ac.uk
7. Department of Comprehensive Care, School of Dental Medicine, Case Western Reserve University, Cleveland, OH 44106, USA; fap17@case.edu
8. Department of General Dentistry, Dentistry School, Universidad San Martín de Porres, Chiclayo 14012, Peru; harbildov@usmp.pe
9. Department of General Dentistry, Dentistry School, Universidad Particular de Chiclayo, Chiclayo 14012, Peru
10. Dental Biomaterials, Preventive and Minimally Invasive Dentistry Departamento de Odontología, Facultad de Ciencias de la Salud, Universidad CEU-Cardenal Herrera C/Del Pozo ss/n, Alfara del Patriarca, 46115 Valencia, Spain; salvatore.sauro@uchceu.es
11. Department of Therapeutic Dentistry, Sechenov University of Moscow, Mozhaisky Val 11, 119435 Moscow, Russia
12. Department of General Dentistry, Medical University of Lodz, 92-213 Lodz, Poland
* Correspondence: monika.lukomska-szymanska@umed.lodz.pl; Tel.: +48-42-675-74-61

Received: 3 April 2020; Accepted: 20 April 2020; Published: 23 April 2020

Abstract: The objective of this systematic review was to determine the influence of surface treatment of glass fiber posts on bond strength to dentine. Laboratory studies were searched in MEDLINE, PubMed, Embase, PubMed Central, Scopus, and Web of Science search engine. All authors interdependently screened all identified articles for eligibility. The included studies were assessed for bias. Because of the considerable heterogeneity of the studies, a meta-analysis was not possible. Twelve articles were found eligible and included in the review. An assessment of the risk of bias in the included studies provided a result that classified the studies as low, medium, and high risk of bias. The available evidence indicated that the coronal region of the root canal bonded better to the glass fiber post than apical regions. Phosphoric acid, hydrogen peroxide, and silane application enhance post's retentiveness. In light of the current evidence, surface treatment strategies increase the bond strength of glass fiber post to dentine. However, recommendations for standardized testing methods and reporting of future clinical studies are required to maintain clinically relevant information and to understand the effects of various surface treatment of glass fiber post and their bond strength with dentine walls of the root canal.

Keywords: glass fiber; composite post; endodontic therapy; etching; post and core technique; silane

1. Introduction

It has been advocated that teeth become more fragile after endodontic treatments, and this increases the failure rate of post-endodontics restorations [1] due to a loss of tooth structure, moisture, and flexibility of the dentine [2]. Therefore, extensive tooth loss after cavity preparation represents a real challenge in the restoration of endodontically treated teeth. Thus, in such a specific clinical scenario, the use of intra-canal posts may be indicated to increase the retention of the core and/or of the coronal restoration [3].

Recently, several different posts and types of core materials are in use, whereby fiber-reinforced posts (FRC) have been used with acceptable results both in clinical practice and in research [4,5]. Due to "dentine-like" elastic modulus and exceptional esthetic properties, FRC has shown superior performance compared to metal posts [2,6]. Moreover, the adhesive bond between FRC and resin cement provides a short term strengthening effect, theoretically creating an endodontic "Monoblock" [7,8]. Nevertheless, bonding posts to root canal dentine can be compromised [4,9]. The limitations are usually related to polymerization effectiveness [9], difficult in creating a water-wet substrate [10–13], reduced number of dentinal tubules [11,14,15], deposition of cementum [10,12] and secondary dentin [13]. As bonding to root dentine is still unpredictable, it may affect the longevity and the clinical performance of the restorations [3,16,17].

The stability of the resin-root dentine interface is also affected by the presence of the endogenous enzymes activated during etching procedures. These are in part responsible for the hybrid layer's degradation and reduction of the longevity of post-endodontists restorations performed with the use of root canal posts [18,19]. Various efforts to enhance the longevity of the interface between root canal dentine and FRC posts have been advocated. Such strategies include adhesive treatment, tribomechanical treatment, sandblasting, as well as a combination of these methods (Co-Jet) [20–22].

Despite the existence of some positive results in vitro [23], the treatment of the surfaces of posts is still considered a technique-sensitive procedure, and there is a lack of long-term clinical studies.

To the best of our knowledge, there is no standard protocol for post-space treatment and FRC surface treatment that can assure long-term performance in both clinical and in vitro scenarios [3]. The wide variety of materials available in the market, associated with the limitations of bonding to root canal dentine, can interfere with the adhesive strategy. Therefore, it is essential to investigate the long-term stability of the interface between fiber posts and adhesive materials to guide clinicians to select the most appropriate materials and treatment protocols in dental practice. The present systematic review aimed at appraising the existing literature on the effect of post-treatment on the bonding of fiber-reinforced composite posts to root dentin.

2. Materials and Methods

This review was carried out following the Preferred Reporting Items for Systematic Reviews and Meta-Analyses (PRISMA) statement guidelines as reported previously [24]. The schematic pattern of the protocol is shown in Figure 1.

2.1. Search Strategy

The initial review carried out with a MEDLINE/PubMed, EMBASE, CENTRAL, Scopus, and Web of Science search with laboratory and clinical trial findings. The structured protocol was adopted for the search and the following keywords were: ("silane" or "treatments" or "tissue conditioning" or "dental etching" or "dental cements" or "bond strength") and ("fiber post" or "glass fiber post" or "post and core technique" or "fiber-reinforced post" or "endodontic post") and ("root dentin" or "dentin" or "tooth root" or "root canal").

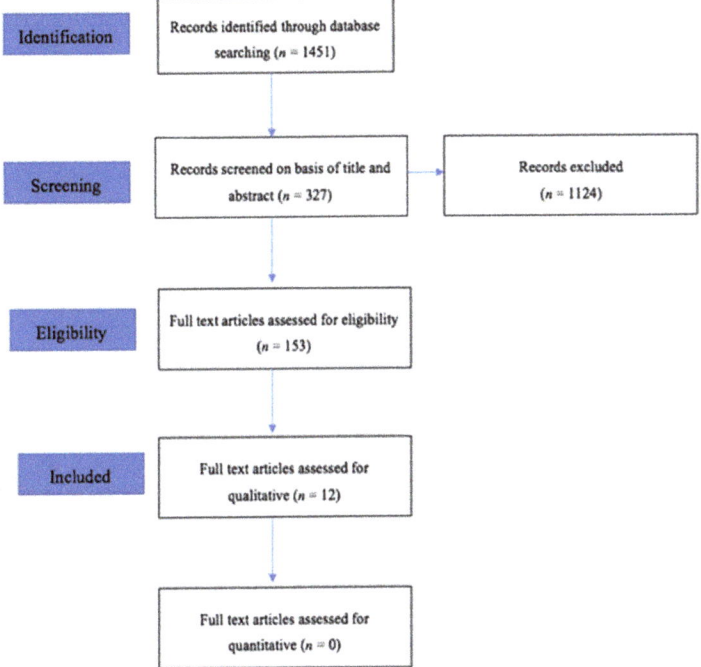

Figure 1. PRISMA flow diagram of the literature search and selection process. PRISMA—Preferred Reporting Items for Systematic Reviews and Meta-Analyses.

2.2. Study Selection

The search of the literature was performed without any restriction on a date and was done up to June 2019. Full texts of papers were obtained from the journals. The inclusion and exclusion criteria for articles are presented in Table 1.

Table 1. The inclusion and exclusion criteria for articles.

Inclusion Criteria	Exclusion Criteria
- In vitro studies carried out on radicular dentin of human extracted teeth. - Surface treatment of radicular dentin was performed after removing the obturating material and prior to restoring with glass fiber posts. - Studies based on glass fiber endodontic posts. - The push-out and pull-out test was performed to evaluate the bond strength.	- In vivo studies and in vitro studies performed on animal models. - Studies testing endodontic posts other than glass fibers, i.e., carbon and metal posts. - Literature not published in peer-reviewed journals. - The grey literature, i.e., the information not reported in the scientific journals. - All papers in a foreign language (not in the English language), where the full text was not available.

2.3. Study Quality Assessment

The title and abstracts of all articles identified by the electronic search were read and assessed by two authors (S.P. and F.A.). Disagreements between the reviewers were resolved by consensus with all the authors. The methodological quality of all selected full-text articles was assessed using the guidelines reported previously [1,24]. The description of the parameters for quality assessment was: randomization of teeth, use of teeth free of caries or restoration, use of materials according to the manufacturer's instructions, use of tooth roots with similar dimensions, endodontic treatment performed by the same operator, description of sample-size calculation, and blinding of the operator of the testing machine. If the parameter was mentioned in the article, a "positive" (+) sign was assigned to that specific parameter; if the information was not provided, the article received a "negative" (−). The searched articles that reported one to three parameters were classified as having a high risk of bias, four or five items as medium risk of bias, and six or seven items as low risk of bias.

After the application of the search strategy, two examiners (L.M. and M.V.) reviewed and performed the selection by consensus to complement the database searches. References in papers were checked and cross-matched with those from the original search. Where additional references were found to meet the inclusion criteria, then these were included in the review. After identifying the eligible studies in the above databases, these studies were imported into Endnote X7 software (Thompson Reuters, Philadelphia, PA, USA) to remove duplicates.

3. Results

After the exclusion of duplicated articles, the electronic search identified 1451 articles from EMBASE, PubMed/Medline, Scopus, and Web of Sciences. However, 327 articles were initially selected based on titles and abstracts, whereby 153 articles appeared as full text. The total number of papers, which met the inclusion criteria for the review, was 12. The flow diagram (Figure 1) depicts the details of the search strategy and Table 2 lists out surface treatment and bond strength analysis of FRC posts.

Table 2. Surface treatment strategies done on the fiber-reinforced (FRC) post.

Serial Number	Pretreatment Technique	BS Analysis	References
1	S	PBS	[22,25–27]
2		RBS	[28,29]
3	AP	PBS	[25,30]
4		RBS	[31]
5		Pull out	[32]
6	AP + S	PBS	[26,27]
7		RBS	[28]
8	HF + S	PBS	[26,33]
9		Pull out	[28]
10	AP + Alc	RBS	[34]
11	AP + BA (Ag/no Ag)	RBS	[31]
12	HF	PBS	[25]
13	Laser	PBS	[26,30]
14		Pull out	[32]
15	H_2O_2	Push out	[21]
16	H_2O_2 + S	PBS	[27,33]
17	AE	PBS	[21,31]
18	AE + BA	RBS	[31]
19	Alc, Alc + primer	RBS	[34]
20	Ethanol + S	RBS	[29]
21	CH_2Cl_2	PBS	[25]

BS—bond strength; PBS—push out bond strength; RBS—retentive bond strength; S—silane; AP—airborne abrasion; Alc—alcohol; BA airborne abrasion application followed by bonding agent application; HF—hydrofluoric acid; H_2O_2—hydrogen peroxide; AE—acid etching; CH_2Cl_2—methylene chloride.

3.1. Pre-Treatment of FRC Post Surface with Silane

Among these studies, six articles [22,25–29] evaluated the effect of surface treatment with silane on the bond strength to dentine. Whereby, four studies [22,25–27] evaluated the push-out bond strength, and the other two assessed the retentive bond strength [28,29]. The push-out bond strength was evaluated in coronal, middle, and apical regions [25–27], whereas, only one study [22] reported the mean values of all three areas together. Only two studies [25,27], compared silane with the control group. All studies concluded that silane alone could not enhance the bond strength of FRC post and resin cement to dentine. Two studies [28,29] evaluated the pull-out bond strength between silane and control groups and reported that the silane treatment of fiber posts did not prevent dislocation.

3.2. Pre-Treatment of FRC Post Surface with Air-Borne Particle Abrasion

The search data showed that alumina particles were used to treat FRC post surface [25,27,28,30,31], whereby the used particle size was mainly 50 μm. Only one study [28] assessed pull-out bond strength by using 30 μm particle size. In a few studies [26–28], silane was applied after air abrasion pre-treatment.

Air-borne particle treatment showed significantly lower push-out bond strength compared to the control group in the coronal and middle levels of the root [25–27]. Nevertheless, in another study [30], the mean results (from the three root regions), showed a higher bond strength than the control group. Two studies performed retentive and pull-out bond strength tests [31,32] and found that air-borne particle treatment group performed better than the control group. In one of those two studies [31], posts received additional aging treatment followed by the bonding agent application.

3.3. Pre-Treatment of FRC Post Surface with Hydrofluoric Acid

In one study [25], push-out bond strength was evaluated in the coronal, middle, and the apical section of the roots after treating the fiber post with hydrofluoric acid, however, found low bond strength values when compared to control group. Other studies [26,33], reported the push-out bond strength values of hydrofluoric acid-treated glass fibers post followed by silane application and showed the non-significant difference in values with the control group. Nevertheless, posts treated with hydrofluoric acid and silane followed by heat treatment showed significantly increased bond strength in comparison to the control group [33]. Another study [28], reported low pull-out bond strength values of hydrofluoric acid-treated glass fiber posts compared to control, silane treated, and sandblasting with silane groups.

3.4. Pre-Treatment of FRC Post Surface with Laser

It was observed in two selected studies [26,30] that Nd:YAG laser with different pulse durations in combination with silane increased the surface roughness, however, no significant difference was found in push-out bond strength between the control and surface treated groups [26]. In particular, the femtosecond laser resulted in lower bond strength when compared to control and air-borne particle abrasion groups [30]. Another study [32] evaluated pull-out bond strength of pre-treated posts using a 1.5-, 3-, and 4.5-W Er:YAG laser and it was found that the 4.5-W laser had higher values compared to air-abrasion and the control group (no treatment).

3.5. Pre-Treatment of FRC Post Surface with Etching Agents

In one of the selected studies [25], it is reported that etching the fiber post surface using CH_2Cl_2 could improve the bond strength values compared to hydrofluoric acid, air-abrasion, silane treated, and the control group. The use of 37% phosphoric acid for the different periods increased the pull-out and push-out bond strength of the posts bonded to root dentine compared to the control group, especially when phosphorous acid was used for 15 s [21,31]. In one of the studies [21], push-out bond strength was evaluated in the coronal, middle, and apical sections of roots after treating the fiber post with 37% phosphoric acid, however, indicating low bond strength values when compared to control group.

3.6. Pre-Treatment of FRC Post Surface with Hydrogen Peroxide (H_2O_2)

The highest bond strength values in both push-out tests were observed after 60 s application of H_2O_2 on glass fiber post in comparison to 15 s, 30 s, and the control group [21]. Pre-treatment of the post with 20% hydrogen peroxide for 20 min before silanization resulted in higher push-out bond strength at all of the three root regions compared to air-abrasion with silane, silane alone, and the control groups [27]. Likewise, treatment of glass fiber posts with 10% hydrogen peroxide for 20 min followed by silane and additional heat treatment showed greater push-out bond strength in all regions of the root compared to the specimens that received no heat treatment [33].

3.7. Pre-Treatment of FRC Post Surface with Alcohol

Two searched studies [29,34] evaluated the pre-treatment of post surfaces with alcohol. One study [29] showed that the post's surface pre-treatment with 21.75% ethanol had lower performance compared to the post-treated with silane, however, performed better in comparison to the non-treated control group. The other study [34] assessed the effects of post's surfaces pre-treatment with 96% alcohol with and without conditioning primer application. It was concluded that the alcohol group and alcohol with primer group exhibited the least retentive forces compared to the air-abrasion group and the combination of all three pre-treated groups.

3.8. Pre-Treatment of FRC and the Artificial Aging Process

Only one study [31], compared the effects of aging on the surface of treated specimens. Groups that underwent phosphorous acid followed by aging showed increased bond strength values when compared to the non-aged groups.

3.9. Risk of Bias Analysis

Five studies [21,28,29,31,34] were classified as high risk, five studies [25,26,30,32,33] were classified as medium risk, and only two studies [22,27] were classified as low risk. The outcome of the risk of bias analysis is presented in Figure 2. It was found that six studies [21,25,28,30,33,34] did not mention randomization of teeth, eight studies [21,26,28,29,31–34] did not mention either experimental procedure was performed by single operator or multiple operators. It was searched that eleven studies [21,22,25–29,31–34] did not mention a description of sample size and two studies [26,34] did not use control group; five studies [21,28–31] did not report about length/dimension of the root; and six studies [21,26,29–31,34] did not report that either sample was used according to manufacturer's instructions or not.

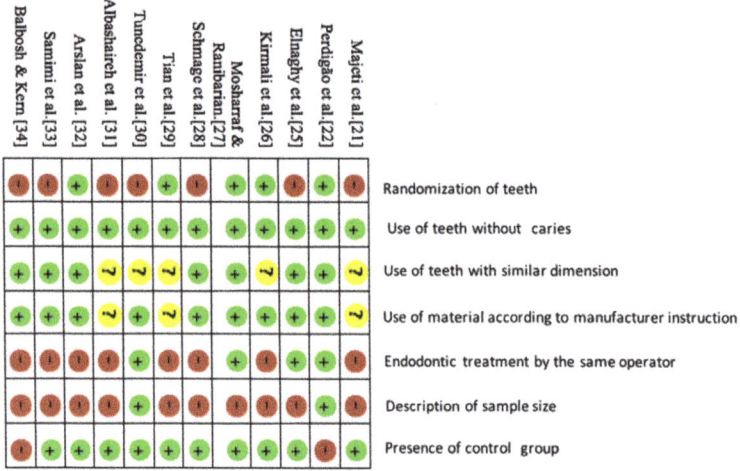

Figure 2. Risk of bias.

4. Discussion

Different strategies have been adopted to enhance the bond strength between the FRC, resin cement, and dentine, as it enables FRC to resist vertical dislodging forces. A high ratio of crosslinking within the polymer matrix makes it unable to reactivate, and negatively affects the bonding of FRC post assembly to dentine [32].

One of the strategies to improve the interfacial bond strength consists of the treatment of the FRC surface before bonding. The plethora of pretreatments of the FRC surface was noted and qualitatively analyzed in this review. Nevertheless, no consensus for surface treatment was seen that can be universally adopted. The bond strength was mainly assessed by push-out and pull-out test methods.

Studies that did pull-out were justified as a more significant methodology to analyze bond strength between fiber posts and root canal dentin [32]. In push-out studies, posts in three regions of roots (coronal, medial, and apical) were tested mainly and least bond strength was observed in the apical region [21,25–27,33]. Whereas, Finite Element Analysis (FEA) data showed homogeneous stress distribution around the post circumference in these regions. The factors related to low bond strength in the apical region could be inefficient removal of smear layer due to lack of accessibility of instruments and irrigating solution, lack of flow of material in constricted space, which results in void and bubble formation, and incomplete curing of material as the light source does not reach that region effectively [35].

It is difficult to relate in vitro condition with actual clinical situation, whereby teeth are subjected to compressive occlusal load along with shear and bending forces. The distribution of these loads and forces are different on anterior and posterior teeth [22,36,37]. The authors could not find any in vitro study in this review search where all these forces and loads were applied on endodontic posts and restoration to simulate the more clinically relevant situation. In this review, only in vitro studies conducted on human teeth were included because inconsistent data exist in the literature regarding bovine teeth. It is inconclusive that bovine teeth should be considered as an appropriate substitute for human teeth or not [38].

Many studies claimed to apply a silane agent on the post to enhance the bonding of post to resin cement and dentine [11,39]. Its application is effortless and easy, making it a widely accepted pre-treatment procedure among researchers. However, the effect on bond strength to dentine is unconvincing [22,25–28,33,40], which might be due to the manual application of silane on the post surface and formed a non-homogenous layer, subsequently led to a weak chemical bond between silane and post [25]. Another reason could be the composition of the resin matrix of FRC posts. It is reported that the methacrylate matrix bonds better to silane solution than epoxy resin containing FRC as depicted in Table 3 [21,22,25,29,40]. The interface between resin cement and post was also considered as a contributing factor for mechanical strength, where limited mechanical interlocking was mentioned between resin cement and untreated post surface [30,31]. Surface treatments such as air-borne abrasion, etching, laser treatment, hydrogen peroxide, etc. altered the surface topography and increased the retention of the FRC post.

Air abrasion is a technique capable of altering the surface topography by partially removing the matrix of the post and creating retention spaces without jeopardizing their mechanical properties [25,26,30,31,41]. However, the reported data showed that the air-abrasion treatment showed significantly lower push-out bond strength compared to the control group in the coronal and middle levels of the root [25]. Nevertheless, in another study [27], a higher push-out bond strength was observed than in the control group. Whereas, a significant reduction of bond strength was found at the apical part when compared with the middle and coronal parts [25,27]. The pull-out bond strength tests [28,32] concluded that air-borne abrasion performed better than the control group. The heterogeneity in the reported results could be due to the differences in particle size, pressure, time, and distance. Distance and time of exposure are critical for the beneficial effect of air-borne abrasion on the FRC post [30–32,34]. Longer abrasion time damages the fibers and reduces the diameter and compromises the fit of the post to the dentine [25–28].

Table 3. Composition of different FRC post used in the selected invitro studies.

Serial Number	Type of Post	Matrix	Filler	References
1	Rebilda post	Dimethacrylate	Glass	[25,26,32]
2	FRC post			[22,33]
3	Hetco fiber post	Epoxy		[27]
4	Radix			[31]
5	Dentin post ER			[28,34]
6	Para post			[22]
7	Glassix posts Nordin			[21]
8	DT light post		Quartz	[22,29,30]

Usually, single surface treatments are not capable of creating strong bond strength between post and resin. Additional etching techniques including mechanical and chemical were rendered to FRC post [25,27]. Micromechanical procedures roughen the post surface and provide access to silane solution and bonding primer agents to have more intimate contact and interaction with fibers [35,41]. Surface activation procedures for the post would be a benefit, and maybe this is one of the reasons that sandblasting followed by silane and primer application performed better than non-etched groups [27,28,31,34]. Other surface treatments, such as 9% hydrofluoric acid resulted in the dissolution of resin matrix at greater depth and also extensively damaged glass fibers within the post, therefore, reduced bond strength values were observed when compared to untreated and other experimental groups [25,26].

Laser parameters in terms of intensity, frequency, wavelength, and ablation rate are the deciding factors to obtain optimal bond strength and roughness values. This could be one of the reasons for the better performance of post irradiated with 4.5-W Er:YAG laser compared to other experimental groups [32]. However, the high power setting of laser etching can damage the fibers and produce inconsistent surface changes [17,23,42].

Many protocols directed towards surface treatments selectively remove the superficial epoxy resin layer but maintain the integrity of the fibers. Most commonly available etchants i.e., 37% phosphoric acid and hydrogen peroxide can facilitate the micromechanical retention of resin cement to the post, achieving the desired results. Two studies have recommended the application of phosphoric acid and hydrogen peroxide for 15 s and 60 s, respectively [21,31]. Aggressive/longer etching can affect the fiber's integrity [21]. Prolonged use of H_2O_2 can jeopardize resin polymerization due to the formation of free radicle and oxygen-rich layer, resulting in low bond strength [14,43]. A newer agent like methylene chloride (CH_2Cl_2) showed enhanced interfacial bond strength between the post and the luting agents. It removed the superficial layer of the resin matrix and exposed the fibers for better retention results [25].

There is a considerable amount of disparity in the retention and bond strength values of surface-treated FRC posts. This heterogeneity of observed values could be a result of not reproducibility of a clinical scenario [28]. The thermo-cycling or aging process is one of the ways to evaluate the bond strength of post in a simulated clinical environment [25]. The average number of thermal cycles that would normally occur in the oral cavity ranges from 4000 to over 10,000 per year [44]. The lack of uniformity was observed in the number of cycles and the period to which specimens were exposed [27,28,31,33,34]. Nevertheless, in a clinical scenario, FRC posts placed in the root, surrounded by dentine and alveolar bone; therefore, they remain unaffected by the temperature changes. It is reported that the thermo-cycling in water weakened the FRC posts and reduced their flexural modulus [45]. This also could be one of the reasons for reduced bond strength when the post was directly exposed or stored in water [28,31,33,34]. Other reasons for inconsistency in results among in vitro studies could be the storage time (24 h and 1 month) of FRC posts, resin, and dentine assembly in distilled water, which is not the true representative of the clinical scenario [22,25–33].

The overall quality of selected in vitro studies for qualitative analysis was poor. Most studies did not justify the sample size selection and did not mention that the endodontic treatment was carried out by the same operator. Another factor that contributed to low evidence of available data is that teeth were not randomly distributed into the groups. The high heterogeneity of the included studies prevented a quantitative analysis of data. Only two studies [22,27] out of 12 had a low risk of bias according to the used quality assessment criteria. Therefore, any general conclusions need to be drawn cautiously. Another limitation was the language restriction placed in the search strategy. There might be good quality in vitro studies in languages other than English which could have influenced the outcome of this review. In addition, the effect of luting cement on bond strength which forms an interface between post and dentine has not been included in this review. The limitation of the present review is that the influence of the shape and design of the post was not taken into account.

Future Consideration

There is a need for good quality in vitro studies which should at least try to simulate clinical conditions. FRC post-pre-treatment compared to other aesthetic posts like zirconia, can be evaluated in future studies. In addition, the effect of dentine conditioning and pre-treatment procedures influencing the bonding assembly of FRC posts can be evaluated in further reviews and research. More studies are required to analyze the efficacy of methylene chloride and ER:YAG laser pre-treatments of FRC post.

5. Conclusions

Based on this systematic review, the following conclusions were drawn:

1. The coronal region of the root canal exhibits higher bond strength to the FRC post than apical regions.
2. Phosphoric acid and hydrogen peroxide are the rapid chairside technique which can be used to increase post retention.
3. Micromechanical treatment followed by silane application can result in optimum bond strength of FRC post to resin cement and dentin.
4. Most of the studies reported that silane application to the epoxy resin-based FRC post did not enhance the bond to dentin.
5. It is not recommended to use hydrofluoric acid to treat the FRC post, as it extensively damages the surface topography of the post.

Author Contributions: Conceptualization, L.M. and M.L.-S.; methodology, L.M. and M.M.d.A.C.V.; software, S.P.; validation, L.M., F.A.P.R., and S.P.; formal analysis, S.P. and H.I.A.V.; investigation, L.M., M.L.-S., M.M.d.A.C.V., A.Z., F.A.P.R., H.I.A.V., and A.S.K.; resources, S.P., H.I.A.V., and F.A.P.R.; data curation, L.M. and M.M.d.A.C.V.; writing—original draft preparation, L.M. and M.M.d.A.C.V.; writing—review and editing, S.S., A.S.K., and M.L.-S.; visualization, S.S.; supervision, S.S.; project administration, L.M.; funding acquisition, M.L.-S. All authors have read and agreed to the published version of the manuscript.

Funding: This research received no external funding.

Conflicts of Interest: The authors declare no conflict of interest.

References

1. Caplan, D.J.; Cai, J.; Yin, G.; White, B.A. Root canal filled versus non-root canal filled teeth: A retrospective comparison of survival times. *J. Public Health Dent.* **2005**, *65*, 90–96. [CrossRef] [PubMed]
2. Rocca, G.T.; Daher, R.; Saratti, C.M.; Sedlacek, R.; Suchy, T.; Feilzer, A.J.; Krejci, I. Restoration of severely damaged endodontically treated premolars: The influence of the endo-core length on marginal integrity and fatigue resistance of lithium disilicate CAD-CAM ceramic endocrowns. *J. Dent.* **2018**, *68*, 41–50. [CrossRef] [PubMed]

3. Bonchev, A.; Radeva, E.; Tsvetanova, N. Fiber Reinforced Composite Posts—A Review of Literature. *Int. J. Sci. Res.* **2017**, *6*, 1887–1893.
4. Schwartz, R.S.; Robbins, J.W. Post placement and restoration of endodontically treated teeth: A literature review. *J. Endod.* **2004**, *30*, 289–301. [CrossRef]
5. Figueiredo, F.E.D.; Martins-Filho, P.R.S.; Faria-E-Silva, A.L. Do metal post–retained restorations result in more root fractures than fiber post–retained restorations? A systematic review and meta-analysis. *J. Endod.* **2015**, *41*, 309–316. [CrossRef]
6. Ferrari, M.; Cagidiaco, M.C.; Goracci, C.; Vichi, A.; Mason, P.N.; Radovic, I.; Tay, F. Long-term retrospective study of the clinical performance of fiber posts. *Am. J. Dent.* **2007**, *20*, 287.
7. Nissan, J.; Dmitry, Y.; Assif, D. The use of reinforced composite resin cement as compensation for reduced post length. *J. Prosthet. Dent.* **2001**, *86*, 304–308. [CrossRef]
8. Bachicha, W.S.; DiFiore, P.M.; Miller, D.A.; Lautenschlager, E.P.; Pashley, D.H. Microleakage of endodontically treated teeth restored with posts. *J. Endod.* **1998**, *24*, 703–708. [CrossRef]
9. Roberts, H.W.; Leonard, D.L.; Vandewalle, K.S.; Cohen, M.E.; Charlton, D.G. The effect of a translucent post on resin composite depth of cure. *Dent. Mater.* **2004**, *20*, 617–622. [CrossRef]
10. Tay, F.R.; Pashley, D.H.; Peters, M.C. Adhesive permeability affects composite coupling to dentin treated with a self-etch adhesive. *Oper. Dent.* **2003**, *28*, 610–621.
11. Goracci, C.; Sadek, F.T.; Fabianelli, A.; Tay, F.R.; Ferrari, M. Evaluation of the adhesion of fiber posts to intraradicular dentin. *Oper. Dent.* **2005**, *30*, 627–635. [PubMed]
12. Hayashi, M.; Okamura, K.; Wu, H.; Takahashi, Y.; Koytchev, E.V.; Imazato, S.; Ebisu, S. The root canal bonding of chemical-cured total-etch resin cements. *J. Endod.* **2008**, *34*, 583–586. [CrossRef] [PubMed]
13. Mazzoni, A.; Marchesi, G.; Cadenaro, M.; Mazzotti, G.; Di Lenarda, R.; Ferrari, M.; Breschi, L. Push-out stress for fibre posts luted using different adhesive strategies. *Eur. J. Oral Sci.* **2009**, *117*, 447–453. [CrossRef]
14. Schwartz, R.S. Adhesive dentistry and endodontics. Part 2: Bonding in the root canal system-the promise and the problems: A review. *J. Endod.* **2006**, *32*, 1125–1134. [CrossRef] [PubMed]
15. Zicari, F.; Couthino, E.; De Munck, J.; Poitevin, A.; Scotti, R.; Naert, I.; Van Meerbeek, B. Bonding effectiveness and sealing ability of fiber-post bonding. *Dent. Mater.* **2008**, *24*, 967–977. [CrossRef]
16. Bitter, K.; Noetzel, J.; Stamm, O.; Vaudt, J.; Meyer-Lueckel, H.; Neumann, K.; Kielbassa, A.M. Randomized clinical trial comparing the effects of post placement on failure rate of postendodontic restorations: Preliminary results of a mean period of 32 months. *J. Endod.* **2009**, *35*, 1477–1482. [CrossRef]
17. Cagidiaco, M.C.; Goracci, C.; Garcia-Godoy, F.; Ferrari, M. Clinical studies of fiber posts: A literature review. *Int. J. Prosthodont.* **2008**, *21*, 328–336.
18. Santos, J.; Carrilho, M.; Tervahartiala, T.; Sorsa, T.; Breschi, L.; Mazzoni, A.; Pashley, D.; Tay, F.; Ferraz, C.; Tjäderhane, L. Determination of matrix metalloproteinases in human radicular dentin. *J. Endod.* **2009**, *35*, 686–689. [CrossRef]
19. Bitter, K.; Perdigão, J.; Exner, M.; Neumann, K.; Kielbassa, A.M.; Sterzenbach, G. Reliability of fiber post bonding to root canal dentin after simulated clinical function in vitro. *Oper. Dent.* **2012**, *37*, 397–405. [CrossRef]
20. de Sousa Menezes, M.; Queiroz, E.C.; Soares, P.V.; Faria-e-Silva, A.L.; Soares, C.J.; Martins, L.R.M. Fiber Post Etching with Hydrogen Peroxide: Effect of Concentration and Application Time. *J. Endod.* **2011**, *37*, 398–402. [CrossRef]
21. Majeti, C.; Veeramachaneni, C.; Morisetty, P.K.; Rao, S.A.; Tummala, M. A simplified etching technique to improve the adhesion of fiber post. *J. Adv. Prosthodont.* **2014**, *6*, 295–301. [CrossRef]
22. Perdigão, J.; Gomes, G.; Lee, I.K. The effect of silane on the bond strengths of fiber posts. *Dent. Mater. Off. Publ. Acad. Dent. Mater.* **2006**, *22*, 752–758. [CrossRef]
23. Monticelli, F.; Toledano, M.; Tay, F.R.; Cury, A.H.; Goracci, C.; Ferrari, M. Post-surface conditioning improves interfacial adhesion in post/core restorations. *Dent. Mater.* **2006**, *22*, 602–609. [CrossRef] [PubMed]
24. Moraes, A.; Sarkis-Onofre, R.; Moraes, R.; Cenci, M.; Soares, C.; Pereira-Cenci, T. Can Silanization Increase the Retention of Glass-fiber posts? A Systematic Review and Meta-analysis of in Vitro Studies. *Oper. Dent.* **2015**, *40*, 567–580. [CrossRef] [PubMed]
25. Elnaghy, A.M.; Elsaka, S.E. Effect of surface treatments on the flexural properties and adhesion of glass fiber-reinforced composite post to self-adhesive luting agent and radicular dentin. *Odontology* **2016**, *104*, 60–67. [CrossRef] [PubMed]

26. Kırmalı, Ö.; Üstün, Ö.; Kapdan, A.; Kuştarcı, A. Evaluation of Various Pretreatments to Fiber Post on the Push-out Bond Strength of Root Canal Dentin. *J. Endod.* **2017**, *43*, 1180–1185. [CrossRef]
27. Mosharraf, R.; Ranjbarian, P. Effects of post surface conditioning before silanization on bond strength between fiber post and resin cement. *J. Adv. Prosthodont.* **2013**, *5*, 126–132. [CrossRef]
28. Schmage, P.; Cakir, F.Y.; Nergiz, I.; Pfeiffer, P. Effect of surface conditioning on the retentive bond strengths of fiberreinforced composite posts. *J. Prosthet. Dent.* **2009**, *102*, 368–377. [CrossRef]
29. Tian, Y.; Mu, Y.; Setzer, F.C.; Lu, H.; Qu, T.; Yu, Q. Failure of fiber posts after cementation with different adhesives with or without silanization investigated by pullout tests and scanning electron microscopy. *J. Endod.* **2012**, *38*, 1279–1282. [CrossRef]
30. Tuncdemir, A.R.; Buyukerkmen, E.B.; Celebi, H.; Terlemez, A.; Sener, Y. Effects of Postsurface Treatments Including Femtosecond Laser and Aluminum-oxide Airborne-particle Abrasion on the Bond Strength of the Fiber Posts. *Niger. J. Clin. Pract.* **2018**, *21*, 350–355.
31. Albashaireh, Z.S.; Ghazal, M.; Kern, M. Effects of endodontic post surface treatment, dentin conditioning, and artificial aging on the retention of glass fiber-reinforced composite resin posts. *J. Prosthet. Dent.* **2010**, *103*, 31–39. [CrossRef]
32. Arslan, H.; Kurklu, D.; Ayrancı, L.B.; Barutcigil, C.; Yılmaz, C.B.; Karatas, E.; Topçuoğlu, H.S. Effects of post surface treatments including Er:YAG laser with different parameters on the pull-out bond strength of the fiber posts. *Lasers Med. Sci.* **2014**, *29*, 1569–1574. [CrossRef] [PubMed]
33. Samimi, P.; Mortazavi, V.; Salamat, F. Effects of heat treating silane and different etching techniques on glass fiber post push-out bond strength. *Oper. Dent.* **2014**, *39*, E217–E224. [CrossRef] [PubMed]
34. Balbosh, A.; Kern, M. Effect of surface treatment on retention of glass-fiber endodontic posts. *J. Prosthet. Dent.* **2006**, *95*, 218–223. [CrossRef]
35. Monticelli, F.; Ferrari, M.; Toledano, M. Cement system and surface treatment selection for fiber post luting. *Med. Oral Patol. Oral Cirugia Bucal* **2008**, *13*, E214–E221.
36. Soares, C.J.; Santana, F.R.; Castro, C.G.; Santos-Filho, P.C.F.; Soares, P.V.; Qian, F.; Armstrong, S.R. Finite element analysis and bond strength of a glass post to intraradicular dentin: Comparison between microtensile and push-out tests. *Dent. Mater. Off. Publ. Acad. Dent. Mater.* **2008**, *24*, 1405–1411. [CrossRef]
37. Zhu, L.; Li, Y.; Chen, Y.-C.; Carrera, C.A.; Wu, C.; Fok, A. Comparison between two post-dentin bond strength measurement methods. *Sci. Rep.* **2018**, *8*, 1–8. [CrossRef]
38. Yassen, G.H.; Platt, J.A.; Hara, A.T. Bovine teeth as substitute for human teeth in dental research: A review of literature. *J. Oral Sci.* **2011**, *53*, 273–282. [CrossRef]
39. Leme, A.A.; Pinho, A.L.; de Gonçalves, L.; Correr-Sobrinho, L.; Sinhoreti, M.A. Effects of silane application on luting fiber posts using self-adhesive resin cement. *J. Adhes. Dent.* **2013**, *15*, 269–274.
40. do Nascimento Rechia, B.C.; Bravo, R.P.; de Oliveira, N.D.; Baratto Filho, F.; Gonzaga, C.C.; Storrer, C.L.M. Influence of different surface treatments of fiberglass posts on the bond strength to dentin. *Braz. J. Oral Sci.* **2016**, *15*, 158–162. [CrossRef]
41. Soares, C.J.; Santana, F.R.; Pereira, J.C.; Araujo, T.S.; Menezes, M.S. Influence of airborne-particle abrasion on mechanical properties and bond strength of carbon/epoxy and glass/bis-gma fiber-reinforced resin posts. *J. Prosthet. Dent.* **2008**, *99*, 444–454. [CrossRef]
42. Nagase, D.Y.; de Freitas, P.M.; Morimoto, S.; Oda, M.; Vieira, G.F. Influence of laser irradiation on fiber post retention. *Lasers Med. Sci.* **2011**, *26*, 377–380. [CrossRef] [PubMed]
43. Kutlu, I.U.; Yanıkoğlu, N.D. The Influences of Various Matrices and Silanization on the Bond Strength between Resin Cores and Glass Fiber Posts. *J. Odontol.* **2018**, *2*, 2.
44. Gale, M.S.; Darvell, B.W. Thermal cycling procedures for laboratory testing of dental restorations. *J. Dent.* **1999**, *27*, 89–99. [CrossRef]
45. Stewardson, D.A.; Shortall, A.C.; Marquis, P.M. The effect of clinically relevant thermocycling on the flexural properties of endodontic post materials. *J. Dent.* **2010**, *38*, 437–442. [CrossRef] [PubMed]

© 2020 by the authors. Licensee MDPI, Basel, Switzerland. This article is an open access article distributed under the terms and conditions of the Creative Commons Attribution (CC BY) license (http://creativecommons.org/licenses/by/4.0/).

Review

Bio-Inductive Materials in Direct and Indirect Pulp Capping—A Review Article

Marta Kunert and Monika Lukomska-Szymanska *

Department of General Dentistry, Medical University of Lodz, 251 Pomorska St., 92-213 Lodz, Poland; marta.kunert@stud.umed.lodz.pl
* Correspondence: monika.lukomska-szymanska@umed.lodz.pl; Tel.: +48-42-675-7461

Received: 5 February 2020; Accepted: 4 March 2020; Published: 7 March 2020

Abstract: The article is aimed at analyzing the available research and comparing the properties of bio-inductive materials in direct and indirect pulp capping procedures. The properties and clinical performances of four calcium-silicate cements (ProRoot MTA, MTA Angelus, RetroMTA, Biodentine), a light-cured calcium silicate-based material (TheraCal LC) and an enhanced resin-modified glass-ionomer (ACTIVA BioACTIVE) are widely discussed. A correlation of in vitro and in vivo data revealed that, currently, the most validated material for pulp capping procedures is still MTA. Despite Biodentine's superiority in relatively easier manipulation, competitive pricing and predictable clinical outcome, more long-term clinical studies on Biodentine as a pulp capping agent are needed. According to available research, there is also insufficient evidence to support the use of TheraCal LC or ACTIVA BioACTIVE BASE/LINER in vital pulp therapy.

Keywords: direct pulp capping; indirect pulp capping; ProRoot MTA; MTA Angelus; retroMTA; biodentine; theraCal LC; ACTIVA BioACTIVE; vital pulp therapy

1. Introduction

The major challenge for the modern approach in restorative dentistry is to induce the remineralization of hypomineralized carious dentine, and therefore, protecting and preserving the vital pulp. Traditionally, deep caries management often resulted in pulp exposure and subsequent root canal treatment. The promotion of biologically-based treatment strategies has been advocated for partial caries removal aimed at avoiding carious pulp exposure. Indeed, recent consensus reports have stated that the complete or nonselective carious removal is now considered overtreatment [1]. Management strategies for the treatment of the cariously exposed pulp are also shifting with avoidance of pulpectomy, claiming the superiority of vital pulp treatment (VPT) techniques such as pulp capping, partial and complete pulpotomy [2]. This constant turn toward a conservative and minimally invasive approach is strongly associated with the fact that the emphasis of the profession is also slowly shifting to care for the senior adult segment of the population [3,4]. Under these circumstances, retaining natural teeth is of paramount importance. All of these changes occur in a strong correlation with development of so-called bioactive dental materials [5,6]. Broadly speaking, the bioactivity of a restorative material usually denotes that it has a biological effect or is biologically active. This characteristic refers to the potential to induce specific and intentional mineral attachment to the dentine substrate [7]. In terms of restorative dentistry, bioactive material is described as "one that forms a surface layer of an apatite-like material in the presence of an inorganic phosphate solution" [8]. Therefore, the remineralization of demineralized dentine is the process of restoring minerals through the formation of inorganic mineral-like matter [9].

Six different bioactive materials have been described and compared: four calcium-silicate cements—ProRoot MTA (Dentsply Sirona, York, PA, USA), MTA Angelus (Angelus, Londrina, Brazil), RetroMTA (BioMTA, Seoul, Korea) and Biodentine (Septodont, Saint-Maur-des-Fossés, France); a light-cured calcium silicate-based material TheraCal LC (Bisco, Schaumburg, IL, USA); and a resin-modified glass-ionomer (RMGIC) with improved resilience and physical properties—ACTIVA BioACTIVE (Pulpdent Corporation, Watertown, MA, USA) [10].

The article is aimed to review the properties of bio-inductive materials applied for direct and indirect pulp capping and to compare their clinical performances.

2. Vital Pulp Treatment (VPT) in Deep Caries and Management of Pulp Exposures

The indirect pulp capping (IPC) procedure is generally used in deep cavity preparations, with or without carious dentine remaining which is in close proximity to the pulp but showing no visible pulp exposure [11,12]. According to the latest (2019) European Society of Endodontology (ESE)-approved definitions and terminology, IPC due to removal of both soft and firm carious dentine until hard dentine is reached, is nowadays considered aggressive, and in many cases may be acknowledged as overtreatment [13].

One-stage selective carious-tissue removal includes the application of a biomaterial onto a dentine barrier after either firm or soft dentine is left only on the pulpal aspect of the cavity, whilst peripheral carious dentine is removed down to hard dentine on the same visit permanent restoration is being placed. On the other hand, stepwise excavation involves re-entry after 6–12 months after application of a biomaterial in an indirect, two-stage, selective carious-tissue removal technique. The first stage involves selective carious removal to soft dentine, to an extent that facilitates proper placement of a temporary restoration, and the second stage includes removal to firm dentine. Eventually, final placement of a permanent restoration is performed. Tooth disinfection with NaOCl or chlorhexidine digluconate 2% solution, as a less aggressive alternative, sterile technique, and rubber dam use, are recommended at all times throughout the management of deep caries whether the pulp is exposed or not [13–18].

According to recent position statement of the ESE, deep caries management should focus on avoidance of pulp exposure by the means of choosing selective one-stage carious-tissue removal or stepwise excavation treatment, rather than redefined indirect pulp capping procedure [13].

Direct pulp capping (DPC) treatment is used when the vital asymptomatic pulp is visibly exposed due to caries or trauma, or due to a misadventure during tooth preparation or caries removal [7]. It includes the application of a biomaterial directly onto the exposed pulp, followed by immediate placement of a permanent restoration. An enhanced protocol should be used, including magnification, a disinfection irrigant and preferably the application of a hydraulic calcium silicate cement, as there is sufficient evidence to substantiate its use in VPT.

In contrast to pulp capping procedures, which does not involve any pulp tissue removal, partial pulpotomy removes 2–3 mm of the pulp tissue at site of exposure. In practice this technique is used for removing the superficial layer of infected tissue in case of asymptomatic carious pulp exposure or when pulp has been exposed to the oral environment. The pulp capping procedures protects the tissue, but may not reverse a superficial inflammatory processes; therefore, it is recommended that 2–3 mm of tissue is removed in a partial pulpotomy procedure. The question about choosing the proper treatment arises when we take into consideration that visual analysis to distinguish between inflamed and noninflamed pulp tissue may not be sufficiently accurate [19]. If hemostasis, as another factor indirectly indicating state of the pulp, cannot be controlled after 5 min, further caries excavation is necessary after exposure; in that case partial or even full pulpotomy may be preferable.

3. Materials Used in Direct and Indirect Pulp Capping

Historically, calcium hydroxide (Ca(OH)$_2$) has been considered the gold standard. Long term clinical observations of calcium hydroxide are incomparable to any other bioactive material, as the first reports of successful pulpal healing using Ca(OH)$_2$ were published between 1934 and 1941 [20]. However, it still has several drawbacks: insufficient adherence to dentinal walls, multiple tunnel defects in the induced dentin bridges [21], poor sealing ability, dissolution over time [22] and lack of antibacterial properties. Long-term clinical studies showed success rates with calcium hydroxide pulp capping on carious exposures to be highly variable, generally unpredictable and often unsuccessful [23]. Indeed, calcium hydroxide no longer seems to be the best possible material of choice [24]. Due to its high basicity, calcium hydroxide in direct contact with the pulp locally destroys a layer of pulp tissue, and thus creates an uncontrolled necrotic zone. This necrotic layer induces an inflammatory reaction which persists in time, or leads to formation of intra-pulpal calcifications [25]. However, it is Ca(OH)$_2$ high solubility that is the major disadvantage of its use as a pulp capping agent. The dissolution of the material within two years after application and the formation of defects in reparative dentin underneath the capping material are responsible for failing to provide a permanent seal against bacterial infection.

Nowadays, calcium hydroxide is being displaced by new generations of materials which result in more predictable clinical outcomes; namely, calcium silicate materials (CSMs) and RMGIC [26,27]. The increasingly predominant role of CSMs could be explained by their high biocompatibility, intrinsic osteoconductive activity and ability to induce regenerative responses in the human body; namely, dentin bridges of improved quality and enhanced sealing of the pulp capped site. Depending on the purpose, CSMs products can be assigned into two clinically relevant groups: restorative cements used in VPT, i.e., ProRoot MTA, MTA Angelus, RetroMTA, Biodentine and TheraCal LC, and endodontic sealers, i.e., BioRoot RCS (Septodont, France). Differences between composition of each material are presented in Table 1.

Table 1. General information on pulp capping agents.

Property	Material					
	MTA			Biodentine	TheraCal LC	ACTIVA BioACTIVE BASE/LINER
	ProRoot MTA	MTA Angelus	RetroMTA			
Release Date	1999	2001	2014	2011	2011	2014
Composition	Powder: tricalcium silicate, icalcium silicate, tricalcium aluminate, bismuth oxide, gypsum Liquid: water	Powder: tricalcium silicate, dicalcium silicate, tricalcium aluminate, silicon oxide, potassium oxide, aluminum oxide, sodium oxide, iron oxide, calcium oxide, bismuth oxide, magnesium oxide, insoluble residues of crystalline silica Liquid: water	Powder: calcium carbonate, silicon dioxide, aluminium oxide, calcium zirconia complex Liquid: water	Powder: tricalcium silicate, dicalcium silicate, calcium oxide, calcium carbonate, zirconium oxide, iron oxide Liquid: calcium chloride, water-soluble polymer, water	Light-curing single paste: resin bis-phenyl glycidyl methacrylate (BisGMA) & polyethylene glycol dimethacrylate (PEGD) modified calcium silicate filled with CaO, calcium silicate particles (type III Portland cement), Sr glass, fumed silica, barium sulphate, barium zirconate	Diurethane dimethacrylate. bis (2-(methacryloyloxy) ethyl) phosphate, barium glass, ionomer glass, polyacrylic acid/maleic acid copolymer, dual-cure chemistry, sodium fluoride, colorants
Color	White	White	White	White	White	Tooth-shade
Mixing	0.5 g pouches of powder + pre-measured unit dose of water (mixed manually)	Powder + liquid (mixed manually)	0.3 g pouches of powder + 3 drops of water (mixed manually)	0.7 g capsule of powder + 5 drops of liquid (30 s; 4000–4200 rpm)	Dispensed directly from a flowable syringe (no mixing)	Two-paste system dispensed directly from an automix syringe
Setting reaction	Hydration reaction					
	MTA + water→calcium hydroxide + calcium silicate hydrate			Tricalcium silicate + water→hydrated calcium silicate gel + calcium hydroxide	Light-cure (20 s)	3 setting mechanisms: - self-cure - light-cure (20 s) - acid-base reaction

3.1. Mineral Trioxide Aggregate

Mineral trioxide aggregate (MTA) is a bioactive cement pioneered by Torabinejad et al. in the early 1990s as an endodontic repair and root-end filling material [28] with favorable physical properties [29]. MTA has proven to induce mineralization beneath exposed pulp and have the potential of maintaining pulp vitality. Hence, the indications for the use of MTA have expanded considerably from its original use, and it has recently became a superior substitute for $Ca(OH)_2$ in many other clinical applications, including direct [14,15,30,31] and indirect pulp capping, perforation repairs in roots or in furcations [32] and the apexification procedure [33–35]. The powder of MTA is a mixture of a purified Portland cement and bismuth oxide to provide radiopacity. The main constituent phases of cement are tricalcium and dicalcium silicate and tricalcium aluminate [36,37].

ProRoot MTA, having been available on the market for two decades, has been extensively studied and proven to be biocompatible. However, the long setting time and also high cost have urged the development of new types of MTA-based materials to overcome these drawbacks (Table 2). The alternative materials must meet the advantages of the ProRoot MTA, and also be more accessible, more cost effective and set in a shorter time. As an alternative, MTA Angelus was developed, offering the advantage of reduced final setting time—24–83 min [38,39], down from the original 228–261 min specific for ProRoot MTA [37,40]. Another of the recently introduced fast-setting calcium silicate cements is Retro MTA, with a final setting time of about 12 min [39]. Although a few studies evaluated and compared the biocompatibility of each of these products with that of Pro Root MTA, those finding are limited and not comprehensive. Characteristics of those materials were collected and presented in Tables 1–3.

Table 2. Key properties of bio-inductive materials used in vital pulp treatment (VPT)—clinical manipulation and performance.

Property		ProRoot MTA	MTA Angelus	RetroMTA	Biodentine	TheraCal LC	ACTIVA BioACTIVE BASE/LINER
Discoloration of tooth structure		+	+	+	-	-	-
Final setting time (min)		261 ± 21 [40] 228.33 ± 2.88 [37]	24.0 [38] 48.3 ± 4 [41,42] 83.66 ± 17.61 [39]	12.66 ± 3.05 [39]	45.0 [43] 85.66 ± 6.03 [37]	Immediate	2.5–3.0
Single visit treatment		-	-	-	+	+	+
Handling		+	+	+	++	+++	+++
Consistency			Granular, initial looseness		Uniform, putty-like	Flowable	Flowable
Cytotoxicity		NA	NA	NA	NA	Observed	Observed
Radiopacity (mm Al)		6.4–8.5 [44]	4.5–5.96 [44]	4.07 ± 0.20 [45] 3.01 ± 0.09 [46]	1.5–4.1 [43,44] 2.79 ± 0.22 [47]	2.17 ± 0.17 [47]	NA
Solubility at 24 h (%)		1.735 ± 0.328 [48] 10.89 ± 0.48 [26]	29.55 ± 2.35 [26]	1.447 ± 0.201 [48]	11.83 ± 0.52 [26]	2.75 ± 1.04 [26]	NA
Bond strength to dentine (MPa)	After 24 h	0 [49]	NA	1.15 ± 0.32 [50]	1.01 ± 0.13 [50]	0.44 ± 0.20 [50] 0.09 ± 0.20 [51]	NA
	After 7 days	0.85 ± 1.42 [49]	NA	NA	9.75 ± 2.19 [49]	NA	NA
	After 14 days	4.96 ± 4.54 [49]	NA	NA	9.34 ± 1.01 [49]	NA	23.7 ± 17.8 [52] after 28 days after DBA application
SBS to composite after 24 h (MPa)	Methacrylate-based composites	8.9 ± 5.7 [53]	11.40 ± 3.19 [54] SE DBA (7th generation)	4.71 ± 2.35 [55] SE DBA (7th generation)	17.7 ± 6.2 [53]	19.3 ± 8.4 [56]	NA
	Silorane-based composites	7.4 ± 3.3 [53]	NA	NA	8.0 ± 3.6 [53]	3.6 ± 2.5 [56]	NA
Clinical success rate in VPT (%)			80–97 [31,57,58] (up to 9–10 years)		73–96	NA	NA
Cost of single package		€45 for 0.5 g	€50 for 1 g	€14 for 0.3 g	€10 for a 0.7 g capsule	€20 for a 1 g syringe	€90 for a 7 mL syringe
Approximate cost per application (€)		22.5	12.5	14	10	5	3.5

NA—not available; SE—self-etch; DBA—dentine bonding agent.

The biocompatibility and sealing ability of MTA result from the dominant calcium ion released from the material which reacts with phosphates in tissue fluid, inducing hydroxyapatite formation (Table 3) [59]. Formation of this layer is a key characteristic responsible for the chemical seal between MTA and the dentinal walls, and cannot be referred to as a genuine bonding process [60].

The first formulation of MTA was grey, but due to the reported discoloration of teeth [61], an altered chemical composition was presented to the market (Table 1). The chemical component of iron is absent in white MTA, although discoloration is still observed and remains one of the major drawbacks of the material [62,63]. Tricalcium and dicalcium silicate react when mixed with water to produce calcium silicate hydrate and calcium hydroxide (Table 1). Despite the high clinical advantages of MTA cement, there were always some limitations which prevented clinicians from using it in on a daily basis. The major ones being a long setting time (up to 284 min) [44,64], handling difficulties, discoloration of the remaining tooth structure [65] and the presence of heavy metals in the powder [66]. The aforementioned properties of MTA are listed in Tables 1 and 2.

When compared with $Ca(OH)_2$ in a randomized clinical trial, confirmatory evidence emerged for the superior performance of MTA as a DPC agent when evaluated in a practice-based research network [31]. In this study, the probability of failure at 24 months was 31.5% for $Ca(OH)_2$ and 19.7% for MTA. A review of the few clinical assessments in 9–10-year observation revealed 92.5–97.96% success for the teeth pulp-capped with MTA [57,58]. Clinical success rates for compared materials are presented in Table 2.

In addition, MTA is less toxic, easier to use in pulp capping procedures and causes less pulpal inflammation than $Ca(OH)_2$ [67]. A histological study confirmed that the application of MTA has a direct effect upon regeneration potential of the dental pulp and is associated with increase in TGF-β1 secretion from pulp cells (Table 3) [68]. This factor directs the progenitor cells' migration to the material-pulp interface and stimulates their differentiation to odontoblastic cells secreting reparative dentin; thus it affects the quality of the induced hard barrier. Further histological study confirmed that the formation of calcified hard tissue after pulp capping with fast-setting MTA (RetroMTA) is not particularly the product of genuine odontoblast differentiation and lacks characteristics of "regular dentine" [69]. These results suggest that the formation of calcified tissues may be more appropriately regarded as a reparative process more than a genuine regeneration response. Hence, it can be concluded that regular dentine was not regenerated, and because of the limited bioactive potential of pulp-capping material (RetroMTA), it cannot be used in regenerative dentistry. Moreover, in contrast to calcium hydroxide, mineral trioxide aggregate (WhiteProRoot®MTA, Dentsply Sirona, York, PA, USA) and Biodentine (BD) exhibit favorable metabolic activity and promote the same desired cellular response, resulting in higher clinical success rate and less tunnel defects [70]. In terms of dentin bridge formation, based on the micro-CT imaging technique, the MTA group shows more a regular, homogenous reparative dentin layer with uniform thickness, when compared with the Biodentine group. These data indicate that both MTA and BD induced hard tissue barriers and that MTA induced dentin with better characteristics (Table 3) [71]. Therefore, MTA is the material of choice in direct pulp capping [70,72].

Table 3. Interfacial properties of bio-active materials.

Property		ProRoot MTA	MTA Angelus	RetroMTA	Biodentine	TheraCal LC	ACTIVA BioACTIVE BASE/LINER
Marginal seal to dentine		○ Chemical and/or micromechanical adhesion ○ Penetration in dentinal tubules				○ Low SBS due to polymerization shrinkage ○ Poor chemical or micromechanical adhesion	○ Poor chemical or micromechanical adhesion due to lack of self-adhesive properties ○ Good seal after DBA application
pH Initially/Endpoint		9.93/8.00 [45] 12.48/11.56 [73] 10.99/7.20 [26]	10.48–9.45 [74] 11.71–10.57 [73] 11.31–8.94 [26]	9.93/7.9 [45]	11.98/11.16 [75] 11.63/9.21 [26]	10.66/9.85 [73] 8.54/8.00 [26]	8.00
Calcium release (ppm)		15.7–27.4 [26]	11.7–55.1 [26]	NA	18.0–95.3 [26]	12.6–34.2 [26]	NA
Pulp/dentine treatment		Rinse with 2.6–5.0% NaOCl	-	-	Hemostasis	Hemostasis	Lightly dry, DBA for higher SBS [52] (E&R DBA not required acc. to manufacturer)
Response of the pulp		○ Non-inflammatory reaction, ○ Increase in TGF-β1, ○ Non-toxic to pulp cells, ○ Favourable odontoblastic layer formation			○ Non-inflammatory reaction, ○ Increase in TGF-β1, ○ Non-toxic to pulp cells, ○ Well-arranged odontoblasts	○ Mild chronic inflammation, ○ Toxic to pulp fibroblasts, ○ Less favourable odontoblastic layer formation	○ Biointeractive, ○ Toxic to pulp cells due to resin component?
Hard tissue barrier quality		○ Regular ○ Homogenous ○ Uniform thickness ○ Lacks characteristics of natural dentine			○ Complete dentin bridge formation ○ Regular ○ Uniform thickness	○ Low quality calcific barrier ○ Inferior dentin bridge formation ○ Reduced dentin bridge thickness	NA
Surface treatment before composite placement	Recommended by manufacturer	37% H$_3$PO$_4$ (15 s) DBA	E&R DBA [76–78]	-	DBA	DBA	ACTIVA BioACTIVE RESTORATIVE or DBA + composite [52]
	Recommended by research	9% HF (90 s), Silane [82]	(after 72 h) 50-µm Al$_2$O$_3$ (15 s, 7 mm distance) [54]		2-SE DBA [79,80]	E&R DBA (higher SBS) [81]	NA
Maturation period		≥7 days [83,84] 1 year [85]			72 h [86] >2 weeks [87,88]	-	NA

NA—not available; E&R—etch and rinse; 2-SE—two-step self-etch; DBA—dentine bonding agent; SBS—shear bond strength.

Other key factors for successful pulp capping procedures are the ability of the bioactive material to seal to the tooth structure, the bond strength between the pulp capping material and restorative properties (Table 2). Recent studies suggest that placement of composite used with a two-step etch and rinse adhesive (E&R) over white MTA performed significantly better than an all-in-one system in terms of bond strength [76]. Other studies concluded that the highest bond strength was obtained when the E&R adhesive was used after 24 h resulting in shear bond strength (SBS) of 7.3 ± 1.49 MPa [77,78]. Supplementary surface treatment protocols were investigated to reliably asses SBS after final composite restoration (Table 3). Additional silanization (Silane, Ultradent, South Jordan, UT, USA) was recommended after treating the surface of ProRoot MTA with 9% hydrofluoric acid (HF) for 90 s before the application of dentin bonding agent (DBA) for higher bond strength [82]. Further research also suggested air abrasion of MTA Angelus surface after 72 h with 50 µm Al_2O_3 particles to achieve higher bond strength to the resin composite compared to specimens treated with 37% phosphoric acid (16.98 ± 4.24 MPa and 11.40 ± 3.19 MPa, respectively) [54]. Although the manufacturers of ProRoot, MTA Angelus and Retro MTA claimed short setting times (165 min, 15 min and 1.5 min, respectively), studies showed adequate setting only after at least 7 days to acquire proper surface properties (Tables 2 and 3) [83,84]. Additionally, evidence exists that MTA continues to mature up to 1 year beyond the setting time with impacts on its mechanical integrity and hence SBS values [85]. This research also puts forward the importance of leaving MTA to mature before the application of the overlying restoration to prevent bacterial infection.

Novel Mineral Trioxide Aggregate Restorative Cements

Alternation in material characteristics has recently led to development of new generation MTA-based cements; namely, Neo MTA Plus (Avalon Biomed Inc., Houston, TX, USA) and the iRoot (Innovative BioCeramix Inc., Vancouver, BC, Canada) products family. Both materials share the same clinical applications in vital pulp treatment, but only a few studies are available to assess their use in clinical practice [89–92]. Neo MTA Plus was developed to be used in pulpotomies without the risk of discoloration due to elimination of bismuth oxide. The radiopacifying agent was replaced by tantalum oxide, which provides a radiopacity value of 3.76 ± 0.13 mm Al and does not exert any effect on hydration [89,92]. What is important is that the final setting time of Neo MTA Plus was proven to be prolonged up to 315 ± 5 min [89]. Moreover, compared to MTA Angelus, NeoMTA Plus showed better apatite formation, higher crystallinity and higher Ca/P, but a lower CO_3/PO_4 ratio, which might result in increased bioactivity [90]. Further in vivo and in vitro studies are required to substantiate such claims.

Addressing the difficult handling of MTA, the iRoot products are available in different consistencies, what could offer an advantage of choosing suitable one for each clinical application. Namely, iRoot BP (IBC, Burnaby, BC, USA) is deposited on preloaded syringes, while the iRoot BP Plus is available in jars and with a thicker consistency, and the iRoot FS was especially developed to set faster with final setting time of 57.0 ± 2.7 min [83]. A systematic review concluded that iRoot BP and iRoot BP Plus are biocompatible materials that enhance human dental pulp cells' proliferation, migration, mineralization and dentinal bridge formation in pulp capping procedures [93]. These findings coincided with those obtained in another study, stating that iRoot BP Plus showed superiority over calcium hydroxide as a pulpotomy agent [94].

There are only a few studies on novel CSMs, as the scientific evidence is in general focused on materials that have been available for a longer time, such as ProRoot MTA and later Biodentine. To determine their clinical efficacy, further studies are needed.

3.2. Biodentine

As a response to the disadvantages of MTA, a new tricalcium silicate-based cement Biodentine (Septodont, France) was released in 2011. This comparatively new biomaterial is claimed to possess properties similar to MTA and is currently being explored for vital pulp therapy procedures. BD was designed as a permanent, biocompatible [95] dentin substitute that could be applied in one session for final composite restoration with the sandwich technique or in the whole volume of the cavity for an observation period before the final restoration. The manufacturer indicates the setting time to be between 9 and 12 min [96]; however, it was proven to set finally after 45 min (Table 2) [43].

BD is available in the form of a capsule containing powder composed of tricalcium silicate, dicalcium silicate, zirconium oxide, calcium carbonate, calcium oxide and iron oxide. A single capsule containing 0.7 g of powder is mixed for 30 s in a mixing device at a speed of 4000–4200 rpm with exactly five drops of liquid containing calcium chloride which act as an accelerator [97]; a hydrosoluble polymer functioning as a water reducing agent; and water (Table 1). The setting accelerator improves its handling properties and strength and mitigates the risk of partial material loss and alteration of the interface when compared to MTA [13,22,45]. Concerning drawbacks of the material, radiopacity is significantly lower than MTA Angelus (Table 2) despite the presence of zirconium oxide [98]. Radiopacity also gradually decreases with time, which causes difficulties in long-term radiographic observations.

The interactions of BD with hard and soft tissues in both the direct and indirect capping procedure lead to a marginal sealing and provide protection to the underlying pulp by inducing tertiary dentin synthesis and remineralization. Based on the calcium (Ca^{2+}) and hydroxide (OH^-) ion release from material, it may be concluded that tricalcium silicate materials such as BD may be preferable for IPC (Table 3) [99].

Marginal sealing is provided by micromechanical retention due to penetration of Biodentine into the dentin tubules forming tag-like structures, and represents similar bond strength to dentin compared to MTA (ProRoot MTA) [49,100]. However, those findings are inconsistent with results of other studies suggesting that Biodentine is superior to MTA in terms of sealing ability [101–103].

When compared to MTA, improvements in BD properties, such as setting time, mechanical qualities and initial cohesiveness, led to widened range of applications, including endodontic repair and vital pulp therapy. On account of its advantages, BD has recently become a preferable agent for both direct and indirect pulp capping procedures. Biodentine, compared with the previous golden standard $Ca(OH)_2$, is mechanically stronger, less soluble and produces a tighter seal [100].

Similarly, another study was conducted to evaluate the clinical response of pulp-dentin complex after DPC with MTA and BD in carious teeth. Over a 6-month observation, MTA and BD showed success rates (subjective symptoms, pulp sensibility tests and radiographic appearance) of 91.7% and 83.3%, respectively [104]. Biodentine and MTA have comparable success rates when used as direct pulp capping or pulpotomy material in permanent mature teeth with carious exposure (Table 2). The success rate of the VPT decreased from overall 96% after 1-year follow-up to 93.8% at the 3-year follow-up (Biodentine—91.7% and MTA—96.0%) [105]. A randomized clinical trial was conducted to investigate the outcome of the DPC of permanent young teeth with Biodentine; it showed no failures after 12 months, while both calcium hydroxide and MTA had a 13.6% failure rate after the same time period [106]. Other studies also support claims of Biodentine's and MTA's superiority over calcium hydroxide in terms of success rate in pulp capping procedures [107,108]. Another study reported that the success rate of DPC with BD is 90.9% in patients younger than 40 and 73.8% in patients 40 or older [109]. It is worth emphasizing that patient's age significantly influences the clinical outcome of this vital pulp therapy and needs to be taken into consideration when choosing successful treatment.

Based on a cell/tissue culture model, it can be concluded that materials used in pulp capping procedures directly affect the regeneration potential of the dental pulp by modulating the secretion of factors such as TGF-β1 [68,110]. Biodentine applied directly onto pulp, induces reparative dentine formation resulting in complete dentin bridge formation, absence of an inflammatory pulp response and layers of well-arranged odontoblasts and odontoblast-like cells observed after 6 weeks [108]. There is ample evidence for the positive effects of BD on vital pulp cells, for stimulating tertiary dentin formation and for early formation of reparative dentin [111]. Biodentine had a similar effect on dentin bridge formation to MTA (Table 3) [71,112,113]. In contrast to another study [46], a clinical trial in adults showed complete dentin bridge formation in 100% of Biodentine cases, compared to 11% and 56% in TheraCal®and ProRoot®MTA, respectively [114]. In terms of quality of induced dentin, MTA exceeded BD, although concerning completion of the dentin bridge, BD had better results. Therefore, those reports are inconclusive in choosing the most reliable pulp capping material. Nevertheless, in the later stages of differentiation and mineralization of dentin bridge, Biodentine had been observed to have a better performance. That is attributed to a notable increase in alkaline phosphatase expression and calcium nodule formation compared to the MTA characteristics [70]. The antibacterial properties of BD and MTA can refer to the high alkaline pH of these materials, which has an inhibitory effect on the growth of microorganism and causes disinfection of dentin (Table 3) [115].

Due to unsatisfactory aesthetics and physical properties of Biodentine, there is a need for an overlying material with final restoration, usually a composite. Yet, despite the possibility of immediate restoration placing after initial set, there is statistically significant increase in SBS when bonding with Biodentine after a 72-h maturation period [86]. Moreover, Biodentine presents a considerably lower SBS to composite than glassionomer cement (GI) even after a 2-week maturation period (Table 3).

Although Biodentine offers many advantages over MTA, it presented significantly lower SBS values than MTA to restorative materials, including composite (Filtek Z250, 3M, Saint Paul, MN, USA), compomer (Dyract XP®) and RMGI cement (Photac-Fil Quick Aplicap, 3M, Saint Paul, MN, USA) [116]. On the other hand, another study showed clinically acceptable scores and higher SBS of Biodentine compared to MTA when used with the methacrylate-based composite [53]. Inferior results obtained with Biodentine might be explained by the deficient intrinsic maturation, which can take over 2 weeks [87,88]. These results highlight the importance of leaving Biodentine and MTA to mature before the application of the overlying restoration for better clinical outcomes.

3.3. TheraCal LC

TheraCal LC (Bisco, Schaumburg, IL, USA) was introduced in 2011 to overcome poor bonding of CSMs to resins in final restorations. TheraCal LC is a light-cured calcium silicate-based material designed as both a direct and indirect pulp capping material that facilitates the immediate placement of final restoration.

A material's ability to remineralize tooth's structure is associated with the resin formula of TheraCal LC that possesses calcium and hydroxide ion release properties. The bioavailability of calcium ions released form TheraCal LC is proven to be in the concentration range for potential stimulatory activity for dental pulp and odontoblasts, although still significantly lower than in the case of Biodentine [117–119]. The hydration process of Theracal LC was found to be incomplete due to the limited moisture diffusion within the material [118]. Thus, no calcium hydroxide is produced, and less calcium ion leaching is recorded, resulting in an inferior remineralization potential compared to Biodentine. The absence of calcium hydroxide in set TheraCal LC suggests that calcium ions released from this material are not in the hydroxide form. Hence, it can be concluded that the presence of a resin matrix modifies the setting mechanism and calcium ion kinetics of TheraCal LC, resulting in lower calcium-releasing ability (Table 3) [120]. In vitro study reported that the CSMs (Biodentine, ProRoot MTA) induced remineralization of artificially demineralized dentine at a definitely higher speed and intensity than TheraCal LC [121].

The material seals the pulp capping site despite contact with dentinal or pulpal fluids, as its solubility is lower than ProRoot MTA, MTA Angelus and Biodentine (Table 2) [26]. Additionally, dentinal fluids play a crucial role in the release of calcium and hydroxide ions supporting the sealing capacity of the induced apatite. TheraCal LC exhibited superior sealing ability and comparable interfacial microleakage to MTA and Biodentine, showing better performance overall [122].

Lack of cytotoxicity and biocompatibility are the significant factors referring to pulp capping agents which directly affect the clinical outcome. Based on evaluation of pulpal responses in dog partial pulpotomy cases, complete dentin bridge was observed only in 33% of specimens [123]. It was found that TheraCal LC produced the least favorable pulpal responses compared to both ProRoot MTA and RetroMTA. Overall, research reported that TheraCal specimens had lower quality calcific barrier formation, extensive inflammation and less favorable odontoblastic layer formation (Table 3).

Those findings were attributed to the presence of acrylic monomer Bis-GMA in the material. However, it should be noted that Bis-GMA was not detected, despite being listed in the safety data sheet provided by the supplier [124]. Presence of resin in the pulp capping agent which may remain unpolymerized is often associated with adverse pulpal reactions that lead to pulp toxicity and inflammation. The study designed to investigate the consequences of adding resins to tricalcium silicates by comparative analysis showed that TheraCal is toxic to pulp fibroblasts and has a higher inflammatory effect and a lower bioactive potential than Biodentine [125]. Those findings coincide with another study which reported that the reparative capacity of TheraCal LC is inferior to Biodentine [126]. TheraCal LC results showed dentin bridge formation with mild chronic inflammation, reduced dentin bridge thickness and a higher inflammatory score, which may be attributed to the hydration properties of TheraCal LC (Table 3). Moreover, despite that the photopolymerization of TheraCal LC is associated with low heat generation, it could still potentially induce adverse pulpal effects when used in pulp capping procedures [127].

TheraCal LC exhibits higher bond strength values than Biodentine when layered with either composite or glassionomer cement [128,129]. To enhance SBS, E&R adhesives are recommended when placing composite restoration on TheraCal LC (Tables 2 and 3) [81].

Although sufficient bioactivity, superior handling properties and superior quality of bonding with the final overlaying restoration could justify the use of TheraCal LC as the IPC agent, further in vitro and in vivo studies are required. Furthermore, TheraCal cannot be recommended for DPC [125].

3.4. ACTIVA BioACTIVE

ACTIVA BioACTIVE-BASE/LINER (Pulpdent, USA) was launched in 2014 claiming the strength, aesthetics and physical properties of composites and increased release and recharge of calcium, phosphate and fluoride in comparison with glassionomer (GI). On the other hand, it has been proven that the fluoride ion release of ACTIVA is lower than that of the conventional GI (Keta Molar Quick Aplicap, 3M ESPE, Saint Paul, MN, USA) and also than that of RMGI cement (VitremerTM, 3M-ESPE, Saint Paul, MN, USA) [130,131]. The results indicate the ACTIVA BioACTIVE does uptake fluoride and re-release it, which could offer decreased incidence of secondary caries. At this time, the ion-releasing ability of these materials has not been proven to protect from caries at restoration margins.

The company markets this product as a "light-cured resin-modified calcium silicate" (RMCS) combining uncompromised attributes of both composite and GI [132]. Despite claimed bioactivity, the manufacturer recommends the use of ACTIVA BioACTIVE products in cases without pulpal involvement and ACTIVA BioACTIVE-BASE/LINER only in cases of IPC.

Comparing to both MTA and Biodentine, ACTIVA BioACTIVE represents a favorable setting time with no delay placing final restoration (Table 2). The material has three setting mechanisms; it cures with low intensity light for 20 s per layer and has both glass-ionomer (acid-base reaction) and composite self-cure setting reactions (Table 1). The anaerobic, self-cure (best under oxygen barrier; e.g., glycerine) setting-time is 3 min. It is also advisable to allow the material to self-cure for 15–20 s before light curing.

The bioactive properties of ACTIVA BioACTIVE products are based on a mechanism whereby the material responds to pH cycles and plays an active role in releasing and recharging of significant amounts of calcium, phosphate and fluoride [131]. Some authors have suggested that ability to release biologically active ions is more accurately termed "biointeractivity" and is a prerequisite for a material to be bioactive [117]. These mineral components are responsible for stimulating the formation of mineralized hard tissue. As calcium ions play a key role in the material-induced proliferation and differentiation of human dental pulp cells, they also stimulate the formation of a connective apatite layer and seal at the material-tooth interface. ACTIVA exhibited the potential to stimulate biomineralization at the same level as MTA, Biodentine and TheraCal LC on the basis of releasing the same amount of Ca and OH ions (ionic supplemented conditions) (data not shown in Table 3) [133].

Despite the fact that ACTIVA BioACTIVE products are RMGICs, the laboratory and clinical findings indicate that the self-adhesive ability of the material is not elucidated [52]. The aforementioned hypothesis was confirmed in a 1-year clinical follow-up of posterior restorations made with ACTIVA BioACTIVE Restorative, indicating a very high initial failure rate [134]. According to the manufacturer, no additional pretreatment of the capped site is required (Table 3). Studies have reported overall higher microleakage of cavities restored with ACTIVA BioACTIVE if no previous etching was performed nor an adhesive applied, compared to cavities restored with resin composite [135,136]. Similar observations were presented in the study where bond strength measurement of ACTIVA BioACTIVE Restorative after 28 days was not possible due to loss of restorations if no pretreatment was performed or if dentine was etched [52]. Under the conditions of this study, the conclusion can be drawn that self-adhesive property of ACTIVA BioACTIVE products is nonexistent; thus, it exhibits an insufficient protection of pulp in VPT (Table 3).

Considering the material's characteristics, the application of ACTIVA BioACTIVE in direct and indirect pulp capping might be justified. Nevertheless, to this day there is insufficient evidence and there have been too few clinical studies to support its reliability in vital pulp therapies. Moreover, the resin in pulp capping materials such as ACTIVA BioACTIVE BASE/LINER would presumably shift the regeneration-inflammation balance towards the latter. As in the other light-cured tricalcium silicate, incomplete resin photopolymerization may lead to free monomers' release, and consequently to pulpal toxicity [137]. However, to accurately assess ACTIVA BioACTIVE's reparative potential or influence on the vital pulp in pulp capping procedures, further in vitro and in vivo studies are necessary. It can be stated that compared with calcium silicates, resin-containing materials are not fully consistent with the spirit of DPC. On the contrary, calcium silicate materials such as Biodentine and MTA shift the balance towards regeneration, resulting in successful clinical outcomes [138]. However, a recent in vivo study concluded that ACTIVA BioACTIVE BASE/LINER exhibited excellent biocompatibility and healing for rat subcutaneous tissues in comparison with CSMs (MTA-HP and iRoot BP Plus) [139]. Nevertheless, further evidence is needed to substantiate such claims.

4. Conclusions

Development of minimally invasive biologically based therapies aimed at preservation of the pulp vitality remains the key theme within contemporary clinical endodontics. The present findings confirm that both MTA and Biodentine are reliable materials in the matter of inducing dentin bridge formation while keeping a vital pulp in both direct and indirect pulp capping procedures [71,104,108,112,140]. This review also reports Biodentine's superiority in relatively easier manipulation, lower cost and faster setting when compared to MTA with comparable or even outstanding clinical outcomes (Table 2). High biocompatibility and excellent bioactivity further go in favor of this dental replacement material, although more long-term clinical studies are needed for a definitive evaluation of Biodentine as a pulp capping agent.

Future in vitro and in vivo studies are necessary to validate the clinical importance of the new generation of light-cured resin-modified calcium silicates; namely, TheraCal LC and ACTIVA BioACTIVE BASE/LINER. In addition, more studies are needed to support use of these materials in VPT other than indirect pulp capping. RMGIC's superior handling properties, quality of bonding with final overlaying restoration and the possibility of immediate restoration placement, might result in more predictable treatments from both a histological and a clinical perspective. Therefore, those materials should constitute the objects of future studies, especially in terms of cytotoxicity, quality of induced dentin bridge and protocols for higher bond strength to tooth structure and final restoration.

Author Contributions: Conceptualization, M.L.-S. and M.K.; writing—original draft preparation, M.L.-S. and M.K.; writing—review and editing, M.L.-S.; supervision M.L.-S. All authors have read and agreed to the published version of the manuscript.

Funding: This research received no external funding.

Conflicts of Interest: The authors declare no conflict of interest.

References

1. Schwendicke, F.; Frencken, J.E.; Bjørndal, L.; Maltz, M.; Manton, D.J.; Ricketts, D.; Van Landuyt, K.; Banerjee, A.; Campus, G.; Doméjean, S.; et al. Managing Carious Lesions: Consensus Recommendations on Carious Tissue Removal. *Adv. Dent. Res.* **2016**, *28*, 58–67. [CrossRef] [PubMed]
2. Bjørndal, L.; Simon, S.; Tomson, P.L.; Duncan, H.F. Management of deep caries and the exposed pulp. *Int. Endod. J.* **2019**, *52*, 949–973. [CrossRef] [PubMed]
3. Tan, S.Y.; Yu, V.S.H.; Lim, K.C.; Tan, B.C.K.; Neo, C.L.J.; Shen, L.; Messer, H.H. Long-term Pulpal and Restorative Outcomes of Pulpotomy in Mature Permanent Teeth. *J. Endod.* **2020**. [CrossRef] [PubMed]
4. Lin, L.M.; Ricucci, D.; Saoud, T.M.; Sigurdsson, A.; Kahler, B. Vital pulp therapy of mature permanent teeth with irreversible pulpitis from the perspective of pulp biology. *Aust. Endod. J.* **2019**. [CrossRef] [PubMed]
5. Stencel, R.; Kasperski, J.; Pakiela, W.; Mertas, A.; Bobela, E.; Barszczewska-Rybarek, I.; Chladek, G. Properties of experimental dental composites containing antibacterial silver-releasing filler. *Materials* **2018**, *11*, 2173. [CrossRef] [PubMed]
6. Barszczewska-Rybarek, I.; Chladek, G. Studies on the curing efficiency and mechanical properties of bis-GMA and TEGDMA nanocomposites containing silver nanoparticles. *Int. J. Mol. Sci.* **2018**, *19*, 3937. [CrossRef] [PubMed]
7. Hench, L.L. Bioceramics. *J. Am. Ceram. Soc.* **2005**, *81*, 1705–1728. [CrossRef]
8. Jefferies, S.R. Bioactive and Biomimetic Restorative Materials: A Comprehensive Review: Part I. *J. Esthet. Restor. Dent.* **2014**, *26*, 14–26. [CrossRef]
9. Cao, C.Y.; Mei, M.L.; Li, Q.L.; Lo, E.C.M.; Chu, C.H. Methods for biomimetic remineralization of human dentine: A systematic review. *Int. J. Mol. Sci.* **2015**, *16*, 4615–4627. [CrossRef]
10. Pameijer, C.H.; Garcia-Godoy, F.; Morrow, B.R.; Jefferies, S.R. Flexural strength and flexural fatigue properties of resin-modified glass ionomers. *J. Clin. Dent.* **2015**, *26*, 23–27.
11. Kim, J.R.; Nosrat, A.; Fouad, A.F. Interfacial characteristics of Biodentine and MTA with dentine in simulated body fluid. *J. Dent.* **2015**, *43*, 241–247. [CrossRef] [PubMed]
12. Costa, C.A.; Hebling, J.; Hanks, C.T. Current status of pulp capping with dentin adhesive systems: A review. *Dent. Mater.* **2000**, *16*, 188–197. [CrossRef]
13. Duncan, H.F.; Galler, K.M.; Tomson, P.L.; Simon, S.; El-Karim, I.; Kundzina, R.; Krastl, G.; Dammaschke, T.; Fransson, H.; Markvart, M.; et al. European Society of Endodontology position statement: Management of deep caries and the exposed pulp. *Int. Endod. J.* **2019**, *52*, 923–934. [PubMed]
14. Tüzüner, T.; Alacam, A.; Altunbas, D.A.; Gokdogan, F.G.; Gundogdu, E. Clinical and radiographic outcomes of direct pulp capping therapy in primary molar teeth following haemostasis with various antiseptics: A randomised controlled trial. *Eur. J. Paediatr. Dent.* **2012**, *13*, 289–292.
15. Thanatvarakorn, O.; Nakajima, M.; Prasansuttiporn, T.; Ichinose, S.; Foxton, R.M.; Tagami, J. Effect of smear layer deproteinizing on resin-dentine interface with self-etch adhesive. *J. Dent.* **2014**, *42*, 298–304. [CrossRef]
16. Abuhaimed, T.S.; Neel, E.A.A. Sodium Hypochlorite Irrigation and Its Effect on Bond Strength to Dentin. *Biomed. Res. Int.* **2017**, *2017*, 1–8. [CrossRef]
17. Mente, J.; Geletneky, B.; Ohle, M.; Koch, M.J.; Friedrich Ding, P.G.; Wolff, D.; Dreyhaupt, J.; Martin, N.; Staehle, H.J.; Pfefferle, T. Mineral Trioxide Aggregate or Calcium Hydroxide Direct Pulp Capping: An Analysis of the Clinical Treatment Outcome. *J. Endod.* **2010**, *36*, 806–813. [CrossRef]
18. Lapinska, B.; Klimek, L.; Sokolowski, J.; Lukomska-Szymanska, M. Dentine surface morphology after chlorhexidine application-SEM study. *Polymers* **2018**, *10*, 905. [CrossRef]
19. Chailertvanitkul, P.; Paphangkorakit, J.; Sooksantisakoonchai, N.; Pumas, N.; Pairojamornyoot, W.; Leela-apiradee, N.; Abbott, P.V. Randomized control trial comparing calcium hydroxide and mineral trioxide aggregate for partial pulpotomies in cariously exposed pulps of permanent molars. *Int. Endod. J.* **2014**, *47*, 835–842. [CrossRef]
20. Fava, L.R.G.; Saunders, W.P. Calcium hydroxide pastes: Classification and clinical indications. *Int. Endod. J.* **1999**, *32*, 257–282. [CrossRef]
21. Cox, C.F.; Sübay, R.K.; Ostro, E.; Suzuki, S.; Suzuki, S.H. Tunnel defects in dentin bridges: Their formation following direct pulp capping. *Oper. Dent.* **1996**, *21*, 4–11. [PubMed]
22. Hilton, T.J. Keys to Clinical Success with Pulp Capping: A Review of the Literature. *Oper. Dent.* **2009**, *34*, 615–625. [CrossRef] [PubMed]

23. Komabayashi, T.; Zhu, Q.; Eberhart, R.; Imai, Y. Current status of direct pulp-capping materials for permanent teeth. *Dent. Mater. J.* **2016**, *35*, 1–12. [CrossRef] [PubMed]
24. Simon, S.; Smith, A.J.; Lumley, P.J.; Cooper, P.R.; Berdal, A. The pulp healing process: From generation to regeneration. *Endod. Top.* **2012**, *26*, 41–56. [CrossRef]
25. Hargreaves, K.M.; Goodis, H.E.; Seltzer, S. *Seltzer and Bender's Dental Pulp*; Quintessence Pub. Co.: Chicago, IL, USA, 2002; ISBN 9780867154153.
26. Gandolfi, M.G.; Siboni, F.; Botero, T.; Bossù, M.; Riccitiello, F.; Prati, C. Calcium silicate and calcium hydroxide materials for pulp capping: Biointeractivity, porosity, solubility and bioactivity of current formulations. *J. Appl. Biomater. Funct. Mater.* **2015**, *13*, 41–60. [CrossRef]
27. Mickenautsch, S.; Yengopal, V.; Banerjee, A. Pulp response to resin-modified glass ionomer and calcium hydroxide cements in deep cavities: A quantitative systematic review. *Dent. Mater.* **2010**, *26*, 761–770. [CrossRef]
28. Torabinejad, M.; Watson, T.F.; Pitt Ford, T.R. Sealing ability of a mineral trioxide aggregate when used as a root end filling material. *J. Endod.* **1993**, *19*, 591–595. [CrossRef]
29. Torabinejad, M.; Hong, C.; McDonald, F.; Pitt Ford, T. Physical and chemical properties of a new root-end filling material. *J. Endod.* **1995**, *21*, 349–353. [CrossRef]
30. Zhu, C.; Ju, B.; Ni, R. Clinical outcome of direct pulp capping with MTA or calcium hydroxide: A systematic review and meta-analysis. *Int. J. Clin. Exp. Med.* **2015**, *8*, 17055–17060.
31. Hilton, T.J.; Ferracane, J.L.; Mancl, L. Comparison of CaOH with MTA for Direct Pulp Capping: A PBRN Randomized Clinical Trial. *J. Dent. Res.* **2013**, *92*, S16–S22. [CrossRef]
32. Baroudi, K.; Samir, S. Sealing Ability of MTA Used in Perforation Repair of Permanent Teeth; Literature Review. *Open Dent. J.* **2016**, *10*, 278–286. [CrossRef] [PubMed]
33. Mc Cabe, P.S. The clinical applications of mineral trioxide aggregate. *J. Ir. Dent. Assoc.* **2003**, *49*, 123–131. [PubMed]
34. Muhamad, A.; Azzaldeen, A.; Hanali, A. Mineral Trioxide Aggregate (MTA) in apexification. *Endodontology* **2013**, *25*, 97–101.
35. Vijayran, M.; Chaudhary, S.; Manuja, N.; Kulkarni, A.U. Mineral trioxide aggregate (MTA) apexification: A novel approach for traumatised young immature permanent teeth. *BMJ Case Rep.* **2013**. [CrossRef] [PubMed]
36. Camilleri, J. The chemical composition of mineral trioxide aggregate. *J. Conserv. Dent.* **2008**, *11*, 141. [CrossRef] [PubMed]
37. Kaup, M.; Schäfer, E.; Dammaschke, T. An in vitro study of different material properties of Biodentine compared to ProRoot MTA. *Head Face Med.* **2015**, *11*, 16. [CrossRef]
38. Vivan, R.R.; Zapata, R.O.; Zeferino, M.A.; Bramante, C.M.; Bernardineli, N.; Garcia, R.B.; Hungaro Duarte, M.A.; Tanomaru Filho, M.; Gomes De Moraes, I. Evaluation of the physical and chemical properties of two commercial and three experimental root-end filling materials. *Oral Surg. Oral Med. Oral Pathol. Oral Radiol. Endod.* **2010**, *110*, 250–256. [CrossRef]
39. Pornamazeh, T.; Yadegari, Z.; Ghasemi, A.; Sheykh-al-Eslamian, S.M.; Shojaeian, S.H. In Vitro cytotoxicity and setting time assessment of calcium-enriched mixture cement, retro mineral trioxide aggregate and mineral trioxide aggregate. *Iran. Endod. J.* **2017**, *12*, 488–492.
40. Choi, Y.; Park, S.J.; Lee, S.H.; Hwang, Y.C.; Yu, M.K.; Min, K.S. Biological effects and washout resistance of a newly developed fast-setting pozzolan cement. *J. Endod.* **2013**, *39*, 467–472. [CrossRef]
41. Tanomaru-Filho, M.; Morales, V.; da Silva, G.F.; Bosso, R.; Reis, J.M.S.N.; Duarte, M.A.H.; Guerreiro-Tanomaru, J.M. Compressive Strength and Setting Time of MTA and Portland Cement Associated with Different Radiopacifying Agents. *ISRN Dent.* **2012**, *2012*, 1–4. [CrossRef]
42. Bortoluzzi, E.A.; Broon, N.J.; Bramante, C.M.; Felippe, W.T.; Tanomaru Filho, M.; Esberard, R.M. The Influence of Calcium Chloride on the Setting Time, Solubility, Disintegration, and pH of Mineral Trioxide Aggregate and White Portland Cement with a Radiopacifier. *J. Endod.* **2009**, *35*, 550–554. [CrossRef] [PubMed]
43. Grech, L.; Mallia, B.; Camilleri, J. Investigation of the physical properties of tricalcium silicate cement-based root-end filling materials. *Dent. Mater.* **2013**, *29*, e20–e28. [CrossRef] [PubMed]
44. Ha, W.N.; Nicholson, T.; Kahler, B.; Walsh, L.J. Mineral trioxide aggregate-A review of properties and testing methodologies. *Materials* **2017**, *10*, 1261. [CrossRef] [PubMed]
45. De Souza, L.C.; Yadlapati, M.; Dorn, S.O.; Silva, R.; Letra, A. Analysis of radiopacity, pH and cytotoxicity of a new bioceramic material. *J. Appl. Oral Sci.* **2015**, *23*, 383–389. [CrossRef]

46. Galarça, A.D.; Da Rosa, W.L.D.O.; Da Silva, T.M.; Da Silveira Lima, G.; Carreño, N.L.V.; Pereira, T.M.; Aguirre Guedes, O.; Borges, A.H.; Da Silva, A.F.; Piva, E. Physical and Biological Properties of a High-Plasticity Tricalcium Silicate Cement. *Biomed. Res. Int.* **2018**, *10*, 1–6. [CrossRef]
47. Corral, C.; Negrete, P.; Estay, J.; Osorio, S.; Covarrubias, C.; De Oliveira Junior, O.B.; Barud, H. Radiopacity and Chemical Assessment of New Commercial Calcium Silicate-Based Cements. *Int. J. Odontostomatol.* **2018**, *12*, 262–268. [CrossRef]
48. Che, J.-L.; Kim, J.-H.; Kim, S.-M.; Choi, N.; Moon, H.-J.; Hwang, M.-J.; Song, H.-J.; Park, Y.-J. Comparison of Setting Time, Compressive Strength, Solubility, and pH of Four Kinds of MTA. *Korean J. Dent. Mater.* **2016**, *43*, 61–72. [CrossRef]
49. Kaup, M.; Dammann, C.H.; Schäfer, E.; Dammaschke, T. Shear bond strength of Biodentine, ProRoot MTA, glass ionomer cement and composite resin on human dentine ex vivo. *Head Face Med.* **2015**, *11*, 14. [CrossRef]
50. Jantarat, J.; Ritsayam, S.; Banomyong, D.; Chaimanakarn, C. Early and 24-hour shear bond strength to dentine of three calcium silicate based pulp capping materials. *Mah. Dent. J.* **2018**, *38*, 177–183.
51. Hong, K. Kwon Dentin Bonding of TheraCal LC Calcium Silicate Containing an Acidic Monomer: An In Vitro Study. *Materials* **2020**, *13*, 293.
52. Benetti, A.R.; Michou, S.; Larsen, L.; Peutzfeldt, A.; Pallesen, U.; van Dijken, J.W.V. Adhesion and marginal adaptation of a claimed bioactive, restorative material. *Biomater. Investig. Dent.* **2019**, *6*, 90–98. [CrossRef] [PubMed]
53. Cantekin, K.; Avci, S. Evaluation of shear bond strength of two resin-based composites and glass ionomer cement to pure tricalcium silicate-based cement (Biodentine®). *J. Appl. Oral Sci.* **2014**, *22*, 302–306. [CrossRef] [PubMed]
54. Nagi, S.M.; Omar, N.; Salem, H.N.; Aly, Y. Effect of different surface treatment protocols on the shear bond strength of perforation repair materials to resin composite. *J. Adhes. Sci. Technol.* **2020**, *34*, 417–426. [CrossRef]
55. Shin, H.; Kim, M.; Nam, O.; Lee, H.; Choi, S.; Kim, K. Shear Bond Strength Comparison of Different Adhesive Systems to Calcium Silicate-based Materials. *J. Korean Acad. Pediatr. Dent.* **2018**, *45*, 445–454. [CrossRef]
56. Cantekin, K. Bond strength of different restorative materials to light-curable mineral trioxide aggregate. *J. Clin. Pediatr. Dent.* **2015**, *39*, 143–148. [CrossRef] [PubMed]
57. Daniele, L. Mineral Trioxide Aggregate (MTA) direct pulp capping: 10 years clinical results. *G. Ital. Endod.* **2017**, *31*, 48–57. [CrossRef]
58. Bogen, G.; Kim, J.S.; Bakland, L.K. Direct pulp capping with mineral trioxide aggregate: An observational study. *J. Am. Dent. Assoc.* **2008**, *139*, 305–315. [CrossRef]
59. Sarkar, N.; Caicedo, R.; Ritwik, P. Physicochemical Basis of the Biologic Properties of Mineral Trioxide Aggregate. *J. Endod.* **2005**, *31*, 97–100. [CrossRef]
60. Parirokh, M.; Torabinejad, M. Mineral Trioxide Aggregate: A Comprehensive Literature Review—Part III: Clinical Applications, Drawbacks, and Mechanism of Action. *J. Endod.* **2010**, *36*, 400–413. [CrossRef]
61. Antunes Bortoluzzi, E.; Sivieri Araújo, G.; Maria Guerreiro Tanomaru, J.; Tanomaru-Filho, M. Marginal Gingiva Discoloration by Gray MTA: A Case Report. *J. Endod.* **2007**, *33*, 325–327. [CrossRef]
62. Asgary, S.; Parirokh, M.; Eghbal, M.; Brink, F. Chemical Differences between White and Gray Mineral Trioxide Aggregate. *J. Endod.* **2005**, *31*, 101–103. [CrossRef] [PubMed]
63. Felman, D.; Parashos, P. Coronal tooth discoloration and white mineral trioxide aggregate. *J. Endod.* **2013**, *39*, 484–487. [CrossRef] [PubMed]
64. Islam, I.; Kheng Chng, H.; Jin Yap, A.U. Comparison of the physical and mechanical properties of MTA and portland cement. *J. Endod.* **2006**, *32*, 193–197. [CrossRef] [PubMed]
65. Salem-Milani, A.; Ghasemi, S.; Rahimi, S.; Ardalan-Abdollahi, A.; Asghari-Jafarabadi, M. The Discoloration effect of White Mineral Trioxide Aggregate (WMTA), Calcium Enriched Mixture (CEM), and Portland Cement (PC) on Human Teeth. *J. Clin. Exp. Dent.* **2017**, *9*, e1397–e1401. [CrossRef]
66. Schembri, M.; Peplow, G.; Camilleri, J. Analyses of heavy metals in mineral trioxide aggregate and Portland cement. *J. Endod.* **2010**, *36*, 1210–1215. [CrossRef]
67. Mente, J.; Hufnagel, S.; Leo, M.; Michel, A.; Gehrig, H.; Panagidis, D.; Saure, D.; Pfefferle, T. Treatment Outcome of Mineral Trioxide Aggregate or Calcium Hydroxide Direct Pulp Capping: Long-term Results. *J. Endod.* **2014**, *40*, 1746–1751. [CrossRef]
68. Laurent, P.; Camps, J.; About, I. BiodentineTM induces TGF-β1 release from human pulp cells and early dental pulp mineralization. *Int. Endod. J.* **2012**, *45*, 439–448. [CrossRef]

69. Dammaschke, T.; Nowicka, A.; Lipski, M.; Ricucci, D. Histological evaluation of hard tissue formation after direct pulp capping with a fast-setting mineral trioxide aggregate (RetroMTA) in humans. *Clin. Oral Investig.* **2019**, *23*, 4289–4299. [CrossRef]
70. Paula, A.; Laranjo, M.; Marto, C.M.; Abrantes, A.M.; Casalta-Lopes, J.; Gonçalves, A.C.; Sarmento-Ribeiro, A.B.; Ferreira, M.M.; Botelho, M.F.; Carrilho, E. Biodentine™ Boosts, WhiteProRoot® MTA Increases and Life® Suppresses Odontoblast Activity. *Materials* **2019**, *12*, 1184. [CrossRef]
71. Kim, J.; Song, Y.-S.; Min, K.-S.; Kim, S.-H.; Koh, J.-T.; Lee, B.-N.; Chang, H.-S.; Hwang, I.-N.; Oh, W.-M.; Hwang, Y.-C. Evaluation of reparative dentin formation of ProRoot MTA, Biodentine and BioAggregate using micro-CT and immunohistochemistry. *Restor. Dent. Endod.* **2016**, *41*, 29. [CrossRef]
72. Moussa, S.A. Mineral Trioxide Aggregate (MTA) vs. Calcium Hydroxide in Direct Pulp Capping—Literature Review. *Online J. Dent. Oral Heal.* **2018**, *1*, 1–6. [CrossRef]
73. Poggio, C.; Lombardini, M.; Colombo, M.; Beltrami, R.; Rindi, S. Solubility and pH of direct pulp capping materials: A comparative study. *J. Appl. Biomater. Funct. Mater.* **2015**, *13*, e181–e185. [CrossRef] [PubMed]
74. Akhavan Zanjani, V.; Tabari, K.; Sheikh-Al-Eslamian, S.M.; Abrandabadi, A.N. Physiochemical properties of experimental nano-hybrid MTA. *J. Med. Life* **2017**, *10*, 182–187. [PubMed]
75. Zeid, S.T.H.A.; Alothmani, O.S.; Yousef, M.K. Biodentine and Mineral Trioxide Aggregate: An Analysis of Solubility, pH Changes and Leaching Elements. *Life Sci. J.* **2015**, *12*, 18–23.
76. Yelamali, S.; Patil, A.C. Evaluation of shear bond strength of a composite resin to white mineral trioxide aggregate with three different bonding systems-An in vitro analysis. *J. Clin. Exp. Dent.* **2016**, *8*, e273–e277. [PubMed]
77. Tyagi, N.; Chaman, C.; Tyagi, S.P.; Singh, U.P.; Sharma, A. The shear bond strength of MTA with three different types of adhesive systems: An in vitro study. *J. Conserv. Dent.* **2016**, *19*, 130–133. [PubMed]
78. Sulwińska, M.; Szczesio, A.; Bołtacz-Rzepkowska, E. Bond strength of a resin composite to MTA at various time intervals and with different adhesive strategies. *Dent. Med. Probl.* **2017**, *54*, 155–160. [CrossRef]
79. Krawczyk-Stuss, M.; Nowak, J.; Bołtacz-Rzepkowska, E. Bond strength of Biodentine to a resin-based composite at various acid etching times and with different adhesive strategies. *Dent. Med. Probl.* **2019**, *56*, 39–44. [CrossRef]
80. Odabaş, M.E.; Bani, M.; Tirali, R.E. Shear Bond Strengths of Different Adhesive Systems to Biodentine. *Sci. World J.* **2013**, *2013*, 1–5. [CrossRef]
81. Karadas, M.; Cantekin, K.; Gumus, H.; Ateş, S.M.; Duymuş, Z.Y. Evaluation of the bond strength of different adhesive agents to a resin-modified calcium silicate material (TheraCal LC). *Scanning* **2016**, *38*, 403–411. [CrossRef]
82. Samimi, P.; Kazemian, M.; Shirban, F.; Alaei, S.; Khoroushi, M. Bond strength of composite resin to white mineral trioxide aggregate: Effect of different surface treatments. *J. Conserv. Dent.* **2018**, *21*, 350. [PubMed]
83. Guo, Y.J.; Du, T.F.; Li, H.B.; Shen, Y.; Mobuchon, C.; Hieawy, A.; Wang, Z.J.; Yang, Y.; Ma, J.; Haapasalo, M. Physical properties and hydration behavior of a fast-setting bioceramic endodontic material. *BMC Oral Health* **2016**, *16*, 1–6. [CrossRef] [PubMed]
84. Nekoofar, M.H.; Aseeley, Z.; Dummer, P.M.H. The effect of various mixing techniques on the surface microhardness of mineral trioxide aggregate. *Int. Endod. J.* **2010**, *43*, 312–320. [CrossRef] [PubMed]
85. Chedella, S.C.V.; Berzins, D.W. A differential scanning calorimetry study of the setting reaction of MTA. *Int. Endod. J.* **2010**, *43*, 509–518. [CrossRef]
86. Ha, H.-T. The effect of the maturation time of calcium silicate-based cement (Biodentine™) on resin bonding: An in vitro study. *Appl. Adhes. Sci.* **2019**, *7*, 1. [CrossRef]
87. Hashem, D.F.; Foxton, R.; Manoharan, A.; Watson, T.F.; Banerjee, A. The physical characteristics of resin composite–calcium silicate interface as part of a layered/laminate adhesive restoration. *Dent. Mater.* **2014**, *30*, 343–349. [CrossRef]
88. Sultana, N.; Nawal, R.; Chaudhry, S.; Sivakumar, M.; Talwar, S. Effect of acid etching on the micro-shear bond strength of resin composite–calcium silicate interface evaluated over different time intervals of bond aging. *J. Conserv. Dent.* **2018**, *21*, 194–197.
89. Siboni, F.; Taddei, P.; Prati, C.; Gandolfi, M.G. Properties of NeoMTA plus and MTA plus cements for endodontics. *Int. Endod. J.* **2017**, *50*, e83–e94. [CrossRef]
90. Zeid, S.T.A.; Alamoudi, N.M.; Khafagi, M.G.; Abou Neel, E.A. Chemistry and Bioactivity of NeoMTA Plus™ versus MTA Angelus® Root Repair Materials. *J. Spectrosc.* **2017**, *2017*, 8736428.

91. Tom As-Catal, C.J.; Collado-Gonz Alez, M.; García-Bernal, D.; Oñate-Sánchez, R.E.; Forner, L.; Llena, C.; an Lozano, A.; Moraleda, J.M.; Rodr ıguez-Lozano, F.J. Biocompatibility of New Pulp-capping Materials NeoMTA Plus, MTA Repair HP, and Biodentine on Human Dental Pulp Stem Cells. *J. Endod.* **2018**, *44*, 126–132. [CrossRef]
92. Camilleri, J. Staining Potential of Neo MTA Plus, MTA Plus, and Biodentine Used for Pulpotomy Procedures. *J. Endod.* **2015**, *41*, 1139–1145. [CrossRef] [PubMed]
93. Mahgoub, N.; Alqadasi, B.; Aldhorae, K.; Assiry, A.; Altawili, Z.; Hong, T. Comparison between iRoot BP Plus (EndoSequence Root Repair Material) and Mineral Trioxide Aggregate as Pulp-capping Agents: A Systematic Review. *J. Int. Soc. Prev. Community Dent.* **2019**, *9*, 542–552. [PubMed]
94. Rao, Q.; Kuang, J.; Mao, C.; Dai, J.; Hu, L.; Lei, Z.; Song, G.; Yuan, G. Comparison of iRoot BP Plus and Calcium Hydroxide as Pulpotomy Materials in Permanent Incisors with Complicated Crown Fractures: A Retrospective Study. *J. Endod.* **2020**. [CrossRef] [PubMed]
95. Fonseca, T.S.; Silva, G.F.; Jm, G.; In, C.P.S. In Vivo evaluation of the inflammatory response and IL-6 immunoexpression promoted by Biodentine and MTA Angelus. *Int. Endod. J.* **2015**, *49*, 1–9.
96. Septodont Biodentine Active Biosilicate Technology Scientific File. In Vitro 2010. Available online: http://www.oraverse.com/bio/img/Biodentine-ScientificFile.pdf (accessed on 20 December 2019).
97. Rajasekharan, S.; Martens, L.C.; Cauwels, R.G.E.C.; Verbeeck, R.M.H. Biodentine^{TM} material characteristics and clinical applications: A review of the literature. *Eur. Arch. Paediatr. Dent.* **2014**, *15*, 147–158. [CrossRef] [PubMed]
98. Tanalp, J.; Karapınar-Kazandağ, M.; Dölekoğlu, S.; Kayahan, M.B. Comparison of the radiopacities of different root-end filling and repair materials. *Sci. World J.* **2013**, *2013*, 594950. [CrossRef] [PubMed]
99. Aksoy, M.K.; Oz, F.T.; Orhan, K. Evaluation of calcium (Ca^{2+}) and hydroxide (OH^-) ion diffusion rates of indirect pulp capping materials. *Int. J. Artif. Organs* **2017**, *40*, 641–646. [CrossRef]
100. About, I. Biodentine: From biochemical and bioactive properties to clinical applications. *G. Ital. Endod.* **2016**, *30*, 81–88. [CrossRef]
101. Mousavi, S.A.; Khademi, A.; Soltani, P.; Shahnaseri, S.; Poorghorban, M. Comparison of sealing ability of ProRoot mineral trioxide aggregate, biodentine, and ortho mineral trioxide aggregate for canal obturation by the fluid infiltration technique. *Dent. Res. J.* **2018**, *15*, 307–312.
102. Sinkar, R.; Patil, S.; Jogad, N.; Gade, V. Comparison of sealing ability of ProRoot MTA, RetroMTA, and Biodentine as furcation repair materials: An ultraviolet spectrophotometric analysis. *J. Conserv. Dent.* **2015**, *18*, 445. [CrossRef]
103. Nifla, F.; Suvarna, N.; KShetty, H.; MoosaKutty, S. Sealing ability of MTA Angelus, Biodentine, Geristore-an ultraviolet spectrophotometric analysis. *Int. J. Adv. Res.* **2019**, *7*, 1085–1090. [CrossRef]
104. Hegde, S.; Sowmya, B.; Mathew, S.; Bhandi, S.H.; Nagaraja, S.; Dinesh, K. Clinical evaluation of mineral trioxide aggregate and biodentine as direct pulp capping agents in carious teeth. *J. Conserv. Dent.* **2017**, *20*, 91–95. [PubMed]
105. Awawdeh, L.; Al-Qudah, A.; Hamouri, H.; Chakra, R.J. Outcomes of Vital Pulp Therapy Using Mineral Trioxide Aggregate or Biodentine: A Prospective Randomized Clinical Trial. *J. Endod.* **2018**, *44*, 1603–1609. [CrossRef] [PubMed]
106. Brizuela, C.; Ormeño, A.; Cabrera, C.; Cabezas, R.; Silva, C.I.; Ramírez, V.; Mercade, M. Direct Pulp Capping with Calcium Hydroxide, Mineral Trioxide Aggregate, and Biodentine in Permanent Young Teeth with Caries: A Randomized Clinical Trial. *J. Endod.* **2017**, *43*, 1776–1780. [CrossRef]
107. Katge, F.A.; Patil, D.P. Comparative Analysis of 2 Calcium Silicate—Based Cements (Biodentine and Mineral Trioxide Aggregate) as Direct Pulp-capping Agent in Young Permanent Molars: A Split Mouth Study. *J. Endod.* **2017**, *43*, 507–513. [CrossRef]
108. Nowicka, A.; Lipski, M.; Parafiniuk, M.; Sporniak-Tutak, K.; Lichota, D.; Kosierkiewicz, A.; Kaczmarek, W.; Buczkowska-Radlińska, J. Response of human dental pulp capped with biodentine and mineral trioxide aggregate. *J. Endod.* **2013**, *39*, 743–747. [CrossRef]
109. Lipski, M.; Nowicka, A.; Kot, K.; Postek-Stefańska, L.; Wysoczańska-Jankowicz, I.; Borkowski, L.; Andersz, P.; Jarząbek, A.; Grocholewicz, K.; Sobolewska, E.; et al. Factors affecting the outcomes of direct pulp capping using Biodentine. *Clin. Oral Investig.* **2018**, *22*, 2021–2029. [CrossRef]
110. About, I. Dentin-pulp regeneration: The primordial role of the microenvironment and its modification by traumatic injuries and bioactive materials. *Endod. Top.* **2013**, *28*, 61–89. [CrossRef]

111. Zanini, M.; Sautier, J.M.; Berdal, A.; Simon, S. Biodentine induces immortalized murine pulp cell differentiation into odontoblast-like cells and stimulates biomineralization. *J. Endod.* **2012**, *38*, 1220–1226. [CrossRef]
112. Mahmoud, S.; El-Negoly, S.; Zaen El-Din, A.; El-Zekrid, M.; Grawish, L.; Grawish, H.; Grawish, M. Biodentine versus mineral trioxide aggregate as a direct pulp capping material for human mature permanent teeth—A systematic review. *J. Conserv. Dent.* **2018**, *21*, 466.
113. Tran, X.V.; Gorin, C.; Willig, C.; Baroukh, B.; Pellat, B.; Decup, F.; Opsahl Vital, S.; Chaussain, C.; Boukpessi, T. Effect of a calcium-silicate-based restorative cement on pulp repair. *J. Dent. Res.* **2012**, *91*, 1166–1171. [CrossRef] [PubMed]
114. Bakhtiar, H.; Nekoofar, M.H.; Aminishakib, P.; Abedi, F.; Naghi Moosavi, F.; Esnaashari, E.; Azizi, A.; Esmailian, S.; Ellini, M.R.; Mesgarzadeh, V.; et al. Human Pulp Responses to Partial Pulpotomy Treatment with TheraCal as Compared with Biodentine and ProRoot MTA: A Clinical Trial. *J. Endod.* **2017**, *43*, 1786–1791. [CrossRef] [PubMed]
115. Kaur, M.; Singh, H.; Dhillon, J.S.; Batra, M.; Saini, M. MTA versus Biodentine: Review of Literature with a Comparative Analysis. *J. Clin. Diagn. Res.* **2017**, *11*, ZG01–ZG05. [CrossRef] [PubMed]
116. Tulumbaci, F.; Almaz, M.E.; Arikan, V.; Mutluay, M.S. Shear bond strength of different restorative materials to mineral trioxide aggregate and Biodentine. *J. Conserv. Dent.* **2017**, *20*, 292–296. [CrossRef] [PubMed]
117. Gandolfi, M.G.; Siboni, F.; Prati, C. Chemical-physical properties of TheraCal, a novel light-curable MTA-like material for pulp capping. *Int. Endod. J.* **2012**, *45*, 571–579. [CrossRef]
118. Camilleri, J.; Laurent, P.; About, I. Hydration of Biodentine, Theracal LC, and a Prototype Tricalcium Silicate–based Dentin Replacement Material after Pulp Capping in Entire Tooth Cultures. *J. Endod.* **2014**, *40*, 1846–1854. [CrossRef]
119. Fathy, S. Remineralization ability of two hydraulic calcium-silicate based dental pulp capping materials: Cell-independent model. *J. Clin. Exp. Dent.* **2019**, *11*, e360–e366. [CrossRef]
120. Camilleri, J. Hydration characteristics of Biodentine and Theracal used as pulp capping materials. *Dent. Mater.* **2014**, *30*, 709–715. [CrossRef]
121. Li, X.; De Munck, J.; Van Landuyt, K.; Pedano, M.; Chen, Z.; Van Meerbeek, B. How effectively do hydraulic calcium-silicate cements re-mineralize demineralized dentin. *Dent. Mater.* **2017**, *33*, 434–445. [CrossRef]
122. Makkar, S.; Kaur, H.; Aggarwal, A.; Vashish, R. A confocal laser scanning microscopic study evaluating the sealing ability of mineral trioxide aggregate, Biodentine and new pulp capping agent—TheraCal. *Dent. J. Adv. Stud.* **2015**, *3*, 20–25.
123. Lee, H.; Shin, Y.; Kim, S.-O.; Lee, H.-S.; Choi, H.-J.; Song, J.S. Comparative Study of Pulpal Responses to Pulpotomy with ProRoot MTA, RetroMTA, and TheraCal in Dogs' Teeth. *J. Endod.* **2015**, *41*, 1317–1324. [CrossRef] [PubMed]
124. Nilsen, B.W.; Jensen, E.; Örtengren, U.; Michelsen, V.B. Analysis of organic components in resin-modified pulp capping materials: Critical considerations. *Eur. J. Oral Sci.* **2017**, *125*, 183–194. [CrossRef] [PubMed]
125. Jeanneau, C.; Laurent, P.; Rombouts, C.; Giraud, T.; About, I. Light-cured Tricalcium Silicate Toxicity to the Dental Pulp. *J. Endod.* **2017**, *43*, 2074–2080. [CrossRef] [PubMed]
126. Kamal, E.; Nabih, S.; Obeid, R.; Abdelhameed, M. The reparative capacity of different bioactive dental materials for direct pulp capping. *Dent. Med. Probl.* **2018**, *55*, 147–152. [CrossRef] [PubMed]
127. Savas, S.; Botsali, M.S.; Kucukyilmaz, E.; Sari, T. Evaluation of temperature changes in the pulp chamber during polymerization of light-cured pulp-capping materials by using a VALO LED light curing unit at different curing distances. *Dent. Mater. J.* **2014**, *33*, 764–769. [CrossRef]
128. Meraji, N.; Camilleri, J. Bonding over Dentin Replacement Materials. *J. Endod.* **2017**, *43*, 1343–1349. [CrossRef]
129. Deepa, V.; Dhamaraju, B.; Bollu, I.; Balaji, T. Shear bond strength evaluation of resin composite bonded to three different liners: TheraCal LC, Biodentine, and resin-modified glass ionomer cement using universal adhesive: An in vitro study. *J. Conserv. Dent.* **2016**, *19*, 166. [CrossRef]
130. Porenczuk, A.; Jankiewicz, B.; Naurecka, M.; Bartosewicz, B.; Sierakowski, B.; Gozdowski, D.; Kostecki, J.; Nasiłowska, B.; Mielczarek, A. A comparison of the remineralizing potential of dental restorative materials by analyzing their fluoride release profiles. *Adv. Clin. Exp. Med.* **2019**, *28*, 815–823. [CrossRef]
131. May, E.; Donly, K.J. Fluoride release and re-release from a bioactive restorative material. *Am. J. Dent.* **2017**, *30*, 305–308.
132. Pulpdent Activa BioActive White Paper. Available online: https://secureservercdn.net/198.71.233.195/91d.e9f.myftpupload.com/wp-content/uploads/2019/12/XF-VWP8-REV10.19.pdf (accessed on 20 December 2019).

133. Jun, S.-K.; Lee, J.-H.; Lee, H.-H. The Biomineralization of a Bioactive Glass-Incorporated Light-Curable Pulp Capping Material Using Human Dental Pulp Stem Cells. *Biomed. Res. Int.* **2017**, *2017*, 2495282. [CrossRef]
134. Van Dijken, J.W.V.; Pallesen, U.; Benetti, A. A randomized controlled evaluation of posterior resin restorations of an altered resin modified glass-ionomer cement with claimed bioactivity. *Dent. Mater.* **2019**, *35*, 335–343. [CrossRef] [PubMed]
135. Kaushik, M.; Yadav, M. Marginal microleakage properties of Activa BioActive Restorative and nanohybrid composite resin using two different adhesives in non carious cervical lesions—An In Vitro study. *J. West. Afr. Coll. Surg.* **2017**, *7*, 1–14. [PubMed]
136. Omidi, B.R.; Naeini, F.F.; Dehghan, H.; Tamiz, P.; Savadroodbari, M.M.; Jabbarian, R. Microleakage of an Enhanced Resin-Modified Glass Ionomer Restorative Material in Primary Molars. *J. Dent.* **2018**, *15*, 205–213.
137. Giraud, T.; Jeanneau, C.; Bergmann, M.; Laurent, P.; About, I. Tricalcium Silicate Capping Materials Modulate Pulp Healing and Inflammatory Activity In Vitro. *J. Endod.* **2018**, *44*, 1686–1691. [CrossRef]
138. Giraud, T.; Jeanneau, C.; Rombouts, C.; Bakhtiar, H.; Laurent, P.; About, I. Pulp capping materials modulate the balance between inflammation and regeneration. *Dent. Mater.* **2019**, *35*, 24–35. [CrossRef]
139. Abou ElReash, A.; Hamama, H.; Abdo, W.; Wu, Q.; Zaen El-Din, A.; Xiaoli, X. Biocompatibility of new bioactive resin composite versus calcium silicate cements: An animal study. *BMC Oral Health* **2019**, *19*, 194. [CrossRef]
140. Nowicka, A.; Wilk, G.; Lipski, M.; Kołecki, J.; Buczkowska-Radlińska, J. Tomographic Evaluation of Reparative Dentin Formation after Direct Pulp Capping with Ca(OH)2, MTA, Biodentine, and Dentin Bonding System in Human Teeth. *J. Endod.* **2015**, *41*, 1234–1240. [CrossRef]

© 2020 by the authors. Licensee MDPI, Basel, Switzerland. This article is an open access article distributed under the terms and conditions of the Creative Commons Attribution (CC BY) license (http://creativecommons.org/licenses/by/4.0/).

Article

Characterization of the Mechanical Properties, Water Sorption, and Solubility of Antibacterial Copolymers of Quaternary Ammonium Urethane-Dimethacrylates and Triethylene Glycol Dimethacrylate

Marta W. Chrószcz-Porębska [1], Izabela M. Barszczewska-Rybarek [1] and Grzegorz Chladek [2,*]

[1] Department of Physical Chemistry and Technology of Polymers, Faculty of Chemistry, Silesian University of Technology, Strzody 9 Str., 44-100 Gliwice, Poland
[2] Department of Engineering Materials and Biomaterials, Faculty of Mechanical Engineering, Silesian University of Technology, 18a Konarskiego Str., 41-100 Gliwice, Poland
* Correspondence: grzegorz.chladek@polsl.pl

Abstract: The use of dental composites based on dimethacrylates that have quaternary ammonium groups is a promising solution in the field of antibacterial restorative materials. This study aimed to investigate the mechanical properties and behaviors in aqueous environments of a series of six copolymers (QA:TEG) comprising 60 wt.% quaternary ammonium urethane-dimethacrylate (QAUDMA) and 40 wt.% triethylene glycol dimethacrylate (TEGDMA); these copolymers are analogous to a common dental copolymer (BG:TEG), which comprises 60 wt.% bisphenol A glycerolate dimethacrylate (Bis-GMA) and 40 wt.% TEGDMA. Hardness (HB), flexural strength (FS), flexural modulus (E), water sorption (WS), and water solubility (SL) were assessed for this purpose. The pilot study of these copolymers showed that they have high antibacterial activity and good physicochemical properties. This paper revealed that QA:TEGs cannot replace BG:TEG due to their insufficient mechanical properties and poor behavior in water. However, the results can help to explain how QAUDMA-based materials work, and how their composition should be manipulated to produce the best performance. It was found that the longer the N-alkyl chain, the lower the HB, WS, and SL. The FS and E increased with the lengthening of the N-alkyl chain from eight to ten carbon atoms. Its further extension, to eighteen carbon atoms, caused a decrease in those parameters.

Keywords: dimethacrylate copolymers; quaternary ammonium methacrylates; urethane-dimethacrylates; photocurable copolymers; mechanical properties; water behavior

1. Introduction

In recent years, a significant increase in the intensity of research into the development of dental restorative materials with antibacterial properties has been observed [1,2]. This results from the fact that teeth and periodontal diseases have become a global problem in the 21st century, and new steps must be taken to keep this issue under control [3–5]. Currently, the most commonly used dental restorative materials are dimethacrylate composites that consist of bisphenol A glycerolate dimethacrylate (Bis-GMA) and its derivatives: urethane-dimethacrylate monomer (UDMA), and triethylene glycol dimethacrylate (TEGDMA) [6]. Their popularity is due to their excellent functional properties, esthetics, and low price. However, they do not protect against secondary caries and dental inflammations, because they have negligible antibacterial activity [7,8]. Their modification with bioactive compounds is perhaps the most reasonable means of giving them antibacterial properties. This can be performed in two ways. The first approach uses the admixing of particles of inorganic or organic compounds that have antibacterial effects [9,10]. The main advantage of this approach is its low price. However, such free particles can leach from the composite, as no covalent bonds exist between them and the matrix. This results in a shorter restoration lifetime and increases composite cytotoxicity. The second approach uses the copolymerization of common dimethacrylates with methacrylate monomers containing quaternary ammonium (QA) groups (QAMs) [11–13]. As these quaternary ammonium groups are positively charged, they interact with negatively charged bacteria cell walls. This causes an electric imbalance inside the bacteria cell, the leaching of cytoplasmic components that are essential for its proper functioning, an increase in the osmotic pressure inside the cell, and, finally, cell lysis, when the existing risk of cytotoxicity for mammalian cell lines is intracellular in origin. Abnormalities in or damage to intracellular biochemical processes, such as intracellular oxidative stress, oxidative DNA damage, and the induction of intrinsic mitochondrial apoptosis, are involved in lieu of membrane disintegration or cell lysis [14]. Covalent bonding between QAM and other dimethacrylates results in better and more stable physicochemical, mechanical, and biocidal properties of chemically modified composites, compared to physically modified composites [15].

First, monomethacrylates with quaternary ammonium groups (mono-QAMs) were produced [16–22]. Composites enriched with their presence showed high antibacterial activity against many strains of bacteria, including *Streptococcus mutans*, *Actinomyces viscosus*, and *Lactobacillus casei* [17–22]. However, it was found that the mono-QAM-repeating units reduced crosslink density in the composite matrix, which caused the deterioration of its mechanical properties, and led to an increase in water sorption and solubility [19,20,23].

Then, dimethacrylates with quaternary ammonium groups (di-QAM) were developed [23–35]. They did not decrease copolymer crosslink density, because they have two methacrylate groups [23]. A series of composite materials enriched with their presence showed high antibacterial activity against many strains of bacteria, such as *S. mutans* [23–25,28–30,32,35], *A. viscosus* [24], *Escherichia coli* [25,27], *Lactobacillus acidophilus* [24], *Streptococcus sanguinis* [24], *Porphyromonas gingivalis* [24], *Prevotella melaninogenica* [24], *Enterococcus faecalis* [24], *Pseudomonas aeruginosa* [25], *Staphylococcus aureus* [24,25,27], and *Bacillus subtillis* [25]. Additionally, the results of these studies showed that the antibacterial activity of composites modified with di-QAMs depended on the number of QA groups in the di-QAM molecule. The more QA groups, the higher the antibacterial activity. Therefore, di-QAM monomers with two QA groups may offer a more promising alternative to those with one QA group, because they achieve an adequate antibacterial effect in the composite at a lower concentration [35]. However, the mechanical and physicochemical properties of copolymers or composites modified with di-QAMs have rarely been examined.

In our previous study [26], we described the synthesis of six novel quaternary ammonium urethane-dimethacrylates (QAUDMAs): namely, the UDMA analogues. They consisted of the trimethylhexamethylene diisocyanate (TMDI) core and two wings. Each wing was terminated with the methacrylate group and contained one quaternary ammonium group substituted with the N-alkyl chain of eight (C8), ten (C10), twelve (C12), fourteen (C14), sixteen (C16), and eighteen (C18) carbon atoms (Figure 1). Novel QAUDMAs show an adequate refractive index, glass transition temperature, and density; however, due to their high viscosity, it is necessary to use a reactive diluent with these monomers for dental applications. Therefore, in next stage, a pilot study on the characterization of the QAUDMA-based polymers was performed for copolymers comprising 60 wt.% QAUDMA and 40 wt.% TEGDMA (Figure 1) (QA:TEGs) [27]. The results of that study showed that QA:TEGs were characterized by a high degree of conversion (DC), a high glass transition temperature (T_{gp}), and low polymerization shrinkage (S); they also showed high antibacterial activity against *S. aureus* and *E. coli*, which justified the next phase of the research. The goal of the current investigation was to enhance our knowledge of the properties of QA:TEGs, focusing on their mechanical properties and behavior in the aqueous environment. To this end, hardness, flexural strength, the flexural modulus, water sorption, and water solubility were determined for six QA:TEGs formulations. This type of research is not widely available for QAM-based polymeric materials. The results of this study provide conclusions about the influence of the QAUDMA chemical structure, and in particular the length of the N-alkyl nitrogen substituent, on the physical and mechanical characteristics of their copolymers. Such knowledge has not previously been attained, and it is essential for understanding how a dimethacrylate copolymer containing QAUDMA repeating units works, and how it should be designed to result in the best performance.

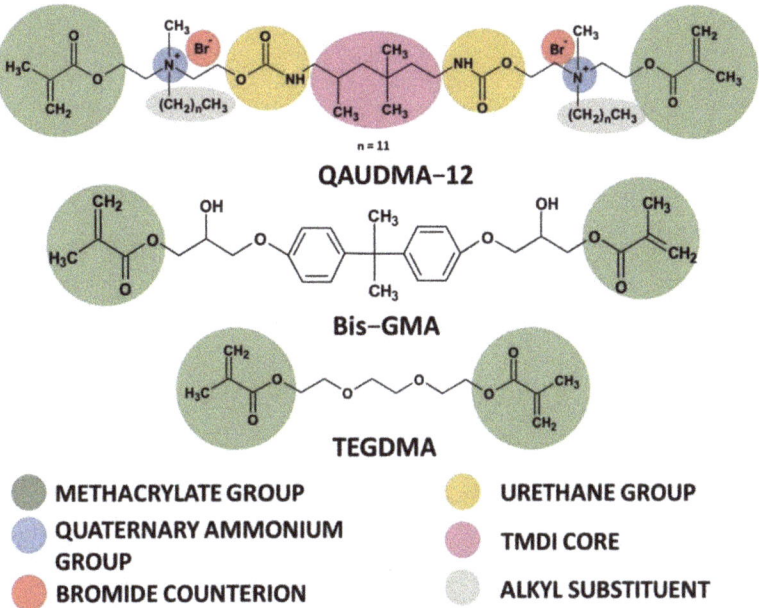

Figure 1. The chemical structure of the QAUDMA, Bis-GMA, and TEGDMA monomers used in this study.

2. Materials and Methods

2.1. Chemicals and Reagents

Alkyl bromides, N-methyldiethanolamine (MDEA), and methyl methacrylate (MMA) were purchased from Acros Organics (Geel, Belgium). Bisphenol A glycerolate dimethacrylate (Bis-GMA), camphorquinone (CQ), dibutyltin dilaurate (DBTDL), 2-dimethylaminoethyl methacrylate (DMAEMA), phenothiazine (PTZ), triethylene glycol dimethacrylate (TEGDMA), and tetramethylsilane (TMS) were purchased from Sigma-Aldrich (St. Louis, MO, USA). Chloroform, methylene chloride, potassium carbonate, and toluene were purchased from POCH S.A. (Gliwice, Poland). Trimethylhexamethylene diisocyanate (TMDI) was purchased from Tokyo Chemical Industry (Tokyo, Japan). All reagents were used as received.

2.2. Monomer Synthesis

QAUDMAs were synthesized in a three-stage process as described in our previous work [26]. First, MMA was transesterificated with MDEA in the presence of a reaction catalyst (K_2CO_3), a polymerization inhibitor (PTZ), and toluene. The product was isolated from the reaction mixture by washing it with distilled water and chloroform. The crude product was vacuum distilled. It was then N-alkylated with alkyl bromides with alkyl chains of 8, 10, 12, 14, 16, and 18 carbon atoms. The quaternized products were subjected to addition with TMDI in the presence of the reaction catalyst (DBTDL), the polymerization inhibitor (PTZ), and methylene chloride. QAUDMAs were isolated from the reaction mixture by evaporating the solvent. The NMR and FT-IR spectra of QAUDMAs are presented in [26].

2.3. Photopolymerization

Six 60 wt.% QAUDMA and 40 wt.% TEGDMA compositions, and one 60 wt.% Bis-GMA and 40 wt.% TEGDMA composition serving as a reference sample (Table 1), were photopolymerized in the presence of the 0.4 wt.% CQ and 1 wt.% DMAEMA initiating system. Polymerizations were performed in square-shaped glass molds with dimensions of 90 mm × 90 mm × 4 mm (length × width × thickness), and disc-like Teflon molds with dimensions of 15 mm × 1.5 mm (diameter × thickness). A UV-VIS lamp with a 280–780 nm wavelength and 2400 mW/cm^2 radiation exitance (Ultra Vitalux 300, Osram, Munich, Germany) was used for irradiation. This curing procedure is described in detail in our previous work [27]. The resulting casts were cut into specimens of dimensions dictated by particular standards, and sanded clean with fine sanding paper (Figure 2).

Table 1. Compositions and structural parameters of the studied 60 wt.% QAUDMA and 40 wt.% TEGDMA (QA:TEG) liquid monomer compositions.

Sample Name	Sample Composition			Structural Properties of the Liquid Monomer Compositions		
	QAUDMA		TEGDMA			
	Number of Carbon Atoms in the N-alkyl Chain	Mole Fraction	Mole Fraction	Molecular Weight (g/mol)	Concentration of Double Bonds (mol/kg)	Degree of Conversion [1] (%)
QA8:TEG	C8	0.31	0.69	496	4.03	84.2
QA10:TEG	C10	0.30	0.71	505	3.96	84.0
QA12:TEG	C12	0.28	0.72	512	3.90	86.0
QA14:TEG	C14	0.27	0.73	520	3.85	88.7
QA16:TEG	C16	0.26	0.74	526	3.80	87.1
QA18:TEG	C18	0.26	0.74	533	3.75	87.1
	Bis-GMA					
BG:TEG	-	0.46	0.54	389	5.14	64.8

[1] taken from [27].

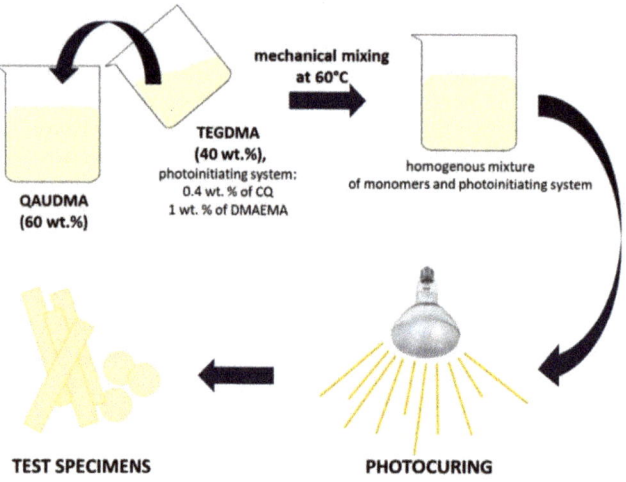

Figure 2. Sample preparation.

2.4. Mechanical Properties

2.4.1. Hardness

Disc-like samples of 40 mm × 4 mm (diameter × thickness) were tested for hardness (*HB*) using VEB Werkstoffprüfmaschinen apparatus (Leipzig, Germany), according to the ISO 2039-1 standard [36].

HB was calculated according to the following formula:

$$HB\ (MPa) = \frac{F_m \frac{0.21}{(h-h_r)+0.21}}{\pi d h_r}, \qquad (1)$$

where:
F_m—the test load;
d—the diameter of the ball intender (d = 5 mm);
h—the immersion depth;
h_r—the reduced depth of immersion (h_r = 0.25 mm).

2.4.2. Flexural Properties

Bars of 80 mm × 10 mm × 4 mm (length × width × thickness) were tested for flexural strength (*FS*) and flexural modulus (*E*) using a universal testing machine (Zwick Z020, Ulm, Germany), according to ISO 178 standards [37].

FS and *E* were calculated according to the following formulas:

$$FS\ (MPa) = \frac{3Pl}{2bd^2}, \qquad (2)$$

$$E\ (MPa) = \frac{P_1 l^3}{4bd^3 \delta}, \qquad (3)$$

where:
P_1—the load at the selected point of the elastic region of the stress-strain plot;
P—the maximum load;
l—the distance between supports (l = 64 mm);
b—the sample width (b = 10 mm);
d—the sample thickness (h = 4 mm);
δ—the deflection of the sample at P_1.

2.5. Water Sorption and Solubility

Disc-like samples of 15 mm × 1.5 mm (diameter × thickness) were tested for water sorption (*WS*) and solubility (*SL*) according to ISO 4049 standards [38].

Samples dried to a constant weight (m_0) were stored in distilled water for seven days at room temperature. After that, the specimens were removed from the water, blotted dry, and weighed (m_1). The samples were then dried again to a constant weight (m_2). Dryings were performed at 100 °C in a conditioning oven. All weightings were performed with an analytical balance (XP Balance, Mettler Toledo, Greifensee, Switzerland) of 0.0001 g accuracy.

WS and *SL* were calculated according to the following formulas:

$$WS\ (\mu g/mm^3) = \frac{m_1 - m_0}{V}, \qquad (4)$$

$$SL\ (\mu g/mm^3) = \frac{m_0 - m_2}{V}, \qquad (5)$$

where:
m_0—the initial mass of the dried samples;
m_1—the mass of the swollen samples;
m_2—the mass of the dried samples after immersion in water;
V—the initial volume of the dried samples.

2.6. Statistical Analysis

The results were expressed as an average value and a corresponding standard deviation (SD) achieved for five specimens in each testing method. A non-parametric Wilcoxon test ($p = 0.05$) was used to determine the statistical significance of the results. The calculations were performed using Statistica 13.1 (TIBCO Software Inc., Palo Alto, CA, USA) software.

3. Results

A series of six copolymers consisting of 60 wt.% QAUDMA and 40 wt.% TEGDMA (QA:TEG) was subject to an investigation that included the measurement of mechanical properties and behavior in water. The copolymer of 60 wt.% Bis-GMA and 40 wt.% TEGDMA served as a reference sample (BG:TEG). The sample names and their compositions are specified in Table 1.

3.1. Mechanical Properties

The summarized results of the mechanical properties HB, FS, and E are given in Table 2.

Table 2. Mechanical properties of the studied copolymers: hardness (HB), flexural strength (FS), and flexural modulus (E). Lower case letters indicate statistically insignificant differences ($p > 0.05$) with a column (non-parametric Wilcoxon test).

Sample Name	HB (MPa)		FS (MPa)		E (MPa)	
	Average	SD	Average	SD	Average	SD
QA8:TEG	51.41 [a,b]	4.32	21.59 [a,b]	0.66	679.0	36.2
QA10:TEG	51.17 [a,c]	6.93	37.37 [c]	2.27	851.6 [a]	47.4
QA12:TEG	50.87 [b,c]	4.08	34.46 [c]	2.18	848.7 [a]	24.7
QA14:TEG	41.60 [d,e]	3.63	28.38	1.38	772.3 [b]	31.1
QA16:TEG	41.21 [d,f]	2.27	20.13 [a]	1.62	753.5 [b]	31.8
QA18:TEG	42.17 [e,f]	1.08	21.75 [b]	1.90	459.4	34.4
BG:TEG	107.56	5.70	51.63	6.76	2800.9	78.9

The HB values of the QA:TEGs ranged from 51.41 to 42.17 MPa. They decreased as the length of the N-alkyl substituent increased. All of the QA:TEGs were characterized by lower HB values than the BG:TEG reference sample (HB = 107.56 MPa). All of these differences were statistically significant. The highest HB value was found for the QA8:TEG, which was lower by 52% in comparison to the BG:TEG reference sample. The differences between QA8:TEG, QA10:TEG, and QA12:TEG were statistically insignificant, as were those between QA14:TEG, QA16:TEG, and QA18:TEG. However, the HB values of the copolymers of the first group were statistically significantly higher, compared to the HB values of the copolymers of the latter group.

The FS values of the QA:TEGs ranged from 37.37 to 20.13 MPa. All of the QA:TEGs were characterized by lower FS values than the BG:TEG reference sample (FS = 51.63 MPa). All of these differences were statistically significant. The highest FS was found for the QA10:TEG (FS = 37.37 MPa), which was 28% lower than the BG:TEG reference sample. The second highest FS value was found for the QA12:TEG (FS = 34.46 MPa). In comparison to QA8:TEG (FS = 21.59 MPa), the FS values of QA10:TEG and QA12:TEG were statistically significantly higher, by 73 and 60%, respectively. They were also statistically significantly higher compared to the FS values of the remaining QA:TEGs. QA16:TEG was characterized by the lowest FS value (FS = 20.13 MPa). In comparison to QA8:TEG, this value was lower by 6%. This value was also slightly lower than the FS value of the QA18:TEG (FS = 21.75 MPa). These differences were statistically insignificant.

The E values of the QA:TEGs ranged from 459.4 to 851.6 MPa. All of the QA:TEGs were characterized by lower E values than the BG:TEG reference sample (E = 2800.9 MPa).

All of these differences were statistically significant. The highest E was found for the QA10:TEG (E = 851.6 MPa), which was 70% lower than the BG:TEG reference sample. The second highest E was found for QA12:TEG (E = 848.9 MPa). In comparison to the QA8:TEG (E = 679.0 MPa), the E values for QA10:TEG and QA12:TEG were statistically significantly higher by 25%. They were also statistically significantly higher than the E values of the remaining QA:TEGs. QA18:TEG was characterized by the lowest E value (E = 459.4 MPa). In comparison to QA8:TEG, this value was lower by 32%. This difference was statistically significant.

3.2. Water Sorption and Solubility

The summarized results related to water sorption (*WS*) and water solubility (*SL*) are given in Figure 3.

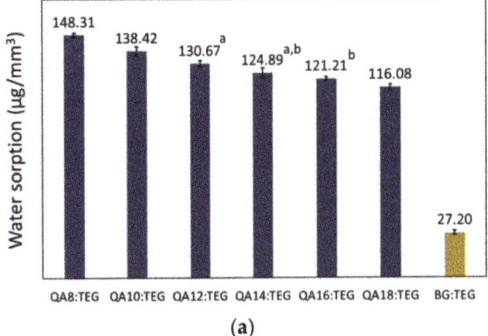
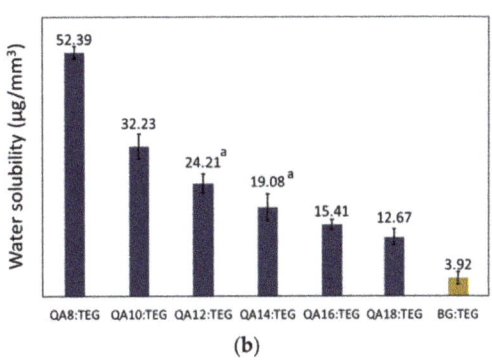

Figure 3. Water sorption (**a**) and water solubility (**b**) of the studied copolymers. Lower case letters indicate statistically insignificant differences ($p > 0.05$, non-parametric Wilcoxon test).

The *WS* values of the QA:TEGs ranged from 116.08 to 148.31 µg/mm^3. All of the QA:TEGs were characterized by higher *WS* values than the BG:TEG reference sample (*WS* = 27.20 µg/mm^3). The *WS* values decreased as the length of the N-alkyl substituent increased. In addition, all of these decreases were statistically significant. The lowest *WS* value was found for QA18:TEG, which was higher by 326% in comparison to the BG:TEG reference sample.

The *SL* values of the QA:TEGs ranged from 12.67 to 52.39 µg/mm^3. All of the QA:TEGs were characterized by higher *SL* values than the BG:TEG reference sample (*SL* = 3.92 µg/mm^3). The *SL* values decreased as the length of the N-alkyl substituent increased. In addition, all of these decreases were statistically significant. The lowest *SL* value was found for QA18:TEG, which was 223% higher than that of the BG:TEG reference sample.

4. Discussion

Dimethacrylate monomers containing quaternary ammonium groups have been recognized as compounds with high antibacterial activity. Therefore, using them to produce novel dental composite matrices represents a potential solution for caries treatment. In recent years, many studies of the development of new methacrylate structures containing quaternary ammonium groups have been conducted. However, the focus of these studies is often limited to the determination of the antimicrobial activity of the polymers.

This work is a continuation of research into the new urethane-dimethacrylate monomers with quaternary ammonium groups and their copolymers. The results of previous works revealed the promising physicochemical characteristics of QAUDMA monomers [26], as well as the structural, physicochemical, and antibacterial properties of their copolymers with TEGDMA [27]. This work was intended to enhance our knowledge of the influence of

QAUDMAs on the mechanical properties and behaviors in water of their copolymers with TEGDMA.

Matrices in dental composite restorative materials often consist of 60 wt.% Bis-GMA and 40 wt.% TEGDMA, where Bis-GMA is responsible for high mechanical performance, and TEGDMA acts as a reactive diluent. Therefore, this work aimed to verify how the complete replacement of Bis-GMA with the new QAUDMAs would affect the copolymers' mechanical properties and behavior in water.

4.1. Mechanical Properties

Mechanical properties are responsible for dental materials' performance and capabilities under particular stresses present in the oral environment. These mechanical properties should be determined to assess the proper functioning and usefulness of dental materials, as well as to evaluate the limitations that result from composition and/or curing parameters (initiation systems, irradiation sources, etc.) [39]. Common mechanical properties that are usually considered are hardness, stiffness, and strength.

4.1.1. Hardness

Hardness (*HB*) is defined as the resistance to permanent surface indentation. Adequate *HB* provides dental restoration materials with suitable resistance to stresses arising from mastication and abrasion [40].

The tested QA:TEGs were characterized by *HB* values lower than that of the BG:TEG reference sample. Such a result suggests that studied QA:TEGs are characterized by insufficient *HB* values; therefore, they cannot be used as matrices in dental composites. The results for *HB* can be interpreted from a structural perspective. The decrease in *HB* values that accompanied the increase in the length of the N-alkyl substituent did not show a correlation with the *DC*, which is a structural parameter that strongly influences poly(dimethacrylate)s' hardness [41]. The QA:TEGs were characterized for *DC* in our previous study [27] and those results are shown in Table 1. As can be seen, the QA:TEGs had high *DC* values, which ranged from 84.0 to 88.7%, and the length of the N-alkyl substituent had no visible influence. These *DC* values can be classified as high, as the *DC* value of the BG:TEG reference sample was 64.8%. As the *DC* did not influence the *HB*, the length of the N-alkyl substituent is probably the key factor affecting *HB*. The precise analysis of hardness showed that QA8:TEG, QA10:TEG, and QA12:TEG had similar *HB* values. A similar situation was observed for QA14:TEG, QA16:TEG, and QA18:TEG. However, the *HB* values of the first group were higher than the *HB* values of the latter group. The lengthening of the N-alkyl substituent from C8 to C12 did not cause a significant decrease in the *HB* values. However, a further increase in its length caused a notable decrease in the *HB* values.

4.1.2. Flexural Strength

Flexural strength (*FS*) is a key factor related to the durability of dental restorative materials. Its value represents the pressure that the material can withstand before breaking. The higher the *FS* value, the higher the stress that the material can withstand [42].

The tested QA:TEGs were characterized by *FS* values lower than that of the BG:TEG reference sample. The initial lengthening of the N-alkyl substituent from C8 to C10 caused a significant increase in the *FS* value. Its further lengthening caused a decrease in *FS* values. This trend did not have any correlation with the *DC* values, which were high for the studied QA:TEGs. The trend of the *FS* values can be explained by the strength of intermolecular interactions between the QAUDMA repeating units in the QA:TEG. This hypothesis can be justified by a comparison of the *FS* values determined for QA:TEGs with the viscosity values of QAUDMAs, which were determined in our previous work (Figure 4) [26].

Viscosity is a common indicator of intermolecular interactions present between monomer molecules. The higher the strength of the molecular interactions, the higher the viscosity, and the more limited the molecular movement [43]. As can be seen in Figure 4, the viscosity of the QAUDMA monomers initially increased with the increase in the length of the N-alkyl substituent from C8 to C10, and its maximum was observed for QA10:TEG. Further lengthening resulted in a decrease in viscosity values. This confirms the hypothesis about the dependency of FS on viscosity.

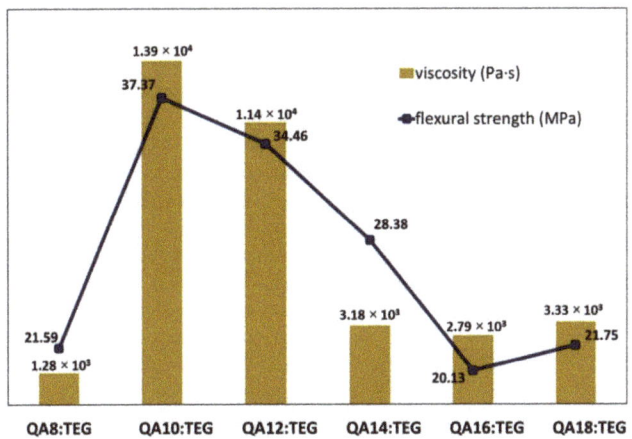

Figure 4. Flexural strength of the QA:TEG copolymers and viscosity of the QAUDMA monomers.

4.1.3. Flexural Modulus

The flexural modulus (E) is a key factor that refers to the stiffness of dental restorative materials. Its value represents the ratio between bending stress and the strain measured in the linear elastic region of a material.

The tested QA:TEGs were characterized by E values lower than that of the BG:TEG reference sample. The initial lengthening of the N-alkyl substituent from C8 to C10 caused a significant increase in the E value. Its further lengthening caused a decrease in E values. As in the case of FS, the trend observed for the E values did not correlate with the DC values, and is related to the strength of the intermolecular interactions between QAUDMA units (Figure 5).

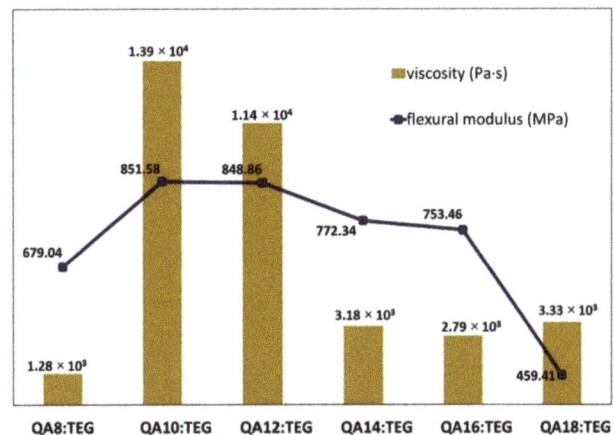

Figure 5. Flexural modulus of the QA:TEG copolymers and viscosity of the QAUDMA monomers.

4.2. Water Sorption and Solubility

Water sorption (WS) and solubility (SL) are two physicochemical factors crucial for the proper functioning of dental restorative materials.

4.2.1. Water Sorption

Excess water absorbed by the dental restorative material usually deteriorates its mechanical properties and has a plasticizing effect on the matrix by decreasing its glass transition temperature [44]. It can also lead to volumetric expansion, resulting in tooth or restoration breakdown [45]. Therefore, the WS of dental materials should be assessed; according to the ISO 4049 standard [38], its value cannot be greater than 40 μg/mm^3. The tested QA:TEGs were characterized by WS values greater than that given in the ISO standard, as well as that of the BG:TEG reference sample. The percentage differences over the value indicated in the ISO standard increased as the length of the N-alkyl chain decreased, and ranged from 192 to 271%. Therefore, none of the six QA:TEGs could be applied as a matrix in dental restorative materials.

The results for WS can be explained with reference to monomer chemical structures and copolymer crosslink density.

Two quaternary ammonium groups in the QAUDMA molecule are probably the main factor responsible for the high WS. This is due to the presence of both positively and negatively charged ions that are prone to absorb water [31].

The QA:TEGs are characterized by lower crosslink density compared to the BG:TEG reference sample. The concentration of methacrylate double bonds in the monomer mixture was used as a parameter to assess the chemical crosslink density in the corresponding copolymer. As can be seen from Table 1, those values decreased as the length of the N-alkyl substituent increased, which means that crosslink density decreases according to the same order. A more detailed analysis of the WS values showed that they had a high linear correlation with the concentration of double bonds on a semi-logarithmic scale (Figure 6).

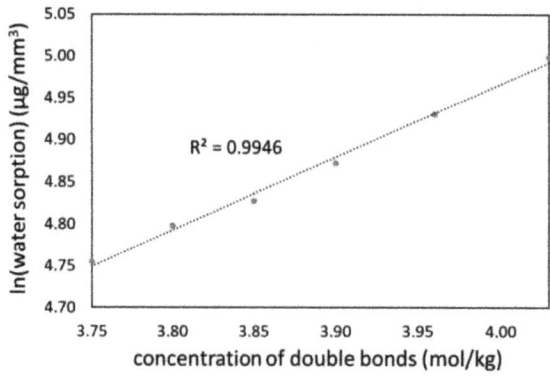

Figure 6. The correlation between the water sorption of the QA:TEG copolymers and the concentration of double bonds in the corresponding monomer compositions. The dotted line - the trendline, the yellow points show the mean values of water sorption.

The relationship between the WS values and the concentration of double bonds is very interesting, as the latter parameter is only diversified by the molecular weight of the N-alkyl chain in QAUDMA. It might be suspected that the increase in the length of the quaternary nitrogen substituent would cause the loosening of the copolymer network structure. Consequently, water would be able to migrate into it more easily, causing an increase in the WS. However, we observed the opposite effect. This can be explained by the following factors. First, the previous study on QA:TEGs shows that the N-alkyl chain takes up less space than suspected, probably due to a coiled conformation [27]. Second, as the length of the N-alkyl chain increases, the quaternary nitrogen structural region gains hydrophobicity. This hypothesis is confirmed by the results achieved in the previous work pertaining to the water contact angle of QA:TEGs. It was found that the hydrophobicity of the QA:TEGs' surface, quantified by the water contact angle values, increased with the increase in the length of the N-alkyl substituent [27]. Third, there is also a hypothesis in the literature that long N-alkyl chains can adopt specific conformations that obscure the quaternary nitrogen atoms, reducing water affinity [46].

QA:TEGs are also physically crosslinked due to the formation of hydrogen bonds with the hydrogen donor of the QAUDMA urethane linkages. Such hydrogen bonds could cause a slight decrease in WS values. Comparing the WS values of the BG:TEG reference sample to the other common dental composition containing the urethane-dimethacrylate monomer (UDMA) (40 wt.% Bis-GMA, 40 wt.% UDMA, 20 wt.% TEGDMA, WS = 25.64 µg/mm^3 [47]), the latter's WS value is 6% lower. The fact that QA:TEGs have much higher WS values than the BG:TEG reference sample indicates that physical crosslinks involving urethane hydrogen bonds were insufficient to reduce water absorption.

4.2.2. Water Solubility

Water solubility (SL) results from the presence of sol fraction [39]; this consists of low molecular weight structures, including monomer molecules, which are not chemically incorporated into the copolymer network. Leaching of sol fraction has an adverse effect on the proper functioning of dental restorative materials, for the following reasons. First, sol fraction is related to incomplete curing [48,49]. Therefore, the mechanical properties of dental restorations might significantly differ from the theoretical level. Second, the elution of the sol fraction may cause the appearance of voids inside the restoration, which would weaken it mechanically [50,51]. Third, the biocompatibility of the filling is reduced. The sol fraction usually has significant cytotoxicity, and therefore its eluting may have harmful effects on surrounding tissues or have a negative impact on organisms [52]. Therefore, the SL of dental materials should be assessed; according to the ISO 4049 standard [38], its value cannot be greater than 7.5 µg/mm^3. The tested QA:TEGs were characterized by SL values greater than that given in the ISO standard, as well as that of the BG:TEG reference sample. The percentage differences over the value indicated in the standard increased as the length of the N-alkyl chain decreased, and ranged from 69 to 599%. Therefore, none of the QA:TEGs could be used as a matrix in dental restorative materials. This result can be explained with reference to monomer chemical structures and molecular weights.

Since the quaternary ammonium groups have a high affinity to water, the QAUDMA sol fraction can easily migrate from the restoration to an aqueous environment. Detailed analysis of the SL values shows that they have a high linear correlation with MW on a semi-logarithmic scale (Figure 7).

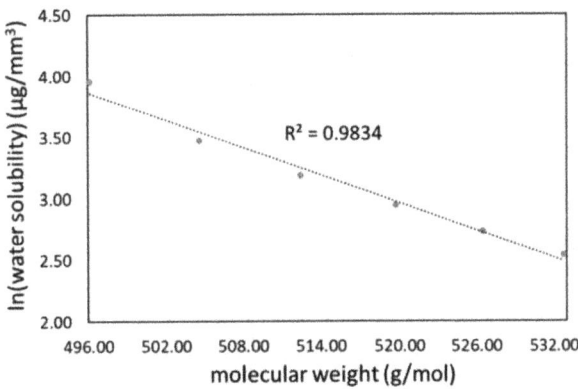

Figure 7. Correlation between the water solubility of the QA:TEG copolymers and the molecular weight of the corresponding monomer compositions. The dotted line-the trendline, the yellow points show the mean values of water solubility.

The MW values of the QA:TEG monomer compositions are high (Table 1). The relationship found for the SL and MW values may indicate that monomers with longer N-alkyl substituents have more difficulty leaching into the aqueous environment, due to the larger size of the monomer molecule. Another factor limiting sol fraction leaching can result from the length of the N-alkyl substituent, which had diverse hydrophobicity. The longer the N-alkyl substituent, the greater the MW, and the higher the hydrophobicity [27,53].

5. Conclusions

A series of six 60 wt.% QAUDMA and 40 wt.% TEGDMA copolymers were characterized for their mechanical properties and behaviors in water. All of the tested properties depended on the N-alkyl substituent length in the QAUDMA repeating unit. Hardness decreased as the length of the N-alkyl chain increased. Flexural strength and modulus initially increased as the length of the N-alkyl chain increased, up to ten carbon atoms. Its further lengthening caused a decrease in those values. The changes in the flexural strength and modulus were similar to the changes in the viscosity of the QAUDMA monomers, which may be attributed to the changes in the strength of intermolecular interactions between monomer units. Water sorption and solubility decreased as the length of the N-alkyl chain increased. Water sorption revealed a correlation with the concentration of double bonds in the QA:TEG monomer compositions, whereas water solubility revealed a correlation with the molecular weight of QA:TEG monomer compositions.

The values of the mechanical properties, water sorption, and solubility obtained for QA: TEGs indicate that their chemical composition is unsuitable for potential matrices of dental restorative composites. As such, additional investigations into biological properties such as cytotoxicity, or more sophisticated antimicrobial tests, are not justified for these materials. Further research into the QAUDMA-based copolymers must be conducted to obtain materials characterized by adequate values of all physicochemical and mechanical properties. This study provided general insight into the influence of the N-alkyl substituent length on hardness, flexural strength, modulus, water sorption, and solubility, which can help to design QAUDMA-based copolymers of suitable performance.

Author Contributions: Conceptualization, I.M.B.-R. and M.W.C.-P.; methodology, I.M.B.-R., M.W.C.-P. and G.C.; investigation, M.W.C.-P. and G.C.; resources, I.M.B.-R. and M.W.C.-P.; data curation, M.W.C.-P. and G.C.; writing—original draft preparation, M.W.C.-P. and I.M.B.-R.; writing—review and editing, I.M.B.-R. and G.C.; visualization, M.W.C.-P.; supervision, I.M.B.-R.; project administration, I.M.B.-R.; funding acquisition, M.W.C.-P. All authors have read and agreed to the published version of the manuscript.

Funding: This research was funded by the Rector's grant for the scientific research and development activities in the Silesian University of Technology, grant number: 04/040/BKM22/0213.

Institutional Review Board Statement: Not applicable.

Informed Consent Statement: Not applicable.

Data Availability Statement: Data supporting the reported results are available from the authors.

Conflicts of Interest: The authors declare no conflict of interest. The funders had no role in the design of the study; in the collection, analyses, or interpretation of data; in the writing of the manuscript; or in the decision to publish the results.

References

1. Cho, K.; Rajan, G.; Farrar, P.; Prentice, L.; Prusty, B.G. Dental Resin Composites: A Review on Materials to Product Realizations. *Compos. Part B Eng.* **2022**, *230*, 109395. [CrossRef]
2. Sun, Q.; Zhang, L.; Bai, R.; Zhuang, Z.; Zhang, Y.; Yu, T.; Peng, L.; Xin, T.; Chen, S.; Han, B. Recent Progress in Antimicrobial Strategies for Resin-Based Restoratives. *Polymers* **2021**, *13*, 1590. [CrossRef] [PubMed]
3. Batchelor, P. Is Periodontal Disease a Public Health Problem? *Br. Dent. J.* **2014**, *217*, 405–409. [CrossRef] [PubMed]
4. Janakiram, C.; Dye, B.A. A Public Health Approach for Prevention of Periodontal Disease. *Periodontoogy* **2020**, *84*, 202. [CrossRef] [PubMed]

5. James, S.L.; Abate, D.; Abate, K.H.; Abay, S.M.; Abbafati, C.; Abbasi, N.; Abbastabar, H.; Abd-Allah, F.; Abdela, J.; Abdelalim, A.; et al. Global, Regional, and National Incidence, Prevalence, and Years Lived with Disability for 354 Diseases and Injuries for 195 Countries and Territories, 1990–2017: A Systematic Analysis for the Global Burden of Disease Study 2017. *Lancet* **2018**, *392*, 1789–1858. [CrossRef]
6. Dursun, E.; Fron-Chabouis, H.; Attal, J.-P.; Raskin, A. Bisphenol A Release: Survey of the Composition of Dental Composite Resins. *Open Dent. J.* **2016**, *10*, 446–453. [CrossRef] [PubMed]
7. Spencer, P.; Ye, Q.; Misra, A.; Goncalves, S.E.P.; Laurence, J.S. Proteins, Pathogens, and Failure at the Composite-Tooth Interface. *J. Dent. Res.* **2014**, *93*, 1243–1249. [CrossRef] [PubMed]
8. Zhang, N.; Melo, M.A.S.; Weir, M.D.; Reynolds, M.A.; Bai, Y.; Xu, H.H.K. Do Dental Resin Composites Accumulate More Oral Biofilms and Plaque than Amalgam and Glass Ionomer Materials? *Materials* **2016**, *9*, 888. [CrossRef]
9. Song, W.; Ge, S. Application of Antimicrobial Nanoparticles in Dentistry. *Molecules* **2019**, *24*, 1033. [CrossRef]
10. Chrószcz, M.; Barszczewska-Rybarek, I. Nanoparticles of Quaternary Ammonium Polyethylenimine Derivatives for Application in Dental Materials. *Polymers* **2020**, *12*, 2551. [CrossRef] [PubMed]
11. Makvandi, P.; Jamaledin, R.; Jabbari, M.; Nikfarjam, N.; Borzacchiello, A. Antibacterial Quaternary Ammonium Compounds in Dental Materials: A Systematic Review. *Dent. Mater.* **2018**, *34*, 851–867. [CrossRef] [PubMed]
12. Ge, Y.; Wang, S.; Zhou, X.; Wang, H.; Xu, H.H.K.; Cheng, L. The Use of Quaternary Ammonium to Combat Dental Caries. *Materials* **2015**, *8*, 3532–3549. [CrossRef] [PubMed]
13. Imazato, S.; Chen, J.-h.; Ma, S.; Izutani, N.; Li, F. Antibacterial Resin Monomers Based on Quaternary Ammonium and Their Benefits in Restorative Dentistry. *Jpn. Dent. Sci. Rev.* **2012**, *48*, 115–125. [CrossRef]
14. Jiao, Y.; Niu, L.; Ma, S.; Li, J.; Tay, F.R.; Chen, J. Quaternary Ammonium-Based Biomedical Materials: State-of-the-Art, Toxicological Aspects and Antimicrobial Resistance. *Prog. Polym. Sci.* **2017**, *71*, 53–90. [CrossRef] [PubMed]
15. Xiao, Y.H.; Chen, J.H.; Fang, M.; Xing, X.D.; Wang, H.; Wang, Y.J.; Li, F. Antibacterial Effects of Three Experimental Quaternary Ammonium Salt (QAS) Monomers on Bacteria Associated with Oral Infections. *J. Oral Sci.* **2008**, *50*, 323–327. [CrossRef] [PubMed]
16. Imazato, S.; Kinomoto, Y.; Tarumi, H.; Torii, M.; Russell, R.R.B.; McCabe, J.F. Incorporation of Antibacterial Monomer MDPB into Dentin Primer. *J. Dent. Res.* **1997**, *76*, 768–772. [CrossRef] [PubMed]
17. Imazato, S.; Kinomoto, Y.; Tarumi, H.; Ebisu, S.; Tay, F.R. Antibacterial Activity and Bonding Characteristics of an Adhesive Resin Containing Antibacterial Monomer MDPB. *Dent. Mater.* **2003**, *19*, 313–319. [CrossRef]
18. Li, F.; Weir, M.D.; Xu, H.H.K. Effects of Quaternary Ammonium Chain Length on Antibacterial Bonding Agents. *J. Dent. Res.* **2013**, *92*, 932–938. [CrossRef] [PubMed]
19. Vidal, M.L.; Rego, G.F.; Viana, G.M.; Cabral, L.M.; Souza, J.P.B.; Silikas, N.; Schneider, L.F.; Cavalcante, L.M. Physical and Chemical Properties of Model Composites Containing Quaternary Ammonium Methacrylates. *Dent. Mater.* **2018**, *34*, 143–151. [CrossRef] [PubMed]
20. Cherchali, F.Z.; Mouzali, M.; Tommasino, J.B.; Decoret, D.; Attik, N.; Aboulleil, H.; Seux, D.; Grosgogeat, B. Effectiveness of the DHMAI Monomer in the Development of an Antibacterial Dental Composite. *Dent. Mater.* **2017**, *33*, 1381–1391. [CrossRef] [PubMed]
21. He, J.; Söderling, E.; Österblad, M.; Vallittu, P.K.; Lassila, L.V.J. Synthesis of Methacrylate Monomers with Antibacterial Effects against S. Mutans. *Molecules* **2011**, *16*, 9755–9763. [CrossRef] [PubMed]
22. Li, F.; Li, F.; Wu, D.; Ma, S.; Gao, J.; Li, Y.; Xiao, Y.; Chen, J. The Effect of an Antibacterial Monomer on the Antibacterial Activity and Mechanical Properties of a Pit-and-Fissure Sealant. *J. Am. Dent. Assoc.* **2011**, *142*, 184–193. [CrossRef] [PubMed]
23. Huang, L.; Yu, F.; Sun, X.; Dong, Y.; Lin, P.T.; Yu, H.H.; Xiao, Y.H.; Chai, Z.G.; Xing, X.D.; Chen, J.H. Antibacterial Activity of a Modified Unfilled Resin Containing a Novel Polymerizable Quaternary Ammonium Salt MAE-HB. *Sci. Rep.* **2016**, *6*, 33858. [CrossRef]
24. Huang, L.; Xiao, Y.H.; Xing, X.D.; Li, F.; Ma, S.; Qi, L.L.; Chen, J.H. Antibacterial Activity and Cytotoxicity of Two Novel Cross-Linking Antibacterial Monomers on Oral Pathogens. *Arch. Oral Biol.* **2011**, *56*, 367–373. [CrossRef]
25. Makvandi, P.; Ghaemy, M.; Mohseni, M. Synthesis and Characterization of Photo-Curable Bis-Quaternary Ammonium Dimethacrylate with Antimicrobial Activity for Dental Restoration Materials. *Eur. Polym. J.* **2016**, *74*, 81–90. [CrossRef]
26. Chrószcz, M.W.; Barszczewska-Rybarek, I.M. Synthesis and Characterization of Novel Quaternary Ammonium Urethane-Dimethacrylate Monomers—A Pilot Study. *Int. J. Mol. Sci.* **2021**, *22*, 8842. [CrossRef] [PubMed]
27. Chrószcz, M.W.; Barszczewska-Rybarek, I.M.; Kazek-Kęsik, A. Novel Antibacterial Copolymers Based on Quaternary Ammonium Urethane-Dimethacrylate Analogues and Triethylene Glycol Dimethacrylate. *Int. J. Mol. Sci.* **2022**, *23*, 4954. [CrossRef]
28. Yanwei, Y.; Li, H.; Yan, D.; Hongchen, Z.; Wei, Z.; Jinghao, B.; Jingjing, W.; Yan, L.; Jing, G.; Jihua, C. In Vitro Antibacterial Activity of a Novel Resin-Based Pulp Capping Material Containing the Quaternary Ammonium Salt MAE-DB and Portland Cement. *PLoS ONE* **2014**, *9*, e112549. [CrossRef]
29. Antonucci, J.M.; Zeiger, D.N.; Tang, K.; Lin-Gibson, S.; Fowler, B.O.; Lin, N.J. Synthesis and Characterization of Dimethacrylates Containing Quaternary Ammonium Functionalities for Dental Applications. *Dent. Mater.* **2012**, *28*, 219–228. [CrossRef]
30. Li, F.; Weir, M.D.; Chen, J.; Xu, H.H.K. Comparison of Quaternary Ammonium-Containing with Nano-Silver-Containing Adhesive in Antibacterial Properties and Cytotoxicity. *Dent. Mater.* **2013**, *29*, 450–461. [CrossRef]
31. Liang, X.; Huang, Q.; Liu, F.; He, J.; Lin, Z. Synthesis of Novel Antibacterial Monomers (UDMQA) and Their Potential Application in Dental Resin. *J. Appl. Polym. Sci.* **2013**, *129*, 3373–3381. [CrossRef]

32. Huang, Q.; Lin, Z.; Liang, X.; Liu, F.; He, J. Preparation and Characterization of Antibacterial Dental Resin with UDMQA-12. *Adv. Polym. Technol.* **2014**, *33*, 21395. [CrossRef]
33. Liang, X.; Söderling, E.; Liu, F.; He, J.; Lassila, L.V.J.; Vallittu, P.K. Optimizing the Concentration of Quaternary Ammonium Dimethacrylate Monomer in Bis-GMA/TEGDMA Dental Resin System for Antibacterial Activity and Mechanical Properties. *J. Mater. Sci. Mater. Med.* **2014**, *25*, 1387–1393. [CrossRef] [PubMed]
34. Huang, Q.T.; He, J.W.; Lin, Z.M.; Liu, F.; Lassila, L.V.J.; Vallittu, P.K. Physical and Chemical Properties of an Antimicrobial Bis-GMA Free Dental Resin with Quaternary Ammonium Dimethacrylate Monomer. *J. Mech. Behav. Biomed. Mater.* **2016**, *56*, 68–76. [CrossRef] [PubMed]
35. Manouchehri, F.; Sadeghi, B.; Najafi, F.; Mosslemin, M.H.; Niakan, M. Synthesis and Characterization of Novel Polymerizable Bis-Quaternary Ammonium Dimethacrylate Monomers with Antibacterial Activity as an Efficient Adhesive System for Dental Restoration. *Polym. Bull.* **2019**, *76*, 1295–1315. [CrossRef]
36. *ISO 2039:2001-1*; Plastics—Determination of Hardness—Part 1: Ball Indentation Method. International Standard Organisation: London, UK, 2001.
37. *ISO 178:2019*; Plastics—Determination of Flexural Properties. International Standard Organisation: London, UK, 2001.
38. *ISO 4049:2019*; Dentistry—Polymer Based Restorative Materials. International Standard Organisation: London, UK, 2001.
39. Barszczewska-Rybarek, I.M. A Guide through the Dental Dimethacrylate Polymer Network Structural Characterization and Interpretation of Physico-Mechanical Properties. *Materials* **2019**, *12*, 4057. [CrossRef]
40. Hardness Testing of Dental Materials and Tooth Substance. Available online: https://niom.no/hardness-testing-of-dental-materials-and-tooth-substance (accessed on 31 May 2022).
41. Barszczewska-Rybarek, I.; Chladek, G. Studies on the Curing Efficiency and Mechanical Properties of Bis-GMA and TEGDMA Nanocomposites Containing Silver Nanoparticles. *Int. J. Mol. Sci.* **2018**, *19*, 3937. [CrossRef]
42. Chung, S.M.; Yap, A.U.J.; Chandra, S.P.; Lim, C.T. Flexural Strength of Dental Composite Restoratives: Comparison of Biaxial and Three-Point Bending Test. *J. Biomed. Mater. Res.* **2004**, *71*, 278–283. [CrossRef]
43. Baker, L.; Bailey, R. Emerging Investigators 2015. *Anal. Methods* **2015**, *7*, 6936. [CrossRef]
44. Ito, S.; Hashimoto, M.; Wadgaonkar, B.; Svizero, N.; Carvalho, R.M.; Yiu, C.; Rueggeberg, F.A.; Foulger, S.; Saito, T.; Nishitani, Y.; et al. Effects of Resin Hydrophilicity on Water Sorption and Changes in Modulus of Elasticity. *Biomaterials* **2005**, *26*, 6449–6459. [CrossRef]
45. Biradar, B.; Biradar, S.; Ms, A. Evaluation of the Effect of Water on Three Different Light Cured Composite Restorative Materials Stored in Water: An In Vitro Study. *Int. J. Dent.* **2012**, *2012*, 640942. [CrossRef] [PubMed]
46. Yudovin-Farber, I.; Beyth, N.; Weiss, E.I.; Domb, A.J. Antibacterial Effect of Composite Resins Containing Quaternary Ammonium Polyethyleneimine Nanoparticles. *J. Nanopart. Res.* **2010**, *12*, 591–603. [CrossRef]
47. Barszczewska-Rybarek, I.M.; Chrószcz, M.W.; Chladek, G. Novel Urethane-Dimethacrylate Monomers and Compositions for Use as Matrices in Dental Restorative Materials. *Int. J. Mol. Sci.* **2020**, *21*, 2644. [CrossRef] [PubMed]
48. Alshali, R.Z.; Silikas, N.; Satterthwaite, J.D. Degree of Conversion of Bulk-Fill Compared to Conventional Resin-Composites at Two Time Intervals. *Dent. Mater.* **2013**, *29*, e213–e217. [CrossRef]
49. Barszczewska-Rybarek, I.; Gibas, M.; Kurcok, M. Evaluation of the Network Parameter in Aliphatic Poly(Urethane Dimethacrylate)s by Dynamic Thermal Analysis. *Polymer* **2000**, *41*, 3129–3135. [CrossRef]
50. Kannurpatti, A.R.; Anseth, J.W.; Bowman, C.N. A Study of the Evolution of Mechanical Properties and Structural Heterogeneity of Polymer Networks Formed by Photopolymerizations of Multifunctional (Meth)Acrylates. *Polymer* **1998**, *39*, 2507–2513. [CrossRef]
51. Gajewski, V.E.S.; Pfeifer, C.S.; Fróes-Salgado, N.R.G.; Boaro, L.C.C.; Braga, R.R. Monomers Used in Resin Composites: Degree of Conversion, Mechanical Properties and Water Sorption/Solubility. *Braz. Dent. J.* **2012**, *23*, 508–514. [CrossRef]
52. Roman, A.; Páll, E.; Moldovan, M.; Rusu, D.; Şoriţău, O.; Feştilă, D.; Lupşe, M. Cytotoxicity of Experimental Resin Composites on Mesenchymal Stem Cells Isolated from Two Oral Sources. *Microsc. Microanal.* **2016**, *22*, 1018–1033. [CrossRef]
53. Law, K.-Y. Definitions for Hydrophilicity, Hydrophobicity, and Superhydrophobicity: Getting the Basics Right. *J. Phys. Chem. Lett.* **2014**, *5*, 686–688. [CrossRef]

Article

The Effect of Liquid Rubber Addition on the Physicochemical Properties, Cytotoxicity, and Ability to Inhibit Biofilm Formation of Dental Composites

Krzysztof Pałka [1,*], Małgorzata Miazga-Karska [2], Joanna Pawłat [3], Joanna Kleczewska [4] and Agata Przekora [2,*]

1. Faculty of Mechanical Engineering, Lublin University of Technology, Nadbystrzycka 36, 20-618 Lublin, Poland
2. Chair and Department of Biochemistry and Biotechnology, Medical University of Lublin, Lublin, Chodźki 1, 20-093 Lublin, Poland; malgorzata.miazga-karska@umlub.pl
3. Institute of Electrical Engineering and Electrotechnologies, Lublin University of Technology, Nadbystrzycka 38A, 20-618 Lublin, Poland; j.pawlat@pollub.pl
4. Arkona Laboratorium Farmakologii Stomatologicznej, Nasutów 99C, 21-025 Niemce, Poland; joanna@arkonadent.com
* Correspondence: k.palka@pollub.pl (K.P.); agata.przekora@umlub.pl (A.P.); Tel.: +48-815384216 (K.P.); +48-814487026 (A.P.)

Abstract: The aim of this study was to evaluate the effect of modification with liquid rubber on the adhesion to tooth tissues (enamel, dentin), wettability and ability to inhibit bacterial biofilm formation of resin-based dental composites. Two commercial composites (Flow-Art–flow type with 60% ceramic filler and Boston–packable type with 78% ceramic filler; both from Arkona Laboratorium Farmakologii Stomatologicznej, Nasutów, Poland) were modified by addition of 5% by weight (of resin) of a liquid methacrylate-terminated polybutadiene. Results showed that modification of the flow type composite significantly ($p < 0.05$) increased the shear bond strength values by 17% for enamel and by 33% for dentine. Addition of liquid rubber significantly ($p < 0.05$) reduced also hydrophilicity of the dental materials since the water contact angle was increased from 81–83° to 87–89°. Interestingly, modified packable type material showed improved antibiofilm activity against *Streptococcus mutans* and *Streptococcus sanguinis* (quantitative assay with crystal violet), but also cytotoxicity against eukaryotic cells since cell viability was reduced to 37% as proven in a direct-contact WST-8 test. Introduction of the same modification to the flow type material significantly improved its antibiofilm properties (biofilm reduction by approximately 6% compared to the unmodified material, $p < 0.05$) without cytotoxic effects against human fibroblasts (cell viability near 100%). Thus, modified flow type composite may be considered as a candidate to be used as restorative material since it exhibits both nontoxicity and antibiofilm properties.

Keywords: resin composite; wettability; biofilm formation; cytotoxicity

1. Introduction

The polymer matrix of dental composites is commonly made of methacrylate resin blends. Unfortunately, due to the brittleness of resin-based composites, they show susceptibility to fracture under the occlusal forces [1]. Most studies focus on the toughening of dental composites by optimization of the resin mixture composition or reinforcement modifications [2–4]. Introduction of elastomers into a matrix of the dental composites is a promising method for their toughening and improvement of their fracture toughness [5–7]. Until recently, rubber modification of resin-based composites (RBC) remained in the realm of the laboratory [6,8]. Our latest research has shown that modification of the composite matrix with an acrylonitrile-free liquid rubber not only increases the fracture toughness [7], but also ensures patient safety due to the use of materials without carcinogenic properties.

The durability and clinical success related to the use of dental restorations highly depend on the strength of adhesive system which provides long-lasting bonds to the hard tooth tissues and the restorative composite. The morphological features of the cavity surface, which are the results of the mechanical and chemical processing, are responsible for the bond strength of the adhesive system. These microirregularities are impregnated by appropriate (meth)acrylate or (meth)acrylamide monomers with both hydrophilic and hydrophobic groups present in the adhesive system. Hydrophilic groups improve the wettability and bond strength to dental hard tissues, while the hydrophobic ones allow the interaction and co-polymerization with the restorative composite. Polymerization of the monomers results in various forms of interpenetrating networks [9]. The presence of liquid rubber in the composition of restoration material may potentially affects the strength of its bonding to the adhesive system.

The cytotoxicity of the resin composites used in dentistry is primarily associated with the release of residual monomers due to incomplete polymerization processes [10]. Over 30 different agents, some of them cytotoxic, have been isolated from cured dental composites, including compounds such as the main monomers, comonomers, various additives and reaction products [11]. Despite an observed increased degree of conversion after modification of the composites with liquid rubber [7], there is still possibility of the release of small amounts of monomers or hydrolytic degradation products. It is clear that consequences other than strong cytotoxicity, e.g., potential carcinogenic effect or inflammation, are also of high importance for the determination of the biological safety of dental materials.

The BisGMA resin, which is the main component of the composites' matrix, has polar hydroxyl groups [12]. In this study, a non-polar liquid rubber [13] was used for the modification of dental composites. Addition of non-polar component may potentially increase the contact angle of the resultant materials, making them more hydrophobic. Importantly, it is known that surface wettability has great impact on bacterial colonization and subsequent biofilm formation on dental materials [14,15]. Other authors have already demonstrated that addition of some components to the dental materials results in increased bacterial biofilm formation, leading to the enamel demineralization [16,17] and even gingival inflammation [18].

In our previous work, it was proven that addition of the acrylonitrile-free liquid rubber as a toughening agent results in improved mechanical properties of the resin composites [7]. It was hypothesized that such a chemical modification of dental materials may also significantly change their physicochemical properties like adhesive properties and wettability. Considering all the above mentioned important issues, the primary purpose of this work was to determine whether applied modification of resin-based materials may influence their adhesive, surface wettability and bacterial colonization properties. Moreover, cytotoxicity tests were performed on eukaryotic cells to evaluate the potential clinical usefulness of the modified composites.

2. Materials and Methods

2.1. Fabrication of Dental Composites

Two commercial composites: Flow-Art–flow type (marked as F) and Boston–packable type (marked as B) (Arkona, Nasutów, Poland) were used as starting research materials. The composites are made of the same components but characterized by different ceramic contents. The detailed chemical composition of the matrix and reinforcement are nor disclosed by the manufacturer. Briefly, the production of the composites was as follows: the matrix was made of Bis-GMA, Bis-EMA, UDMA and TEGDMA dimethacrylate resins (Sigma-Aldrich Chemicals, Munich, Germany) and it was modified with the 5 wt.% addition of a liquid methacrylate-terminated polybutadiene Hypro®2000X168LC VTB (CAS 68649-04-7, CVC Thermoset Specialties, Moorestown, NJ, USA) having vinyl reactive functional groups. The mixture was supplemented in each case with a photoinitiator (camphorquinone), co-initiator (2-dimethylaminoethyl methacrylate, DMAEMA) and an inhibitor (butylated hydroxytoluene, BHT) (all additives were purchased from Sigma-Aldrich Chemicals). The composite reinforcement was a mixture of pyrogenic silica, Ba-Al-B-Si glass and titanium dioxide added in various proportions: 60% by weight and 78% by weight in the case of flow (F) and packable (B) type of composite, respectively. The composition and the manufacturing of the composites modified with liquid rubber were claimed in the Polish patent application no. P.427219 [19]. Briefly, the base resins were premixed with the liquid rubber and then the first batch of the ceramic phase was introduced into the mixer in order to disperse the copolymer and obtain its homogeneous distribution in the whole volume of the matrix. Subsequently, the second batch of fillers (ceramic phase and other additives) was introduced into the mixer and the mixing process was continued using a vacuum to remove air from the composite. All materials were cured for 20 s in a stainless steel using led lamp (1350 mW/cm^2 intensity). Curing was performed according to ISO 4049:2019 standard [20]. All tested materials used in the research were prepared by Arkona Laboratorium Farmakologii Stomatologicznej (Nasutów, Poland). Modified flow (F) and packable (B) composites were marked as FM and BM, respectively.

2.2. Wettability Determination

Wettability (surface water contact angle–CA) was determined using a Krüss DSA25E goniometer (Krüss Scientific Instruments, Hamburg, Germany) equipped with CCD camera. Specimens for the wettability test were prepared in a form of discs (15 mm in diameter, 1 mm thick). Water contact angle was studied through the sessile drop method (1 µL of ultrapure water droplet was dosed at 0.16 mL/min flow rate) using static contact angle measurements [21]. The experiments were performed at room temperature (22 °C) after 24 h polymerization period of the composites (index "1" next to the sample symbol). The measurements were repeated for the samples that were additionally incubated for 24 h in pure deionized water (index "2" next to the sample symbol) to assess whether humidified oral cavity environment may influence composites' wettability. The value of the water contact angle, characteristic to selected surface, was obtained by averaging the mean contact angles (in fifteen measurements performed for 3 independent samples, n = 3). The unpaired t test and the Statistica software (TIBCO Software Inc., Palo Alto, CA, USA) were used to determine statistically significant results between unmodified samples and corresponding modified composites (B was compared to BM and F was compared to FM). Samples incubated in deionized water were not compared to untreated ones (without incubation in water).

2.3. Assessment of the Shear Bond Strength to the Tooth Tissues

The shear bond strength (SBS) was examined according to ISO 29022:2013 standard [22]. The use of the SBS methodology was applied to assess the impact of the presence of liquid rubber in the composites on their adhesive properties.

Twenty human molars without caries were used in this study after obtaining informed consent and approval of the Bioethics Committee of Medical University of Lublin (Lublin, Poland, KE-0254/339/2016). Each tooth was recovered after testing and used four times to obtain the required number of repetitions. In total, each material was tested ten times (n = 10) using enamel or dentin [23]. At the same time, this approach allowed us to maintain the randomness of using the teeth to make the results more reliable. Teeth were sectioned by a low speed diamond saw in order to reveal appropriate tissues while constantly keeping the teeth moist. Then, teeth tissues were mounted in cold-curing resin in cylindrical polycarbonate holders, so as to expose the facial enamel or dentin. After the specimens were mounted, their surfaces were abraded on silicon carbide abrasive paper (P600 grit size). Immediately after the abrasive treatment of the surface, it was etched with orthophosphoric acid for 30 s, then rinsed with a strong stream of water and dried. Then the bonding agent was applied (Masterbond, Arkona) and cured for 10 s using a led curing lamp of intensity 1350 mW/cm^2. The Masterbond is a single bottle, etch-and-rinse adhesive type, which combines the primer and adhesive action.

Teflon molds with central cylindrical cavity (2.38 mm in diameter, 2 mm thick) were used for the specimens and filled with the appropriate composite followed by polymerization for 20 s. Specimens were then stored for 24 h at 37 °C in distilled water prior to the testing and were then loaded in a universal testing machine (Autograph AG-X plus, Shimadzu, Kyoto, Japan) at a constant crosshead speed of 0.5 mm/min until fracture. The SBS was then calculated as the stress corresponding to maximum load force divided by the area of the bonded surface. The unpaired t test and Statistica software were used to determine statistically significant results between unmodified samples and corresponding modified composites (B was compared to BM and F was compared to FM). Results obtained with dentin were not compared to the results obtained with enamel.

2.4. Antibiofilm Activity

2.4.1. Bacterial Culture

Evaluation of biofilm formation on B, BM, F, FM samples was conducted using a modified Tu et al. method [24]. Bacterial adhesion assays were performed using *Streptococcus mutans* PCM 2502 and *Streptococcus sanguinis* PCM 2335 strains (the strains were obtained from the Polish Collection of Microorganisms PCM, Institute of Immunology and Experimental Therapy, Polish Academy of Sciences, Wroclaw, Poland) as a model of primary colonizer in biofilm formation on dental materials. In our tests reference bacteria species were used to minimize confounding variables. Initially, bacteria were cultured under anaerobic conditions for 48 h at 37 °C on BHI agar (BioMaxima S.A., Lublin, Poland). Each bacterial strain was diluted in BHI broth (BioMaxima S.A.) to get 1.5×10^8 CFU/mL of bacteria for biofilm assays.

2.4.2. Seeding of the Dental Composites with Bacteria

Eighteen disc samples (three for each experiment) with a diameter of approximately 12 mm and a height of 1 mm were sterilized by immersing the specimens for 10 s in 70% ethanol and drying at room temperature. Dry samples were rinsed twice with 200 µL of phosphate buffered saline (PBS, Pan-Biotech, Aidenbach, Germany) and then placed in the wells of 12-well polystyrene plates. Next, 1000 µL of BHI broth were added to each well. Simultaneously, other plates with composite discs were filled up with 1000 µL of BHI broth with 0.25% sucrose, which is a main sugar responsible for biofilm formation on the dental materials. Sucrose was added to broth to determine the changes in bacterial biofilm formation on the surface of B, BM, F, FM samples dependent on the absence or presence of this sugar in the medium. Finally, an amount of 20 µL of bacterial inoculum (1.5×10^8 CFU/mL) was transferred to each well. In the case of mono-species biofilm tests it was *S. sanguinis* or *S. mutans*. For mixed-species biofilm assays it was 10 µL of *S. sanguinis* (1.5×10^8 CFU/mL) and 10 µL of *S. mutans* (1.5×10^8 CFU/mL) added together. Sterility controls (only BHI or BHI+sucrose) and positive biofilm growth controls formed on the polystyrene surface (BHI or BHI + sucrose with bacterial inoculum) were included in all experiments. Plates were anaerobically incubated at 37 °C for 48 h.

2.4.3. Quantitative Biofilm Determination

Biofilm determination was performed according to the procedures described in [25,26]. After 48 h incubation of composite samples with bacterial strains, the medium was discarded and the samples were rinsed twice with 200 µL of fresh medium to leave only bacteria attached to abiotic surfaces. Remaining cells that were attached to the discs, were subsequently stained for 10 min. with CV (1 mL of 0.1% crystal violet) at room temperature, what allowed to visualize biofilm architecture. Discs were transferred to fresh wells, washed twice with 500 µL of sterile water to remove any CV that were not bound to bacteria. Finally, discs with biofilm were individually placed for 15 min. into tubes with 1000 µL of 20% acetic acid to allow the dye to solubilize at room temperature and sonicated (3 min.) to disperse the biofilm. Obtained CV/acetic acid solutions were transferred (200 µL) to a new 96-well plate to measure the optical density (OD at 590 nm) of each sample and additionally positive and negative controls. OD determination (200 µL from 1000 µL of one sample) was repeated five times and the average value was calculated. For each group of materials, a negative control (sterility disc control) consisting of the disc sample immersed in proper broth and positive growth controls (*S. mutans* or *S. sanguinis* or mixed *S. mutans* with *S. sanguinis* with proper broth in wells of polystyrene plates) were included. The OD value obtained for the positive controls was considered as equal to 100% biofilm formation. The results obtained with the experiment were checked for statistically significant differences ($p < 0.05$, n = 3) compared to the positive control of biofilm formation according to the unpaired *t* test (GraphPad Prism 5, Version 5.03 GraphPad Software, Inc., San Diego, CA, USA). Moreover, biofilm formation was compared between unmodified samples and corresponding modified composites (B was compared to BM and F was compared to FM). Results obtained with different bacterial species were not compared to each other.

2.4.4. Qualitative Biofilm Determination by Confocal Microscopy

The aim of the experiment was to visualize the viability and possible adhesion of bacterial cells to the modified materials (B, BM, F, FM) or control polystyrene surface. For this purpose, a Viability/Cytotoxicity Assay kit for Bacteria Live and Dead Cells (Biotium, Fremont, CA, USA) was used. The kit contains dyes that stain live cells (green fluorescence provided by DMAO dye) and dead bacteria (red fluorescence provided by EthD-III dye and green fluorescence provided by DMAO). Based on conducted staining it was possible to evaluate the structure and viability of the biofilm formed on the abiotic surfaces [27,28]. After 48 h of incubation with bacteria suspension (see Section 2.4.2.), the disc samples were subjected to confocal microscopy observation. Firstly, tested discs were gently rinsed with BHI medium (200 µL) to eliminate planktonic bacteria that were loosely attached to the samples and to leave only the biofilm on the surfaces. Then, samples were placed in the wells and stained using fluorescent dyes' solution prepared in PBS (Pan-Biotech). The staining procedure was carried out according to the manufacturer instructions. After 15 min of incubation at room temperature, bacterial biofilm on the disc surfaces was visualized using confocal laser scanning microscope (CLSM, Olympus Fluoview equipped with FV1000, Olympus, Tokyo, Japan). The mono-species biofilm positive controls or mixed-species biofilm positive control were grown simultaneously on the bottom of polystyrene wells under the same anaerobic conditions (37 °C for 48 h).

2.5. Cytotoxicity Evaluation

2.5.1. Eukaryotic Cell Culture

In vitro cell culture experiments were conducted with the use of normal human fibroblast cell line (BJ, ATCC, CRL-2522^{TM}), which is a model cell line commonly applied for cytotoxicity tests on new biomaterials and medical devices. The BJ cells were cultured in EMEM medium (ATCC-LGC Standards, Teddington, UK) with the following supplementation: 10% fetal bovine serum (FBS, Pan-Biotech) and penicillin-streptomycin solution (Sigma-Aldrich Chemicals, Warsaw, Poland). The cells were maintained in an incubator at 37 °C (humidified atmosphere of 95%, 5% CO_2, 95% air).

2.5.2. Quantitative Evaluation of Cytotoxicity

Cytotoxicity of the composites was evaluated in direct contact with the eukaryotic cells [29]. Prior to the cell seeding, tested materials were sterilized in the same manner as described in Section 2.4.2. Then the samples were placed in a 24-multiwell plate and presoaked for 20 min. in the complete culture medium at 37 °C. Cell culture-treated glass coverslip in the form of disc (13 mm in diameter) served as a control nontoxic material (negative control of cytotoxicity). 5×10^4 of BJ cells were seeded on top surface of the tested materials in 500 µL of the complete culture medium. After 48-h incubation at 37 °C, viability of BJ cells was assessed by colorimetric WST-8 test (Sigma-Aldrich Chemicals, Poland), which was performed according to the manufacturer protocol. The experiment was carried out in triplicate–3 independent samples of B, BM, F, FM, and control material were tested (n = 3). Cells cultured on the control material were considered to reveal 100%

viability. Viability of cells grown on the tested materials was expressed in % and calculated based on the absorbance values obtained with cells grown on the control material and cells cultured on the tested samples. The results obtained with WST-8 assay were checked for statistically significant differences ($p < 0.05$) among tested groups (all groups were compared to each other, including control cells) using one-way ANOVA test followed by Tukey's multiple comparison test (GraphPad Prism 5, Version 5.03).

2.5.3. Qualitative Evaluation of Cytotoxicity

The experiment was performed in direct contact with tested materials according to protocols described previously [30,31]. BJ fibroblasts were seeded on top surface of the materials in the same manner as described above (Section 2.5.2.). Upon 48-h incubation, cells grown on the top surface of the sample discs as well the cells which flawed down and were cultured on the polystyrene around the tested materials were stained using calcein-AM (green fluorescence of live cells) and propidium iodide (red fluorescence of dead cells). The dyes were the components of Live/Dead Double Staining Kit (Sigma-Aldrich Chemicals, Poland). Stained cells (their viability and morphology) were visualized using CLSM (Olympus Fluoview equipped with FV1000).

3. Results and Discussion

3.1. Wettability Determination

Average values of water contact angle (CA) for the tested materials are presented in Figure 1. The pairs of B-F and BM-FM samples had similar water CA values (CA < 90°), indicating their hydrophilic character. Modification of RBC by using liquid rubber significantly ($p < 0.05$) increased the CA value from 81–83° to 87–89°.

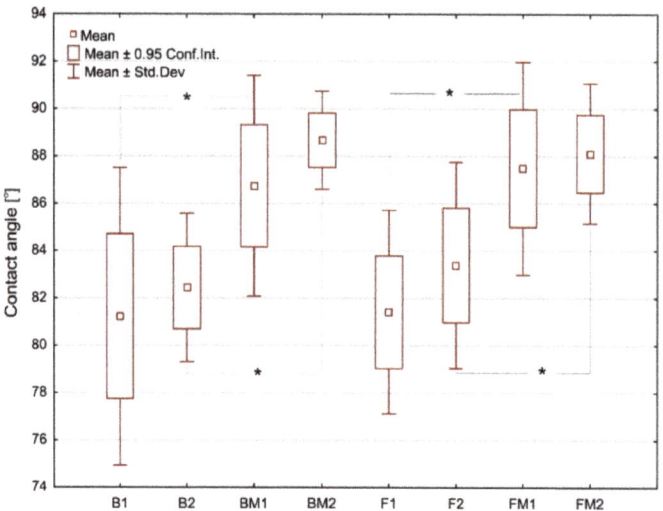

Figure 1. The surface water contact angles determined for tested materials. Asterisks (*) point at statistically significant differences ($p < 0.05$) according to the unpaired t test.

Thus, the introduced modification of dental composites slightly reduced their hydrophilicity. After immersion of the materials in deionized water (for 24 h), a slight increase of water CA values was observed for all samples, however without statistical significance. Wettability of the biomaterials' surface may have impact on eukaryotic cells' and bacterium adhesion and thus biocompatibility and biofilm formation on the dental materials, respectively [32–35]. Results of these studies showed that modification of RBC with liquid rubber reduced their wettability (hydrophilicity). Moreover, humidified oral cavity may further reduce hydrophilicity of fabricated dental composites. The presence of the aquatic environment as well as mechanical interactions, can form–by releasing nonpolar compounds–a thin film on the surface with strong hydrophobic properties. This thin film may provide a strong repulsive force that prevents bacteria adhesion. Similar phenomenon was observed by Rüttermann et al. [36].

3.2. Shear Bond Strength

Results of SBS tests are presented in Figure 2. Modification of flow type composite (FM sample) resulted in significant increase ($p < 0.05$) of SBS value for both kinds of tooth tissues. The 17% increase in the SBS value for enamel was observed, whereas SBS value for dentine increased by over 33%. In the case of modified packable composite (BM sample), an upward trend was observed however without statistical significance.

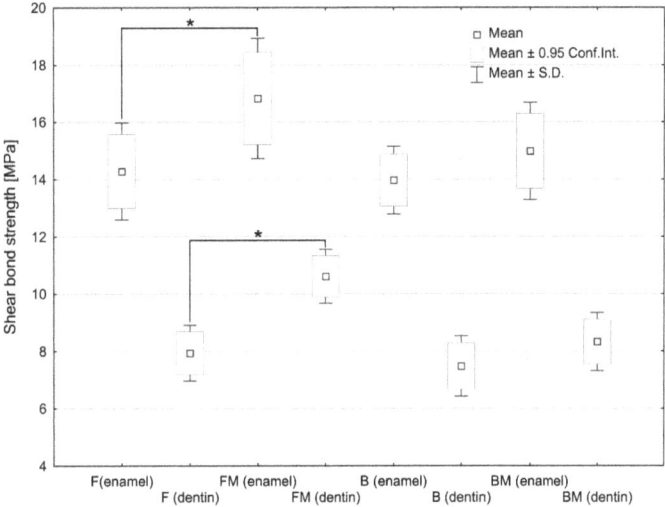

Figure 2. The shear bond strength estimated for tested materials depending on type of the tooth tissue. Asterisks (*) point at statistically significant differences ($p < 0.05$) according to the unpaired t test.

The test showed a significant increase in SBS value in the case of flow type dental composite after modification with liquid rubber (FM specimen). It was assumed that observed increase may be related to the reduced viscosity after liquid rubber modification of resin matrix, which was demonstrated in our previous work [7]. An upward trend, however without statistical significance, was observed for modified packable composite (BM). It was noticed that flow type composites showed almost equal SBS values as packable ones. In the case of flowable material a higher SBS value can be explained by better penetration to microirregularities of the substrate surface. Moreover, improved conversion degree obtained after the modification [7] may enhance the connection with the bonding agent.

With a similar chemical composition of the matrix, it can be expected that the SBS value will be comparative to the matrix volume fraction in the composite. It should also be taken into account that in the case of the packable composite, it is possible to form porosity by closing the air bubbles. This fact will facilitate cracking according to Griffith's theory. However, the SBS values estimated for both packable composites (B and BM) were almost equal to the values obtained for both flow composites (F and FM). The high SBS value obtained for the packable composite may be the result of its higher elastic modulus demonstrated in our previous study [7], leading to more uniform distribution of the stress over the bonded interface.

3.3. Antibiofilm Activity

To evaluate the effects of the B, BM, F, FM materials on biofilm formation, composite discs were added to planktonic bacteria (separately *S. mutans*, *S. sanguinis* for mono-species biofilm assay or *S. mutans* with *S. sanguinis* for mixed-species biofilm) and incubated for 48 h. Next, crystal violet staining was used to quantitatively detect biofilm. Figure 3 clearly shows strong reduction of biofilm formation on all tested materials compared to positive control grown on polystyrene. Flow type materials (F, FM) were more resistant (but the results were not statistically significant) to bacterial colonization (regardless of the species and the type of medium) than packable type composites (B, BM) having a higher content of ceramics. Importantly, modified FM material had the most resistant surface to biofilm adhesion among all tested samples. Considering % of positive controls (biofilm formation = 100%), it was less than 7% of mono- or mixed-species biomass grown (both in BHI medium and enriched with sugar) on the FM disc. The inhibition of biofilm formation followed the trend: FM (less than 7%), F disc (less than 12.79%), BM (less than 15.01%), and B (less than 18.9%) against the positive control.

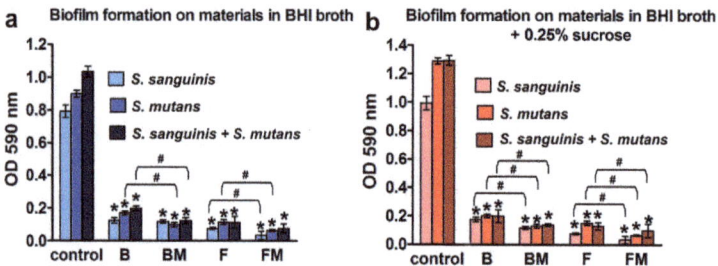

Figure 3. Quantification of the adhered bacteria to the B, BM, F, and FM materials compared to the control surface (polystyrene): (**a**) biofilm formation on materials in BHI broth; (**b**) biofilm formation on materials in BHI broth with 0.25% sucrose; results are shown as mean values ± SD from triplicate experiments; * indicates statistically significant results ($p < 0.05$) compared to the control, # indicates statistically significant results ($p < 0.05$) between unmodified and corresponding modified composite (the unpaired *t* test).

Figure 3a,b show that addition of 0.25% sucrose to the medium increased (compared to clear BHI broth) the growth of biofilm but mainly in the case of controls grown on polystyrene (increase by 43.3% for *S. mutans*, 25.2% for *S. sanguinis*, and 25.1% for mixed species). Slight increase in biofilm formation in the medium with sucrose was observed also on B material (increase by 1.94% for *S. sanguinis* and by 0.43% for mixed species). Importantly, in the case of both modified dental biomaterials (BM and FM), this extra sugar did not promote biofilm formation.

Results obtained with the quantitative assay were next confirmed by biofilm determination with the use of CLSM. The CLSM images presented in Figure 4a,b show *S. mutans* and *S. sanguinis* mono- and mixed-species biofilm on the materials surface with viable (stained green) and dead colonies (stained yellow-red). Surface of the disc without biofilm is visible as a non-stained area (no fluorescent signal—black color).

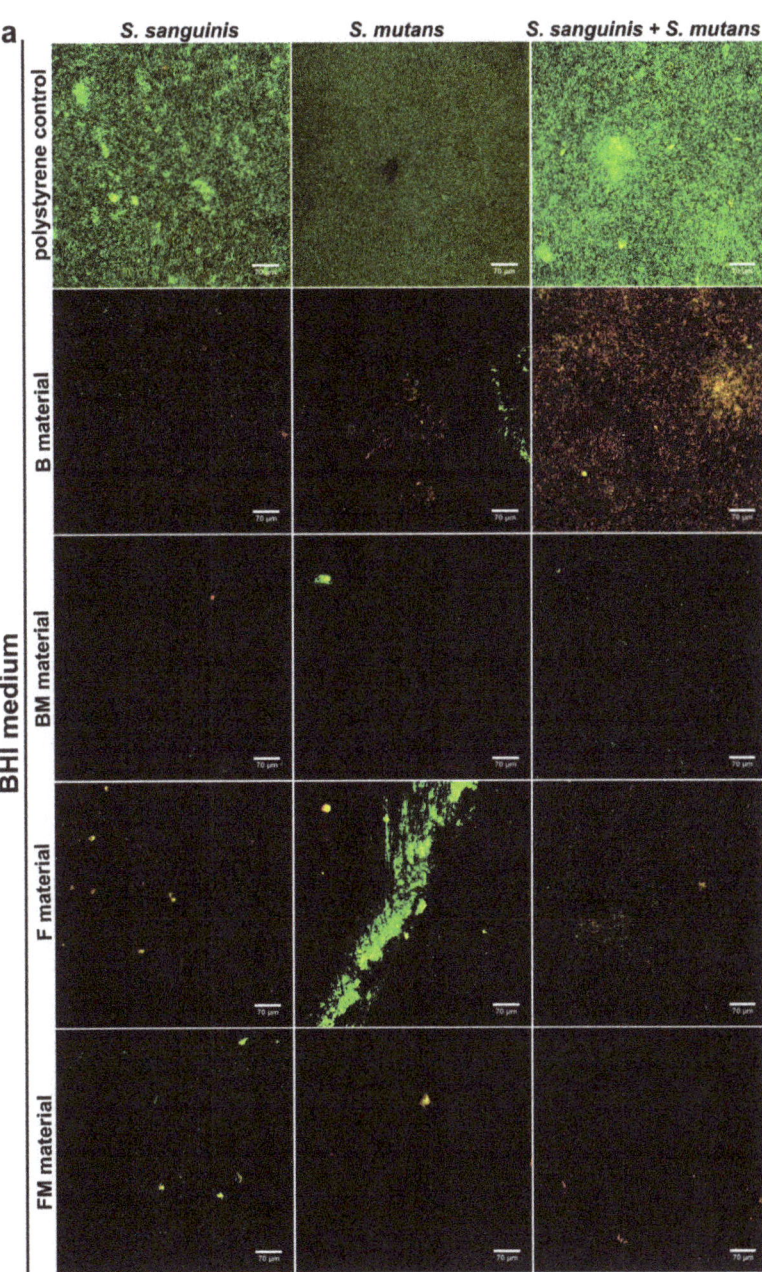

Figure 4. *Cont.*

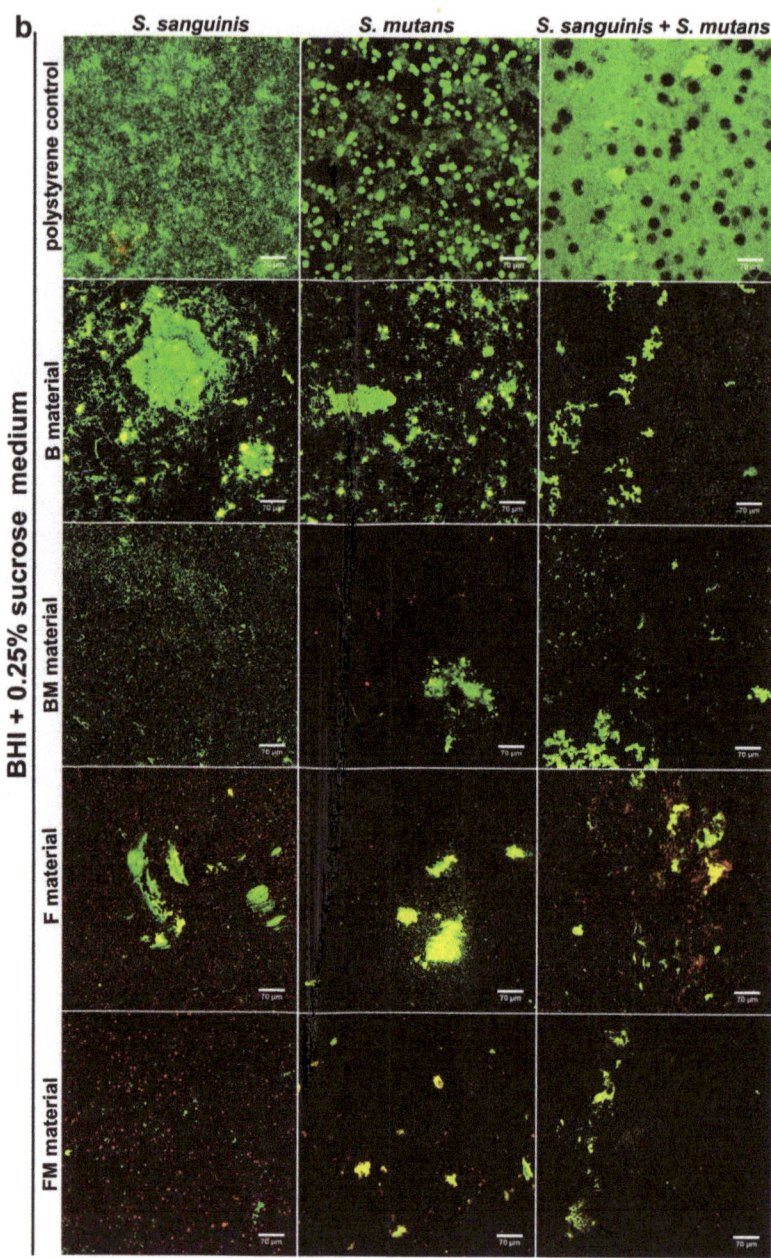

Figure 4. CLSM images of biofilm formed by *S. sanguinis*, *S. mutans*, and mixed species: *S. sanguinis* and *S. mutans* on B, BM, F, FM composite materials in BHI medium (**a**) and in BHI + 0.25% sucrose medium (**b**). Magn. 400×, scale bar = 70 μm (green fluorescence—viable cells, yellow and red fluorescence—dead cells).

The control group demonstrated viable bacteria after 48 h of incubation in BHI and BHI with sucrose, which formed green fluorescent colonies, proving that bacteria were viable during the period of evaluation. Whereas dead bacteria (red colonies) and some small green colonies (formed by viable bacteria) were observed on the tested composite discs. The images clearly demonstrate weaker adhesion of the bacteria to the modified FM, BM samples as well as to the unmodified F and B discs in comparison to the control (Figure 4a,b). However, the weakest bacteria adhesion was observed on the FM composite that contained liquid rubber and lower content of ceramics compared to packable type composites. Therefore the images confirmed the quantitative tests, where mono- and mixed-species biofilms showed limited growth on FM, F, BM and B disc surfaces in comparison with control surfaces. Importantly, the FM composite had the most resistant surface to bacterial adhesion.

Periodontal diseases and caries are caused by oral microflora changes and are the most common bacterial diseases occurring in the human and animal oral cavity. Almost 80% of bacterial infections of tissue or implants are associated with biofilm formation. Despite various findings concerning the physical and biochemical parameters of the biofilm-forming bacteria and the surface characteristics of the implant, the need for an ideal material still exists. Within these studies the antibiofilm properties of new dental materials were tested against caries bacteria. The effect of various types of material surfaces (B, BM, F, FM) and sucrose additions on biofilm formation were determined.

There are contrary opinions in the available literature concerning biofilm formation ability of main oral cavity bacteria (*S. sanguinis* and *S. mutans*) dependent on the surface wettability of the dental materials. It was observed that *S. sanguinis* bacteria better adhere to the ceramic materials compared to titanium regardless of the surface hydrophilicity [33]. Interestingly, some papers have described better adhesion and biofilm formation by *S. mutans* on hydrophilic surfaces [37] while majority of the authors proved that *S. mutans* bacteria have hydrophobic character of cell membrane and thus better adhere to the hydrophobic materials [32,38]. Moreover, it was observed that surface topography and chemistry are the key factors responsible for bacterial biofilm formation, whereas wettability is the second-rate issue [33,34].

Unfortunately, there is no clear factor enabling prediction of the probability of bacterial adhesion to the surface of dental materials. However, there are many features of materials' surface affecting the adhesion of bacteria to the implants, e.g., material chemical composition, its specific texture and topography as well as wettability and physicochemical properties of the surface. In fact, multiple factors are working simultaneously affecting bacterial adhesion. According to available literature it is known that surface roughness and character (hydrophilic or hydrophobic) of oral cavity bacteria wall–rather than material wettability–are the major agents responsible for bacterial adhesion [34,39,40]. This is in agreement with our results which demonstrated that samples modified with liquid rubber disc (FM) were the most resistant to bacterial biofilm formation among all tested samples, although the modification reduced the hydrophilic character. Thus, it is possible that modifications of composites influenced also their surface topography providing improved antibacterial protection. Importantly, all investigated samples had hydrophilic character, which according to many authors limits *S. mutans* and *S. sanguinis* bacteria adhesion [33,41]. Since unmodified samples revealed slightly higher hydrophilicity compared to BM and FM discs, it may be inferred that slightly reduced biofilm formation on modified composites most likely resulted from their different topography compared to unmodified materials.

3.4. Evaluation of Materials' Cytotoxicity

Direct cytotoxicity tests performed after 48-h culture of fibroblasts on the surfaces of the materials demonstrated lack of cytotoxicity only for B (viability = 83%), F (viability = 95%), and FM (viability = 104%) materials, whereas modification of B composite (BM sample) resulted in a significant reduction of cell viability to 37% (Figure 5).

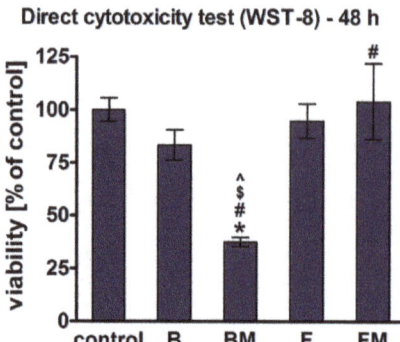

Figure 5. Cytotoxicity of the materials assessed by WST-8 test after direct 48 h contact of BJ fibroblasts with the materials (*significantly different results compared to the negative control of cytotoxicity, #significantly different results compared to the B material, $significantly different results compared to the FM material, ^significantly different results compared to the F material, according to one-way ANOVA test followed by Tukey's multiple comparison test, $p < 0.05$).

Live/dead staining of fibroblasts cultured directly on the materials showed that B and BM surfaces were not favorable to cell adhesion and growth (Figure 6). Only apoptotic (green and red fluorescence) and dead cells (red fluorescence) were found on the surfaces of these materials. F sample allowed attachment of small amounts of cells, whereas FM material was very supportive to cell adhesion and survival, which is consistent with WST-8 direct cytotoxicity results (Figure 5). Importantly, there was also great number of apoptotic and dead cells around the BM sample, proving cytotoxicity of this material. Cells grown around B, F, and FM materials were viable and had typical fibroblastic morphology. Thus, in the case of B and F samples slight reduction in cell viability (to 83% and 95%, respectively) observed in quantitative direct contact test (WST-8) primarily resulted from hindered fibroblast adhesion to the surfaces of these materials.

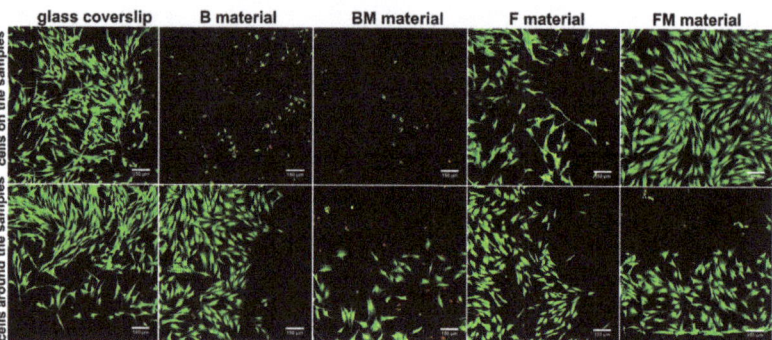

Figure 6. Qualitative evaluation of materials' cytotoxicity by live/dead double staining of BJ fibroblasts cultured for 48 h on the materials' surfaces, and grown around the samples (viable cells—green fluorescence, nuclei of dead cells—red fluorescence, apoptotic cells—green and red florescence); magn. 100×, scale bar = 150 μm.

Within these studies cytotoxicity was assessed in direct contact with the tested composites. WST-8 quantitative assay showed that fibroblasts revealed the highest viability on the flow type samples (FM and F) which showed also the highest ability to inhibit biofilm formation, proving their potential in biomedical applications. Moreover, CLSM observation demonstrated that FM surface was the most supportive to cell adhesion and growth among all test samples.

It is well known that hydrophilic surfaces of the biomaterials, revealing low water CA, promote the adsorption of cell adhesion proteins (e.g., fibronectin, laminin), allowing for more effective eukaryotic cell adhesion [21,35,42]. Interestingly, wettability tests showed higher hydrophilicity of the B and F samples compared to modified BM and FM composites. However, the B and BM samples did not allow for attachment of any BJ cell and F composite hindered cell adhesion. Therefore, surface hydrophilicity was not the key factor responsible for biocompatibility of fabricated composites. It should be noted that also other features of materials' surface influence cell adhesion, e.g., surface chemistry, charge or roughness, especially at nanometric scale [43,44]. All composites showed flat and rather smooth surfaces, thus sensitive analysis such as atomic force microscopy (AFM) should be performed to evaluate differences in nanoroughness between the samples. Nevertheless, it may be assumed that slightly lower amount of BJ fibroblasts attached to the F sample compared to modified FM composite most likely resulted from its smoother surface. The only difference between B and BM composites was the presence of additional component in BM material (5% of a liquid methacrylate-terminated polybutadiene), which improved biocompatibility of F sample. Thus, the reason of cytotoxic effect of only BM sample is hard to be explained without detailed analysis of surface chemical composition with the use of ATR-FTIR or XPS, which are planned to be performed in our future studies.

4. Conclusions

Results of the studies described herein revealed that the modification of packable type resin-based material (B) with liquid rubber reduced its hydrophilicity, improved its antibiofilm activity against *Steptococcus mutans* and *Streptococcus sanguinis*, but made the resultant BM material cytotoxic against eukaryotic cells. Interestingly, the introduction of the same modification to a flow type resin-based material (F)—which had lower content of the ceramic filler compared to the packable type–allowed us to obtain a novel dental material (FM) with higher adhesion to the tooth tissues, reduced hydrophilicity, significantly improved ability to inhibit bacterial biofilm formation and nontoxicity against human fibroblasts. Based on the obtained results it may be concluded that the fabricated FM composite can be a candidate to be used as restorative material since it exhibits nontoxicity and possesses antibiofilm properties, which are the best among all tested composites.

Author Contributions: Conceptualization, K.P. and A.P.; methodology, K.P., A.P., M.M.-K., J.P.; software, K.P.; validation, K.P., A.P. and J.K.; formal analysis, K.P.; investigation, K.P., M.M.-K., J.P. and A.P.; resources, K.P.; data curation, K.P.; writing—original draft preparation, K.P., M.M.-K. and A.P.; writing—review and editing, K.P. and A.P.; visualization, K.P., M.M.-K. and A.P.; supervision, A.P. and J.K.; project administration, K.P. All authors have read and agreed to the published version of the manuscript.

Funding: The paper was supported by Ministry of Science and Higher Education in Poland within statutory activity of Lublin University of Technology (M/KIM/FN25/2021). Studies conducted by AP and MMK was supported by Ministry of Science and Higher Education in Poland within statutory activity of Medical University of Lublin (DS3/2021 project). The paper was also partially developed using the equipment purchased within the agreement No. POPW.01.03.00-06-010/09-00 Operational Programme Development of Eastern Poland 2007–2013, Priority Axis I, Modern Economy, Operations 1.3. Innovations Promotion.

Institutional Review Board Statement: The study was conducted according to the guidelines of the Declaration of Helsinki, and approved by the Bioethics Committee of Medical University of Lublin (Lublin, Poland, KE-0254/339/2016, 15 December 2016).

Informed Consent Statement: Human teeth were used in this study after obtaining informed consent.

Data Availability Statement: The raw/processed data required to reproduce these findings can be obtained from the corresponding authors (k.palka@pollub.pl and agata.przekora@umlub.pl) upon reasonable request.

Conflicts of Interest: The authors declare no conflict of interest.

References

1. Ilie, N.; Hickel, R.; Valceanu, A.S.; Huth, K.C. Fracture toughness of dental restorative materials. *Clin. Oral Investig.* **2012**, *16*, 489–498. [CrossRef] [PubMed]
2. Fugolin, A.P.P.; Pfeifer, C.S. New resins for dental composites. *J. Dent. Res.* **2017**, *96*, 1085–1091. [CrossRef] [PubMed]
3. Chen, M.-H. Update on dental nanocomposites. *J. Dent. Res.* **2010**, *89*, 549–560. [CrossRef] [PubMed]
4. Moszner, N.; Salz, U. Composites for dental restoratives. In *Polymers for Dental and Orthopedic Applications*; Shalaby, W.S., Salz, U., Eds.; CRC Press: Boca Raton, FL, USA, 2006; pp. 13–68. ISBN 9781420003376.
5. Lee, V.A.; Cardenas, H.L.; Rawls, H.R. Rubber-toughening of dimethacrylate dental composite resin. *J. Biomed. Mater. Res. Part B Appl. Biomater.* **2010**, *94*, 447–454. [CrossRef] [PubMed]
6. Kerby, R.E.; Tiba, A.; Knobloch, L.A.; Schricker, S.R.; Tiba, O. Fracture toughness of modified dental resin systems. *J. Oral Rehabil.* **2003**, *30*, 780–784. [CrossRef]
7. Palka, K.; Kleczewska, J.; Sasimowski, E.; Belcarz, A.; Przekora, A. Improved fracture toughness and conversion degree of resin-based dental composites after modification with liquid rubber. *Materials* **2020**, *13*, 1–13. [CrossRef]
8. Mante, F.K.; Wadenya, R.O.; Bienstock, D.A.; Mendelsohn, J.; LaFleur, E.E. Effect of liquid rubber additions on physical properties of Bis-GMA based dental resins. *Dent. Mater.* **2010**, *26*, 164–168. [CrossRef]
9. Salz, U.; Bock, T. Testing adhesion of direct restoratives to dental hard tissue—A review. *J. Adhes. Dent.* **2010**, *12*, 343–371. [CrossRef]
10. Ausiello, P.; Cassese, A.; Miele, C.; Beguinot, F.; Garcia-Godoy, F.; Di Jeso, B.; Ulianich, L. Cytotoxicity of dental resin composites: An in vitro evaluation. *J. Appl. Toxicol.* **2013**, *33*, 451–457. [CrossRef] [PubMed]
11. Pelka, M.; Distler, W.; Petschelt, A. Elution parameters and HPLC-detection of single components from resin composite. *Clin. Oral Investig.* **1999**, *3*, 194–200. [CrossRef] [PubMed]
12. Cornelio, R.B.; Wikant, A.; Mjosund, H.; Kopperud, H.M.; Haasum, J.; Gedde, U.W.; Örtengren, U.T. The influence of bis-EMA vs bis GMA on the degree of conversion and water susceptibility of experimental composite materials. *Acta Odontol. Scand.* **2014**, *72*, 440–447. [CrossRef]
13. Xu, S.A. Miscibility and phase separation of epoxy/rubber blends. In *Handbook of Epoxy Blends*; Parameswaranpillai, J., Hameed, N., Pionteck, J., Woo, E.M., Eds.; Springer International Publishing AG: Berlin/Heidelberg, Germany, 2017; pp. 68–100.
14. Azam, M.T.; Khan, A.S.; Muzzafar, D.; Faryal, R.; Siddiqi, S.A.; Ahmad, R.; Chauhdry, A.A.; Rehman, I.U. Structural, surface, in vitro bacterial adhesion and biofilm formation analysis of three dental restorative composites. *Materials* **2015**, *8*, 3221–3237. [CrossRef]
15. Cazzaniga, G.; Ottobelli, M.; Ionescu, A.; Garcia-Godoy, F.; Brambilla, E. Surface properties of resin-based composite materials and biofilm formation: A review of the current literature. *Am. J. Dent.* **2015**, *28*, 311–320.
16. Magno, A.F.F.; Enoki, C.; Ito, I.Y.; Matsumoto, M.A.N.; Faria, G.; Nelson-Filho, P. In-vivo evaluation of the contamination of super slick elastomeric rings by Streptococcus mutans in orthodontic patients. *Am. J. Orthod. Dentofac. Orthop.* **2008**, *133*, S104–S109. [CrossRef] [PubMed]
17. Thurnheer, T.; Belibasakis, G.N. Effect of sodium fluoride on oral biofilm microbiota and enamel demineralization. *Arch. Oral Biol.* **2018**, *89*, 77–83. [CrossRef]
18. Larsen, T.; Fiehn, N.E. Dental biofilm infections—An update. *Apmis* **2017**, *125*, 376–384. [CrossRef] [PubMed]
19. Pałka, K.; Kleczewska, J.; Kalbarczyk, G. Light-curing dental composite modified with liquid rubber and the method for its production. Polish Patent Application P.427219, 2018.
20. *Dentistry-Polymer-Based Restorative Materials*; ISO 4049:2019; International Organization for Standardization: Geneva, Switzerland, 2019.
21. Przekora, A.; Benko, A.; Blazewicz, M.; Ginalska, G. Hybrid chitosan/β-1,3-glucan matrix of bone scaffold enhances osteoblast adhesion, spreading and proliferation via promotion of serum protein adsorption. *Biomed. Mater.* **2016**, *11*. [CrossRef]
22. *Dentistry—Adhesion—Notched-Edge Shear Bond Strength Test*; ISO 29022:2013; International Organization for Standardization: Geneva, Switzerland, 2013.
23. De Munck, J.; Van Landuyt, K.; Peumans, M.; Poitevin, A.; Lambrechts, P.; Braem, M.; Van Meerbeek, B. A critical review of the durability of adhesion to tooth tissue: Methods and results. *J. Dent. Res.* **2005**, *84*, 118–132. [CrossRef]
24. Tu, Y.; Ling, X.; Chen, Y.; Wang, Y.; Zhou, N.; Chen, H. Effect of *S. Mutans* and *S. Sanguinis* on growth and adhesion of *P. Gingivalis* and their ability to adhere to different dental materials. *Med. Sci. Monit.* **2017**, *23*, 4539–5445. [CrossRef] [PubMed]
25. Merritt, J.H.; Kadouri, D.E.; O'Toole, G.A. Growing and analyzing static biofilms. In *Current Protocols in Microbiology*; John Wiley & Sons, Inc.: Hoboken, NJ, USA, 2005; Volume 1, Unit 1B.1.
26. O'Toole, G.A. Microtiter dish biofilm formation assay. *J. Vis. Exp.* **2011**, 2437. [CrossRef]
27. Khajotia, S.S.; Smart, K.H.; Pilula, M.; Thompson, D.M. Concurrent quantification of cellular and extracellular components of biofilms. *J. Vis. Exp.* **2013**, e50639. [CrossRef]

28. Pączkowski, P.; Puszka, A.; Miazga-Karska, M.; Ginalska, G.; Gawdzik, B. Synthesis, characterization and testing of antimicrobial activity of composites of unsaturated polyester resins with wood flour and silver nanoparticles. *Materials* **2021**, *14*, 1122. [CrossRef]
29. Przekora, A.; Palka, K.; Ginalska, G. Biomedical potential of chitosan/HA and chitosan/β-1,3-glucan/HA biomaterials as scaffolds for bone regeneration-A comparative study. *Mater. Sci. Eng. C* **2016**, *58*. [CrossRef] [PubMed]
30. Przekora, A.; Ginalska, G. Addition of 1,3-β-d-glucan to chitosan-based composites enhances osteoblast adhesion, growth, and proliferation. *Int. J. Biol. Macromol.* **2014**, *70*. [CrossRef]
31. Przekora, A.; Palka, K.; Ginalska, G. Chitosan/β-1,3-glucan/calcium phosphate ceramics composites-novel cell scaffolds for bone tissue engineering application. *J. Biotechnol.* **2014**, *182–183*, 46–53. [CrossRef] [PubMed]
32. Brambilla, E.; Ionescu, A.; Mazzoni, A.; Cadenaro, M.; Gagliani, M.; Ferraroni, M.; Tay, F.; Pashley, D.; Breschi, L. Hydrophilicity of dentin bonding systems influences in vitro *Streptococcus mutans* biofilm formation. *Dent. Mater.* **2014**, *30*, 926–935. [CrossRef] [PubMed]
33. Wassmann, T.; Kreis, S.; Behr, M.; Buergers, R. The influence of surface texture and wettability on initial bacterial adhesion on titanium and zirconium oxide dental implants. *Int. J. Implant Dent.* **2017**, *3*, 32. [CrossRef] [PubMed]
34. Rupp, F.; Scheideler, L.; Rehbein, D.; Axmann, D.; Geis-Gerstorfer, J. Roughness induced dynamic changes of wettability of acid etched titanium implant modifications. *Biomaterials* **2004**, *25*, 1429–1438. [CrossRef]
35. Ercan, B.; Webster, T. Cell response to nanoscale features and its implications in tissue regeneration: An orthopedic perspective. In *Nanotechnology and Tissue Engineering: The Scaffold*; Laurencin, C., Nair, L., Eds.; CRC Press: Boca Raton, FL, USA, 2017; pp. 151–155. ISBN 9781138076587.
36. Rüttermann, S.; Trellenkamp, T.; Bergmann, N.; Raab, W.H.M.; Ritter, H.; Janda, R. A new approach to influence contact angle and surface free energy of resin-based dental restorative materials. *Acta Biomater.* **2011**, *7*, 1160–1165. [CrossRef]
37. Buergers, R.; Schneider-Brachert, W.; Hahnel, S.; Rosentritt, M.; Handel, G. Streptococcal adhesion to novel low-shrink silorane-based restorative. *Dent. Mater.* **2009**, *25*, 269–275. [CrossRef]
38. Kim, D.H.; Kwon, T.-Y. In vitro study of *Streptococcus mutans* adhesion on composite resin coated with three surface sealants. *Restor. Dent. Endodont.* **2017**, *42*, 39. [CrossRef] [PubMed]
39. Van Brakel, R.; Cune, M.; Van Winkelhoff, A.; De Putter, C.; Verhoeven, J.; Van Der Reijden, W. Early bacterial colonization and soft tissue health around zirconia and titanium abutments: An in vivo study in man. *Clin. Oral Implant. Res* **2011**, *22*, 571–577. [CrossRef] [PubMed]
40. Al-Ahmad, A.; Wiedmann-Al-Ahmad, M.; Faust, J.; Bächle, M.; Follo, M.; Wolkewitz, M.; Hannig, C.; Hellwig, E.; Carvalho, C.; Kohal, R. Biofilm formation and composition on different implant materials in vivo. *J. Biomed. Mater. Res. Part B Appl. Biomater.* **2010**, *95*, 101–109. [CrossRef] [PubMed]
41. Mitchell, L. Decalcification during orthodontic treatment with fixed appliances—An overview. *Br. J. Orthod.* **1992**, *19*, 199–205. [CrossRef]
42. Chang, H.-I.; Wang, Y. Cell response to surface and architecture of tissue engineering scaffolds. In *Regenerative Medicine and Tissue Engineering–Cells and Biomaterials*; Eberli, D., Ed.; InTech Open Access Publisher: London, UK, 2011.
43. Chen, H.; Liu, Y.; Jiang, Z.; Chen, W.; Yu, Y.; Hu, Q. Cell-scaffold interaction within engineered tissue. *Exp. Cell Res.* **2014**, *323*, 346–351. [CrossRef]
44. Anselme, K.; Bigerelle, M. On the relation between surface roughness of metallic substrates and adhesion of human primary bone cells. *Scanning* **2014**, *36*, 11–20. [CrossRef]

Article

An In Vitro Study on the Antimicrobial Properties of Essential Oil Modified Resin Composite against Oral Pathogens

Barbara Lapinska [1], Aleksandra Szram [1], Beata Zarzycka [2], Janina Grzegorczyk [2], Louis Hardan [3], Jerzy Sokolowski [1] and Monika Lukomska-Szymanska [1,*]

[1] Department of General Dentistry, Medical University of Lodz, 92-213 Lodz, Poland; barbara.lapinska@umed.lodz.pl (B.L.); ola.wilzak@gmail.com (A.S.); jerzy.sokolowski@umed.lodz.pl (J.S.)
[2] Department of Microbiology and Laboratory Medical Immunology, Medical University of Lodz, 92-213 Lodz, Poland; beata.zarzycka@umed.lodz.pl (B.Z.); janina.grzegorczyk@umed.lodz.pl (J.G.)
[3] Department of Restorative Dentistry, Dental School, Saint Joseph University, 11072180 Beirut, Lebanon; louis.hardan@usj.edu.lb
* Correspondence: monika.lukomska-szymanska@umed.lodz.pl; Tel.: +85-42-675-74-61

Received: 17 August 2020; Accepted: 28 September 2020; Published: 1 October 2020

Abstract: Modifying the composition of dental restorative materials with antimicrobial agents might induce their antibacterial potential against cariogenic bacteria, e.g., *S. mutans* and *L. acidophilus*, as well as antifungal effect on *C. albicans* that are major oral pathogens. Essential oils (EOs) are widely known for antimicrobial activity and are successfully used in dental industry. The study aimed at evaluating antibacterial and antifungal activity of EOs and composite resin material (CR) modified with EO against oral pathogens. Ten EOs (i.e., anise, cinnamon, citronella, clove, geranium, lavender, limette, mint, rosemary thyme) were tested using agar diffusion method. Cinnamon and thyme EOs showed significantly highest antibacterial activity against *S. mutans* and *L. acidophilus* among all tested EOs. Anise and limette EOs showed no antibacterial activity against *S. mutans*. All tested EOs exhibited antifungal activity against *C. albicans*, whereas cinnamon EO showed significantly highest and limette EO significantly lowest activity. Next, 1, 2 or 5 µL of cinnamon EO was introduced into 2 g of CR and microbiologically tested. The modified CR showed higher antimicrobial activity in comparison to unmodified one. CR containing 2 µL of EO showed the best antimicrobial properties against *S. mutans* and *C. albicans*, while CR modified with 1 µL of EO showed the best antimicrobial properties against *L. acidophilus*.

Keywords: essential oils; oral pathogens; antibacterial activity; *S. mutans*; *L. acidophilus*; *C. albicans*; antifungal activity

1. Introduction

Resin composites are the most commonly used dental restorative materials. They are composed of organic matrix and inorganic filler and their properties can be modeled with addition of specific components. The literature provides data on various modifications of dental composites and adhesives performed to enhance their physico-chemical, mechanical and antimicrobial properties [1–5]. Antibacterial activity of monomers, such as 12-methacryloyloxydodecylpyridinium bromide (MDPB), has been widely investigated [5]. Among antibacterial agents introduced into the composition of dental resin materials, most commonly described in the literature are nanoparticles, such as silver, gold, titanium dioxide, zinc oxide or calcium phosphate, as well as fluoride-containing filler and fluoride compounds [6–12]. Essential oils (EOs) could be promising alternative to contribute to the antimicrobial effect of resin composite materials [13,14].

Essential oils are natural, volatile complex compounds characterized by the odor of their corresponding aromatic plants [15]. There is no systematic chemical nomenclature for chemical compounds found in EOs. However, the scientific names are based on their properties or prominent sources (e.g., limonene, pinene and thymol) [16–18]. They exhibit hydrophobic nature and often lower density in comparison to water and are generally lipophilic. Moreover, EOs are soluble in organic solvents, but immiscible with water [19].

EOs are plant products that for decades have been used in traditional healing worldwide. EOs are biosynthesized as secondary metabolites such as bark (cinnamon), buds (clove), flowers (jasmine, rose, violet and lavender), fruits (star anise), herbs, leaves (thyme, eucalyptus and salvia), twigs, wood (sandal), rhizome and roots (ginger), seeds (cardamom) and zest (citrus) [19]. EOs represent a small fraction of plant composition (less than 5% of the vegetal dry matter) and comprise mainly hydrocarbon terpenes (monoterpenes and sesquiterpenes) and terpenoids (isoprenoids). The chemical components of EOs may be produced through either the methylerythritol or the mevalonate or the shikimic acid pathway [19]. Over 100 different components in various ratios (1%–70%) can be found in a single type of EO.

EOs exhibit different biological and pharmacological activities, such as antibacterial, antifungal, antiviral, antimutagenic, antiprotozoal, anti-inflammatory, antidiabetic, antinociceptive, antiphlogistic and antioxidant properties [20–30].

The combination of several EOs may lead to an additive or antagonistic effect against pathogens [31]. The enhanced antibacterial activity of EO mixture in comparison to individual products may result from the synergic effect of EO compounds. This effect relies either on inhibiting common biological pathway in microorganisms, suppressing the protective enzymes, or modifying the functions of the cellular wall [32]. EOs consist of different chemical compounds which may have different antimicrobial modes of action. Therefore, the possibility of antimicrobial resistance is minimized [17].

The mechanism of action of EOs against microorganisms has not been completely understood so far. EOs owe the antimicrobial properties to their volatile components, including terpenoids and phenolic compounds [33]. EO phenolic compounds are known to penetrate through the microbial membrane (formatting pores) leading to the leakage of ions and cytoplasmatic content and finally to cellular breakdown [17,34].

In oral hygiene and dentistry, essential oils are used as components of mouthwashes (i.e., Cool Mint, Listerine Antiseptic, Johnson&Johnson, Skillman, NJ, USA), toothpastes, antiseptic solutions and temporary filing materials (eugenol-based products, i.e., zinc oxide-eugenol cement) [35,36]. Incorporating essential oils into adhesive systems may contribute to the decrease in occurrence of secondary caries due to its antimicrobial activity reported in an *in vitro* microcosm dental biofilm model [14]. The main oral pathogens, *Streptococcus mutans* and *Lactobacillus acidophilus* are crucial in caries development. *S. mutans* plays main role in early demineralization of dental hard tissues, while *L. acidophilus* is pivotal in caries development. Various attempts has been made to enhance antibacterial properties of dental materials, involving the addition of silver-releasing filler [6,7], calcium fluoride [8,12] or amorphous calcium phosphate [9] into the composition of dental resin materials or adhesives. Studies reported that incorporation of essential oil into dental composite structure do not significantly compromise the mechanical properties [13,37], while it could improve its antibacterial activity and thus reduce the risk of secondary caries.

Yeasts, such as *Candida albicans*, are found in oral cavity as structural component of dental plaque biofilm, but more recently it has been recognized as part of cariogenic microbiota [38–41]. *C. albicans* is capable of producing acids that might demineralize dental hard tissues. According to Nikawa et al. [42], *C. albicans* possesses the ability to dissolve hydroxyapatite to a greater extent (approximately 20-fold) when compared with *S. mutans*. Furthermore, the presence of *streptococci* may promote *C. albicans* colonization of dental hard tissues [43]. Studies suggest symbiotic fungal-bacterial relationship between *S. mutans* and *C. albicans* within the biofilm that prevents from killing or inhibiting each other [44]. However, Maijala et al. [45] claimed that the role of *C. albicans* in cariogenic process is highly overestimated. Incorporating essential oils into dental materials composition seems like a promising alternative that would allow for enhancement of antimicrobial activity of dental restorative materials. In terms of potential anticariogenic effect, it would be favorable to investigate antimicrobial activity of various EOs against major cariogenic pathogens, such as *S. mutans*, *L. acidophilus* and *C. albicans*, in the same study, in homogeneous conditions. That would help to select the most active EOs in order to further incorporate them into dental materials composition to enhance their clinical performance. Thus, the primary aim of this study was to assess which of the different essential oils has the highest antimicrobial activity against oral pathogens (*S. mutans*, *L. acidophilus* and *C. albicans*). Next, the most effective essential oil would be selected to incorporate into resin material and the secondary aim of the study was to evaluate antimicrobial activity against *S. mutans*, *L. acidophilus* and *C. albicans* of the modified resin composite material.

2. Materials and Methods

This study used the following ten commercially available essential oils (dr Beta, Pollena Aroma, Nowy Dwór Mazowiecki, Poland): anise, cinnamon, citronella, clove, geranium, lavender, limette, mint, rosemary and thyme. The composition of tested EOs was presented in Table 1, based on data obtained from previous studies analyzing the EOs' composition by gas chromatography with flame-ionization and mass spectroscopic detection (GC-FID-MS) [46–52] or data from European Pharmacopoeia [53].

Table 1. Characteristics of essential oils used in the study.

Essential Oil (Name of EO in INCI)	Composition
Star anise (*Illicium Verum Oil*)	*trans*-anethole (86.0%–93.0%), linalool (0.2%–2.5%), estragole (0.5%–6.0%), α-terpineol (<0.3%), *cis*-anethole (0.1%–0.5%), anisaldehyde (0.1%–0.5%), foeniculin (0.1%–3.0%) [53]
Cinnamon (*Cinnamomum Zeylanicum Bark Oil*)	cinnamaldehyde (76.8%), methoxycinnamaldehyde (11.7%), cinnamyl acetate (3.2%), cumarin (1.5%), benzaldehyde (1.1%) [48,49]
Citronella (*Cymbopogon Winterianus Oil*)	citronellal (36.2%), geraniol (22.4%), citronellol (14.1%), limonene (3.5%), elemol (3.3%), citronellyl acetate (3.2%) [51]
Clove (*Eugenia Caryophyllus Oil*)	eugenol (85.3%), β-caryophyllene (10.6%), α-humulene (2.0%) [47,49]
Geranium (*Pelargonium Graveolens Oil*)	citronellol (26.7%), geraniol (13.4%), nerol (8.7%), citronellyl formate (7.1%), isomenthone (6.3%), linalool (5.2%), 10-*epi*-γ-eudesmol (4.4%), geranyl formate (2.5%), menthone (1.6%), β-caryophyllene (1.5%), geranyl butyrate, *cis*-rose oxide (1.4%), geranial (1.1%), β-baurobonene (1.1%) [47–49,52]
Lavender (*Lavandula Angustifolia Oil*)	linalool (34.1%), linalyl acetate (33.3%), lavandulil acetate (3.2%), β-ocymene (3.2%), β-caryophyllene (2.7%), cineole (2.5%), terpinen-4-ol (2.5%), myrecene (2.4%), α-terpineol (1.8%) [48,49]
Limette (*Citrus aurantifolia oil*)	linalyl acetate (48.06%), linalool (26.88%), α-terpineol (5.74%), geranyl acetate (3.92%), geraniol (3.05%), geranial (2.44%) [50]
Mint (*Mentha Piperita Oil*)	menthol (30.0%–55.0%), menthone (14.0%–32.0%), cineole (3.5%–14.0%), menthyl acetate (2.8%–10.0%), isomenthone (1.5%–10.0%), menthofuran (1.0%–9.0%), limonene (1.0%–5.0%), isopulegol (<0.2%), pulegone (<4.0%), carvone (<1.0%) [53]
Rosemary (*Rosmarinus Officinalis Oil*)	1,8-cineole (46.4%), camphor (11.4%), α-pinene (11.0%), β-pinene (9.2%), camphene (5.2%), β-caryophyllene (3.5%), borneol (3.1%), αa-terpineol (1.8%), *p*-cymene (1.3%), myrecene (1.2%) [47,49]
Thyme (*Thymus Vulgaris Oil*)	thymol (38.1%), *p*-cymene (29.1%), γ-terpinene (5.2%), linalool (3.7%), β-Caryophyllene (3.1%), carvacrol (2.3%) [46,47,49]

Legend: INCI = International Nomenclature of Cosmetic Ingredients.

2.1. Microbiological Studies of Essential Oils

Microbiological studies were performed on three reference strains: *Streptococcus mutans* ATCC 25175 (Oxoid, Basingstoke, UK), *Lactobacillus acidophilus* ATCC 4356 (Oxoid, Basingstoke, UK) and *Candida albicans* ATTC 10231 (Biocorp, Warsaw, Poland). The colonies were stored in Microbank system (Viabank, Medical Wire&Equipment, Corsham, UK) in the temperature of −30 °C until the experiment was performed. The study protocol was described in detail in previous research paper [8].

Antimicrobial activity of essential oils was tested using agar diffusion test. After 18 h of cultivation, the suspension has been prepared with the turbidity of the 0.5 McFarland standard and inoculated on Mueller–Hinton II Agar medium (Becton Dickinson, Franklin Lakes, NJ, USA) for *S. mutans*, on RPMI 1640 + $NaHCO_3$ + L-Glutamine + phenol red medium (Biocorp, Warsaw, Poland) for *C. albicans* and media consisting of 90% IST (Oxoid, Basingstoke, UK) agar and 10% MRS (Oxoid, Basingstoke, UK) agar adjusted to pH 6.7 for *L. acidophilus*.

An automatic micropipette (Proline® Plus 2–20 µL, Sartorius Biohit Liquid Handling Oy, Helsinki, Finland) was used to apply 6 µL of tested essential oil on filter paper discs (Oxoid, Basingstoke, UK). Chlorhexidine digluconate (CHX) aqueous solution (0.2%) served as a positive control. Filter paper discs (6 mm in diameter) were incubated for 20 min in room temperature in order to ensure the homogenous absorption of tested essential oil. Blank discs were used as negative control.

Next, sterile filter paper discs with tested oils, CHX and blank ones were placed directly on the inoculated agar surface. Special care was taken to ensure uniform contact of the paper disc with the media surface. The cultures were incubated for 18 h at temperature of 35 °C: for *S. mutans* in CO_2 enriched conditions—GENbox CO_2 (bioMerieux S.A., Marcy l'Etoile, France), for *L. acidophilus* in anaerobic conditions; GENbox anaer (bioMerieux S.A., Marcy l'Etoile, France), for *C. albicans*—in aerobic conditions. After the removal of paper discs, the inhibition growth zones were measured (without subtracting disc diameter). For each tested EO and CHX, twelve filter paper discs were used to measure inhibition growth zone of every tested strain.

2.2. Microbiological Studies of Composite Resin Material Modified with Essential Oil

The chosen essential oil was introduced into flowable bulk-fill composite resin (CR) material (SDR flow, Dentsply Sirona, Konstanz, Germany) and mixed mechanically until obtaining uniform and homogenous consistency. The essential oils and dimetacrylate resins possess hydrophobic features hence they can be easily mixed to obtain homogeneous material. The material was modified with the essential oil that exhibited the highest antimicrobial activity. The concentrations of the essential oil were chosen as follows: Group 1: 1 µL of EO in 2 g of CR; Group 2: 2 µ of EO in 2 g of CR; Group 3: 5 µL of EO in 2 g CR.

Disc-shaped (3 mm of height and 6 mm in diameter) samples of composite resin material modified with essential oil were performed. Each sample was light-cured with halogen lamp (Megalux Soft-start, Mega-PHYSIC GmbH & Co. KG, Rastatt, Germany) according to the manufacturer's instruction (i.e., 20 s). To evaluate antimicrobial activity against *S. mutans*, *L. acidophilus* and *C. albicans* of essential oil modified composite resin (EO-CR) eluate method was used.

The samples were placed in 2.5 mL of 0.95% NaCl solution and incubated for 24 h in temperature of 35 °C. Next, after removing tested samples from the eluate, serial dilutions of the tested microbial strains were prepared (10^{-1}, 10^{-2}, 10^{-3}, 10^{-4}, 10^{-5} and 10^{-6}) by the introduction of 200 µL of the strains into 1.8 mL of eluate. Strains were incubated for 18 h.

The control group was a sample of composite resin material, not modified with essential oil, that was incubated in the same conditions as the study groups samples.

After the incubation, to evaluate bacterial susceptibility, 100 µL of the control and 100 µL of bacteria dilution (of each dilution) in eluate were cultivated as follows: *S. mutans* on MH agar medium (Becton-Dickinson, Franklin Lakes, NJ, USA) in CO_2-enriched conditions—GENbox CO_2 (bioMerieux S.A., Marcy l'Etoile, France); *L. acidophilus* in anaerobic conditions on GENbox anaer medium (bioMerieux S.A., Marcy l'Etoile, France), and *C. albicans* in aerobic conditions on RPMI

1640 medium (Thermo Fisher Scientific, Waltham, MA, USA). The strains were incubated for 24 h in temperature of 35 °C.

Upon the cultivation period, bacterial colonies in the studied samples and the control group were counted. The experiment was repeated twelve times for each EO-CR group and for the control group.

2.3. Statistical Analysis

The descriptive analysis of numerical variables encompasses the calculation of the mean (M) along with standard deviations (SD) values. The statistical analysis of the significance consisted of the following: Shapiro–Wilk W test for normality; Levene's tests for equality of variances; One-way analysis of variance; Kruskal–Wallis equality-of-populations rank test; Post-hoc multiple comparison tests; Zero-inflated Poisson regression with robust standard errors. A level of $p < 0.05$ was deemed statistically significant. The statistical analyses of were carried out using Stata®/Special Edition, release 14.2 (StataCorp LP, College Station, TX, USA). The post-hoc statistical power was calculated using post-hoc power analysis calculator (https://clincalc.com/stats/Power.aspx) and a statistical power of 98.56% was found.

3. Results

3.1. Antimicrobial Activity of Essential Oils (Inhibition Growth Zone)

Figure 1 shows a measurement of representative inhibition growth zone of tested EO. The inhibition growth zones of tested microbes measured for each essential oil were presented in Figures 2–4. All tested essential oils, with exception to anise and limette EOs, were found to possess antibacterial activity against *S. mutans* (Figure 2). The diameter of the inhibition zone of *S. mutans* ranged from 0 mm for anise and limette essential oils up to 40 mm for cinnamon essential oil.

The cinnamon oil showed significantly highest antibacterial activity among all ten tested essential oils. Next, it was the thyme EO that exhibited significantly higher activity than anise, citronella, clove, geranium, lavender, limette, mint and rosemary EOs, but significantly lower than the cinnamon EO. Clove and lavender EOs exhibited antibacterial activity comparable to the one of 0.2% CHX. Citronella, geranium and mint showed significantly lower activity than other EOs, with exception to anise and limette EOs (Table A1). The latter showed the lowest antibacterial activity against *S. mutans* among all EOs tested.

All tested essential oils were found to possess antibacterial activity against *L. acidophilus* (Figure 3).

The diameter of the inhibition zone of *L. acidophilus* bacteria ranged from 8 to 40 mm. Again, significantly highest antibacterial activity among all tested essential oils showed cinnamon and thyme EOs, followed by anise and citronella EOs. Geranium, mint EOs and CHX showed significantly higher activity than lavender, limette and rosemary EOs, but significantly lower—than anise, cinnamon, citronella, clove and thyme EOs (Table A2). Lavender and rosemary EOs exhibited the significantly lowest antibacterial activity.

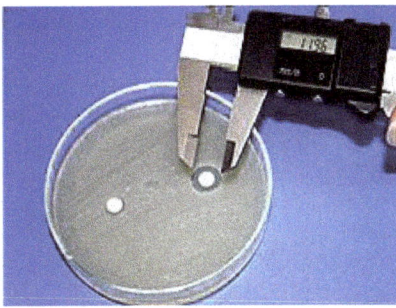

Figure 1. Representative figure of measurement of growth inhibition zone.

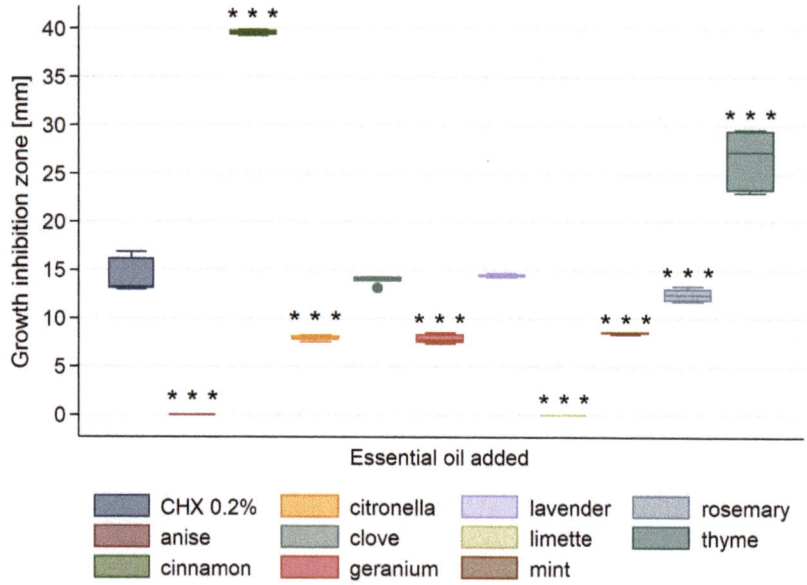

Figure 2. Antibacterial activity of tested essential oils against *S. mutans*. *** $p < 0.001$ versus positive control (0.2% CHX).

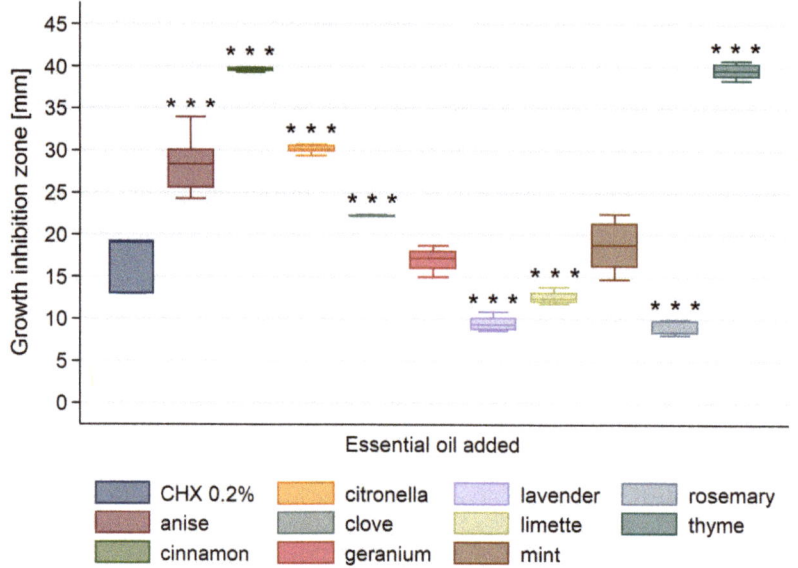

Figure 3. Antibacterial activity of tested essential oils against *L. acidophilus*. *** $p < 0.001$ versus positive control (0.2% CHX).

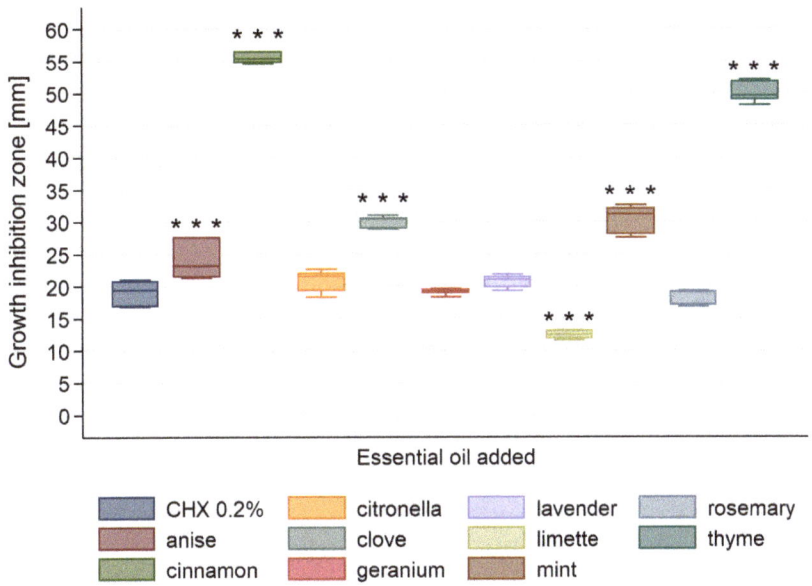

Figure 4. Antifungal activity of tested essential oils against *C. albicans*. *** $p < 0.001$ versus positive control (0.2% CHX).

All tested essential oils were found to possess antifungal activity (Figure 4). The diameter of the inhibition zone of *C. albicans* ranged from 12 to 56 mm.

The significantly highest antifungal activity among all tested essential oils showed cinnamon EO, followed by thyme EO. Clove and mint EOs showed significantly higher activity than other EOs (with exception to cinnamon and thyme EOs). Citronella, geranium and lavender EOs exhibited significantly lower activity than anise, cinnamon, clove, mint and thyme EOs (Table A3). CHX possessed similar antifungal activity as citronella and lavender EOs. Limette exhibited the significantly lowest antifungal activity among all tested EOs.

3.2. Antimicrobial Activity of Composite Resin Modified with Essential Oil

The highest antimicrobial activity against all tested pathogens showed cinnamon EO, hence it was introduced into composite resin material. Next, the modified material was tested for antimicrobial activity against oral pathogens, i.e., *S. mutans*, *L. acidophilus* and *C. albicans*.

For all tested microbes, the essential oil modified composite resins showed statistically significant different CFU than for the control group regardless of the EO concentration (Figures 5–7). Antimicrobial activity of EO-CRs was significantly higher than that of unmodified CR. Furthermore, Fisher's post-hoc test revealed, that for each tested oral pathogen, the differences in CFU between the study groups were statistically significant.

As for *S. mutans*, the significantly highest antibacterial activity showed 2 µL/2 g EO-CR, followed by 1 µL/2 g EO-CR and 5 µL/2 g EO-CR ($p < 0.001$). Whereas, for *L. acidophilus* the least CFU were noted for 1 µL/2 g EO-CR, followed by 2 µL/2 g EO-CR and 5 µL/2 g EO-CR ($p < 0.001$).

As far as antifungal activity against *C albicans* was concerned, the highest activity showed 2 µL/2 g EO-CR and the lowest 5 µL/2 g EO-CR ($p < 0.001$).

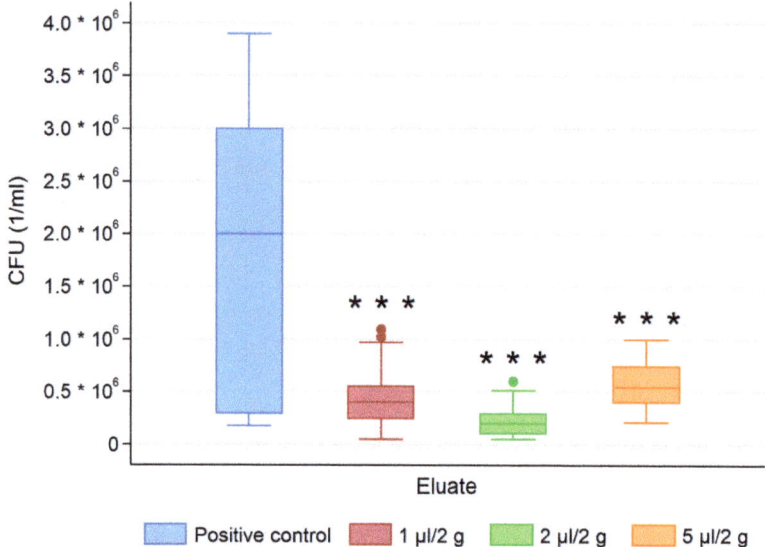

Figure 5. Colony forming units (CFU) of *S. mutans* for essential oil modified composite resins and the control group. *** $p < 0.001$ versus control.

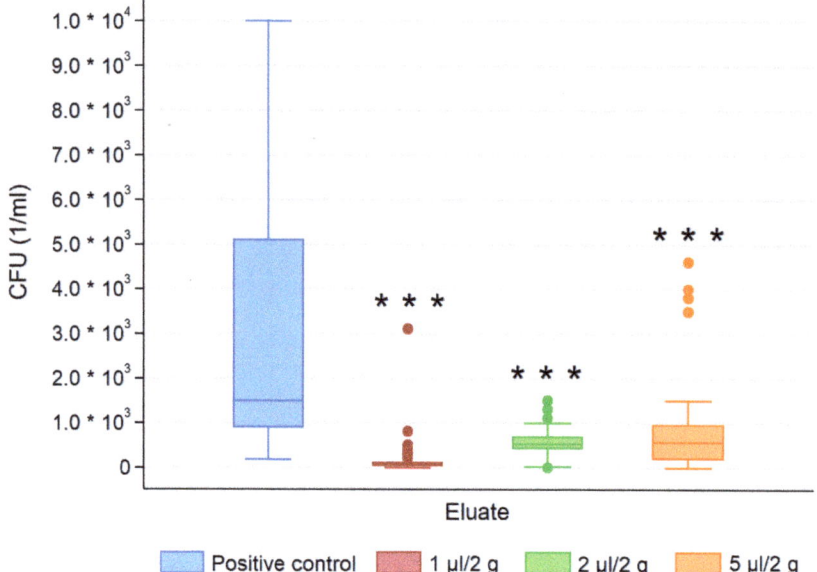

Figure 6. Colony forming units (CFU) of *L. acidophilus* for essential oil modified composite resins and the control group. *** $p < 0.001$ versus control.

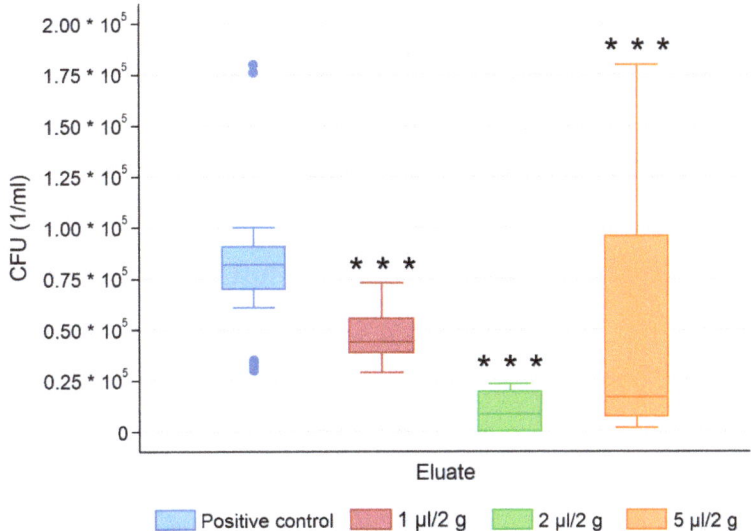

Figure 7. Colony forming units (CFU) of *C. albicans* for essential oil modified composite resins and the control group. *** $p < 0.001$ versus control.

4. Discussion

Essential oils have been used in many walks of life, including dentistry. Researchers constantly search for new possibilities of application of effective formulas into dental products. EOs seem to be the promising ingredients of future oral care products and dental materials used both by patients and dentists. The present study investigated antibacterial activity of ten essential oils against three cariogenic pathogens: *S. mutans*, *L. acidophilus* and *C. albicans*. Such great variety of essential oils tested in one study seemed advantageous as the experiment was performed in the same standardized conditions and allowed for reliable assessment and comparison of EOs' antimicrobial properties. As far as oral pathogens are concerned, most of the previous studies described only a few essential oils [13,14,37,54,55] against one or two most cariogenic bacteria in one setting [56,57].

In the present study, among ten tested essential oils, the most prominent antimicrobial activity exhibited two EOs: cinnamon and thyme. The other EO that showed both significant antibacterial and antifungal effect was clove oil. These results confirmed other findings that EOs possessed potent antibacterial activity and antifungal properties against oral pathogens, including cariogenic bacteria [54,56–58]. The study used *S. mutans* and *L. acidophilus*, due to their undisputable involvement in the carious process. The former one is responsible for the initiation of the process and the latter for its development [59–61]. Given their proven cariogenic activity, *S. mutans* as well as *Lactobacillus spp.*, have been used in the present study. In addition, *C. albicans* is considered to play a supportive role in cariogenic process [42]. Similarly to other studies [62], current study used 0.2% chlorhexidine digluconate aqueous solution as a positive control due to its proved antimicrobial and antifungal activity [63,64].

The composition of EOs determines their antibacterial potential. The highest activity of EOs is provided by thymol, eugenol and carvacrol content, followed by alcohol-containing EOs, with alcohols such as citronellol, geraniol, linalool, menthol, terpinen-4-ol and α-terpineol. Another bioactive group comprise of EOs that contain either ketones, e.g., camphor, carvone, menthone, or thujene or aldehyde groups, i.e., cinnamaldehyde, as well as those with other functional groups, such as anethole and cineole.

Cinnamon essential oil has high percentage of aldehydes (cinnamon aldehydes), that possess antifungal, anti-inflammatory and disinfectant qualities [65]. The effectiveness of cinnamon EO and cinnamon aldehyde against *S. mutans*, *S. mitis*, *S. salivarius*, *A. actinomycetemcomitans*, *P. gingivalis* and *Fusobacterium nucleatum* was reported by Zainal-Abidin et al. [66]. Other studies also confirmed antibacterial activity of cinnamon [47,67] and clove [47,68,69] essential oils against *S. mutans*. High antibacterial activity of clove EOs depends on its aromatic compound content: eugenol (85.3%). Eugenol was reported to have antiseptic, antimicrobial, anesthetic, analgesic, antioxidant, anti-inflammatory and cardiovascular activities [70]. In the present study, cinnamon EO showed significantly higher antimicrobial activity than clove EO, which is consistent with other study [69]. Clove and cinnamon were found to inhibit fungal growth at a concentration of 6% [57]. The results of the present study are consistent with other findings [71] reporting cinnamon oil to have the most potential antibacterial properties. Another study [72] proved cinnamon essential oil to possess the highest antibacterial activity against *S. mutans* among other nine oils (eucalyptol, lime, clove, mint, vinegar, cedar and citrus grass). In addition, Arora and Kaur [73] observed the antimicrobial activity of clove EO against *C. albicans*. It was confirmed by the present study, in which clove EOs exhibited significantly higher activity against *C. albicans* and *L. acidophilus* than CHX, whereas no significant difference between in activity of clove EO and CHX against *S. mutans* was found.

Thyme EO was reported to show antimicrobial activity against oral pathogens due to high content of thymol (38.1%) and *p*-cymene (29.1%) [47,74]. Phenolic compound—thymol, the main component of thyme EO—is reported to disintegrate the outer membrane of Gram-negative bacteria and make bacterial cytoplasmic membrane more permeable to ATP [75]. Another constituent of thyme EO—carvacrol—is proved to exhibit antibacterial potential against *S. mutans* and *C. albicans* [54,62]. Carvacrol emulsion might be also a promising alternative to NaOCl in irrigation of dental root canal system and eradication of intracanal bacteria: *E. feacalis* [76]. Studies proved also a potent antimicrobial activity of thymol against *S. mutans* and *C. albicans* [54,57], as well as against *Porphyromonas gingivalis* and *A. actinomycetemcomitans*, which play a role in development of periodontal disease [77]. That was confirmed by the present study. Thyme EO exhibited the significantly highest antimicrobial activity against *C. albicans* and *S. mutans*, whereas the antibacterial activity against *L. acidophilus* was significantly higher than of other EOs, but lower than that of cinnamon EO. Another study stated that clove, thyme, cinnamon and peppermint EOs are potent antimicrobial phenols [17].

Other EOs tested in the study, such as citronella, geranium, lavender, limette, mint, rosemary presented medium antibacterial activity that is associated with the content of citronellol and geraniol, linalool and linalyl acetate, 1,8-cineole, camphor and α-pinene [74]. As for anise EO, it showed no antibacterial activity against *S. mutans*, whereas its activity against *L. acidophilus* and antifungal activity were high. Antifungal potential of this EO can be attributed to high content of *trans*-anethole, which can interact with fungal plasma [78].

The positive correlation between antibacterial activity of EOs and high content of certain components was reported only for few EOs (e.g., mint, thyme and oregano). For others (e.g., limette and lavender), it is most likely that their antibacterial potential is the result of synergistic effect of the components, since some of those major components exhibit higher antibacterial effect than the EO itself [79].

The present study proved that cinnamon and thyme followed by clove EOs exhibited significantly higher or equal antimicrobial properties against oral pathogens than CHX. These findings would be the introduction to further investigations aiming at the incorporation of these oils into oral care products, i.e., tooth pastes, mouth rinses. The antimicrobial potential of these EOs might be also used to enhance the antibacterial properties of dental materials such as dental resin materials, temporary dressings, disinfectants or root canal filling materials. Furthermore, extracting the most active components from EOs and introducing them into dental products (e.g., restorative materials) composition might be promising line of research. The abundance literature reported that there is great need for development of dental materials with antibacterial properties [4,8,10,12,55,80]. The results of previous studies on antibacterial properties of dental materials seemed promising and suggested that introducing antimicrobial agents into the composition of dental materials might improve their antibacterial potential without deteriorating the physico-mechanical performance [8,12,37].

Given the highest antimicrobial activity obtained in the present study, cinnamon essential oil was used to incorporate, in three different concentrations, into composite resin material composition. Based on preliminary experiments performed, the tested concentration of the EO in composite resin was established as 1, 2 or 5 µL of EO in 2 g of CR. The best antimicrobial properties against *S. mutans* and *C. albicans* were achieved for composite resin containing this essential oil in concentration 2 µL/2 g, whereas against *L. acidophilus* in concentration 1 µL/2 g. Ideally, the composite resin material would present antimicrobial effect and possess very good mechanical properties. The addition of antibacterial or antifungal agents should not change the mechanical performance of the resin material. The current experiment showed that the addition of 2 µL of cinnamon essential oil into 2 g of composite material allows for limiting microbial growth of tested oral pathogens in comparison to unmodified material. This composition might be optimal in terms of antimicrobial properties due to mild influence on polymerization process and enabling release of active compounds into environment.

Still, the present study has some limitations. First of all, the study used single-species model with isolated strains of specific oral pathogens tested in in vitro conditions, without saliva involvement, whereas oral cavity is complex environment holding variety of pathogens interacting in formation of oral biofilm on hard dental tissues. Therefore, these findings must be confirmed in further microbiological studies.

Next, mechanical properties of composite resins modified with essential oils should be tested if considering such materials for clinical application. One study [13] tested mechanical properties of composite resin material modified with cinnamon EO, such as hardness, tensile and flexural strength. The results of the study provided inconsistent data on the proper concentration of the EO in the CR to obtain desirable mechanical performance of the EO-CR material. However, the addition of cinnamon EO to composite material did not adversely affect all the mechanical properties. CR material showed significantly higher flexural strength when modified with 1 µL of cinnamon EO (in 2 g of the material) than non-modified CR. In contrast, non-modified CR showed significantly higher hardness (HV1) and tensile strength values in comparison to modified CR. As far as tensile strength of EO-CR material was concerned, the addition of 2 µL of cinnamon EO (in 2 g of the material) allowed for obtaining significantly highest results. On the contrary, the addition of high amount of EO (5 µL/2 g) significantly deteriorated all tested mechanical properties. Still, such EO-modified bulk-fill material could be clinically used in pediatric dentistry as a final filling in primary teeth or in permanent teeth as a temporary filling, as a liner or in two-step bulk restorative technique in deep cavities. Furthermore, class V cavities, with minimum occlusal loading could be restored with such composite material. Still, long-term performance of such restorations and their aesthetic features must be evaluated.

Moreover, long term study should be performed to evaluate possible allergic reaction to essential oil modified composite resin material as well as the cytotoxic effect of EOs released from EO-CRs. Study showed that EOs present cytotoxic effects on living cells and the severity depends on their type and concentration [16]. Hence, further studies should be conducted to evaluate the potential cytotoxicity and long-term antimicrobial effect of essential oils incorporated into the dental restorative materials.

Since the present study tested only one restorative material, the results cannot be translated to other composites resin materials due to some variation in their composition.

5. Conclusions

The study showed that all ten tested essential oils possess antibacterial activity against *L. acidophilus* and antifungal activity against *C. albicans*. Only two essential oils, anise and limette were ineffective towards *S. mutans*. Among tested essential oils, the cinnamon and thyme showed overall the highest antibacterial and antifungal activity against oral pathogens used in the study. Composite resin modified with cinnamon essential oil showed antimicrobial effect regardless of the EO concentration.

Considering these preliminary results, essential oils seem promising alternative to other antibacterial agents incorporated into resin composite and further studies should be conducted to further evaluate the antimicrobial effect of dental composites modified with essential oils, as well as their mechanical properties.

Author Contributions: Conceptualization, M.L.-S. and A.S.; methodology, B.Z.; software, B.Z.; validation, A.S., B.Z. and J.G.; formal analysis, A.S.; investigation, A.S.; resources, J.S.; data curation, A.S. and B.L.; writing—original draft preparation, A.S., M.L.-S. and B.L.; writing—review and editing, B.L., L.H. and M.L.-S.; visualization, B.L.; supervision, M.L.-S.; project administration, M.L.-S.; funding acquisition, J.S. All authors have read and agreed to the published version of the manuscript.

Funding: This research received no external funding.

Conflicts of Interest: The authors declare no conflict of interest.

Appendix A

Table A1. Levels of statistical significance (p) in Fisher's post-hoc test of EOs activity against *S. mutans*.

Essential Oil	Anise	Citronella	Cinnamon	Clove	Geranium	Lavender	Limette	Mint	Rosemary	Thyme
Anise								NS	NS	NS
Citronella	NS							0.688		
Cinnamon	<0.001	<0.001				<0.001		<0.001	<0.001	
Clove	<0.001	<0.001	<0.001			0.157		<0.001		
Geranium	0.905					<0.001	NS	0.604	<0.001	<0.001
Lavender	<0.001			NS				<0.001	NS	
Limette	NS	NS	NS			NS				
Mint										
Rosemary	<0.001	<0.001		0.181		0.940	NS	<0.001		
Thyme	<0.001	<0.001	<0.001	<0.001		<0.001		<0.001	<0.001	

NS = not significant.

Table A2. Levels of statistical significance (p) in Fisher's post-hoc test of EOs activity against *L. acidophilus*.

Essential oil	Anise	Cinnamon	Citronella	Clove	Geranium	Lavender	Limette	Mint	Rosemary	Thyme
Anise								<0.001	<0.001	<0.001
Cinnamon	<0.001							<0.001	<0.001	
Citronella				<0.001		<0.001		<0.001		
Clove	<0.001		<0.001			<0.001		<0.001	<0.001	
Geranium			<0.001	<0.001		<0.001	<0.001	0.209	<0.001	<0.001
Lavender	<0.001	0.544	<0.001					<0.001	<0.001	
Limette			<0.001	<0.001		<0.001	<0.001	<0.001	<0.001	
Mint										
Rosemary	<0.001		<0.001	<0.001		0.557	<0.001	<0.001	<0.001	
Thyme	0.734		<0.001	<0.001		<0.001	<0.001	<0.001	<0.001	

Table A3. Levels of statistical significance (p) in Fisher's post post-hoc test of EOs activity against *C. albicans*

Essential oil	Anise	Cinnamon	Citronella	Clove	Geranium	Lavender	Limette	Mint	Rosemary	Thyme
Anise		<0.001	<0.001	<0.001	<0.001	<0.001	<0.001	<0.001	<0.001	<0.001
Cinnamon			<0.001	<0.001		<0.001		<0.001	<0.001	
Citronella								<0.001		
Clove			<0.001			<0.001		0.486		
Geranium		<0.001	0.030	<0.001		0.001	<0.001	<0.001	0.648	<0.001
Lavender			0.252					<0.001		
Limette		<0.001	<0.001	<0.001		<0.001		<0.001	<0.001	
Mint										
Rosemary			0.009	<0.001		<0.001		<0.001		
Thyme		<0.001	<0.001	<0.001		<0.001	<0.001	<0.001	<0.001	

References

1. Beyth, N.; Farah, S.; Domb, A.J.; Weiss, E.I. Antibacterial dental resin composites. *React. Funct. Polym.* **2014**, *75*, 81–88. [CrossRef]
2. Lukomska-Szymanska, M.; Konieczka, M.; Zarzycka, B.; Lapinska, B.; Grzegorczyk, J.; Sokolowski, J. Antibacterial activity of commercial dentine bonding systems against E. faecalis-flow cytometry study. *Materials* **2017**, *10*, 481. [CrossRef] [PubMed]
3. Lapinska, B.; Konieczka, M.; Zarzycka, B.; Sokolowski, K.; Grzegorczyk, J.; Lukomska-Szymanska, M. Flow cytometry analysis of antibacterial effects of universal dentin bonding agents on streptococcus mutans. *Molecules* **2019**, *24*, 532. [CrossRef] [PubMed]
4. Imazato, S. Bio-active restorative materials with antibacterial effects: New dimension of innovation in restorative dentistry. *Dent. Mater. J.* **2009**, *28*, 11–19. [CrossRef]
5. Cocco, A.R.; Rosa, W.L.D.O.D.; Silva, A.F.; Lund, R.G.; Piva, E. A systematic review about antibacterial monomers used in dental adhesive systems: Current status and further prospects. *Dent. Mater.* **2015**, *31*, 1345–1362. [CrossRef]
6. Barszczewska-Rybarek, I.; Chladek, G. Studies on the curing efficiency and mechanical properties of bis-GMA and TEGDMA nanocomposites containing silver nanoparticles. *Int. J. Mol. Sci.* **2018**, *19*, 3937. [CrossRef]
7. Stencel, R.; Kasperski, J.; Pakiela, W.; Mertas, A.; Bobela, E.; Barszczewska-Rybarek, I.; Chladek, G. Properties of experimental dental composites containing antibacterial silver-releasing filler. *Materials* **2018**, *11*, 1031. [CrossRef]
8. Lukomska-Szymanska, M.; Zarzycka, B.; Grzegorczyk, J.; Sokolowski, K.; Poltorak, K.; Sokolowski, J.; Lapinska, B. Antibacterial properties of calcium fluoride-based composite materials: *In Vitro* study. *BioMed Res. Int.* **2016**, *2016*, 1–7. [CrossRef]
9. Drzewiecka, K.; Kleczewska, J.; Krasowski, M.; Lapinska, B.; Sokolowski, J. The Influence of amorphous calcium phosphate addition on mechanical properties of the experimental light-cured dental composite. *Dent. Med. Probl.* **2016**, *53*, 34–40. [CrossRef]
10. Pokrowiecki, R.; Zareba, T.; Mielczarek, A.; Opalińska, A.; Wojnarowicz, J.; Majkowski, M.; Lojkowski, W.; Tyski, S. Evaluation of biocidal properties of silver nanoparticles against cariogenic bacteria. *Med. Dosw. Mikrobiol.* **2013**, *65*, 197–206.
11. Marović, D.; Šariri, K.; Demoli, N.; Ristić, M.; Hiller, K.-A.; Škrtić, D.; Rosentritt, M.; Schmalz, G.; Tarle, Z. Remineralizing amorphous calcium phosphate based composite resins: The influence of inert fillers on monomer conversion, polymerization shrinkage, and microhardness. *Croat. Med. J.* **2016**, *57*, 465–473. [CrossRef] [PubMed]
12. Lukomska-Szymanska, M.; Kleczewska, J.; Nowak, J.; Prylinski, M.; Szczesio, A.; Podlewska, M.; Sokolowski, J.; Lapinska, B. Mechanical properties of calcium fluoride-based composite materials. *BioMed Res. Int.* **2016**, *2016*, 1–8. [CrossRef]
13. Szram, A.; Sokolowski, J.; Nowak, J.; Domarecka, M.; Lukomska-Szymanska, M. Mechanical properties of composite material modified with essential oil. *Inż. Mater.* **2017**, *1*, 49–53. [CrossRef]
14. Peralta, S.L.; Carvalho, P.H.A.; van de Sande, F.H.; Pereira, C.M.P.; Piva, E.; Lund, R.G. Self-etching dental adhesive containing a natural essential oil: Anti-biofouling performance and mechanical properties. *Biofouling* **2013**, *29*, 345–355. [CrossRef] [PubMed]
15. Noorizadeh, H.; Farmany, A.; Noorizadeh, M. Quantitative structure-retention relationships analysis of retention index of essential oils. *Quím. Nova* **2011**, *34*, 242–249. [CrossRef]
16. Bakkali, F.; Averbeck, S.; Averbeck, D.; Idaomar, M. Biological effects of essential oils—A review. *Food Chem. Toxicol.* **2008**, *46*, 446–475. [CrossRef]
17. Burt, S. Essential oils: Their antibacterial properties and potential applications in foods—Review. *Int. J. Food Microbiol.* **2004**, *94*, 223–253. [CrossRef]
18. Carson, C.F.; Hammer, K.A. Chemistry and bioactivity of essential oils. In *Lipids and Essential Oils as Antimicrobial Agents*; John Wiley & Sons, Ltd.: Hoboken, NJ, USA, 2010; pp. 203–238. ISBN 9780470741788.
19. El Asbahani, A.; Miladi, K.; Badri, W.; Sala, M.; Addi, E.A.; Casabianca, H.; El Mousadik, A.; Hartmann, D.; Jilale, A.; Renaud, F.; et al. Essential oils: From extraction to encapsulation. *Int. J. Pharm.* **2015**, *483*, 220–243. [CrossRef]

20. Mazzarrino, G.; Paparella, A.; Chaves-López, C.; Faberi, A.; Sergi, M.; Sigismondi, C.; Compagnone, D.; Serio, A. Salmonella enterica and Listeria monocytogenes inactivation dynamics after treatment with selected essential oils. *Food Control* **2015**, *50*, 794–803. [CrossRef]
21. Chang, Y.; McLandsborough, L.; McClements, D.J. Fabrication, stability and efficacy of dual-component antimicrobial nanoemulsions: Essential oil (thyme oil) and cationic surfactant (lauric arginate). *Food Chem.* **2015**, *172*, 298–304. [CrossRef]
22. Saad, N.Y.; Muller, C.D.; Lobstein, A. Major bioactivities and mechanism of action of essential oils and their components. *Flavour Fragr. J.* **2013**, *28*, 269–279. [CrossRef]
23. Koul, O.; Singh, R.; Kaur, B.; Kanda, D. Comparative study on the behavioral response and acute toxicity of some essential oil compounds and their binary mixtures to larvae of Helicoverpa armigera, Spodoptera litura and Chilo partellus. *Ind. Crops Prod.* **2013**, *49*, 428–436. [CrossRef]
24. Olmedo, R.; Nepote, V.; Grosso, N.R. Antioxidant activity of fractions from oregano essential oils obtained by molecular distillation. *Food Chem.* **2014**, *156*, 212–219. [CrossRef]
25. Ebada, S.S.; Lin, W.; Proksch, P. Bioactive sesterterpenes and triterpenes from marine sponges: Occurrence and pharmacological significance. *Mar. Drugs* **2010**, *8*, 313–346. [CrossRef] [PubMed]
26. Sobral, M.V.; Xavier, A.L.; Lima, T.C.; De Sousa, D.P. Antitumor activity of monoterpenes found in essential oils. *Sci. World J.* **2014**, *2014*, 35. [CrossRef] [PubMed]
27. Grigore, A.; Paraschiv, I.; Colceru-Mihul, S.; Bubueanu, C.; Draghici, E.; Ichim, M. Chemical composition and antioxidant activity of Thymus vulgaris L. volatile oil obtained by two different methods. *Helios* **2010**, *15*, 5436–5443.
28. Sonboli, A.; Esmaeili, M.A.; Gholipour, A.; Kanani, M.R. Composition, cytotoxicity and antioxidant activity of the essential oil of Dracocephalum surmandinum from Iran. *Nat. Prod. Commun.* **2010**, *5*, 341–344. [CrossRef]
29. Sá, R.D.C.D.S.E.; Andrade, L.N.; De Sousa, D.P. A review on anti-inflammatory activity of monoterpenes. *Molecules* **2013**, *18*, 1227–1254. [CrossRef]
30. Solomakos, N.; Govaris, A.; Koidis, P.; Botsoglou, N. The antimicrobial effect of thyme essential oil, nisin, and their combination against Listeria monocytogenes in minced beef during refrigerated storage. *Food Microbiol.* **2008**, *25*, 120–127. [CrossRef]
31. Bassolé, I.H.N.; Juliani, H.R. Essential oils in combination and their antimicrobial properties. *Molecules* **2012**, *17*, 3989–4006. [CrossRef]
32. Ambrosio, C.M.S.; de Alencar, S.M.; de Sousa, R.L.M.; Moreno, A.M.; Da Gloria, E.M. Antimicrobial activity of several essential oils on pathogenic and beneficial bacteria. *Ind. Crops Prod.* **2017**, *97*, 128–136. [CrossRef]
33. Cosentino, S.; Tuberoso, C.I.G.; Pisano, B.; Satta, M.; Mascia, V.; Arzedi, E.; Palmas, F. In-vitro antimicrobial activity and chemical composition of Sardinian Thymus essential oils. *Lett. Appl. Microbiol.* **1999**, *29*, 130–135. [CrossRef] [PubMed]
34. Bajpai, V.K.; Baek, K.H.; Kang, S.C. Control of Salmonella in foods by using essential oils: A review. *Food Res. Int.* **2012**, *45*, 722–734. [CrossRef]
35. Haas, A.N.; Wagner, T.P.; Muniz, F.W.M.G.; Fiorini, T.; Cavagni, J.; Celeste, R.K. Essential oils-containing mouthwashes for gingivitis and plaque: Meta-analyses and meta-regression. *J. Dent.* **2016**, *55*, 7–15. [CrossRef] [PubMed]
36. Dagli, N.; Dagli, R.; Mahmoud, R.; Baroudi, K. Essential oils, their therapeutic properties, and implication in dentistry: A review. *J. Int. Soc. Prev. Community Dent.* **2015**, *5*, 335. [CrossRef] [PubMed]
37. Peralta, S.L.; Valente, L.L.; Bueno, A.S.; Piva, E.; Lund, R.G. Antibacterial and mechanical properties of one experimental adhesive containing essential oil. *Dent. Mater.* **2011**, *27*, e15. [CrossRef]
38. Moalic, E.; Gestalin, A.; Quinio, D.; Gest, P.E.; Zerilli, A.; Le Flohic, A.M. The extent of oral fungal flora in 353 students and possible relationships with dental caries. *Caries Res.* **2001**, *35*, 149–155. [CrossRef]
39. Sedgley, C.M.; Samaranayake, L.P.; Chan, J.C.Y.; Wei, S.H.Y. A 4-year longitudinal study of the oral prevalence of enteric gram–negative rods and yeasts in Chinese children. *Oral Microbiol. Immunol.* **1997**, *12*, 183–188. [CrossRef]
40. Jacob, L.S.; Flaitz, C.M.; Nichols, C.M.; Hicks, M.J. Role of dentinal carious lesions in the pathogenesis of oral candidiasis in HIV infection. *J. Am. Dent. Assoc.* **1998**, *129*, 187–194. [CrossRef]
41. Shen, S.; Samaranayake, L.; Yip, H.; Dyson, J. Bacterial and yeast flora of root surface caries in elderly, ethnic Chinese. *Oral Dis.* **2002**, *8*, 207–217. [CrossRef]

42. Nikawa, H.; Yamashiro, H.; Makihira, S.; Nishimura, M.; Egusa, H.; Furukawa, M.; Setijanto, D.; Hamada, T. In vitro cariogenic potential of Candida albicans. *Mycoses* **2003**, *46*, 471–478. [CrossRef] [PubMed]
43. Jenkinson, H.F.; Lala, H.C.; Shepherd, M.G. Coaggregation of Streptococcus sanguis and other streptococci with Candida albicans. *Infect. Immun.* **1990**, *58*, 1429–1436. [CrossRef] [PubMed]
44. Kim, D.; Sengupta, A.; Niepa, T.H.R.; Lee, B.-H.; Weljie, A.; Freitas-Blanco, V.S.; Murata, R.M.; Stebe, K.J.; Lee, D.; Koo, H. Candida albicans stimulates Streptococcus mutans microcolony development via cross-kingdom biofilm-derived metabolites. *Sci. Rep.* **2017**, *7*, 41332. [CrossRef] [PubMed]
45. Maijala, M.; Rautemaa, R.; Järvensivu, A.; Richardson, M.; Salo, T.; Tjäderhane, L. Candida albicans does not invade carious human dentine. *Oral Dis.* **2007**, *13*, 279–284. [CrossRef]
46. Sienkiewicz, M.; Łysakowska, M.; Denys, P.; Kowalczyk, E. The antimicrobial activity of thyme essential oil against multidrug resistant clinical bacterial strains. *Microb. Drug Resist.* **2012**, *18*, 137–148. [CrossRef]
47. Grzesiak, B.; Głowacka, A.; Krukowski, H.; Lisowski, A.; Lassa, H.; Sienkiewicz, M. The in vitro efficacy of essential oils and antifungal drugs against prototheca zopfii. *Mycopathologia* **2016**, *181*, 609–615. [CrossRef]
48. Sienkiewicz, M.; Glowacka, A.; Kowalczyk, E.; Wiktorowska-Owczarek, A.; Jóźwiak-Będenista, M.; Łysakowska, M. The biological activities of cinnamon, geranium and lavender essential oils. *Molecules* **2014**, *19*, 20929–20940. [CrossRef]
49. Sienkiewicz, M.; Łysakowska, M.; Kowalczyk, E.; Szymańska, G.; Kochan, E.; Krukowska, J.; Olszewski, J.; Zielińska-Bliźniewska, H. The ability of selected plant essential oils to enhance the action of recommended antibiotics against pathogenic wound bacteria. *Burns* **2017**, *43*, 310–317. [CrossRef]
50. Gniewosz, M.; Kraśniewska, K.; Kosakowska, O.; Pobiega, K.; Wolska, I. Chemical compounds and antimicrobial activity of petitgrain (*Citrus aurantium* L. var. amara) essential oil. *Herba Pol.* **2018**, *63*, 18–25. [CrossRef]
51. Budzyńska, A.; Sadowska, B.; Lipowczan, G.; Maciąg, A.; Kalemba, D.; Różalska, B. Activity of selected essential oils against *Candida* spp. strains. Evaluation of New aspects of their specific pharmacological properties, with special reference to lemon balm. *Adv. Microbiol.* **2013**, *3*, 317–325. [CrossRef]
52. Sienkiewicz, M.; Poznańska-Kurowska, K.; Kaszuba, A.; Kowalczyk, E. The antibacterial activity of geranium oil against Gram-negative bacteria isolated from difficult-to-heal wounds. *Burns* **2014**, *40*, 1046–1051. [CrossRef] [PubMed]
53. The European Pharmacopoeia (Ph. Eur.) 10th Edition. Available online: https://www.edqm.eu/en/european-pharmacopoeia-ph-eur-10th-edition (accessed on 5 September 2020).
54. Botelho, M.A.; Nogueira, N.A.P.; Bastos, G.M.; Fonseca, S.G.C.; Lemos, T.L.G.; Matos, F.J.A.; Montenegro, D.; Heukelbach, J.; Rao, V.S.; Brito, G.A.C. Antimicrobial activity of the essential oil from Lippia sidoides, carvacrol and thymol against oral pathogens. *Braz. J. Med. Biol. Res.* **2007**, *40*, 349–356. [CrossRef] [PubMed]
55. Łysakowska, M.E.; Sienkiewicz, M.; Banaszek, K.; Sokołowski, J.; McPhee, D.J. The sensitivity of endodontic Enterococcus spp. strains to geranium essential oil. *Molecules* **2015**, *20*, 22881–22889. [CrossRef] [PubMed]
56. Galvão, L.C.D.C.; Furletti, V.F.; Bersan, S.M.F.; Da Cunha, M.G.; Ruiz, A.; De Carvalho, J.E.; Sartoratto, A.; Rehder, V.L.G.; Figueira, G.M.; Duarte, M.C.T.; et al. Antimicrobial activity of essential oils against Streptococcus mutans and their antiproliferative effects. *Evid. Based Complement. Altern. Med.* **2012**, *2012*. [CrossRef]
57. Radwan, I.A.; Abed, A.H.; Abeer, M.R.; Ibrahim, M.A.; Abdallah, A.S. Effect of thyme, clove and cinnamon essential oils on Candida albicans and moulds isolated from different sources. *Am. J. Anim. Vet. Sci.* **2014**, *9*, 303–314. [CrossRef]
58. Swamy, M.K.; Akhtar, M.S.; Sinniah, U.R. Antimicrobial properties of plant essential oils against human pathogens and their mode of action: An updated review. *Evid. Based Complement. Altern. Med.* **2016**, *2016*. [CrossRef]
59. Jokstad, A. Secondary caries and microleakage. *Dent. Mater.* **2016**, *32*, 11–25. [CrossRef]
60. Cummins, D. The development and validation of a new technology, based upon 1.5% arginine, an insoluble calcium compound and fluoride, for everyday use in the prevention and treatment of dental caries. *J. Dent.* **2013**, *41*, S1–S11. [CrossRef]
61. Kulshrestha, S.; Khan, S.; Hasan, S.; Khan, M.E.; Misba, L.; Khan, A.U. Calcium fluoride nanoparticles induced suppression of Streptococcus mutans biofilm: An *in vitro* and in vivo approach. *Appl. Microbiol. Biotechnol.* **2015**, *100*, 1901–1914. [CrossRef]

62. Freires, I.A.; Denny, C.; Benso, B.; De Alencar, S.M.; Rosalen, P.L. Antibacterial activity of essential oils and their isolated constituents against cariogenic bacteria: A systematic review. *Molecules* **2015**, *20*, 7329–7358. [CrossRef]
63. Lapinska, B.; Klimek, L.; Sokolowski, J.; Lukomska-Szymanska, M. Dentine surface morphology after chlorhexidine application—SEM study. *Polymers* **2018**, *10*, 905. [CrossRef] [PubMed]
64. Lukomska-Szymanska, M.; Sokolowski, J.; Lapinska, B. Chlorhexidine—Mechanism of action and its application to dentistry. *J. Stomatol.* **2017**, *70*, 405–417. [CrossRef]
65. Szczepanski, S.; Lipski, A. Essential oils show specific inhibiting effects on bacterial biofilm formation. *Food Control* **2013**, *36*, 224–229. [CrossRef]
66. Zainal-Abidin, Z. Anti-bacterial activity of cinnamon oil on oral pathogens. *Open Conf. Proc. J.* **2014**, *4*, 12–16. [CrossRef]
67. Almeida, L.; De Paula, J.F.; De Almeida, R.V.D.; Williams, D.W.; Hebling, J.; Cavalcanti, Y.W. Efficacy of citronella and cinnamon essential oils on Candida albicans biofilms. *Acta Odontol. Scand.* **2016**, *74*, 393–398. [CrossRef]
68. Moon, S.E.; Kim, H.Y.; Cha, J.D. Synergistic effect between clove oil and its major compounds and antibiotics against oral bacteria. *Arch. Oral Biol.* **2011**, *56*, 907–916. [CrossRef]
69. Gupta, C.; Kumari, A.; Garg, A.P.; Catanzaro, R.; Marotta, F. Comparative study of cinnamon oil and clove oil on some oral microbiota. *Acta Biomed.* **2011**, *82*, 197–199.
70. Pramod, K.; Ansari, S.H.; Ali, J. Eugenol: A natural compound with versatile pharmacological actions. *Nat. Prod. Commun.* **2010**, *5*, 1999–2006. [CrossRef]
71. Prabuseenivasan, S.; Jayakumar, M.; Ignacimuthu, S. In vitro antibacterial activity of some plant essential oils. *BMC Complement. Altern. Med.* **2006**, *6*, 39. [CrossRef]
72. Chaudhari, L.K.D.; Jawale, B.A.; Sharma, S.; Sharma, H.; Kumar, H.S.C.M.; Kulkarni, P.A. Antimicrobial activity of commercially available essential oils against Streptococcus mutans. *J. Contemp. Dent. Pract.* **2012**, *13*, 71–74. [CrossRef]
73. Arora, R.; Rao, M.H. Comparative evaluation of the antibacterial effects of four dentine bonding systems: An *in vitro* study. *J. Conserv. Dent.* **2013**, *16*, 466–470. [CrossRef] [PubMed]
74. Wińska, K.; Mączka, W.; Łyczko, J.; Grabarczyk, M.; Czubaszek, A.; Szumny, A. Essential oils as antimicrobial agents—Myth or real alternative? *Molecules* **2019**, *24*, 2130. [CrossRef] [PubMed]
75. Lambert, R.J.W.; Skandamis, P.N.; Coote, P.J.; Nychas, G.J.E. A study of the minimum inhibitory concentration and mode of action of oregano essential oil, thymol and carvacrol. *J. Appl. Microbiol.* **2001**, *91*, 453–462. [CrossRef] [PubMed]
76. Nosrat, A.; Bolhari, B.; Shariffian, M.R.; Aligholi, M.; Mortazavi, M.S. The effect of Carvacrol on Enterococcus faecalis as a final irrigant Archive of SID. *Int. Endod. J.* **2009**, *4*, 96–100.
77. Fani, M.; Kohanteb, J. *In Vitro* antimicrobial activity of thymus vulgaris essential oil against major oral pathogens. *J. Evid. Based Complement. Altern. Med.* **2017**, *22*, 660–666. [CrossRef] [PubMed]
78. Neto, A.C.D.R.; Navarro, B.B.; Canton, L.; Maraschin, M.; Di Piero, R.M. Antifungal activity of palmarosa (*Cymbopogon martinii*), tea tree (*Melaleuca alternifolia*) and star anise (*Illicium verum*) essential oils against Penicillium expansum and their mechanisms of action. *LWT* **2019**, *105*, 385–392. [CrossRef]
79. Soković, M.; Glamočlija, J.; Marin, P.D.; Brkić, D.; van Griensven, L.J.L.D. Antibacterial effects of the essential oils of commonly consumed medicinal herbs using an *in vitro* model. *Molecules* **2010**, *15*, 7532–7546. [CrossRef]
80. Nedeljkovic, I.; Teughels, W.; De Munck, J.; Van Meerbeek, B.; Van Landuyt, K.L. Is secondary caries with composites a material-based problem? *Dent. Mater.* **2015**, *31*, e247–e277. [CrossRef]

© 2020 by the authors. Licensee MDPI, Basel, Switzerland. This article is an open access article distributed under the terms and conditions of the Creative Commons Attribution (CC BY) license (http://creativecommons.org/licenses/by/4.0/).

Article

Antibacterial and Antibiofilm Properties of Three Resin-Based Dental Composites against *Streptococcus mutans*

Simonetta D'Ercole [1,†], Francesco De Angelis [1,*,†], Virginia Biferi [1], Chiara Noviello [1], Domenico Tripodi [1], Silvia Di Lodovico [2], Luigina Cellini [2,‡] and Camillo D'Arcangelo [1,‡]

1. Department of Medical, Oral and Biotechnological Sciences, "G. d'Annunzio" University of Chieti–Pescara, 66100 Chieti, Italy; simonetta.dercole@unich.it (S.D.); virginia.biferi@unich.it (V.B.); chiara96sgr@gmail.com (C.N.); tripodi@unich.it (D.T.); cdarcang@unich.it (C.D.)
2. Department of Pharmacy, "G. d'Annunzio" University of Chieti–Pescara, 66100 Chieti, Italy; silvia.dilodovico@unich.it (S.D.L.); l.cellini@unich.it (L.C.)
* Correspondence: fda580@gmail.com; Tel.: +39-(0)85-4549652
† These authors contributed equally to this work.
‡ These authors contributed equally to this work.

Abstract: Antibacterial and antibiofilm properties of restorative dental materials may improve restorative treatment outcomes. The aim of this in vitro study was to evaluate *Streptococcus mutans* capability to adhere and form biofilm on the surface of three commercially available composite resins (CRs) with different chemical compositions: GrandioSO (VOCO), Venus Diamond (VD), and Clearfil Majesty (ES-2). Disk-shaped specimens were manufactured by light-curing the CRs through two glass slides to maintain a perfectly standardized surface topography. Specimens were subjected to Planktonic OD_{600nm}, Planktonic CFU count, Planktonic MTT, Planktonic live/dead, Adherent Bacteria CFU count, Biomass Quantification OD_{570nm}, Adherent Bacteria MTT, Concanavalin A, and Scanning Electron Microscope analysis. In presence of VOCO, VD, and ES2, both Planktonic CFU count and Planktonic OD_{600nm} were significantly reduced compared to that of control. The amount of Adherent CFUs, biofilm Biomass, metabolic activity, and extracellular polymeric substances were significantly reduced in VOCO, compared to those of ES2 and VD. Results demonstrated that in presence of the same surface properties, chemical composition might significantly influence the in vitro bacterial adhesion/proliferation on resin composites. Additional studies seem necessary to confirm the present results.

Keywords: composite resins; *Streptococcus mutans*; biofilm formation; dental caries; bacterial adhesion

1. Introduction

The use of composite resin (CR)-based materials in restorative dentistry considerably grew in recent years, mainly due to their latest improvements in mechanical properties and esthetic features [1–4]. CRs can be successfully employed for small- and medium-sized adhesive restorations, with direct techniques, and even when dealing with wide restorations and extensive rehabilitations, following an indirect approach [5,6]. However, oral cavities may present many challenges to their physical and chemical stability, mainly due to an extremely moist environment and the frequent temperature/pH variations [7]. Acids released by acidogenic bacteria such as *Streptococcus mutans*, together with the degradative attack pursued by enzymes present in saliva, can progressively compromise the CR functional and esthetic features over time, which may even result in interfacial gaps at the margins of restoration [8].

Many research efforts into studying the bacterial biofilm formation on composite resins were conducted. Biofilms adhering to CRs contain microorganisms, such as streptococci and lactobacilli, that can survive in a highly acidic environment [9]. The amount of *S. mutans* in biofilm is particularly relevant as it plays a key role in determining its cariogenic potential [10]. Many studies showed that CR surface is susceptible to bacterial adhesion, mainly in the marginal areas of restorations [11,12]. Moreover, the adhesion phase is strongly influenced by surface characteristics such as surface roughness (Ra) and surface free energy [13]. An Ra value below 0.2 µm is considered to have a low effect on bacterial adhesion in vivo [14]. Thus, optimizing surface properties of dental composite resins through an accurate finishing and polishing can contribute to restoration success by preventing biofilm formation [15–17].

Beyond surface roughness, other parameters seem to be able to significantly affect bacterial adhesion on composite resins. Aside from the *ad hoc* inclusion of antibacterial additives, it could be interesting to better investigate if even for general composites, the inherent CR formulation could play a relevant role in biofilm formation phases [18,19]. For instance, there are controversial results in literature concerning the correlation between biofilm formation and filler particle size [18,20,21]: several studies reported that micro-hybrid composites might lead to higher bacterial adhesion than nanohybrids [22], while other studies observed no differences [21]. Likewise, concerning the effect of the organic matrix components on bacteria, a study by Van Landuyt et al. showed that in presence of ethylene glycol dimethacrylate (DEGDMA), triethylene glycol dimethacrylate (TEGDMA), or diethylene glycol dimethacrylate (EGDMA), the count of *Streptococcus sanguinis* and *Streptococcus sobrinus* were similarly decreased compared to that of the broth control after 6 h and 9 h incubation periods, respectively, with no significant differences among the tested monomers [23]. As well, Takahashi et al. reported no differences between the effect of TEGDMA and DEGDMA on CFU numbers (CFU/mL) of *S. sobrinus* and *S. sanguinis* [24]. On the other hand, other results suggest that different monomers might differently affect cell proliferation [25,26].

Based on the above-mentioned contrasting findings, the aim of the present in vitro study was to analyze the *S. mutans* capability to adhere and form biofilm on the surface of three commercially available CRs with different chemical compositions. The null hypothesis was that there are no differences in terms of antibacterial and antibiofilm properties among chemically different resin composites if they show the same standardized surface topography.

2. Materials and Methods

The list of the composite resins included in the experimental design, together with their composition, is given in Table 1.

Table 1. Composite resins included in the study.

Group	Material	Manufacturer	Lot Number	Composition
VOCO	GrandioSO Shade A2 (Nanohybrid)	Voco GmbH (Cuxhaven, Germany)	1847313	89% (w/w) fillers (1 µm glass ceramic filler, 20–40 nm silicon dioxide fillers), Bis-GMA *, Bis-EMA *, TEGDMA *.
VD	Venus DiamondShade A2 (Nanohybrid)	Kulzer GmbH (Hanau, Germany)	K010070	80–82% (w/w) fillers (5 nm–20 µm barium aluminum fluoride glass fillers), TCD-UA *, UDMA *, TEGDMA *.
ES-2	Clearfil Majesty ES-2 Classic Shade A2 (Nanohybrid)	Kuraray (Chiyoda, Tokyo, Japan)	7D008	78% (w/w) fillers (0.37 µm–1.5 µm silanated barium glass fille, pre-polymerized organic fillers), Bis-GMA *, hydrophobic aromatic dimethacrylate.

* Bis-GMA = Bisphenol A-glycidyl methacrylate; Bis-EMA = Bisphenol A-glycidyl methacrylate ethoxylated; TEGDMA = Triethylene glycol dimethacrylate; TCD-UA = Tricyclodecane-urethane acrylate; UDMA = Urethane Dimethacrylate.

2.1. Realization of Composite Disks

Disk-shaped specimens were prepared by positioning the uncured material in a polyvinylsiloxane mold, with a diameter of 4 mm and a height of 2 mm, resulting in a total surface area of 50.27 mm². To achieve a perfectly smooth and standardized surface topography, the samples were inserted between two glass slides and stuck with a paper clip for 20 s to extrude the excess material. They were then light-cured for 20 s from the upper surface and for 20 s from the lower surface, using a light-emitting-diode curing unit (Celalux 3, VOCO, Cuxhaven, Germany) with 8 mm diameter tip and an output power of 1300 mW/cm^2. All disks were washed in an ultrasonic bath and were not subjected to any further surface treatment.

2.2. Saliva Collection

Human saliva samples were taken from healthy volunteers with age >18 years, according to a previously described protocol. The Ethics Committee of University "G. d'Annunzio", Chieti–Pescara, Italy (approval code SALI, N. 19 of the 10 September 2020) approved the collection and the use of saliva [27]. The healthy volunteers refrained from oral hygiene for 24 h, did not have any active oral caries, periodontitis, dental care in progress, or antibiotics therapy for at least three months prior to the beginning of the study.

Saliva was pooled, mixed, centrifuged (16.000× g for 1 h at 4 °C), and filtered from microorganisms by filters with pore diameters of 0.8, 0.45, and 0.2 µm. Saliva samples were considered sterile if no growth could be detected in both aerobic and anaerobic atmospheres after incubation for 24–48 h at 37 °C [27]. Sterile saliva was collected in sterile tubes and kept frozen until needed for the study.

2.3. Microbial Strain

Streptococcus mutans CH02, a clinical strain isolated from caries which was sourced from the private collection of the microbiology laboratory of the University "G. d'Annunzio", Chieti–Pescara, was used in this experimental study [28]. The frozen (−80 °C) strain was recovered in Brain Heart Infusion broth (BHI, Oxoid, Milan, Italy) overnight at 37 °C under anaerobic condition. Then, the broth culture was diluted 1:10 in BHI broth (OXOID) containing 1% (w/v) sucrose and refreshed for 2 h at 37 °C in a shaking thermostatic water bath (160 rpm). Bacterial suspension was prepared using a spectrophotometer (Eppendorf, Milan, Italy) to obtain an optical density of OD_{600} = 0.12, corresponding to 9×10^6 CFU/mL. This bacterial suspension was used for the experiment.

2.4. Experimental Design

All composite disk specimens were placed in 96-well polystyrene microtiter plates and sterilized through top and bottom surface exposure to ultraviolet UV light for 40 min [29]. Then, the sterile specimens were inoculated for 2 h in saliva at 37 °C in a shaking incubator with slight agitation to form the protein pellicle layer on the surface and to provide bacterial adhesion. Biofilms were grown on each composite disk, coated with saliva, by inoculation of 200 µL of standardized *S. mutans* CH02 bacterial suspension and incubation at 37 °C for 24 h under anaerobic condition, and then for another 24 h in aerobic atmosphere, in accordance with a previous study [28].

Negative controls, consisting of noninoculated composite disks, were also prepared. After incubation, the planktonic microbial growth was carefully removed from each well and analyzed for:

(i). Total mass amount by measuring the planktonic optical density (OD_{600nm});
(ii). Planktonic CFU count of the bacterial cells (CFU/mL);
(iii). Planktonic bacterial metabolic activity by MTT assay;
(iv). Planktonic bacterial viability assay by live/dead staining.

The disks were rinsed three times with phosphate-buffered saline (PBS) to remove unbound bacteria, and the bacterial load on each disk was assessed for:

(i). Adherent CFU count for the quantification of cultivable cells;
(ii). Adherent OD for biofilm biomass evaluation, using Hucker's crystal violet staining method (OD_{570nm});
(iii). Adherent biofilm metabolic activity by MTT assay;

(iv). Extracellular polymeric substances (EPS) of the biofilm matrix by the Concanavalin A assay;
(v). Biofilm morphology by SEM evaluation.

Each evaluation was performed in triplicate for three independent experiments.

2.5. Planktonic Optical Density Detection

The effects of composite disks on the growth in the biofilm supernatant was determined. The planktonic phase, coming from the *S. mutans* CH02 biofilm formation, was removed from each well and transferred to wells of a new 96-well polystyrene flat-bottomed microtiter plates to evaluate the total mass amount by determining the OD_{600nm} with ELISA reader (SAFAS, Munich, Germany). For the detection, the planktonic phases coming from 39 disks (10 tests and 3 negative controls for each different CR material) in triplicate, for a total of 117 disks, were analyzed.

2.6. Planktonic CFU Count

To evaluate the ability of the different composite disks to influence the growth and viability of *S. mutans* CH02, the count of CFU/mL of planktonic bacteria was determined. The planktonic bacterial phase, removed from the wells containing the resin disks, was vortexed, diluted, and spread on Tryptic Soy Agar (TSA) plates and incubated for 24–48 h in anaerobic conditions at 37 °C. Then, the CFU/mL count was performed. For the detection, the planktonic phases coming from 30 disks (10 for each different CR material) in triplicate, for a total of 90 disks, were analyzed.

2.7. Planktonic MTT Assay

The planktonic growth was placed in a 96-well flat bottom microtiter plate and incubated with 20 μL of MTT (3-(4,5-dimethylthiazol-2-yl)-2,5-diphenyltetrazolium bromide) labeling reagent (Sigma–Aldrich Chemical Co., St. Louis, MO, USA) at concentration of 5 mg/mL for 2 h, as indicated by the manufacturer. After incubation, 100 μL of Dimethyl Sulfoxide (DMSO, Sigma–Aldrich Chemical Co., St. Louis, MO, USA) was added at each well and incubated for 10 min at the dark. Medium incubated with composite disks served as negative control. The assay is based on the metabolic conversion of water soluble MTT compound to a colored insoluble formazan derivate. Viable cells with active metabolism convert MTT into formazan; however, dead cells lose this ability [30]. The optical density reading was measured spectrophotometrically at 570 nm by ELISA reader (SAFAS, Munich, Germany). For the detection, the planktonic phases coming from 39 disks (10 tests and 3 negative controls for each different CR material) in triplicate, for a total of 117 disks, were analyzed.

2.8. Planktonic Bacterial Viability Assay

For the evaluation of planktonic cells viability, the cells were observed at fluorescent Leica 4000 DM microscope (Leica Microsystems, Milan, Italy). The planktonic bacterial phase was removed from each well and stained using the live/dead BacLight staining (Invitrogen, Milan, Italy), as previously reported [31]. The microscopic observation allowed the differentiation between live and dead cells based on the relative green and red fluorescence from SYTO 9 (500 nm) and propidium iodide (635 nm) staining. Ten fields of view, randomly chosen for each disk, were examined. Ten fields of view, randomly chosen for each planktonic phase coming from 9 disks (3 for each different CR material) in triplicate, for a total of 27 disks, were examined.

2.9. Adherent Bacteria CFU Count

The number of adhered viable bacteria on the surface of the specimens was determined to evaluate the ability of *S. mutans* to colonize the different composite disks, as previously described [32].

Briefly, after cultivating for 48 h, the adherent viable cells were washed with PBS to remove nonadherent cells. The disks were placed in a sterile test tube containing 1 mL PBS. Then, each test tube was placed in a 4 kHz ultrasonic water bath (Euronda, Italy) for 4 min followed by vortex mixing for 2 min to detach the bacteria adhering to the surface of each disk. The live/dead analysis was performed to confirm that all detached cells were viable and disaggregated. Then, serial 10-fold dilutions were carried out, plated on TSA plates, and incubated overnight at 37 °C, followed by counting of CFU/mL. For this detection, 30 disks (10 for each different CR material) were analyzed in triplicate, for a total of 90 disks.

2.10. Biomass Quantification by Optical Density (OD_{570nm})

After 48 h (biofilm formation), adherent viable biomass assessment was performed. A crystal violet (CV) staining was used to evaluate relative biofilm biomass formed by *S. mutans* CH02 on composite surfaces.

The disks were washed three times with PBS, fixed by air drying, stained with crystal violet 0.1% (Sigma–Aldrich, Milan, Italy) for 1 min and washed with PBS. After drying, bound CV was eluted with ethanol for reading. After 10 min, the composite disks were removed and the biofilm formation was quantified by measuring absorbance at 570 nm with a microplate reader (SAFAS, Munich, Germany). The absorbance of the eluted stain is proportional to the concentration of biofilm biomass formed on the sample surface [33]. For this detection, 39 disks (10 tests and 3 negative controls for each different CR material) were analyzed in triplicate, for a total of 117 disks.

2.11. Adherent Bacteria MTT Assay

To evaluate the metabolic activity of the biofilm formed by *S. mutans* CH02 on composite surfaces, the disks were incubated with 100 µL of MTT (3-(4,5-dimethylthiazol-2-yl)-2,5-diphenyltetrazolium bromide) labeling reagent (Sigma–Aldrich Chemical Co. St. Louis, MO, USA) at the concentration of 5 mg/mL for 2 h, as indicated by the manufacturer. A total of 100 µL of supernatant from each composite disk was transferred to a 96-well flat bottom microtiter plate, and the optical density reading of formazan derivative was read at 570 nm using an ELISA reader spectrophotometer (SAFAS, Munich, Germany). Disks incubated with media alone served as negative control. For this detection, 39 disks (10 tests and 3 negative controls for each different CR material) were analyzed in triplicate, for a total of 117 disks.

2.12. Concanavalin Assay

To analyze the extracellular polymeric substances (EPS) of the biofilms matrix, rhodamine-labeled Concanavalin A (rhodamine-conA) (Vector Laboratories, Burlingame, CA, USA), was used for its ability to bind to d−(+)−glucose and d−(+)−mannose groups on EPS.

The sessile bacterial population, adherent on composite disks, was washed with 1 mL of PBS twice, stained with 500 µL of the rhodamine-conA (10 µg/mL) solution, and incubated in the dark at room temperature for 30 min. Then, the excess staining solution was removed, and the stained specimens were rinsed with 1 mL of PBS and examined under fluorescence Leica 4000 DM microscope (Leica Microsystems, Milan, Italy). The fluorescence microscopy was used to obtain images using an excitation of 514 nm and an emission wavelength of 600 ± 50 nm. For this detection, 9 disks (3 for each different CR material) were analyzed in triplicate, for a total of 27 disks.

2.13. Scanning Electron Microscope (SEM) Analysis

After 48 h of in vitro biofilm formation, 5 specimens from each group were fixed for 1 h in 2.5% glutaraldehyde, dehydrated in six ethanol washes (10%, 25%, 50%, 75%, and 90% for 20 min and 100% for 1 h), and then dried overnight in a bacteriological incubator at 37 °C. Then, they were coated with gold (Emitech K550, Emitech Ltd., Ashford, Kent, UK) and observed carefully under a SEM (EVO 50 XVP LaB6, Carl Zeiss SMT Ltd., Cambridge, UK) at 15 kV, under 500×, 1000×, and 2000× magnifications. The representative micrographs of the biofilm on the specimens' surface were recorded, and their descriptive analysis was performed.

2.14. Statistical Analysis

Means and standard deviations for data collected following the quantitative experiments (Planktonic OD_{600nm}, Planktonic CFU count, Planktonic MTT, Adherent Bacteria CFU count, Biomass Quantification by OD_{570nm}, Adherent Bacteria MTT) were calculated in each group. Statistical analysis was performed using SPSS for Windows version 21 (IBM SPSS Inc, Chicago, IL, USA), by means of the analysis of variance (ANOVA) and Tukey tests for posthoc intergroup comparisons. Homogeneity of variances and normality of the data sets were respectively confirmed by means of Levene's and Kolmogorov–Smirnov tests. p-values less than 0.05 were considered significant.

3. Results

The results of the quantitative tests performed on the three composites examined in the study are summarized in Table 2.

Table 2. *Streptococcus mutans* detection on three resin composites investigated.

Streptococcus Mutans CH02	VOCO	VD	ES2	CTRL+
Planktonic OD_{600nm} (SD)	0.3054 [b] (0.0567)	0.2931 [b] (0.0540)	0.3117 [b] (0.0532)	0.3978 [a] (0.0491)
Planktonic CFU count ($\times 10^5$ CFU/mL) (SD)	583.0 [b] (53.4)	281.0 [c] (26.4)	273.0 [c] (14.1)	1900.0 [a] (353.3)
Planktonic MTT (SD)	0.451 [a] (0.094)	0.549 [a] (0.136)	0.448 [a] (0.085)	0.442 [a] (0.120)
Adherent Bacteria CFU count ($\times 10^3$ CFU) (SD)	150.0 [b] (29.4)	252.0 [a] (39.7)	248.3 [a] (47.8)	
Biomass Quantification OD_{570nm} (SD)	0.4775 [b] (0.1548)	0.6364 [b] (0.2376)	1.6040 [a] (0.2075)	
Adherent Bacteria MTT (SD)	0.003 [b] (0.004)	0.040 [b] (0.013)	0.035 [b] (0.021)	0.367 [a] (0.274)

VOCO, VD and ES2 are the experimental groups, based on the materials described in Table 1. CTRL+ indicates the positive control group. Same superscript letters indicate not statistically significant differences.

3.1. Planktonic Optical Density Detection

The planktonic *S. mutans* OD_{600} mean values obtained in presence of VOCO, VD, and ES2 composite disks and control are shown in Table 2. In respect to the control, statistically significant ($p < 0.05$) OD_{600} values were recorded in presence of all tested composite disks. A major percentage of OD_{600} reduction in respect to the control was obtained in presence of VD; instead, a slightly decreased OD_{600} reduction was shown in presence of VOCO and ES2, with no statistically significant differences.

3.2. Planktonic CFU Count

As shown in Table 2, in presence of VOCO, the amount of CFU/mL was 5.8×10^7 $\pm 5.3 \times 10^7$ with a percentage of reduction in respect to the control of 69.47%. A similar trend was recorded in presence of VD and ES2 with $2.8 \times 10^7 \pm 2.6 \times 10^6$ CFU/mL and $2.7 \times 10^7 \pm 1.4 \times 10^6$ CFU/mL, respectively. In presence of VD and ES2, the percentage of CFU/mL reduction in respect to the control was about 85%. All CFU/mL results were significantly different in respect to the control ($p < 0.05$), and the CFU/mL obtained with VOCO were also significantly different in respect to the CFU/mL obtained with VD and ES2 ($p < 0.05$).

3.3. Planktonic MTT Assay

The planktonic metabolic activity showed the major value obtained with VD (0.549 ± 0.136) in respect to the VOCO (0.451 ± 0.094), ES2 (0.448 ± 0.085), and control (0.442 ± 0.120). In general, similar metabolic activity values were obtained in presence of all tested composite disks and control with no statistical significance ($p > 0.05$) (Table 2). All recorded cells were metabolically active.

3.4. Planktonic Bacterial Viability Assay

Typical live/dead images of planktonic *S. mutans* cells after 48 h of incubation on composite disks are shown in Figure 1. The live/dead images showed remarkable green viable cells in all conditions. In fact, no difference in terms of killing action was observed when VOCO, VD, and ES2 were compared to the control. A reduced number of cells was recorded in presence of the composite disks, when compared to the control.

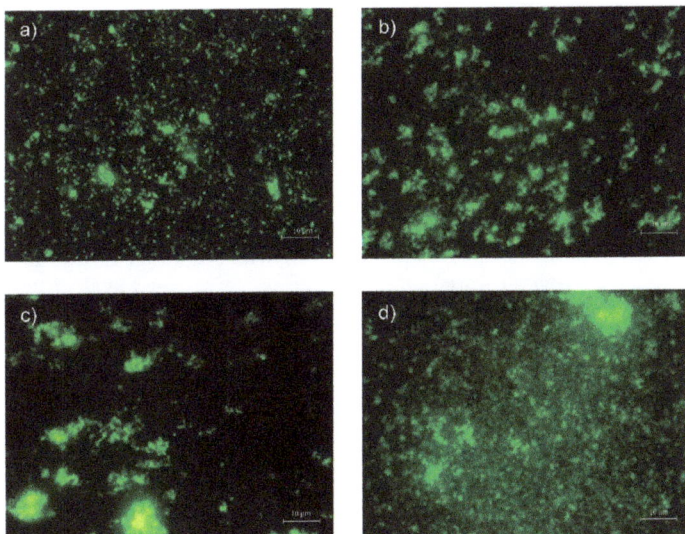

Figure 1. Representative live/dead images of *Streptococcus mutans* in planktonic phase of (**a**) control, (**b**) VOCO, (**c**) VD, and (**d**) ES2.

3.5. Adherent Bacteria CFU Count

Similar numbers of CFU/mL adherent on VD and ES2 were obtained. The lowest CFU/mL value was recorded in presence of VOCO ($1.5 \times 10^5 \pm 2.9 \times 10^5$), with statistically significant differences in respect to VD and ES2 (Table 2) ($p < 0.05$).

3.6. Biomass Quantification by Optical Density (OD_{570nm})

The best antibiofilm biomass effect was obtained with VOCO (0.4775 ± 0.1548), with a significant reduction in respect to ES2 ($p < 0.05$). A statistically significant difference between VD and ES2 (0.6364 ± 0.2376) was also recorded (Table 2), with VD showing a relevant increase in biomass biofilm compared to that of ES2.

3.7. Adherent Bacteria MTT Assay

Significant metabolic activity reductions were obtained in presence of all tested composite disks in respect to the control ($p < 0.05$) (Table 2). No difference between VOCO, VD, and ES2 disks was detected. Few cells were detected on the composite disks corresponding to the low metabolic activity in respect to the control.

3.8. Concanavalin Assay

The polysaccharides matrix production by *S. mutans* biofilms on different composites disks are plotted in Figure 2. A major production of carbohydrates was displayed on ES2 disks in respect to the other samples (c). As shown in Figure 2a,b, in presence of VOCO and VD a scarce EPS matrix was detected, with no significant differences.

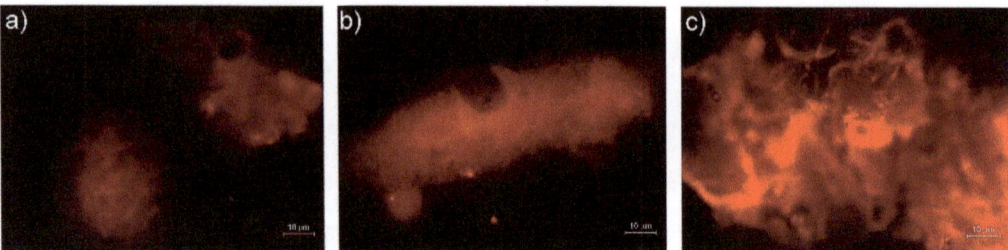

Figure 2. Concanavalin A assay. (**a**) VOCO shows a compact matrix, less than ES2, correlated with quantification of biomass produced by *Streptococcus mutans* biofilm; (**b**) VD shows a compact matrix, less than ES2, correlated with quantification of biomass produced by *Streptococcus mutans* biofilm; and (**c**) ES2 shows increased production of sugars from biofilm matrix.

3.9. SEM Analysis

Representative SEM images of the *S. mutans* biofilm formation on the surface of the disk-shaped specimens are shown in Figure 3a–c. After 48 h of in vitro biofilm formation, the presence of the *S. mutans* cells was noted in all groups, but biofilm formation time was not sufficient to coat entirely the surface with *S. mutans* cells. Large adherent aggregates were observed in the ES2 group, which had the highest quantity of bacteria on its surface (Figure 3c), whereas small aggregates were found on VD e VOCO groups, with an apparently reduced biofilm formation (Figure 3a,b).

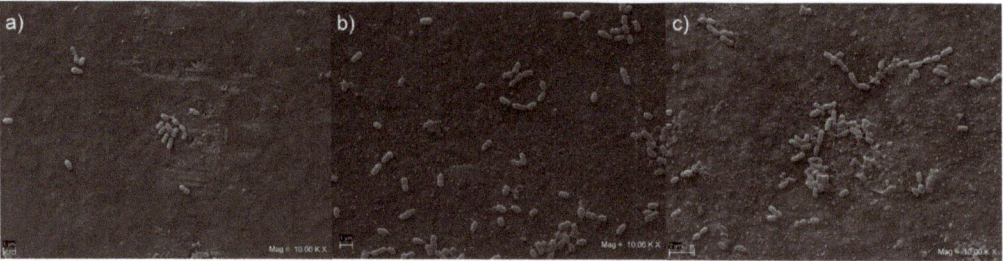

Figure 3. Representative SEM images (10.00 KX) of *Streptococcus mutans* biofilm formed on disk-shaped specimens from VOCO (**a**), VD (**b**), and ES2 (**c**) groups.

4. Discussion

In this study, the behavior of *Streptococcus mutans* exposed to three commercially available, nanohybrid, dental CR-based materials, coated with human saliva, was quantitatively assessed through the evaluation of total microbial population, the viable count, and the metabolic activity in planktonic and sessile growth mode. To better analyze the potential influence of different chemical formulations on biofilm formation, all samples were subjected to the same standardized light-curing protocol through two glass slides and inside polyvinylsiloxane molds, to predictably achieve the smoothest surface possible with the least amount of surface roughness. This allowed to remove the effects of a potentially confounding variable, the surface topography, which is also hypothetically able to affect biofilm adhesion [34,35]. Based on the achieved results, the null-hypothesis tested had to be rejected: statistically significant differences were observed when comparing the antibacterial and antibiofilm properties of chemically different CR materials.

Focusing on Planktonic Optical Density assay, a statistically significant reduction emerged in planktonic cells exposed to all the nanohybrid CRs, compared to that of the control. Kim et al. showed that bis-GMA inhibited the planktonic growth of *S. mutans* in media containing glucose, fructose, or mannose [36]. However, since in our research bis-GMA was present just in two out of the three tested resins, it is possible to suppose that the *S. mutans* decrease can be determined even by other monomers, or their combination.

The Planktonic Optical Density reduction was in accordance with the results of the planktonic CFU assay, which showed for all resins a statistically significant reduced planktonic CFU count compared to that of the control. Furthermore, the planktonic microbial cell populations detected by CFU count, when studied for their viability, displayed a green color suggesting a bacteriostatic effect for all composites. Such behavior was more relevant for ES2 and VD supposing a major capability of *S. mutans* to adhere to these materials with a reduction in planktonic counterpart, which is in line with the findings of the Adherent Bacteria CFU count test.

Regarding the Adherent Bacteria count, a significant increase was observed for ES2 and VD compared to that of VOCO. Indeed, for VOCO, results showed a significant anti-adhesive effect with a reduction in Adherent CFU count, biomass quantification, and presence of EPS matrix. This material expressed the best performances compared to that of the other composites with significance in terms of antiadhesive action and a general reduction in biomass quantification and MTT detection. Of course, antiadhesive and antibacterial properties should be cautiously pondered, taking into account also any potential cytotoxicity due to the monomers included in (and released from) adhesive dental materials [37,38]. Several in vitro studies demonstrated that the potential cytotoxicity of the CR organic components is mainly due to the residues of free methacrylate monomers following the phase of polymerization, which may trigger the production of prostaglandin E2 (PGE2), the expression of cyclooxygenase 2 (COX2), and a proinflammatory activation through the increase in interleukin-1β (IL-1β), IL-6, and nitric oxide (NO) [39,40]. The capability of resin monomers to influence cellular physiology and adaptive cell responses by increasing ROS production was also reported [41,42]. In this regard, a recent study showed reduced cytotoxic and genotoxic effects for VOCO, compared to that of VD, toward human gingival fibroblasts, somehow strengthening the clinical relevance of the present findings [43].

A lower biomass production of biofilms and polysaccharides results in thinner biofilm thickness, lower biofilm matrix barrier for protection, and less carbohydrate amount accumulation. This could be clinically beneficial, as the resulting biofilm would be more susceptible to the buffering action of saliva, fluoride ions, and antibacterial agents.

Interestingly, the MTT assay on adherent cells led to concordant results: metabolic activity underwent a significant reduction following the exposure to all the tested composites, but it appeared relatively and slightly increased in presence of ES2 and VD, compared to that of VOCO. As observed by Aqawi et al., the metabolic activity-related reduction in preformed biofilm could be associated with the modification of membrane polarization [44].

As already reported by Kim et al., cells grown under bis-GMA showed significantly increased surface hydrophobicity, which could potentially enhance the ability of *S. mutans* to adhere to hydrophobic surfaces [36]. However, in the present study, both bis-GMA free (VD)- and bis-GMA-based (ES2) composites led to a significant increase in adherent *S. mutans* CFU count, compared to that of a third bis-GMA based (VOCO) material. Thus, as already pointed out concerning the Planktonic Optical Density results, also when dealing with the adherent CFU count, all the composite components might simultaneously interact; thus, bis-GMA should not be seen as the only monomer able to modulate the adherent properties of bacteria. Further studies would be needed to clarify such effects properly and comprehensively, with respect to the complex formulation of commercial CR-based materials. Also, the filler amount/size/shape might represent another possible factor influencing bacterial adherence to resin composite surface, as suggested by previous papers [21,22,45]. VOCO had a higher proportion of filler particles than ES2. VOCO contained approximately 89 wt% of filler particles, while ES2 contained 78 wt%. From this point of view, the present results seem in line with Ikeda et al., who observed resin composites with a higher filler content showing a reduced biofilm retention [20,46].

The presence of bacterial biofilm, in association with a gap at the margin between the composite and the tooth structure, may lead to the development of secondary caries [47]. This is, in fact, one of the main reasons for restorations replacement [48]. For this reason, the present research focused on the adhesive and biofilm-forming capabilities of S. mutans toward CRs with different chemical compositions, as this microorganism can be isolated in almost all carious lesions [10]. S. mutans tends to accumulate more on composites than on enamel or other restorations [49], and its presence on tooth surfaces is usually followed by caries after 6 to 24 months [50]. Then, the inhibition of S. mutans biofilm formation is a key goal for preventing dental caries.

Based on the present results, VOCO showed a strong reduction in biofilm growth in terms of CFU count, biofilm biomass, metabolic activity, and polysaccharide production compared to those of VD and ES2. These features could indicate VOCO as a more promising restorative material, even though it also allowed microorganism growth to a certain extent. The limitation of this study is related to the use of a single clinical strain of S. mutans and a static model. The interesting results herein obtained suggest carrying out further studies, including a multispecies dynamic biofilm model, to better evaluate the susceptibility of adhesion of microorganisms to resin-based materials.

Author Contributions: Conceptualization, F.D.A., S.D., L.C., D.T. and C.D.; methodology, F.D.A., S.D., L.C. and C.D.; formal analysis, F.D.A., S.D., V.B., C.N., D.T., S.D.L., L.C. and C.D.; investigation, S.D., V.B., C.N. and S.D.L.; resources, S.D., L.C. and C.D.; data curation, F.D.A.; writing—original draft preparation, F.D.A., S.D., V.B., C.N. and S.D.L.; writing—review and editing, F.D.A., S.D., V.B., C.N., D.T., S.D.L., L.C. and C.D.; visualization, F.D.A. and S.D.; supervision, L.C. and C.D.; project administration, F.D.A., S.D., L.C. and C.D.; funding acquisition, S.D., L.C. and C.D. All authors have read and agreed to the published version of the manuscript.

Funding: This research received no external funding.

Institutional Review Board Statement: The study was conducted according to the guidelines of the Declaration of Helsinki and approved by the Ethics Committee of University "G. d'Annunzio", Chieti–Pescara, Italy (approval code SALI, N. 19 of the 10 September 2020).

Informed Consent Statement: Informed consent was obtained from all subjects involved in the study.

Data Availability Statement: The data presented in this study are available on request from the corresponding author.

Conflicts of Interest: The authors declare no conflict of interest.

References

1. Thomaidis, S.; Kakaboura, A.; Mueller, W.D.; Zinelis, S. Mechanical properties of contemporary composite resins and their interrelations. *Dent. Mater.* **2013**, *29*, e132–e141. [CrossRef] [PubMed]
2. De Angelis, F.; D'Arcangelo, C.; Maliskova, N.; Vanini, L.; Vadini, M. Wear properties of different additive restorative materials used for onlay/overlay posterior restorations. *Oper. Dent.* **2020**, *45*, E156–E166. [CrossRef] [PubMed]
3. Gurgan, S.; Koc Vural, U.; Miletic, I. Comparison of mechanical and optical properties of a newly marketed universal composite resin with contemporary universal composite resins: An in vitro study. *Microsc. Res. Tech.* **2022**, *85*, 1171–1179. [CrossRef]
4. D'Arcangelo, C.; Vanini, L.; Rondoni, G.D.; Pirani, M.; Vadini, M.; Gattone, M.; De Angelis, F. Wear properties of a novel resin composite compared to human enamel and other restorative materials. *Oper. Dent.* **2014**, *39*, 612–618. [CrossRef] [PubMed]
5. D'Arcangelo, C.; Vadini, M.; Buonvivere, M.; De Angelis, F. Safe clinical technique for increasing the occlusal vertical dimension in case of erosive wear and missing teeth. *Clin. Case Rep.* **2021**, *9*, e04747. [CrossRef] [PubMed]
6. D'Arcangelo, C.; Vanini, L.; Casinelli, M.; Frascaria, M.; De Angelis, F.; Vadini, M.; D'Amario, M. Adhesive cementation of indirect composite inlays and onlays: A literature review. *Compend. Contin. Educ. Dent.* **2015**, *36*, 570–577; quiz 78.

7. Frassetto, A.; Breschi, L.; Turco, G.; Marchesi, G.; Di Lenarda, R.; Tay, F.R.; Pashley, D.H.; Cadenaro, M. Mechanisms of degradation of the hybrid layer in adhesive dentistry and therapeutic agents to improve bond durability—A literature review. *Dent. Mater.* **2016**, *32*, e41–e53. [CrossRef]
8. Kusuma Yulianto, H.D.; Rinastiti, M.; Cune, M.S.; de Haan-Visser, W.; Atema-Smit, J.; Busscher, H.J.; van der Mei, H.C. Biofilm composition and composite degradation during intra-oral wear. *Dent. Mater.* **2019**, *35*, 740–750. [CrossRef]
9. Auschill, T.M.; Arweiler, N.B.; Brecx, M.; Reich, E.; Sculean, A.; Netuschil, L. The effect of dental restorative materials on dental biofilm. *Eur. J. Oral Sci.* **2002**, *110*, 48–53. [CrossRef]
10. Forssten, S.D.; Björklund, M.; Ouwehand, A.C. Streptococcus mutans, caries and simulation models. *Nutrients* **2010**, *2*, 290–298. [CrossRef]
11. Hannig, M. Transmission electron microscopy of early plaque formation on dental materials in vivo. *Eur. J. Oral Sci.* **1999**, *107*, 55–64. [CrossRef] [PubMed]
12. Hahnel, S.; Henrich, A.; Rosentritt, M.; Handel, G.; Bürgers, R. Influence of artificial ageing on surface properties and streptococcus mutans adhesion to dental composite materials. *J. Mater. Sci. Mater. Med.* **2010**, *21*, 823–833. [CrossRef] [PubMed]
13. Song, F.; Koo, H.; Ren, D. Effects of material properties on bacterial adhesion and biofilm formation. *J. Dent. Res.* **2015**, *94*, 1027–1034. [CrossRef]
14. Bollen, C.M.; Lambrechts, P.; Quirynen, M. Comparison of surface roughness of oral hard materials to the threshold surface roughness for bacterial plaque retention: A review of the literature. *Dent. Mater.* **1997**, *13*, 258–269. [CrossRef]
15. Jung, M.; Wehlen, O.; Klimek, J. Finishing and polishing of indirect composite and ceramic inlays in-vivo: Occlusal surfaces. *Oper. Dent.* **2004**, *29*, 131–141.
16. Lutz, F.; Setcos, J.C.; Phillips, R.W. New finishing instruments for composite resins. *J. Am. Dent. Assoc.* **1983**, *107*, 575–580. [CrossRef]
17. Kurt, A.; Cilingir, A.; Bilmenoglu, C.; Topcuoglu, N.; Kulekci, G. Effect of different polishing techniques for composite resin materials on surface properties and bacterial biofilm formation. *J. Dent.* **2019**, *90*, 103199. [CrossRef]
18. Cazzaniga, G.; Ottobelli, M.; Ionescu, A.C.; Paolone, G.; Gherlone, E.; Ferracane, J.L.; Brambilla, E. In vitro biofilm formation on resin-based composites after different finishing and polishing procedures. *J. Dent.* **2017**, *67*, 43–52. [CrossRef]
19. Ionescu, A.; Brambilla, E.; Wastl, D.S.; Giessibl, F.J.; Cazzaniga, G.; Schneider-Feyrer, S.; Hahnel, S. Influence of matrix and filler fraction on biofilm formation on the surface of experimental resin-based composites. *J. Mater. Sci. Mater. Med.* **2015**, *26*, 5372. [CrossRef]
20. Ikeda, M.; Matin, K.; Nikaido, T.; Foxton, R.M.; Tagami, J. Effect of surface characteristics on adherence of *S. mutans* biofilms to indirect resin composites. *Dent. Mater. J.* **2007**, *26*, 915–923. [CrossRef]
21. Motevasselian, F.; Zibafar, E.; Yassini, E.; Mirzaei, M.; Pourmirhoseni, N. Adherence of streptococcus mutans to microhybrid and nanohybrid resin composites and dental amalgam: An in vitro study. *J. Dent.* **2017**, *14*, 337–343.
22. Pereira, C.A.; Eskelson, E.; Cavalli, V.; Liporoni, P.C.; Jorge, A.O.; do Rego, M.A. Streptococcus mutans biofilm adhesion on composite resin surfaces after different finishing and polishing techniques. *Oper. Dent.* **2011**, *36*, 311–317. [CrossRef] [PubMed]
23. Nedeljkovic, I.; Yoshihara, K.; De Munck, J.; Teughels, W.; Van Meerbeek, B.; Van Landuyt, K.L. No evidence for the growth-stimulating effect of monomers on cariogenic streptococci. *Clin. Oral Investig.* **2017**, *21*, 1861–1869. [CrossRef]
24. Takahashi, Y.; Imazato, S.; Russell, R.R.; Noiri, Y.; Ebisu, S. Influence of resin monomers on growth of oral streptococci. *J. Dent. Res.* **2004**, *83*, 302–306. [CrossRef] [PubMed]
25. Kim, K.; Kim, J.N.; Lim, B.S.; Ahn, S.J. Urethane dimethacrylate influences the cariogenic properties of streptococcus mutans. *Materials* **2021**, *14*, 1015. [CrossRef] [PubMed]
26. Khalichi, P.; Cvitkovitch, D.C.; Santerre, J.P. Effect of composite resin biodegradation products on oral streptococcal growth. *Biomaterials* **2004**, *25*, 5467–5472. [CrossRef]
27. Petrini, M.; Giuliani, A.; Di Campli, E.; Di Lodovico, S.; Iezzi, G.; Piattelli, A.; D'Ercole, S. The bacterial anti-adhesive activity of double-etched titanium (dae) as a dental implant surface. *Int. J. Mol. Sci.* **2020**, *21*, 8315. [CrossRef]
28. D'Ercole, S.; Di Giulio, M.; Grande, R.; Di Campli, E.; Di Bartolomeo, S.; Piccolomini, R.; Cellini, L. Effect of 2-hydroxyethyl methacrylate on streptococcus spp. Biofilms. *Lett. Appl. Microbiol.* **2011**, *52*, 193–200. [CrossRef]
29. Yaghmoor, R.B.; Xia, W.; Ashley, P.; Allan, E.; Young, A.M. Effect of novel antibacterial composites on bacterial biofilms. *J. Funct. Biomater.* **2020**, *11*, 55. [CrossRef]
30. Peralta, S.L.; Leles, S.B.; Dutra, A.L.; Guimarães, V.; Piva, E.; Lund, R.G. Evaluation of physical-mechanical properties, antibacterial effect, and cytotoxicity of temporary restorative materials. *J. Appl. Oral. Sci.* **2018**, *26*, e20170562. [CrossRef]
31. Di Giulio, M.; D'Ercole, S.; Zara, S.; Cataldi, A.; Cellini, L. Streptococcus mitis/human gingival fibroblasts co-culture: The best natural association in answer to the 2-hydroxyethyl methacrylate release. *Apmis* **2012**, *120*, 139–146. [CrossRef] [PubMed]
32. D'Ercole, S.; Cellini, L.; Pilato, S.; Di Lodovico, S.; Iezzi, G.; Piattelli, A.; Petrini, M. Material characterization and streptococcus oralis adhesion on polyetheretherketone (peek) and titanium surfaces used in implantology. *J. Mater. Sci. Mater. Med.* **2020**, *31*, 84. [CrossRef] [PubMed]
33. D'Ercole, S.; Di Fermo, P.; Di Giulio, M.; Di Lodovico, S.; Di Campli, E.; Scarano, A.; Tripodi, D.; Cellini, L.; Petrini, M. Near-infrared nir irradiation and sodium hypochlorite: An efficacious association to counteract the enterococcus faecalis biofilm in endodontic infections. *J. Photochem. Photobiol. B* **2020**, *210*, 111989. [CrossRef] [PubMed]

34. Mena Silva, P.A.; Garcia, I.M.; Nunes, J.; Visioli, F.; Leitune, V.C.B.; Melo, M.A.; Collares, F.M. Myristyltrimethylammonium bromide (mytab) as a cationic surface agent to inhibit streptococcus mutans grown over dental resins: An in vitro study. *J. Funct. Biomater.* **2020**, *11*, 9. [CrossRef]
35. Cazzaniga, G.; Ottobelli, M.; Ionescu, A.; Garcia-Godoy, F.; Brambilla, E. Surface properties of resin-based composite materials and biofilm formation: A review of the current literature. *Am. J. Dent.* **2015**, *28*, 311–320.
36. Kim, K.; An, J.S.; Lim, B.S.; Ahn, S.J. Effect of bisphenol a glycol methacrylate on virulent properties of streptococcus mutans ua159. *Caries Res.* **2019**, *53*, 84–95. [CrossRef]
37. Trubiani, O.; Caputi, S.; Di Iorio, D.; D'Amario, M.; Paludi, M.; Giancola, R.; Di Nardo Di Maio, F.; De Angelis, F.; D'Arcangelo, C. The cytotoxic effects of resin-based sealers on dental pulp stem cells. *Int. Endod. J.* **2010**, *43*, 646–653. [CrossRef]
38. Trubiani, O.; Cataldi, A.; De Angelis, F.; D'Arcangelo, C.; Caputi, S. Overexpression of interleukin-6 and -8, cell growth inhibition and morphological changes in 2-hydroxyethyl methacrylate-treated human dental pulp mesenchymal stem cells. *Int. Endod. J.* **2012**, *45*, 19–25. [CrossRef]
39. Kuan, Y.H.; Huang, F.M.; Lee, S.S.; Li, Y.C.; Chang, Y.C. Bisgma stimulates prostaglandin e2 production in macrophages via cyclooxygenase-2, cytosolic phospholipase a2, and mitogen-activated protein kinases family. *PLoS ONE* **2013**, *8*, e82942. [CrossRef]
40. Huang, F.M.; Chang, Y.C.; Lee, S.S.; Yeh, C.H.; Lee, K.G.; Huang, Y.C.; Chen, C.J.; Chen, W.Y.; Pan, P.H.; Kuan, Y.H. Bisgma-induced cytotoxicity and genotoxicity in macrophages are attenuated by wogonin via reduction of intrinsic caspase pathway activation. *Environ. Toxicol.* **2016**, *31*, 176–184. [CrossRef]
41. Lottner, S.; Shehata, M.; Hickel, R.; Reichl, F.X.; Durner, J. Effects of antioxidants on DNA-double strand breaks in human gingival fibroblasts exposed to methacrylate based monomers. *Dent. Mater.* **2013**, *29*, 991–998. [CrossRef] [PubMed]
42. Gallorini, M.; Petzel, C.; Bolay, C.; Hiller, K.A.; Cataldi, A.; Buchalla, W.; Krifka, S.; Schweikl, H. Activation of the nrf2-regulated antioxidant cell response inhibits hema-induced oxidative stress and supports cell viability. *Biomaterials* **2015**, *56*, 114–128. [CrossRef] [PubMed]
43. De Angelis, F.; Mandatori, D.; Schiavone, V.; Melito, F.P.; Valentinuzzi, S.; Vadini, M.; Di Tomo, P.; Vanini, L.; Pelusi, L.; Pipino, C.; et al. Cytotoxic and genotoxic effects of composite resins on cultured human gingival fibroblasts. *Materials* **2021**, *14*, 5225. [CrossRef] [PubMed]
44. Aqawi, M.; Sionov, R.V.; Gallily, R.; Friedman, M.; Steinberg, D. Anti-biofilm activity of cannabigerol against streptococcus mutans. *Microorganisms* **2021**, *9*, 2031. [CrossRef]
45. Uçtaşli, M.B.; Arisu, H.D.; Omürlü, H.; Eligüzeloğlu, E.; Ozcan, S.; Ergun, G. The effect of different finishing and polishing systems on the surface roughness of different composite restorative materials. *J. Contemp. Dent. Pract.* **2007**, *8*, 89–96.
46. Ono, M.; Nikaido, T.; Ikeda, M.; Imai, S.; Hanada, N.; Tagami, J.; Matin, K. Surface properties of resin composite materials relative to biofilm formation. *Dent. Mater. J.* **2007**, *26*, 613–622. [CrossRef]
47. Ferracane, J.L. Models of caries formation around dental composite restorations. *J. Dent. Res.* **2017**, *96*, 364–371. [CrossRef]
48. Opdam, N.J.; van de Sande, F.H.; Bronkhorst, E.; Cenci, M.S.; Bottenberg, P.; Pallesen, U.; Gaengler, P.; Lindberg, A.; Huysmans, M.C.; van Dijken, J.W. Longevity of posterior composite restorations: A systematic review and meta-analysis. *J. Dent. Res.* **2014**, *93*, 943–949. [CrossRef]
49. Svanberg, M.; Mjör, I.A.; Orstavik, D. Mutans streptococci in plaque from margins of amalgam, composite, and glass-ionomer restorations. *J. Dent. Res.* **1990**, *69*, 861–864. [CrossRef]
50. Balakrishnan, M.; Simmonds, R.S.; Tagg, J.R. Dental caries is a preventable infectious disease. *Aust. Dent. J.* **2000**, *45*, 235–245. [CrossRef]

Article

An Evaluation of the Properties of Urethane Dimethacrylate-Based Dental Resins

Agata Szczesio-Wlodarczyk [1,*], Monika Domarecka [2], Karolina Kopacz [3,4], Jerzy Sokolowski [2] and Kinga Bociong [2]

1. University Laboratory of Materials Research, Medical University of Lodz, ul. Pomorska 251, 92-213 Lodz, Poland
2. Department of General Dentistry, Medical University of Lodz, ul. Pomorska 251, 92-213 Lodz, Poland; monika.domarecka@umed.lodz.pl (M.D.); jerzy.sokolowski@umed.lodz.pl (J.S.); kinga.bociong@umed.lodz.pl (K.B.)
3. "DynamoLab" Academic Laboratory of Movement and Human Physical Performance, Medical University of Lodz, ul. Pomorska 251, 92-216 Lodz, Poland; kkopacz@mum.edu.pl
4. Department of Health Sciences, Medical University of Mazovia, Ludwika Rydygiera 8, 01-793 Warszawa, Poland
* Correspondence: agata.szczesio@umed.lodz.pl

Citation: Szczesio-Wlodarczyk, A.; Domarecka, M.; Kopacz, K.; Sokolowski, J.; Bociong, K. An Evaluation of the Properties of Urethane Dimethacrylate-Based Dental Resins. *Materials* **2021**, *14*, 2727. https://doi.org/10.3390/ma14112727

Academic Editor: Grzegorz Chladek

Received: 21 April 2021
Accepted: 18 May 2021
Published: 21 May 2021

Publisher's Note: MDPI stays neutral with regard to jurisdictional claims in published maps and institutional affiliations.

Copyright: © 2021 by the authors. Licensee MDPI, Basel, Switzerland. This article is an open access article distributed under the terms and conditions of the Creative Commons Attribution (CC BY) license (https://creativecommons.org/licenses/by/4.0/).

Abstract: Most of the dental materials available on the market are still based on traditional monomers such as bisphenol A-glycidyl methacrylate (Bis-GMA), urethane dimethacrylate (UDMA), triethyleneglycol dimethacrylate (TEGDMA), and ethoxylated bisphenol-A dimethacrylate (Bis-EMA). The interactions that arise in the monomer mixture and the characteristics of the resulting polymer network are the most important factors, which define the final properties of dental materials. The use of three different monomers in proper proportions may create a strong polymer matrix. In this paper, fourteen resin materials, based on urethane dimethacrylate with different co-monomers such as Bis-GMA or Bis-EMA, were evaluated. TEGDMA was used as the diluting monomer. The flexural strength (FS), diametral tensile strength (DTS), and hardness (HV) were determined. The impacts of material composition on the water absorption and dissolution were evaluated as well. The highest FS was 89.5 MPa, while the lowest was 69.7 MPa. The median DTS for the tested materials was found to range from 20 to 30 MPa. The hardness of the tested materials ranged from 14 to 16 HV. UDMA/TEGDMA matrices were characterized by the highest adsorption values. The overall results indicated that changes in the materials' properties are not strictly proportional to the material's compositional changes. The matrices showed good properties when the composite contained an equal mixture of Bis-GMA/Bis-EMA and UDMA or the content of the UDMA monomer was higher.

Keywords: dental resins; UDMA; Bis-GMA; Bis-EMA; TEGDMA; mechanical properties; hardness; water absorption; water dissolution

1. Introduction

One of the most important dental achievements in the last century was the introduction of resin matrix composites as a restoration material [1,2]. The organic matrix is typically based on dimethacrylate resins [3–5], while fillers primarily consist of silicon, quartz, borosilicates, zirconium, and aluminum oxides. Inorganic components have different sizes, shapes, and morphologies [6,7]. In dentistry, rear restorations (class I or II according to Black [7]) require composites that have good mechanical properties, while frontal restorations (classes IV and V) require excellent aesthetics. Thus far, no dental composite that meets all these requirements has appeared on the market [2]. Therefore, there is great interest in identifying dental composites with improved esthetics, and antibacterial, physical, and mechanical properties.

The most commonly used monomer in dental composites is bisphenol A-glycidyl methacrylate resin (Bis-GMA). Due to its high viscosity, which is caused by strong intermolecular interactions and the formation of hydrogen bonds between macromolecules, low-viscosity monomers are needed to dilute the polymer matrix and obtain desirable properties [8]. Typical diluent substances include dimethacrylate monomers such as triethylene glycol dimethacrylate (TEGDMA), ethylene glycol dimethacrylate (EGDMA), ethylene diglycol dimethacrylate (DEGDMA), 2-hydroxyethyl methacrylate (HEMA), and 1,10-decanediol dimethacrylate (DDDMA or D3MA) [9]. Free-radical chain polymerization of the organic phase, most commonly initiated by photo initiators or by a chemical initiator and co-initiator, leads to the formation of a cross-linked network bound by esters, urethanes, amide bonds, and van der Waals interactions [10]. Table 1 summarizes the selected properties of popular dimethacrylate homopolymers [11,12].

Table 1. Selected properties of A-glycidyl methacrylate (Bis-GMA), triethyleneglycol dimethacrylate (TEGDMA), urethane dimethacrylate (UDMA), and ethoxylated bisphenol-A dimethacrylate (Bis-EMA)—the most popular monomers used in dental composites.

Monomer	Molecular Weight (g/mol)	Viscosity (Pa·s)	Flexural Strength (MPa)	Flexural Modulus (GPa)	Water Sorption ($\mu g/mm^3$)	Solubility ($\mu g/mm^3$)
Bis-GMA	512	1200 [a]	72.4 [b]	1 [b]	51.2 [b]	9.5 [b]
TEGDMA	286	0.01 [a]	99.1 [b]	1.7 [b]	28.8 [b]	27.5 [b]
UDMA	470	23 [a]	133.8 [b]	1.8 [b]	42.3 [b]	20.4 [b]
Bis-EMA	540	0.9 [a]	87.3 [b]	1.1 [b]	21.3 [b]	2.1 [b]

[a]—taken from [12]; [b]—taken from [11].

The aim of the present study was to determine the flexural strength (FS), diametral tensile strength (DTS), and hardness (HV) of polymer matrices based on urethane dimethacrylate (UDMA) with TEGDMA, Bis-GMA, and ethoxylated bisphenol-A dimethacrylate (Bis-EMA) as co-monomers. The dynamic water absorbency was also studied. The null hypothesis was: there is no effect of compositional changes in more complex resin systems on the three-point bending flexural strength (TPB), diametral tensile strength (DTS), Vickers hardness (HV), and the dynamics of water absorbency of the materials.

2. Materials and Methods

In this study, flexural strength, diametral tensile strength, and hardness were determined. Due to teeth anatomy and the nature of jaw mechanics, loads applied at restoration will cause various stresses, e.g., tensile and shear stresses. Therefore, it is justified to conduct a broader evaluation of mechanical properties, which will determine the behavior of the material under challenging mechanical conditions. Tensile loading is considered the most appropriate. However, it is a difficult test to conduct for dental materials. Diametral tensile strength is proposed as a substitute method. During this test (DTS) tensile, compressive and shear stresses are developed. A flexural strength test is recommended by ISO 4049 [13] for all restorative materials. The hardness is easy to test and may indicate some wear resistance of the materials [14,15]. High water absorption values may indicate that the material will be more susceptible to hydrolytic degradation [16].

Before mechanical tests, samples were immersed in water and stored at 37 °C for 24 h. Such a protocol is in accordance with ISO 4049 [13]. In order to ensure the completion of the post-cure polymerization processes, an interval of 24 h from the preparation of samples was used [17]. The impact of material composition on the water absorption and dissolution was evaluated.

2.1. Materials

The monomers used in the study are described in Table 2. Fourteen different resin mixtures were prepared according to the weight percentage of selected monomers (Table 3). Each mixture contained camphorquinone (an initiator (CAS 10373-78-1), <1 wt.%) and

N,N-dimethylaminoethyl methacrylate (CAS 2867-47-2). After mixing, the resins were stored for a week prior to the study. The materials were cured for 20 s. Increments of 2 mm in thickness were polymerized. To ensure consistent irradiance values, the light curing units (Mini L.E.D, Satelec, France) were calibrated with a radiometer system (Digital Light Meter 200, Rolence Enterprice Inc., Taoyuan, Taiwan).

Table 2. Monomers used in the study.

Monomer	Abbreviation	Manufacturer	Purity	Viscosity at 25 °C
Bis-GMA	G	Esstech, Inc., Essington, PA, USA	97%	718,641 cps
TEGDMA	T		99.8%	—
UDMA	U		98.4%	9387 cps
Bis-EMA	E		98.9%	911 cps

Table 3. Composition of tested matrices.

Matrix Signature	UDMA Content (wt.%)	TEGDMA Content (wt.%)	Bis-EMA Content (wt.%)	Bis-GMA Content (wt.%)
U/T 80/20	80	20	—	—
U/T 70/30	70	30	—	—
U/T 60/40	60	40	—	—
U/T 50/50	50	50	—	—
E/T 80/20	—	20	80	—
U/E/T 70/10/20	70	20	10	—
U/E/T 60/20/20	60	20	20	—
U/E/T 50/30/20	50	20	30	—
U/E/T 40/40/20	40	20	40	—
G/T 80/20	—	20	—	80
U/G/T 70/10/20	70	20	—	10
U/G/T 60/20/20	60	20	—	20
U/G/T 50/30/20	50	20	—	30
U/G/T 40/40/20	40	20	—	40

2.2. Flexural Strength

Flexural strength (FS) was determined using the three-point bending test (Appendix A, Figure A1). Rectangular samples (dimensions: 2 mm × 2 mm × 25 mm) were used for the tests. For each study group, seven samples were tested. Measurements were carried out using a Zwick Roell Z020 universal strength machine (Zwick-Roell, Ulm, Germany). The traverse speed was 1 mm/min. During the test, the modulus of elasticity in bending was also determined.

2.3. Diametral Tensile Strength

The tests were performed on samples in the form of a cylinder (6 mm in diameter and 3 mm in height) (Appendix A, Figure A2). The DTS was measured on nine samples from each study group using a Zwick Roell Z020 universal strength machine (Zwick-Roell, Ulm, Germany). The traverse speed was 2 mm/min. The DTS values were calculated using Equation (1):

$$DTS = \frac{2F}{\pi dh} \ (MPa) \qquad (1)$$

F—force that caused the destruction of the sample [N],
d—diameter of the sample [mm],
h—height of the sample [mm].

2.4. Hardness

The hardness of the tested materials was measured using the Vickers method using a Zwick ZHV2-m hardness tester (Zwick-Roell, Ulm, Germany) (Appendix A, Figure A3). The applied load was 1000 g and the penetration time was 10 s. Nine measurements were performed on three out of nine DTS samples for each study group.

2.5. Dynamic Absorbency

In order to determine the dynamic absorbency, the samples were prepared using a silicone mold (15 mm in diameter and 1 mm in width). Tested materials were applied in one layer and cured with an LED light lamp (Mini L.E.D., Acteon, Norwich, France) in nine zones partially overlapping, in accordance with ISO 4049 recommendations [13]. Five samples were prepared for each dental composite. The samples were weighed (RADWAG AS 160/C/2, Poland) immediately after preparation, on 30 consecutive days, and then after 60, 90, and 120 days. The absorbency was calculated according to Equation (2):

$$A = \frac{m_i - m_0}{m_0} \times 100\% \qquad (2)$$

A—the absorbency of water,
m_0—the initial mass of the sample,
m_i—the mass of the sample after storage in water for a specified (i) period of time.

After 120 days, the specimens were dried to a constant weight using a protocol similar to the dissolution test from standard 4049 [13]. The weight loss (dissolution) in water was calculated according to Equation (3), this being the absolute value:

$$D = \left| \frac{m_0 - m_z}{m_0} \times 100\% \right| \qquad (3)$$

D—the dissolution in water,
m_0—the initial mass of the sample,
m_z—the constant mass of the sample after drying.

2.6. Statistical Analysis

The obtained data were processed with the use of Statistica 13.1 (Statsoft, Kraków, Poland). For statistical analysis, elements of descriptive statistics were used. The Shapiro–Wilk test was used to confirm normality. As the data were found to be nonconsistent with a normal distribution, the data were then analyzed using the Kruskal–Wallis test with the multiple comparisons of mean ranks. The accepted level of significance was α = 0.05.

3. Results

The obtained results are presented in Figures 1–5 and Table 4.

3.1. Flexural Strength

The highest median value of the three-point flexural strength was 89.5 MPa (UDMA/Bis-GMA/TEGDMA 40/40/20 wt.%), while the lowest was 69.7 MPa (Bis-EMA/TEGDMA

80/20 wt.%) (Figure 1). These values were significantly different (Kruskal–Wallis test; p-value = 0.0042). Based on the multiple comparisons of mean ranks for all groups, statistically significant differences were found between UDMA/Bis-GMA/TEGDMA 40/40/20 wt.% and UDMA/TEGDMA 80/20 wt.% (p-value = 0.044215); UDMA/Bis-GMA/TEGDMA 40/40/20 wt.% and Bis-EMA/TEGDMA 80/20 wt.% (p-value = 0.04118). UDMA/Bis-GMA/TEGDMA 40/40/20 wt.% demonstrated a higher FS value than UDMA/TEGDMA 80/20 wt.% or Bis-EMA/TEGDMA 80/20 wt.% did. An analysis of the median FS depending on the matrix composition is presented in Figure 1.

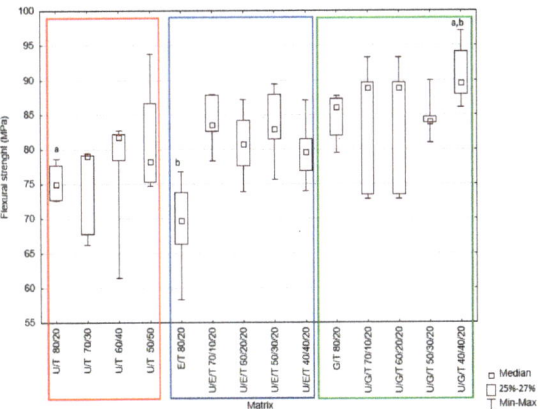

Figure 1. Box-and-whisker plot of three-point bending flexural strength (FS). For variables with the same letter (a,b), the difference is statistically significant ($p \leq 0.05$).

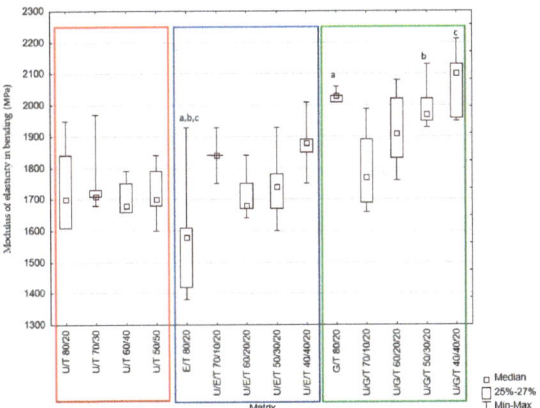

Figure 2. Box-and-whisker plot of modulus of elasticity in bending. For variables with the same letter (a,b,c), the difference is statistically significant ($p \leq 0.05$). The modulus of elasticity in bending (FM) was also determined (Figure 2). Median FM values ranged from 1.58 GPa (Bis-EMA/TEGDMA 80/20 wt.%) to 2.1 GPa (UDMA/Bis-GMA/TEGDMA 40/40/20 wt.%) (Figure 2). These differences were statistically significant (Kruskal–Wallis test; p-value = 0.0000). Based on the multiple comparisons of mean ranks for all groups, statistically significant differences were also found between Bis-EMA/TEGDMA 80/20 wt.% and Bis-GMA/TEGDMA 80/20 wt.% (p-value = 0.010261); Bis-EMA/TEGDMA 80/20 wt.% and UDMA/Bis-GMA/TEGDMA 50/30/20 wt.% (p-value = 0.044215); Bis-EMA/TEGDMA 80/20 wt.% and UDMA/Bis-GMA/TEGDMA 40/40/20 wt.% (p-value = 0.044215).

In the case of Bis-EMA/TEGDMA 80/20 wt.%, the FM value was lower than those in Bis-GMA/TEGDMA 80/20 wt.%, UDMA/Bis-GMA/TEGDMA 50/30/20 wt.%, and UDMA/Bis-GMA/TEGDMA 40/40/20 wt.%. The analysis of the median modulus of elasticity in bending, depending on the matrix composition, is presented in Figure 2.

3.2. Diametral Tensile Strength

Median DTS values ranged from 30.1 MPa (UDMA/Bis-EMA/TEGDMA 40/40/20 wt.%.) to 46.8 MPa (Bis-EMA/TEGDMA 80/20 wt.%.). This difference was statistically significant (Kruskal–Wallis test; p-value = 0.0001). Based on the multiple comparisons of mean ranks for all groups, statistically significant differences were also found between: Bis-EMA/TEGDMA 80/20 wt.%. and UDMA/Bis-EMA/TEGDMA 50/30/20 wt.% (p-value = 0.001177); Bis-EMA/TEGDMA 80/20 wt.%. and UDMA/Bis-EMA/TEGDMA 40/40/20 wt.% (p-value = 0.000668); Bis-GMA/TEGDMA 80/20 wt.%. and UDMA/Bis-EMA/TEGDMA 50/30/20 wt.% (p-value = 0.013824); Bis-GMA/TEGDMA 80/20 wt.%. and UDMA/Bis-EMA/TEGDMA 40/40/20 wt.% (p-value = 0.008395).

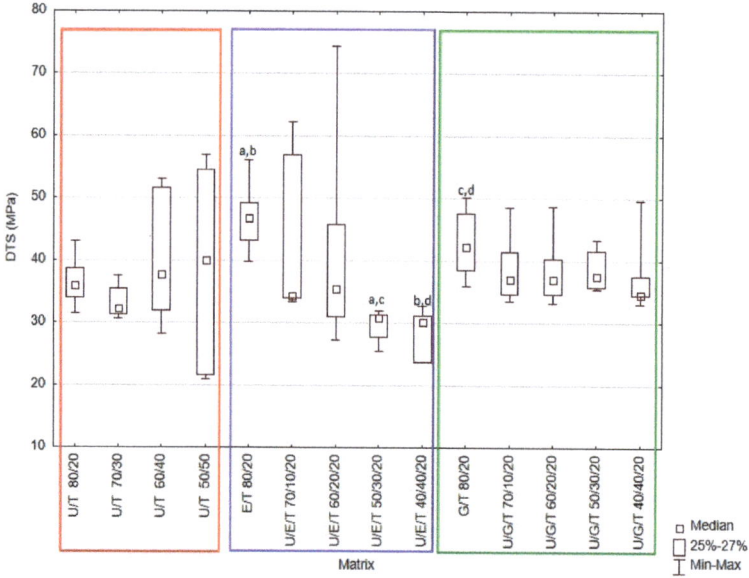

Figure 3. Box-and-whisker plot of diametral tensile strength (DTS). For variables with the same letter (a–d), the difference is statistically significant ($p \leq 0.05$).

Bis-EMA/TEGDMA 80/20 wt.% and Bis-GMA/TEGDMA 80/20 wt.% demonstrated higher DTS values than UDMA/Bis-EMA/TEGDMA 50/30/20 wt.% and UDMA/Bis-EMA/TEGDMA 40/40/20 wt.% did. An analysis of the median DTS with regard to material is presented in Figure 3.

3.3. Hardness

The highest median Vickers hardness (HV) value was 16 (-) (UDMA/TEGDMA 80/20 wt.% and UDMA/Bis-GMA/TEGDMA 70/10/20 wt.%), while the lowest was 14 (-) (UDMA/Bis-EMA/TEGDMA 70/10/20 wt.% and Bis-GMA/TEGDMA 80/20 wt.%.) (Figure 4). These differences were statistically significant (Kruskal–Wallis test; p-value = 0.0000). Most of the tested resin matrices had a hardness of 15. Based on the multiple comparisons of mean ranks for all groups, statistically significant differences were found between: UDMA/TEGDMA 80/20 wt.% vs. UDMA/Bis-EMA/TEGDMA 70/10/20 wt.% (p-value = 0.000292); UDMA/TEGDMA 80/20 wt.% vs. UDMA/Bis-EMA/TEGDMA

60/20/20 wt.% (*p*-value = 0.022403); UDMA/Bis-EMA/TEGDMA 70/10/20 wt.% vs. UDMA/Bis-GMA/TEGDMA 70/10/20 wt.% (*p*-value = 0.011175); UDMA/Bis-EMA/TEGDMA 70/10/20 wt.% vs. UDMA/Bis-GMA/TEGDMA 40/40/20 wt.% (*p*-value = 0.014966); Bis-GMA/TEGDMA 80/20 wt.% vs. UDMA/TEGDMA 80/20 wt.% (*p*-value = 0.000081); Bis-GMA/TEGDMA 80/20 wt.%. vs. UDMA/Bis-GMA/TEGDMA 70/10/20 wt.% (*p*-value = 0.003782); Bis-GMA/TEGDMA 80/20 wt.%. vs. UDMA/Bis-GMA/TEGDMA 40/40/20 wt.% (*p*-value = 0.005057).

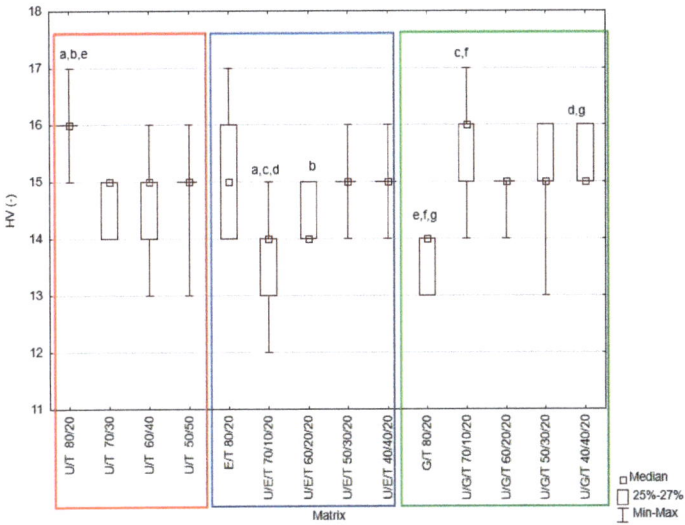

Figure 4. Box-and-whisker plot of Vickers hardness (HV) of tested materials. For variables with the same letter (a–g), the difference is statistically significant ($p \leq 0.05$).

3.4. Water Absorbency Dynamic Study

The matrices based on UDMA and TEGDMA had the highest absorbency. Bis-EMA-TEGDMA 80/20 wt.% showed the lowest water absorbency (Figure 5, Table 4). The highest dissolution values were observed for matrices with Bis-GMA (Table 4).

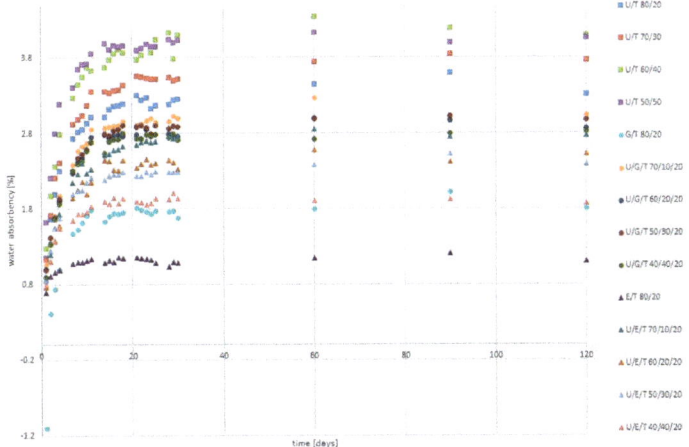

Figure 5. Dynamic water absorbency of tested matrices.

Table 4. The mean values of the water sorption, after 120 days, and dissolution (weight loss, absolute value) of the tested matrices with standard deviations (SD).

	Sorption after 120 Days (wt.%)	SD	Dissolution (wt.%)	SD
U/T 80/20	3.3092	0.1843	0.3465	0.1336
U/T 70/30	3.7575	0.2674	0.4832	0.0759
U/T 60/40	4.0989	0.1561	0.4488	0.0221
U/T 50/50	4.0548	0.0803	0.3997	0.0580
E/T 80/20	1.1094	0.0916	0.1646	0.0896
U/E/T 70/10/20	2.7707	0.0089	0.5118	0.0775
U/E/T 60/20/20	2.5219	0.0562	0.4014	0.0775
U/E/T 50/30/20	2.3936	0.0703	0.3160	0.1450
U/E/T 40/40/20	1.8683	0.0778	0.3243	0.0796
G/T 80/20	2.9333	0.1986	0.5991	0.0759
U/G/T 70/10/20	3.0348	0.1340	0.6208	0.1615
U/G/T 60/20/20	2.8643	0.0728	0.5860	0.0851
U/G/T 50/30/20	2.9782	0.2060	0.6064	0.1655
U/G/T 40/40/20	2.8182	0.1324	0.6192	0.0604

4. Discussion

The properties of the polymer matrix depend on its composition. The most popular base monomer used in dental composites is Bis-GMA, with a molecular weight (MW) of 512 g/mol; the compound comprises a stiff bisphenol A core and hydroxyl groups that are able to form strong hydrogen bonds [18]. Hence, it demonstrates a high viscosity (1200 Pa·s) [12]. An alternative base monomer with a flexible aliphatic core, and, hence, a lower viscosity (23 Pa·s), is UDMA (MW = 470 g/mol) [12]. It also has two urethane links, which are able to form hydrogen bonds, but these interactions are not as strong as in Bis-GMA. Still, this hydrogen bond, formed by the urethane proton donor group, is strong enough to increase the mechanical properties of dental composites [19]. Due to the more flexible nature, UDMA demonstrates a higher degree of conversion than Bis-GMA does and a higher morphological homogeneity [19–21].

Bis-EMA has a similar structure to Bis-GMA, being based on a stiff bisphenol A core; however, as it lacks two pendant hydroxyl groups and has longer ethoxylated linkages, Bis-EMA (MW = 540 g/mol) is more flexible and mobile, with a lower viscosity (0.9 Pa·s) than Bis-GMA [12]. Therefore, it demonstrates a higher overall conversion [12,22]. Finally, TEGDMA (MW = 286 g/mol) is a flexible, low-viscosity (0.01 Pa·s) diluent monomer [12], which is used to obtain a higher degree of conversion and filler homogenization. However, it is characterized by a high hydrophilicity and greater susceptibility to cyclization and polymerization shrinkage [23].

It is important to underline that the final properties result not only from the characteristics of individual monomers, but above all, from the interactions that arise in the monomer mixture and the characteristics of the resulting polymer network [24]. Recent research has examined new monomers based on methacrylate, urethanes, or new resin systems [19,20,25–29]. However, only a few commercial materials, such as Venus Diamond (Heraeus Kulzer), Kalore (GC Corp), and SDR (Dentsply Sirona), are based on such modified monomers [30,31]. Most of the restoration materials on the market are based on conventional resins (Bis-GMA, TEGDMA, UDMA, and Bis-EMA) [29]. The most extensively researched resins are Bis-GMA/TEGDMA and UDMA/TEGDMA and, more rarely, Bis-EMA/TEGDMA mixtures [12,18,22,23,32–39].

The present study examined the effect of the addition of TEGDMA, Bis-GMA, and Bis-EMA on selected properties of the matrix based on UDMA resin. The null hypothesis can be rejected due to changes in the evaluated properties, along with the modification of resins composition. Increasing the amount of TEGDMA monomer in the U/T mixtures did not affect the modulus of elasticity. However, it resulted in a slight increase in the flexural strength (FS) value, but this was not statistically significant; this may be due to the polymer network demonstrating a greater conversion and crosslink density [32]. Studies have shown that the addition of TEGDMA reduces the viscosity of systems based on Bis-GMA or UDMA, thus allowing a higher degree of conversion. However, excessively high amounts of diluent monomer (TEGDMA) result in the deterioration of properties of the tested matrices probably due to primary cyclization. The FS values for mixtures based on a molar fraction of UDMA or Bis-GMA of approx. 0.7 were approximately 140 MPa for a molar fraction of TEGDMA of approximately 0.3, but close to 110 MPa for a molar fraction of TEGDMA of more than 0.6 [32]. Lower-viscosity resins are more likely to demonstrate TEGDMA primary cyclization. Cyclization leads to a reduction in the effective cross-linking density and heterogeneity in the polymer due to microgel formation [38]. Composites based on Bis-GMA and 66% TEGDMA were characterized by high conversion values, but lower shrinkage values than would be expected, probably due to more severe primary cyclization. Although primary cyclization increases the conversion, it can compromise network formation and reduce the crosslinking density [39]. It is possible to achieve a high conversion with a relatively low TEGDMA content in UDMA-based composites (compared to Bis-GMA), due to their lower viscosity and structure characteristics [36]. A compromise between the degree of conversion and the desired properties (leaching and mechanical strength) was achieved for the systems with base monomers, Bis-GMA or UDMA, with molar fractions between 0.375 and 0.625 [32].

The addition of Bis-GMA in U/B/T mixtures was found to increase the flexural modulus, which has an impact on the value of the three-point bending strength (Figures 1 and 2). The highest median values (FS = 89.5 MPa) were observed for the UDMA/Bis-GMA/TEGDMA 40/40/20 wt.% matrix, while the lowest FS (69.7 MPa) and FM (1.6 GPa) values were demonstrated by the Bis-EMA/TEGDMA 80/20 wt.% matrix. The addition of UDMA had a positive effect on the tested properties, with the resulting polymer network being characterized by a greater stiffness and resistance to three-point bending. The addition of Bis-GMA and UDMA in the mixtures U/G T and U/E/T, respectively, increased intermolecular interactions, mostly hydrogen bonding, which is considered to be one of the most important factors influencing the strength and modulus of crosslinked dimethacrylate systems [11,24,40]. High values of FS in mixtures containing Bis-EMA may indicate a high conversion. This monomer is less viscous than Bis-GMA and the systems can therefore achieve a higher degree of conversion [41]. Additionally, the introduction of UDMA to the U/E/T mixtures resulted in the formation of a denser polymer network. UDMA resins have a higher reactivity than Bis-EMA resins due to the greater flexibility of their molecular structure, possible hydrogen abstraction, and chain transfer reaction mechanism [18,22]. The FS values of unfilled resins based on UDMA/TEGDMA ranged from 44 to 78 MPa, while those of matrices based on Bis-GMA/TEGDMA were between 51 and 66 MPa [37]. The ISO 4049 demands a flexural strength of at least 80 MPa for restorative materials in occlusion-bearing areas. Resins obtained in this study will allow the requirements of the standard to be met after using a filler system.

The second most frequently defined mechanical property for polymeric dental materials is diametral tensile strength (DTS). This property allows the tensile strength to be indirectly examined [42]. This value for dental composites varies from 30 to 55 MPa [43]. However, it should be emphasized that the DTS value increases with filler content. The DTS of unfilled Bis-GMA/TEGDMA (75 wt.%/25 wt.%) resin was previously found to be 21.9 MPa [44]. The median DTS for tested materials was found to range from 20 to 30 MPa, with the highest value being observed for the Bis-EMA/TEGDMA 80/20 wt.% matrix (approximately 46 MPa). The DTS test assumes a negligible deformation before fracture. Some distortions may have appeared in the tests on resin matrices as they demonstrate greater plastic deformation than filled materials [33,45]. The materials containing Bis-EMA and TEGDMA tended to display greater deformations. In addition, the E/T 80/20 wt.% matrix had the lowest FM and modulus of elasticity, which suggests that it should also have a lower DTS value. The smallest dispersion of values was observed for the U/G/T matrices, which may be due to the formation of a stiffer and brittle system that was more suitable for this type of test.

Of the studied properties, hardness is one of the most sensitive to changes in the degree of conversion [46]. This value increases with the degree of conversion [47]. In the present study, the highest hardness value was demonstrated by the UDMA/TEGDMA 80/20 wt.% matrix (16 HV) and the lowest by the UDMA/Bis-EMA/TEGDMA 70/10/20 wt.% and Bis-GMA/TEGDMA 80/20 wt.% matrices (14 HV) (Figure 4). Previous studies have indicated that U/T matrices have higher hardness values than U/G matrices do; UDMA has a lower viscosity and is more flexible than Bis-GMA, which leads to a higher conversion and denser polymer network [37]. However, like other properties of dental resins, hardness also depends on intramolecular interactions and the polymer structure [24]. The presence of aromatic rings and urethane bonds increases hardness values [48].

When dental resins are soaked in water and oral fluids, unreacted monomers and small oligomers are eluted, and water is absorbed by the resin matrix. The absorbed water occupies the space between polymer chains or it is bonded with the polymer. This process is controlled by diffusion and requires a few weeks to complete [49,50]. Our dynamic absorbency testing showed that the fastest mass increase due to water sorption occurs during the first month (Figure 5). The UDMA/TEGDMA mixtures showed a more rapid increase than UDMA/Bis-EMA/TEGDMA. In addition, the highest values were observed for the U/T and U/G/T mixtures (Table 4).

Hydrophobicity of the monomer is one of the most important factors that allows water sorption to be predicted. In our study, the highest values of water absorbency were observed for U/T. The values did not change significantly within the selected formulations. High values were also observed for mixtures with Bis-GMA. Due to the presence of urethane linkages in UDMA, ether linkages in TEGDMA, and hydroxyl groups in Bis-GMA, monomers have a hydrophilic nature and will more easily cause water to penetrate into the polymer network. The smallest values of water absorbency were observed by formulations with the Bis-EMA monomer. This monomer decreases water sorption and solubility due to its hydrophobic character [11,51]. An additional factor that significantly affects sorption and solubility values is the degree of conversion, and the characteristic of the polymer network. Homogenous networks with high cross-linking densities and small levels of porosity or microvoids have reduced solvent uptake and swelling [50,52–54]. This factor may explain the higher sorption values for U/T matrices than for formulations with the addition of a more hydrophilic monomer such as Bis-GMA. The higher addition of TEGDMA in the U/T matrices could result in a more cross-linked network, which may create a more heterogeneous polymer structure [35]. Additionally, the structure may be disturbed by the occurrence of a cyclization process of TEGDMA monomer [18]. The more heterogeneous the structure, the larger the spaces created between the polymer clusters (microporous), which can accommodate a larger amount of water [35]. The high sorption values for U/T matrices can be explained by the higher flexibility of the network in comparison with U/B/T formulations (Figure 2). This permits the higher swelling of polymer chains by water [35]. Solubility and sorption cause the hygroscopic expansion, plasticization, and hydrolytic degradation of resins, thus weakening the mechanical properties over time [16]. Therefore, knowledge of the behavior of the resin under the influence of the aquatic environment is also a very important element in assessing its properties.

Little research has been performed into more complex resin systems. One of the most extensive works was published in 1998 by Asmussen and Peutzfeldt, but in this study, the resins were loaded with silanized glass filler (78 wt.%) [33]. Blends based on TEGDMA/UDMA/Bis-GMA showed good properties, when the composite contained an equal mixture of Bis-GMA and UDMA or the content of UDMA monomer was higher. Flexural strength values were found to be 159 ± 18, 164 ± 18, and 167 ± 12 MPa for composites based on TEGDMA/UDMA/Bis-GMA with monomer contents (mol%) of 30/40/30, 30/30/40, and 30/20/50, respectively. The modulus values for these materials were 10.2 ± 0.4, 9.1 ± 1.2, and 8.0 ± 0.7 GPa, respectively. The tensile strength was similar for all matrices (approximately 55 MPa) [33]. Similar studies found that medium-viscosity resin (TEGDMA/UDMA/Bis-GMA 30:33:33 wt.%) provided optimum mechanical properties, and that the viscosity should be adjusted to achieve a balance between efficient conversion and the best mechanical properties [34]. Another similar study based on only five experimental groups also suggested that Bis-GMA:TEGDMA:UDMA (30:35:35 mol%) resin showed promising properties [55].

It should be noted that this work had some limitations. A fairly wide spectrum of tests were carried out that allowed for the exclusion of matrices, which did not meet certain strength criteria. However, for a more complete evaluation, the viscosity of the formulations, the degree of conversion, and the polymer network structure should be determined. Currently selected formulations (both unfilled and filled systems) are under evaluation using the aging protocol to assess the behavior of these materials in complex oral environments.

5. Conclusions

Certain relationships were observed regarding the influence of individual components on the properties of polymer matrices. However, they are not directly proportional to the compositional changes. Matrices with compositions of UDMA/Bis-GMA/TEGDMA 70/10/20 wt.% and 40/40/20 wt.%, and UDMA/Bis-EMA/TEGDMA 40/40/20 wt.% matrices are characterized with a good flexural strength (FS), modulus of elasticity (ME), hardness (HV), diametral tensile strength (DTS), and satisfactory water absorption and dissolution values. The use of three different monomers in proper proportions may create a stronger polymer matrix. Tested formulations after filling should meet the requirements of standard 4049 on the minimum flexural strength for restoration materials. The DTS and hardness values were also promising.

In addition to the degree of conversion—which can be partially controlled by obtaining medium-viscous systems—an important issue is secondary bonds such as hydrogen bonding and van der Waals forces. These interactions may improve the mechanical properties by increasing the polymer network density. The addition of such monomers as UDMA and Bis-GMA, which, due to their chemical structure, are capable of producing such interactions, may result in the creation of a material with a higher strength.

Author Contributions: Conceptualization, A.S.-W., M.D., K.B. and J.S.; methodology, A.S.-W., K.B. and J.S.; formal analysis, A.S.-W. and K.K.; investigation, A.S.-W.; data curation, A.S.-W.; writing—original draft preparation, A.S.-W., M.D. and K.K.; writing—review and editing, A.S.-W., K.B. and J.S.; visualization, A.S.-W. and K.K.; supervision, K.B. and J.S.; project administration A.S.-W. and K.B. All authors have read and agreed to the published version of the manuscript.

Funding: This research received no external funding.

Institutional Review Board Statement: Not applicable.

Informed Consent Statement: Not applicable.

Data Availability Statement: Data available in a publicly accessible repository Zenodo at 10.5281/zenodo.4772497, https://zenodo.org/record/4772497#.YKaRMrUzZPY.

Acknowledgments: We would like to offer our thanks to Arkona–Dental Laboratory of Pharmacology (Nasutów, Poland) for providing resins for the study.

Conflicts of Interest: The authors declare no conflict of interest. The funders had no role in the design of the study; in the collection, analyses, or interpretation of data; in the writing of the manuscript, or in the decision to publish the results.

Appendix A

Figure A1. The three-point bending test.

Figure A2. Diametral tensile strength test.

Figure A3. Vickers hardness test.

References

1. Ferracane, J.L. Current trends in dental composites. *Crit. Rev. Oral Biol. Med.* **1995**, *6*, 302–318. [CrossRef]
2. Iftikhar, S.; Jahanzeb, N.; Saleem, M.; ur Rehman, S.; Matinlinna, J.P.; Khan, A.S. The trends of dental biomaterials research and future directions: A mapping review. *Saudi Dent. J.* **2021**. [CrossRef]
3. Karabela, M.M.; Sideridou, I.D. Synthesis and study of properties of dental resin composites with different nanosilica particles size. *Dent. Mater.* **2011**, *27*, 825–835. [CrossRef] [PubMed]
4. Pratap, B.; Gupta, R.K.; Bhardwaj, B.; Nag, M. Resin based restorative dental materials: Characteristics and future perspectives. *Jpn. Dent. Sci. Rev.* **2019**, *55*, 126–138. [CrossRef] [PubMed]
5. Yadav, R.; Kumar, M. Dental restorative composite materials: A review. *J. Oral Biosci.* **2019**, *61*, 78–83. [CrossRef] [PubMed]
6. Kim, K.-H.; Ong, J.L.; Okuno, O. The effect of filler loading andmorphology on the mechanical properties of contemporary composites. *J. Prosthet. Dent.* **2002**, *87*, 642–649. [CrossRef] [PubMed]
7. Mount, G.J.; Rory Hume, W. A new cavity classification. *Aust. Dent. J.* **1998**, *43*, 153–159. [CrossRef]
8. Sandner, B.; Baudach, S.; Davy, K.W.M.; Braden, M.; Clarke, R.L. Synthesis of BISGMA derivatives, properties of their polymers and composites. *J. Mater. Sci. Mater. Med.* **1997**, *8*, 39–44. [CrossRef]
9. Cramer, N.B.; Stansbury, J.W.; Bowman, C.N. Recent Advances and Developments in Composite Dental Restorative Materials. *J. Dent. Res.* **2011**, *90*, 402–416. [CrossRef]
10. Peutzfeldt, A. Resin composites in dentistry: The monomer systems. *Eur. J. Oral Sci.* **1997**, *105*, 97–116. [CrossRef]
11. Gajewski, V.E.S.; Pfeifer, C.S.; Fróes-Salgado, N.R.G.; Boaro, L.C.C.; Braga, R.R. Monomers used in resin composites: Degree of conversion, mechanical properties and water sorption/solubility. *Braz. Dent. J.* **2012**, *23*, 508–514. [CrossRef] [PubMed]
12. Sideridou, I.; Tserki, V.; Papanastasiou, G. Effect of chemical structure on degree of conversion in light-cured dimethacrylate-based dental resins. *Biomaterials* **2002**, *23*, 1819–1829. [CrossRef]
13. Polski Komitet Organizacyjny. *Stomatologia-Materiały na Bazie Żywic Syntetycznych do Wypełnień, Odbudowy i Cementowania*; PN-EN ISO 4049:2003; Polski Komitet Organizacyjny: Warsaw, Poland, 2003.
14. Ilie, N.; Hilton, T.J.; Heintze, S.D.; Hickel, R.; Watts, D.C.; Silikas, N.; Stansbury, J.W.; Cadenaro, M.; Ferracane, J.L. Academy of Dental Materials guidance—Resin composites: Part I—Mechanical properties. *Dent. Mater.* **2017**, *33*, 880–894. [CrossRef]
15. Ferracane, J.L. Resin-based composite performance: Are there some things we can't predict? *Dent. Mater.* **2013**, *29*, 51–58. [CrossRef] [PubMed]
16. Szczesio-Wlodarczyk, A.; Sokolowski, J.; Kleczewska, J.; Bociong, K. Ageing of dental composites based on methacrylate resins-A critical review of the causes and method of assessment. *Polymers* **2020**, *12*, 882. [CrossRef] [PubMed]

17. Par, M.; Lapas-Barisic, M.; Gamulin, O.; Panduric, V.; Spanovic, N.; Tarle, Z. Long Term Degree of Conversion of two Bulk-Fill Composites. *Acta Stomatol. Croat.* **2016**, *50*, 292–300. [CrossRef]
18. Pfeifer, C.S.; Shelton, Z.R.; Braga, R.R.; Windmoller, D.; MacHado, J.C.; Stansbury, J.W. Characterization of dimethacrylate polymeric networks: A study of the crosslinked structure formed by monomers used in dental composites. *Eur. Polym. J.* **2011**, *47*, 162–170. [CrossRef]
19. Barszczewska-Rybarek, I.M.; Chrószcz, M.W.; Chladek, G. Novel urethane-dimethacrylate monomers and compositions for use as matrices in dental restorative materials. *Int. J. Mol. Sci.* **2020**, *21*, 2644. [CrossRef]
20. Fugolin, A.P.; de Paula, A.B.; Dobson, A.; Huynh, V.; Consani, R.; Ferracane, J.L.; Pfeifer, C.S. Alternative monomer for Bis-GMA-free resin composites formulations. *Dent. Mater.* **2020**, *36*, 884–892. [CrossRef]
21. Pomes, B.; Derue, I.; Lucas, A.; Nguyen, J.F.; Richaud, E. Water ageing of urethane dimethacrylate networks. *Polym. Degrad. Stab.* **2018**, *154*, 195–202. [CrossRef]
22. Dickens, S.H.; Stansbury, J.W.; Choi, K.M.; Floyd, C.J.E. Photopolymerization kinetics of methacrylate dental resins. *Macromolecules* **2003**, *36*, 6043–6053. [CrossRef]
23. Gonalves, F.; Kawano, Y.; Pfeifer, C.; Stansbury, J.W.; Braga, R.R. Influence of BisGMA, TEGDMA, and BisEMA contents on viscosity, conversion, and flexural strength of experimental resins and composites. *Eur. J. Oral Sci.* **2009**, *117*, 442–446. [CrossRef] [PubMed]
24. Barszczewska-Rybarek, I.M. Structure-property relationships in dimethacrylate networks based on Bis-GMA, UDMA and TEGDMA. *Dent. Mater.* **2009**, *25*, 1082–1089. [CrossRef]
25. Martim, G.C.; Pfeifer, C.S.; Girotto, E.M. Novel urethane-based polymer for dental applications with decreased monomer leaching. *Mater. Sci. Eng. C* **2017**, *72*, 192–201. [CrossRef] [PubMed]
26. Fu, J.; Liu, W.; Hao, Z.; Wu, X.; Yin, J.; Panjiyar, A.; Liu, X.; Shen, J.; Wang, H. Characterization of a low shrinkage dental composite containing bismethylene spiroorthocarbonate expanding monomer. *Int. J. Mol. Sci.* **2014**, *15*, 2400–2412. [CrossRef] [PubMed]
27. Wu, J.; Xie, X.; Zhou, H.; Tay, F.R.; Weir, M.D.; Melo, M.A.S.; Oates, T.W.; Zhang, N.; Zhang, Q.; Xu, H.H.K. Development of a new class of self-healing and therapeutic dental resins. *Polym. Degrad. Stab.* **2019**, *163*, 87–99. [CrossRef]
28. Ding, Y.; Li, B.; Wang, M.; Liu, F.; He, J. Bis-GMA Free Dental Materials Based on UDMA/SR833s Dental Resin System. *Adv. Polym. Technol.* **2016**, *35*, 396–401. [CrossRef]
29. Fugolin, A.P.P.; Pfeifer, C.S. New Resins for Dental Composites. *J. Dent. Res.* **2017**, *96*, 1085–1091. [CrossRef]
30. De Oliveira, D.C.R.; Rovaris, K.; Hass, V.; Souza-Júnior, E.J.; Haiter-Neto, F.; Sinhoreti, M.A.C. Effect of low shrinkage monomers on physicochemical properties of dental resin composites. *Braz. Dent. J.* **2015**, *26*, 272–276. [CrossRef]
31. Ilie, N.; Hickel, R. Investigations on a methacrylate-based flowable composite based on the SDR^{TM} technology. *Dent. Mater.* **2011**, *27*, 348–355. [CrossRef]
32. Floyd, C.J.E.; Dickens, S.H. Network structure of Bis-GMA- and UDMA-based resin systems. *Dent. Mater.* **2006**, *22*, 1143–1149. [CrossRef] [PubMed]
33. Asmussen, E.; Peutzfeldt, A. Influence of UEDMA, BisGMA and TEGDMA on selected mechanical properties of experimental resin composites. *Dent. Mater.* **1998**, *14*, 51–56. [CrossRef]
34. Musanje, L.; Ferracane, J.L. Effects of resin formulation and nanofiller surface treatment on the properties of experimental hybrid resin composite. *Biomaterials* **2004**, *25*, 4065–4071. [CrossRef]
35. Sideridou, I.; Tserki, V.; Papanastasiou, G. Study of water sorption, solubility and modulus of elasticity of light-cured dimethacrylate-based dental resins. *Biomaterials* **2003**, *24*, 655–665. [CrossRef]
36. Gonçalves, F.; Pfeifer, C.C.S.; Stansbury, J.W.; Newman, S.M.; Braga, R.R. Influence of matrix composition on polymerization stress development of experimental composites. *Dent. Mater.* **2010**, *26*, 697–703. [CrossRef] [PubMed]
37. Nicolae, L.C.; Shelton, R.M.; Cooper, P.R.; Martin, R.A.; Palin, W.M. The Effect of UDMA/TEGDMA Mixtures and Bioglass Incorporation on the Mechanical and Physical Properties of Resin and Resin-Based Composite Materials. *Conf. Pap. Sci.* **2014**, *2014*, 1–5. [CrossRef]
38. Elliott, J.E.; Lovell, L.G.; Bowman, C.N. Primary cyclization in the polymerization of bis-GMA and TEGDMA: A modeling approach to understanding the cure of dental resins. *Dent. Mater.* **2001**, *17*, 221–229. [CrossRef]
39. Gonçalves, F.; Pfeifer, C.S.; Ferracane, J.L.; Braga, R.R. Contraction Stress Determinants in Dimethacrylate Composites. *J. Dent. Res.* **2008**, *87*, 367–372. [CrossRef]
40. Lemon, M.T.; Jones, M.S.; Stansbury, J.W. Hydrogen bonding interactions in methacrylate monomers and polymers. *J. Biomed. Mater. Res.* **2007**, *83*, 734–746. [CrossRef]
41. Barszczewska-Rybarek, I.M. A guide through the dental dimethacrylate polymer network structural characterization and interpretation of physico-mechanical properties. *Materials* **2019**, *12*, 4057. [CrossRef] [PubMed]
42. Penn, R.; Craig, R.; Tesk, J. Diametral tensile strength and dental composites. *Dent. Mater.* **1987**, *3*, 46–48. [CrossRef]
43. Bona, D.A.; Benetti, P.; Borba, M.; Cecchetti, D. Flexural and diametral tensile strength of composite resins. *Restor. Dent. Braz. Oral Res.* **2008**, *22*, 84–89. [CrossRef] [PubMed]
44. Lin, C.T.; Lee, S.Y.; Keh, E.S.; Dong, D.R.; Huang, H.M.; Shih, Y.H. Influence of silanization and filler fraction on aged dental composites. *J. Oral Rehabil.* **2000**, *27*, 919–926. [CrossRef] [PubMed]
45. Zidan, O.; Asmussen, E.; Jørgensen, K.D. Tensile strength of restorative resins. *Eur. J. Oral Sci.* **1980**, *88*, 285–290. [CrossRef]

46. Barszczewska-Rybarek, I.; Chladek, G. Studies on the curing efficiency and mechanical properties of bis-GMA and TEGDMA nanocomposites containing silver nanoparticles. *Int. J. Mol. Sci.* **2018**, *19*, 3937. [CrossRef]
47. Ferracane, J.L. Correlation between hardness and degree of conversion during the setting reaction of unfilled dental restorative resins. *Dent. Mater.* **1985**, *1*, 11–14. [CrossRef]
48. Barszczewska-Rybarek, I.M. Characterization of urethane-dimethacrylate derivatives as alternative monomers for the restorative composite matrix. *Dent. Mater.* **2014**, *30*, 1336–1344. [CrossRef]
49. Braden, M.; Causton, E.E.; Clarke, R.L. Diffusion of Water in Composite Filling Materials. *J. Dent. Res.* **1976**, *55*, 730–732. [CrossRef]
50. Ferracane, J.L. Hygroscopic and hydrolytic effects in dental polymer networks. *Dent. Mater.* **2006**, *22*, 211–222. [CrossRef]
51. Putzeys, E.; De Nys, S.; Cokic, S.M.; Duca, R.C.; Vanoirbeek, J.; Godderis, L.; Van Meerbeek, B.; Van Landuyt, K.L. Long-term elution of monomers from resin-based dental composites. *Dent. Mater.* **2019**, *35*, 477–485. [CrossRef]
52. Kalachandra, S.; Kusy, R.P. Comparison of water sorption by methacrylate and dimethacrylate monomers and their corresponding polymers. *Polymer* **1991**, *32*, 2428–2434. [CrossRef]
53. Kalachandra, S.; Turner, D.T. Water sorption of polymethacrylate networks: Bis-GMA/TEGDM copolymers. *J. Biomed. Mater. Res.* **1987**, *21*, 329–338. [CrossRef] [PubMed]
54. Venz, S.; Dickens, B. NIR-spectroscopic investigation of water sorption characteristics of dental resins and composites. *J. Biomed. Mater. Res.* **1991**, *25*, 1231–1248. [CrossRef]
55. Pfeifer, C.S.; Silva, L.R.; Kawano, Y.; Braga, R.R. Bis-GMA co-polymerizations: Influence on conversion, flexural properties, fracture toughness and susceptibility to ethanol degradation of experimental composites. *Dent. Mater.* **2009**, *25*, 1136–1141. [CrossRef] [PubMed]

Article

Assessment of the Potential Ability to Penetrate into the Hard Tissues of the Root of an Experimental Preparation with the Characteristics of a Dental Infiltratant, Enriched with an Antimicrobial Component—Preliminary Study

Małgorzata Fischer [1],*, Małgorzata Skucha-Nowak [1], Bartosz Chmiela [2] and Anna Korytkowska-Wałach [3]

[1] Unit of Dental Propedeutics, Department of Conservative Dentistry with Endodontics, Faculty of Medical Sciences in Zabrze, Medical University of Silesia, 40-055 Katowice, Poland; mskucha-nowak@sum.edu.pl

[2] Department of Materials Technologies, Faculty of Materials Engineering, Silesian University of Technology, 40-019 Katowice, Poland; bartosz.chmiela@polsl.pl

[3] Department of Organic Chemistry, Bioorganic Chemistry and Biotechnology, Faculty of Chemistry, Silesian University of Technology, 44-100 Gliwice, Poland; anna.korytkowska-walach@polsl.pl

* Correspondence: skorusmalgorzata@gmail.com; Tel.: +48-32-2828-7942

Abstract: Infiltration is a method of penetration with a low viscosity resin that penetrates deep into demineralised tooth tissue and fills the intergranular spaces, hence reducing porosity. Carious lesions initially located at the enamel–cement junction are usually found in elderly patients. Those spots are predisposed to bacterial adhesion originating both from biofilm and from gingival pocket bacteria. The aim of this study was to evaluate the penetration of an experimental preparation, which has the characteristics of a dental infiltrant, enriched with an antibacterial component, into the decalcified root cement tissues of extracted human teeth in elderly patients. An experimental preparation with the characteristics of a dental infiltrant was prepared, applied, and polymerised on the surface of extracted, previously decalcified human teeth. The control sample was Icon (DMG, Hamburg, Germany). The ability of the preparations to penetrate deep into the root cement was evaluated using scanning electron and light microscopy. The study showed that an experimental preparation could potentially be used for treatment of early carious lesions within the tooth root in elderly patients, among others, as it penetrates deep into demineralised tissues. More research is needed.

Keywords: root caries; infiltration; microstomatology; dental materials; antimicrobial properties; polymers

1. Introduction

Despite extensive development ongoing in medicine, dental caries is still one of the most common diseases of the oral cavity. As a result, demineralization and proteolytic tissue breakdown occur. The carious process involves both tissues located in the coronal part, such as enamel and dentin, and those located in the root part, such as root cementum and root dentin [1–5].

Microinvasive dentistry (also known as microstomatology) is a modern concept based on early detection of carious lesions at the molecular level. This new branch in restorative dentistry makes it possible to stop the development of the disease quickly and to restore the normal structure and function of tooth tissue [6–9]. In order to assess the degree of demineralisation, diagnostic equipment characterised by a wide range of sensitivity and high specificity was used [10–15].

One of the basic principles of microstomatology is the remineralisation of hard dental tissues. Preparations used for that purpose contain fluoride compounds, such as sodium fluoride, acidulated phosphate fluoride (APF), tin fluoride, and amine fluoride [16–18].

Fluoridisation strengthens tooth enamel against the action of bacteria and acids metabolised by them, reduces tooth hypersensitivity in the cervical region, and prevents the formation of carious lesions in areas with previously placed fillings. However, fluoridation preparations have disadvantages. Fluoride compounds applied to tooth surfaces are not mechanically resistant, especially during oral hygiene procedures. These preparations are also not able to penetrate deep into demineralised tooth tissue. Another disadvantage is the lack of ability to inhibit the development of microorganisms on their surface, which would facilitate long-term inhibition of carious lesion development. Moreover, allergy to any of its ingredients is also a contraindication for the use of fluoride preparations.

Infiltration is an answer to the disadvantages of fluoride preparations—it is also a method that does not require surgical interference in the hard tissues of the tooth. This method comprises deep penetration into the demineralised tooth tissues, while filling the intercrystalline spaces with low viscosity resin and reducing the tissue porosity. As a result, the window for penetration of bacterial toxins into the hard tissues and the bacteria producing them is closed. The result is inhibition of the decalcification process and caries development [19,20]. An infiltrant is a substance based on polymeric resins, capable of penetrating tissues due to capillary forces. According to Manji [21,22], dental infiltrants are used to treat lesions on smooth vestibular and tangential surfaces at the white spot stage, with a maximum radiological depth of up to 1/3 of the outer dentin layer. The advantages of that method include mechanical stabilisation of demineralised tissue, permanent closure of superficial micropores, and inhibition of the progression of lesions. In this way, we minimise the risk of secondary caries and delay surgical intervention, while at the same time achieving aesthetic benefits [23,24]. The use of infiltrants also reduces the risk of gingivitis and the occurrence of tooth hypersensitivity in the cervical region [25]. The requirements for infiltrants to be used in treatment of carious lesions include: a high surface tension, a low density, hydrophilicity, a lack of toxicity to the host, a resistance to chemical and mechanical effects, the ability to polymerise to a solid state, the similarity of the infiltrant colour to that of the tooth, a lack of interaction with food and drugs, and the ability to inhibit multiplication of microorganisms on their surface. The last feature mentioned is very important as it reduces the risk of secondary caries. The only commercially available preparation of such kind is Icon (DMG, Hamburg, Germany), launched in 2009. It is used for the treatment of early enamel caries lesions located on the vestibular and proximal surfaces. However, this preparation does not meet all the requirements for infiltrants, as it does not have a component responsible for bacteriostaticity [25]. This fact prompted the authors of the study to undertake research towards the synthesis of an experimental preparation enriched with the missing component. The developed experimental preparation was enriched with a PMMAn-MTZ monomer, which has adhesive properties. Metronidazole acts on Gram-negative anaerobic microorganisms and Gram-positive anaerobic bacteria (organisms living under anaerobic or low oxygen conditions). Synthesis of PMMAn with metronidazole provides a component that potentially inhibits microbial proliferation on the surface of the hard tooth tissue. PMMAn and metronidazole are linked by an ester bond, which is hydrolysed when exposed to water (saliva). In this way, metronidazole can interact with the experimental preparation to inhibit bacterial growth on the tooth surface continuously [26]. According to the instructions of the manufacturer, the commercial preparation Icon is mainly used for caries within the enamel; therefore, in the opinion of the authors of this study, the potential of infiltrants is not fully exploited. An attempt was made to apply the synthesised infiltrant to the surface of decalcified root cement.

Root cement is the mineralized hard tissue of the tooth that covers the outer surface of the tooth root. Thickness of root cement is locally variable, being thickest in the apical region (150–200 micrometres) and thinnest in the cervical region (20–50 micrometres) [27,28].

Carious lesions initially located at the enamel–cement junction are usually found in older patients [29,30]. As life expectancy increases, there is no apparent correlation with the number of retained healthy teeth [27]. Modern dentistry faces the challenge of improving oral health. With age, there is an increase in the number of general comorbidities and medications taken by patients, as well as reduced manual dexterity [31]. All these factors contribute to the development of root caries. In the population, we also observed generalized senile periodontal atrophy, characterized by bone loss of the alveolar process and the alveolar part of the mandible, as well as gingival recession.

As a result, the clinical crowns of the teeth are elongated, while the roots and cementum are exposed [32]. The root surface exposed to the oral environment is much more sensitive to mechanical and chemical destruction than the highly mineralised enamel. This process is initially asymptomatic, manifesting itself as a small brown demarcated lesion (grade 1 according to Billings) below a well-mineralised surface layer [33]. Demineralisation of the root cement is similar to that observed in enamel (destruction of apatite crystals occurs subsurface, before bacterial penetration) [29]. Some differences are also apparent in caries of enamel and root cement. Demineralisation in root cement starts much faster than in enamel, already at the plaque pH of 6–6.5 [34]. In addition, the root surface contains less fluoride than the superficial layer of enamel, which is related to its lower exposure to fluoride contained in saliva and preparations used to maintain oral hygiene [35]. Bacterial invasion occurs at an earlier stage of root caries than in case of enamel caries. The accumulation of bacteria at the cemento–dentin interface causes rapid decomposition of collagen fibres. The presence of traumatic nodes in the oral cavity is not without significance in the formation of decalcifications in the cervical region in case of elderly patients. Bite loads acting eccentrically to the long axis of the tooth cause a loss of connection between hydroxyapatite crystals, which subsequently leads to their disruption and the formation of tissue loss in the cervical region [36]. An oral environment rich in carious bacteria and poor in saliva adversely affects the structure of the exposed cement, leading to the formation of carious defects.

There are few reports in the available literature on the use of infiltrants enriched with an antibacterial component [37]. The authors of the study focused on the potential attempt to apply infiltrants on the surface of the root cement in order to achieve the intended therapeutic effect that is minimally invasive for the hard tissues of the tooth.

The aim of this study was to assess the potential degree of penetration of the experimental preparation valuate, which reached deep into the decalcified root cement tissues of extracted human teeth, with the characteristics of a dental infiltrant enriched with an antibacterial component in adult patients, including the elderly.

This study is a pioneering and pilot study in this field. The authors based their work on the assessment of the degree of penetration of the experimental preparation based on the SEM observations.

2. Materials and Methods

The material for the study included 12 extracted human teeth of adult patients: molars and premolars removed for prosthetic and periodontitis, with naturally exposed root cement to a depth of 6 mm, i.e., the limit above which the treatment of choice is invasive, surgical treatment. The teeth with a preserved anatomical crown and root were included in the study, whereas the teeth with a loss of root cement during extraction were excluded. The teeth were kept in a prepared 0.5% chloramine solution until the start of the study. Before the beginning of the study, the teeth were cleaned of sediment and calculus, as well as hard and soft tissue debris with use of dental curettes (LM-Instruments, Parainem, Finland). The teeth were then polished with SuperPolish paste (Kerr Dental, Bioggio, Italy), with an RDA of 9.8 and without fluoride, with an Opti Shine toothbrush (Kerr, Bioggio, Italy) using 2000 r/min. The teeth were rinsed three times with distilled water and left for 24 h.

2.1. Demineralization of Tooth Root Cement

A solution was prepared to demineralise the hard tissues of the tooth, the composition of which is shown in Table 1 [25,38].

Table 1. Composition of the solution for demineralization of hard tooth tissues.

Component	Molar Concentration	Quantity
$CaCl_2*2H_2O$	3 mM	0.441 g
KH_2PO_4	3 mM	0.408 g
CH_3COOH	50 mM	2.88 mL
MHDP	6 µM	1.416×10^{-3} g

The teeth were divided into two groups. Then, 1 L of demineralisation solution was poured into one container in which 6 teeth (the test group) were placed. pH = 6.2 of the solution was then determined using acetic acid and potassium hydroxide, depending on the fluctuation of the pH value [39]. The container was placed in an incubator at 37 °C for a period of 5 weeks. The conditions in the incubator mimicked the conditions in the oral cavity to initiate the process of initial caries formation. During the entire 5-week period, the pH was measured daily using a CP 411 pH-meter (Elmetron, Gliwice, Poland), with adjustments of high pH values with acetic acid and low pH values with potassium hydroxide to maintain a constant level of pH = 6. The remaining 6 teeth, which constituted the control group, were kept in distilled water. After 5 weeks, the teeth of both the test and control groups were rinsed three times with distilled water and thoroughly dried.

2.2. Preparation of the Experimental Solution

In cooperation with the Department of Organic Chemistry, Bioorganic Chemistry and Biotechnology of the Silesian University of Technology (Gliwice, Poland), an experimental solution was prepared, which consisted of the components listed in Table 2 [40].

Table 2. Composition and percentage content of the experimental solution.

Component	Quantity (g)	Content (%)
TEGDMA	3.75	75
HEMA	1.25	25
PMMAn-MTZ [1]	0.05	1 *
DMAEMA [2]	0.05	1 *
CQ [3]	0.025	0.5 *

* ratio to total mass of monomers; TEGDMA (triethylene glycol dimethacrylate, Fluka, Buchs, Switzerland); HEMA (2-hydroxyethyl methacrylate, Acros, New Jersey, USA); [1] PMMAn (2-(7-methyl-1,6-dioxo-2,5-dioxa-7-octenyl) trimellitic anhydride); MTZ (metronidazole, Acros, New Jersey, USA); [2] DMAEMA (N,N-dimethylaminoethyl methacrylate, Merck, Darmstadt, Germany); [3] CQ (camphorquinone, Aldrich, St. Louis, Mo, USA).

An experimental solution with the characteristics of a dental infiltrant was prepared with a mass of 5 g, and it was placed in a dark bottle and mixed thoroughly. DMAEMA in the preparation is the component responsible for polymerisation. PMMAn was enriched with metronidazole to obtain the antimicrobial component. In order to facilitate observation under the microscope and to be able to assess the penetration of the preparations into the dental hard tissue, a dye called eosin (WarChem, Warsaw, Poland) (0.25 mL) was added to both experimental and Icon preparations, as shown in Figure 1.

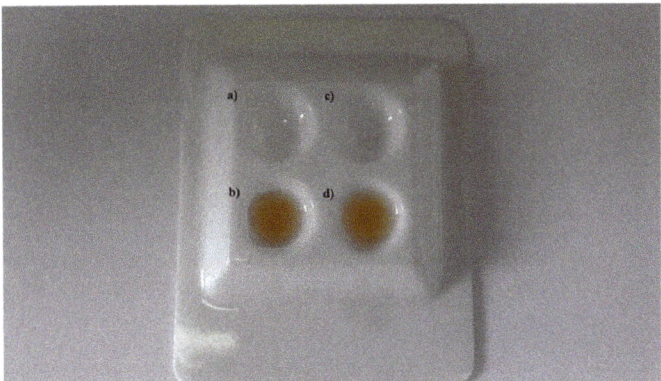

Figure 1. Observation of change in colour of the polymerised experimental preparation (**a**) before the addition of eosin and (**b**) after the addition of eosin, as well as of Icon solution (**c**) before the addition of eosin and (**d**) after the addition of eosin.

2.3. Tooth Infiltration Procedure

Two zones were delineated on the surface of both decalcified teeth (study group) and non-decalcified teeth (control group). Each zone was marked with colour varnish on the surfaces of tooth crowns. The blue zone marked teeth infiltrated with the Icon preparation containing eosin, and the red zone marked teeth infiltrated with the experimental preparation with eosin, as shown in Figure 2.

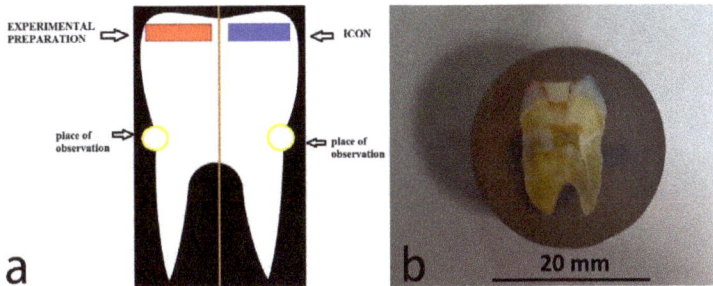

Figure 2. (**a**) Graphically marked tooth zones and areas for microscopic observation; (**b**) a photo of a resin-embedded tooth imaged with a stereoscopic microscope. The red side shows the zone with the applied experimental preparation, and the blue side shows the tooth with the applied commercial preparation Icon.

Both preparations were applied according to the instructions for use supplied with the commercial preparation.

During polymerisation, a C01-C Premium Plus wireless LED polymerisation lamp (Premium Plus International Limited, Bournemouth, UK) was used, emitting light in the wavelength range of 440–480 nm. The full mode was used for the study, with the radiation power of 1200 mW/cm^2.

2.4. Preparation of Samples for Microscopic Analysis

The teeth were cut along the long axis, perpendicularly to the traced zones. Tooth surfaces were polished with R100 felt discs (Stoddard, London, UK) at 4000 r/min with SeptoDiamond fluoride-free diamond paste (Septodont, Warsaw, Poland). The teeth were then cold-fused in the Metalogis Opti-Mix two-component methylmethacrylate-based resin, (Metalogis s.c., Warsaw, Poland). They were then cut longitudinally, using a Buehler precision cutter (Buehler Holding A.G., Uzwil, Switzerland), after which the cross-sections were made. The cross-sectional surfaces were sanded with waterproof sandpaper of decreasing gradations: P320, P500, P800, and P1000 Metalogis DEMPAX (Metalogis s.c, Warsaw, Poland); and then polished with Struers DP Suspension diamond slurries: 9 μm, 3 μm, 1 μm, 0.25 μm (Struers A/S, Ballerup, Denmark).

2.5. Microscope Observations

Observations of longitudinal sections of teeth were made with a Hitachi S-3400N scanning electron microscope (Hitachi Ltd., Tokyo, Japan) and with an Olympus GX71 light microscope (Olympus, Tokyo, Japan) at 50× magnification.

3. Results

The first stage of microscope observation was to determine the anatomical structures visible in the Hitachi S-34000N scanning electron microscope, which are shown in Figure 3.

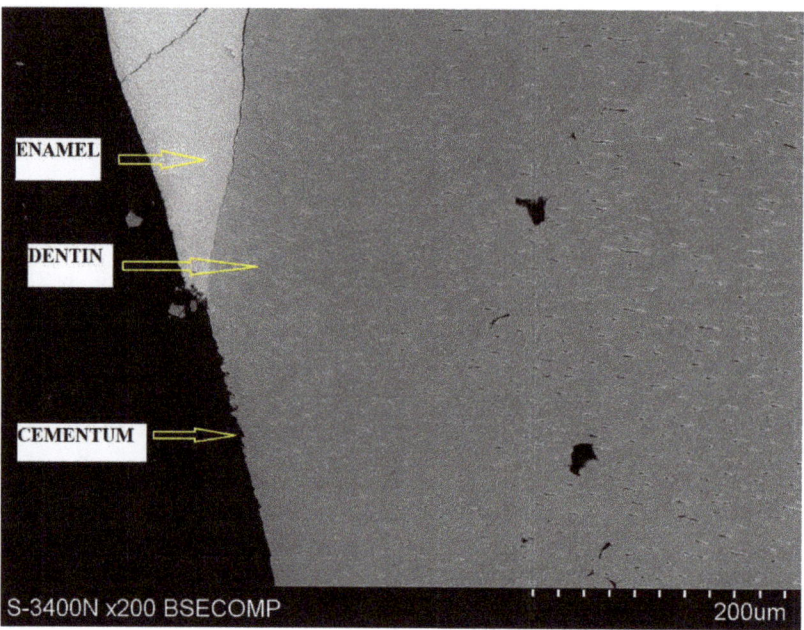

Figure 3. A photograph showing anatomical tooth structures.

With the use of a scanning microscope, the thickness of root cement in the cervical region was determined to be 7.47 μm, as shown in Figure 4. It was also possible to see the expansion of root cement, with an increasing thickness to 55.6 μm towards the root apex, as shown in Figure 5.

Figure 4. Scanning electron microscope image showing the thickness of root cement in the cervical region.

Figure 5. Scanning electron microscope image showing the thickness of root cement below the cervical region, in the apical direction.

The second stage of microscopic observation was to analyse the degree of penetration of the experimental preparation and the Icon preparation. Both the resin-embedded teeth of the study group and the control group were observed under an Olympus GX71 light microscope (Tokyo, Japan) at 50× magnification. The results of the observations are shown in Figures 6 and 7.

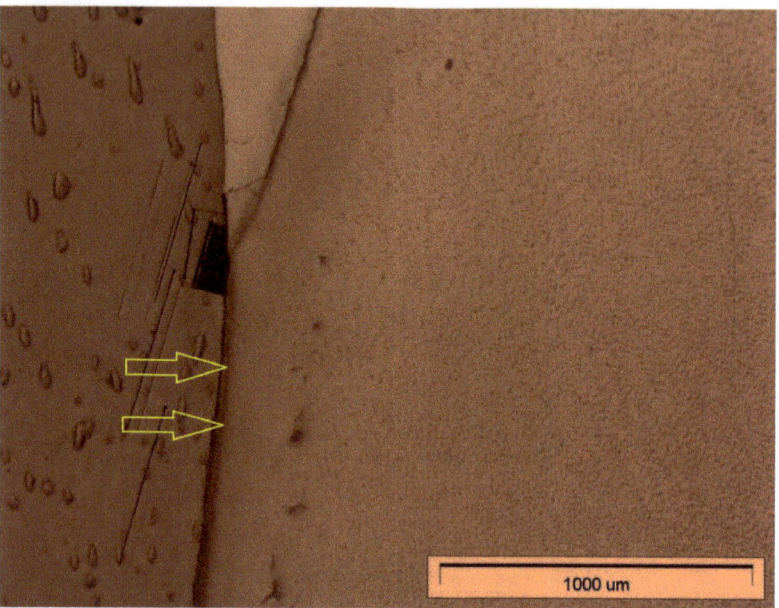

Figure 6. Visible penetration of the experimental preparation in the cervical area of a tooth. The arrows indicate the direction of penetration.

The light microscope observations revealed penetration of both the experimental preparation and the commercial preparation Icon deep into the decalcified hard tissue in the area of the root. The study showed that both preparations penetrated not only the decalcified root cement but that it also reached the root dentine.

Observation under a light microscope made it possible to assess the degree of penetration of the experimental preparation and the commercial preparation Icon in decalcified teeth. In both cases, penetration of the infiltrant into root cement is a desirable phenomenon, as these preparations should be able to penetrate demineralised tooth root tissues.

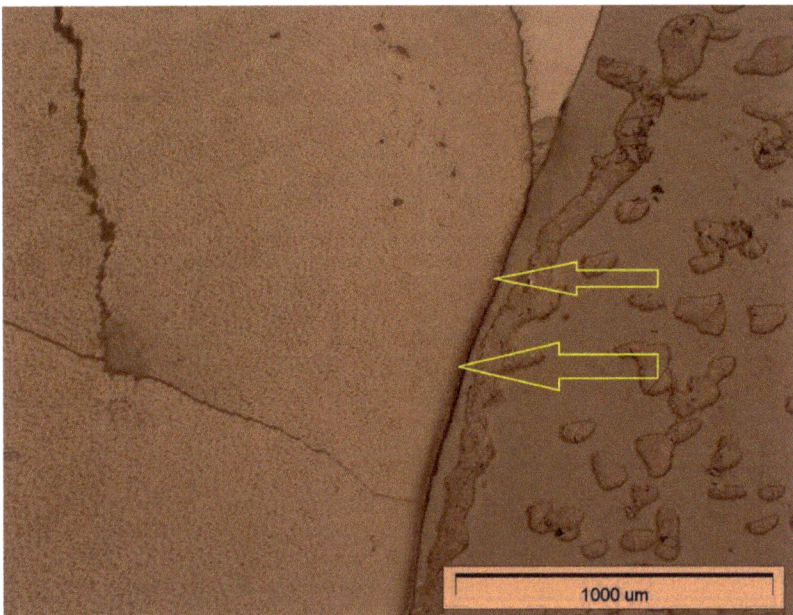

Figure 7. Visible penetration of Icon in the cervical region. The arrows indicate the direction of penetration.

4. Discussion

Contemporary dentistry is faced with a challenge to maintain oral health by preserving as much healthy tooth tissue as possible, to treat caries and periodontal disease early, and to improve the function of the masticatory apparatus in elderly patients. Based on US statistical data from 2020, it is estimated that the group of people aged 65 and over constitutes 17% of the population [41]. Statistical data of European countries, e.g., Poland, also show that the group of people aged 65 and over is steadily growing. The participation of this age group in the total population (old age index) in 2020 was 18.6%, compared to 1990, when it was 10% [42]. Konopka et al. in their study noted that, in recent years, as life expectancy increases in the studied age groups of the population, the number of retained dentition increases [43]. While considering the aging process of the population, it should be noted that geriatric dentistry will constitute a significant percentage of cases covered by prevention and treatment. With age, a number of physiological changes occur, which in case of the dentition are usually atrophic in nature. In many cases, the dental or periodontal age of a patient does not correspond to their biological age. The coexistence of systemic diseases, the deterioration of oral hygiene associated with reduced manual dexterity and poorer vision, as well as past or active periodontal diseases significantly increase the likelihood of developing dental root caries [44,45]. There are also a number of other behavioural and sociodemographic factors that contribute to the occurrence of root caries [46]. Moreover, it is important to mention that, with age, the salivary glands become fibrotic and fatty, which causes a reduction in saliva secretion in the oral cavity and the development of caries [47]. Senile bone atrophy of the maxillary alveolar process and the alveolar part of the mandible together with gingival recession affects apparent elongation of the clinical crowns and exposure of root cement. Barczak et al. in their study estimated that the incidence of periodontal disease in the elderly occurs in 98–100% of patients [27]. Katz RV et al., while evaluating teeth with recessions, proved that as many as one in nine cases had root caries (11.4%) [32]. Moreover, Slavkin noted that more than 50% of the people in the age group of 65 and older had a diagnosis of root caries [48]. Hellyer et al.

determined that the majority of active root carious lesions were located within 1 mm of the gingival margin, while inactive lesions were located above or 1 mm from the gingival margin [49]. It should be mentioned that not only elderly patients but also adolescents who undergo orthodontic treatment, especially with brackets located in the cervical region of teeth and at the same time on a high-sugar diet, are at risk of caries in the cervical region and in the root cement [47].

Until recently, the treatment of primary carious lesions was based on two basic mechanisms: prophylactic and curative, i.e., the use of an appropriate diet and fluoride prophylaxis—the effect of exogenous fluoride compounds (tooth brushing, rubbing), as well as the action of endogenous compounds (their oral intake). These measures were applied before any signs of tooth decay occurred, in order to prevent demineralization, or after a white spot was observed. Scribante et al., in their study, confirmed the correctness of enamel remineralization with hydroxyapatite and sodium fluoride, which was also after teeth whitening [50]. The exogenous use of fluoride compounds is aimed at remineralization and is therefore a form of treatment. The gap between preventive and interventional ("surgical") treatment has been bridged by micro-invasive dentistry with its promising treatment of early carious lesions, based on the infiltration technique [8].

Most studies involving the process of root tissue infiltration have assessed the degree of microhardness. Zhou et al., in their study, evaluated the effect of infiltration on root caries induced by *Streptococcus mutans* biofilm. The researchers demonstrated that the microhardness of the tooth surface and resistance to caries increased significantly after the application of Icon infiltrant [51]. Yazkan and Gurdogan et al. in their study confirmed that the application of Icon infiltrant increased the microhardness of the enamel surface [52,53]. Additionally, Gurdogan et al. in their study found that Icon increased surface roughness, when compared to the control group [53]. Zhou et al. also noted that Icon resin causes an increase in surface roughness compared to a decalcified surface without application of the infiltrant. That research may suggest that Icon resin application tends to increase plaque accumulation on the tooth surface at the application site [51]. M-Skucha Nowak et al. in their previous study pointed out that the commercial preparation Icon does not meet one of the main characteristics of a dental infiltrant, namely, it does not have a component responsible for bacteriostaticity [54]. In previous studies, the authors demonstrated that the inclusion of metronidazole in the experimental formulation allows for targeted activity against both anaerobic and oxygen-deficient organisms. The PMMAn—MTZ component demonstrates strong antimicrobial activity against *Bacteroides*, *Fusobacterium*, *Eubacterium*, *Clostridium*, *Peptococcus*, *Peptostreptococcus*, and protozoa [26]. Both the authors and a researcher, Zhang, assessed that the inclusion of an antibacterial component in the composition of a dental infiltrant could potentially inhibit the development of caries at the site of application [26,55]. Clinically, this is of great importance because the exposed surface of root cement provides a site of plaque retention. The authors, in their previous study, also demonstrated that an experimental formulation with a PMMAn-MTZ component at the 2-fold dilution applied to an early carious lesion in close proximity to the gingival mucosa showed similar cytotoxicity to the commercial formulation Icon after 24 h, while, at the 4-fold dilution, the experimental formulation showed less cytotoxicity than Icon, suggesting that application of the experimental formulation does not require isolation of the treatment field [26]. Neres et al. focused on the abrasion resistance of Icon on the enamel surface by subjecting it to 10,000 brushing cycles, proving that when the application of the preparation is performed according to the manufacturer's recommendations in the form of two layers of infiltrant, the preparation will not be abraded [56]. These results allow us to conclude that the application of infiltrants to the exposed root surface subjected to regular hygienic procedures will allow the preparation to persist and protect against the development of caries.

Numerous researchers have infiltrated the enamel surface. Skucha-Nowak et al., while applying the experimental preparation on the enamel surface, proved that the preparation penetrates the pores to a depth of 30–40 µm. The study obtained values similar to those described by Subramaniam et al. [25,57]. The authors of the present study attempted to evaluate the ability of the experimental preparation to penetrate deep into the root cement. By evaluating the degree of penetration of the experimental preparation and the commercial preparation Icon in decalcified teeth, the degree of penetration in both groups was observed, which is a desirable phenomenon, as these preparations should penetrate the demineralised tissues of the root. The phenomenon of infiltration shows that preparation reaches into pores or cavities on the tooth surface. The literature reports that infiltration to a depth of 60 μm is sufficient to prevent demineralisation of enamel tissues [58]. It is known that, depending on the type of tooth, thickness of the enamel can vary from 2.6 mm on the chewing surface to 2.4 mm in molars and premolars, respectively. The thinnest layer was found in the cervical region of the tooth root and was 0.1 mm. In comparison, the thickness of root cement was 150–200 micrometres in the periapical region, while, in the cervical region, it was about 20–50 micrometres. Based on these values, it can be estimated that it is sufficient for the infiltrant to penetrate to a depth of 2.3% of enamel thickness. The authors of this study performed a pioneering study based on a subjective exclusion in the SEM image. In the future, the use of X-ray microtomography (μ-CT) will allow clinicians to accurately analyse the depth of penetration of the experimental preparation into hard tissues [59,60]. This method can provide valuable information in a non-invasive manner, without cutting the samples. These studies need to be continued in order to thoroughly assess the penetration of the experimental preparation based on a statistical study. The results obtained in the study allow for a preliminary assessment that resin infiltration can be used not only on the enamel surface but also within the root. Conservative treatment of elderly patients should be carefully planned. Routine diagnostics combined with treatment of early carious lesions on the root surface may hinder caries progression, resulting in improved oral health by preserving the maximum number of teeth in that group of patients.

The obtained microscopic results showed that infiltrants not only penetrate decalcified root cement but also penetrate the surface area of root dentine.

5. Conclusions

The pioneering, preliminary study based on microscopic viewing has proven that the experimental preparation and the features of a dental infiltrate enriched with an antimicrobial component in the form of metronidazole can potentially be used to treat early carious lesions within the tooth root. This study proved that it penetrates into demineralized tissues. There is a need to continue research.

Author Contributions: Conceptualization, M.F. and M.S.-N.; methodology, M.F., M.S.-N., B.C. and A.K.-W.; software, M.F., B.C. and M.S.-N.; validation, M.F. and M.S.-N.; formal analysis, M.F., M.S.-N., A.K.-W. and B.C.; investigation, M.F., M.S.-N. and A.K.-W.; resources, M.F. and M.S-N.; data curation, M.F. and M.S.-N.; writing original draft preparation, M.F. and M.S.-N.; writing—review and editing, M.F. and M.S.-N.; visualization, M.S.-N.; project administration, M.S.-N.; funding acquisition, M.F. and M.S.-N. All authors have read and agreed to the published version of the manuscript.

Funding: Financing of the work is based on a contract PCN-1-I34/N/0/K and KNW-2-I44/D/9/N.

Institutional Review Board Statement: In order to conduct this research, an application was filled with the Bioethics Committee of the Medical University of Silesia in Katowice. The Committee issued an approving motion no. KNW/0022/KB/258/1/17 on the day of 14 November 2017.

Informed Consent Statement: Not applicable.

Data Availability Statement: Data is contained within the article.

Acknowledgments: We would like to thank Mirosław Gibas for initiating research in this field.

Conflicts of Interest: The authors declare no conflict of interest.

Sample Availability: Samples of the compounds are available from the authors.

References

1. Frecken, J. Caries Epidemiology and Its Challenges. *Monogr. Oral Sci.* **2018**, *27*, 11–23.
2. Conrads, G.; About, I. Pathophysiology of Dental Caries. *Monogr. Oral Sci.* **2018**, *27*, 1–10.
3. Machiulskiene, V.; Campus, G.; Carvalho, J.C.; Dige, I.; Ekstrand, K.R.; Jablonski-Momeni, A.; Maltz, M.; Manton, D.J.; Martignon, S.; Martinez-Mier, A.; et al. Terminology of Dental Caries and Dental Caries Management: Consensus Report of a Workshop Organized by ORCA and Cariology Research Group of IADR. *Caries Res.* **2020**, *54*, 7–14. [CrossRef]
4. Mathur, V.P.; Dhillon, J.K. Dental Caries: A Disease Which Needs Attention. *Indian J. Pediatr.* **2018**, *3*, 202–206. [CrossRef] [PubMed]
5. Wójcicka, A.; Zalewska, M.; Czerech, E.; Jabłoński, R.; Grabowska, S.Z.; Maciorkowska, E. Próchnica Wieku Rozwojowego Chorobą Cywilizacyjną. *Przegl Epidemiol.* **2012**, *66*, 705–711. [PubMed]
6. Giacaman, R.A.; Muñoz-Sandoval, C.; Neuhaus, K.W.; Fontana., M.; Chałas, R. Evidence-based strategies for the minimally invasive treatment of carious lesions: Review of the literature. *Adv. Clin. Exp. Med.* **2018**, *7*, 1009–1016. [CrossRef] [PubMed]
7. Wierichs, R.J.; Meyer-Lueckel, H. Systematic review on noninvasive treatment of root caries lesions. *J. Dent. Res.* **2015**, *2*, 261–271. [CrossRef]
8. Tanasiewicz, M.; Skucha-Nowak, M.; Skorus, M.; Nowak, M. Stomatologia minimalnie inwazyjna. *TPS* **2018**, *3*, 23–26.
9. Kaczmarek, U. Minimal intervention Dentistry—Review of literature. *Czas. Stomatol.* **2007**, *6*, 367–376.
10. Dawett, B.; Young, S.; Deery, C.; Banerjee, A. Minimally Invasive Selective Caries Removal put into Practice. *Dent. Update* **2020**, *47*, 10. [CrossRef]
11. Showkat, N.; Singh, G.; Singla, K.; Sareen, K.; Chowdhury, C.; Jindal, L. Minimal Invasive Dentistry: Literature Review. *J. CMRO* **2020**, *9*, 631–636.
12. Ekstrand, K.R.; Gimenez, T.; Ferreira, F.R.; Mendes, F.M.; Braga, M.M. The International Caries Detection and Assessment System—ICDAS: A Systematic Review. *Caries Res.* **2018**, *5*, 406–419. [CrossRef] [PubMed]
13. Litzenburger, F.; Schäfer, G.; Hickel, R.; Kühnisch, F.; Heck, K. Comparison of novel and established caries diagnostic methods: A clinical study on occlusal surfaces. *BMC Oral Health* **2021**, *21*, 79. [CrossRef]
14. Mitchell, C.; Zaku, H.; Milgrom, P.; Mancl, L.; Prince, D.B. The accuracy of laser fluorescence (DIAGNOdent) in assessing caries lesion activity on root surfaces, around crown margins, and in furcations in older adults. *BDJ Open* **2021**, *23*, 1–5.
15. Abogazalah, N.; Ando, M. Alternative methods to visual and radiographic examinations for approximal caries detection. *J. Oral Sci.* **2017**, *59*, 315–322. [CrossRef]
16. Berczyński, P.; Gmerek, A.; Buczkowska-Radlińska, J. Remineralizing methods in early caries Lesions—Review of the liteerature. *Pom. J. Life Sci.* **2015**, *61*, 68–72. [CrossRef]
17. Arifa, M.K.; Ephraim, R.; Rajamani, T. Recent Advances in Dental Hard Tissue Remineralization: A Review of Literature. *Int J Clin. Pediatr. Dent.* **2019**, *2*, 139–144. [CrossRef]
18. Sivapriya, E.; Sridevi, K.; Periasamy, R.; Lakshminarayanan, L.; Pradeepkumar, A.R. Remineralization ability of sodium fluoride on the microhardness of enamel, dentin, and dentinoenamel junction: An in vitro study. *J. Conserv. Dent.* **2017**, *2*, 100–104.
19. Chen, Y.; Chen, D.; Lin, H. Infiltration and sealing for managing non-cavitated proximal lesions: A systematic review and meta-analysis. *BMC Oral Health* **2021**, *21*, 13. [CrossRef]
20. Skucha-Nowak, M.; Fischer, M.; Nowak, M.; Łopaciński, M.; Tanasiewicz, M. Infiltracja odwapnionego szkliwa jako sposób leczenia próchnicy. *Med. Trib. Stomatol.* **2019**, *4*, 5–10.
21. Zakizade, M.; Davoudi, A.; Akhavan, A.; Shirban, F. Effect of Resin Infiltration Technique on Improving Surface Hardness of Enamel Lesions: A Systematic Review and Meta-analysis. *J. Evid. Based Dent. Pract.* **2020**, *2*, 101405. [CrossRef]
22. Kajka-Hawryluk, K.; Furmaniak, K.; Gromak-Zaremba, J.; Szopiński, K. Bitewing radiography in modern pediatric dentistry. *Nowa Stomatol.* **2015**, *2*, 73–80. [CrossRef]
23. Ozyurt, E.; Arisu, H.D.; Turkoz, E. In Vitro Comparison of the Effectiveness of a Resin Infiltration Systemand a Dental Adhesive System in Dentinal Tubule Penetration. *Clin. Exp. Health Sci.* **2019**, *9*, 253–260.
24. Diago, A.M.; Cadenaro, M.; Ricchiuto, R.; Banchelli, F.; Spinas, E.; Checchi, V.; Giannetti, L. Hypersensitivity in Molar Incisor Hypomineralization: Superficial Infiltration Treatment. *Appl. Sci.* **2021**, *11*, 1823. [CrossRef]
25. Skucha-Nowak, M.; Machorowska-Pieniążek, A.; Tanasiewicz, M. Assesing the Penetrating Abilities of Experimental Preparation with Dental Infiltrant Features Using Optical Microscope: Preliminary Study. *Adv. Clin. Exp. Med.* **2016**, *25*, 961–969. [CrossRef] [PubMed]
26. Fischer, M.; Mertas, A.; Czuba, Z.P.; Skucha-Nowak, M. Study of cytotoxic properties of an experimental preparation 2 with features of a dental infiltrant. *Materials* **2021**, *14*, 2442. [CrossRef]
27. Barczak, K.; Palczewska-Komsa, M.; Buczkowska-Radlińska, J. Physiological and pathological changes in the teeth and periodontal tissues related to age. *GERIATRIA* **2016**, *10*, 98–104.
28. Yamamoto, T.; Hasegawa, T.; Yamamoto, T.; Hongo, H.; Amizuka, N. Histology of human cementum: Its structure, function, and development. *Jpn. Dent. Sci. Rev.* **2016**, *52*, 63–74. [CrossRef]
29. Prymas, A.; Wędrychowicz-Welman, A.; Mania-Końsko, A. Root Caries Treatment –Clinical Case. *Dent. Med. Probl.* **2006**, *1*, 135–138.
30. Gernhardt, C.R. Wurzelkaries—Ein Problem im Alter Eine Übersicht über Ätiologie, Epidemiologie und das klinische Erscheinungsbild kariöser Läsionen im Wurzelbereich. *Art Dent.* **2018**, *67*, 22–29.

31. Dinakaran, S.; Gopinathan, A.S. Root caries: A geriatric challenge. *Dent. Med. Probl.* **2017**, *4*, 403–408. [CrossRef]
32. Katz, R.V.; Hazen, S.P.; Chilton, N.W.; Mumma, R.D. Prevalence and intraoral distribution of root caries in an adult population. *Caries Res.* **1982**, *16*, 265–271. [CrossRef]
33. Składnik-Jankowska, J.; Pregiel, B.; Wrzyszcz-Kowalczyk, A.; Kaczmarek, U. Management of Dental Root Caries Using Ozone. *Dent. Med. Probl.* **2005**, *2*, 273–279.
34. Shay, K. Root caries in the older patient. *Dent. Clin. N. Am.* **1997**, *41*, 763–793.
35. Ravald, N.; Birkhed, D. Factors associated with active and inactive root caries in patients with periodontal disease. *Caries Res.* **1991**, *25*, 377–384. [CrossRef] [PubMed]
36. Hryncewicz, M.; Tropak, K. Non-carious lesions—abfraction, abrasion, attrition, erosion. Review of literature. *Borgis Nowa Stomatol.* **2014**, *1*, 46–52.
37. Collares, F.M.; Garcia, I.M.; Bohns, F.R.; Melo, M.A.; Branco Leitune, V.C. Guanidine hydrochloride polymer additive to 563 undertake ultraconservative resin infiltrant against Streptococcus mutans. *Eur. Polym. J.* **2020**, *133*, 109746. [CrossRef]
38. Skucha-Nowak, M.; Mertas, A.; Tanasiewicz, M. Using an Electron Scanning Microscope to Assess the Penetrating Abilities of an Experimental Preparation with Features of a Dental Infiltrant: Preliminary Study. *Adv. Clin. Exp. Med.* **2016**, *6*, 1293–1301. [CrossRef] [PubMed]
39. Rusyan, E. Etiology and modifying factors of dental erosion. *Borgis—Nowa Stomatol.* **2003**, *1*, 33–36.
40. Skucha-Nowak, M.; Tanasiewicz, M.; Gibas, M.; Twardawa, H. Analysis of the composition of preparations used as a barrier 544 to protect tissues of the patient against the influence of the environment in the oral cavity. *Pol. J. Environ. Stud.* **2013**, *22*, 53–57.
41. Vespa, J.; Medina, L.; Armstrong, D.M. *Demographic Turning Points for the United States: Population Projections for 2020 to 2060 Population Estimates and Projections*; U.S. Department of Commerce, U.S. Census Bureau: Washington, DC, USA, 2018. Available online: http://www.census.gov/ (accessed on 15 March 2018).
42. Statistics Poland. *Ludność. Stan I Struktura Oraz Ruch Naturalny W Przekroju Terytorialnym W 2020 R. Stan W Dniu 31 XII. Population. Size and Structure and Vital Statistics in Poland by Territorial Division in 2020. As of 31 December*; Demographic Surveys Department: Warszawa, Poland, 2021.
43. Konopka, T.; Zawada, Ł.; Kobierzycka, A.; Chrzęszczyk, D. Periodontal Condition in 35–44 and 65–74 Year-Old Residents from Lower Silesia Region. *Dent. Med. Probl.* **2015**, *4*, 447–454. [CrossRef]
44. Gati, D.; Vieira, A.R. Elderly at greater risk for root caries: A look at the multifactorial risks with emphasis on genetics susceptibility. *Int. J. Dent.* **2011**, *2011*, 647168. [CrossRef]
45. Gavriilidou, N.N.; Belibasakis, G.N. Root caries: The intersection between periodontal disease and dental caries in the course of ageing. *Br. Dent. J.* **2019**, *12*, 1063–1067. [CrossRef] [PubMed]
46. Theophilus, L.V.; Kida Minja, I.; Lembariti, B.S. Root Caries Prevalence and Associated Socio-Behavioral and Clinical Factors Among Elderly Patients Attending Selected Public Dental Clinics in Dar Es Salaam, Tanzania. *J. Dent. Oral Sci.* **2021**, *1*, 11–12.
47. Strassler, H.E. Cervical Caries—Treatment Options Based Upon Etiology of the Lesion. *Inside Dentistry* 2005, *1*, 1.
48. Slavkin, H.C. Maturity and oral health: Live longer and better. *J. Am. Dent. Assoc.* **2000**, *6*, 805–808. [CrossRef]
49. Hellyer, P.H.; Beighton, D.; Heath, P.; Lynch, E.J. Root caries in older people attending a general dental practice in East Sussex. *Br. Dent. J.* **1990**, *6*, 201–206. [CrossRef]
50. Scribante, A.; Poggio, C.; Gallo, S.; Riva, A.; Cuocci, A.; Carbone, M.; Arciola, C.R.; Colombo, M. In Vitro Re-Hardening of Bleached Enamel Using Mineralizing Pastes: Toward Preventing Bacterial Colonization. *Materials* **2020**, *4*, 818. [CrossRef] [PubMed]
51. Zhou, Y.; Matin, K.; Shimada, Y.; Sumi, Y.; Tagami, J. Evaluation of resin infiltration on demineralized root surface: An in vitro study. *Dent. Mater. J.* **2017**, *2*, 195–204. [CrossRef]
52. Yazkan, B.; Ermis, B. Effect of resin infiltration and microabrasion on the microhardness, surface roughness and morphology of incipient carious lesions. *Acta Odontol. Scand.* **2018**, *76*, 473–481. [CrossRef]
53. Gurdogan, E.B.; Ozdemir-Ozenen, D.; Sandalli, N. Evaluation of Surface Roughness Characteristics Using Atomic Force Microscopy and Inspection of Microhardness Following Resin Infiltration with Icon®. *JERD* **2017**, *3*, 201–208. [CrossRef]
54. Skucha-Nowak, M. Attempt to assess the infiltration of enamel made with experimental preparation using a scanning electron microscope. *Open Med. Former. Cent. Eur. J. Med.* **2015**, *1*, 238–248. [CrossRef]
55. Zhang, K.; Wang, S.; Zhou, X. Effect of Antibacterial Dental Adhesive on Multispecies Biofilms Formation. *J. Dent. Res.* **2015**, *4*, 622–629. [CrossRef] [PubMed]
56. Neres, E.; Moda, M.D.; Chiba, E.K.; Briso, A.; Pessan, J.P.; Fagundes, T.C. Microhardness and Roughness of Infiltrated White Spot Lesions Submitted to Different Challenges. *Oper. Dent.* **2017**, *4*, 428–435. [CrossRef] [PubMed]
57. Subramaniam, P.; Babu, G.; Lakhotia, D. Evaluation of penetration depth of a commercially available resin infiltrate into artificially created enamel lesions: An in vitro study. *J. Conserv. Dent.* **2014**, *2*, 146–149. [CrossRef] [PubMed]
58. Kielbassa, A.M.; Müller, A.; Gerhard, C.R. Closing the gap between oral higiene and minimally invasive infiltrationtechnique of incipient (proximal) enamel lesions. *Quintessence Int.* **2009**, *40*, 663–681. [PubMed]

59. Tosco, V.; Vitiello, F.; Furlani, M.; Gatto, M.L.; Monterubbianesi, R.; Giuliani, A.; Orsini, G.; Putignano, A. Microleakage Analysis of Different Bulk-Filling Techniques for Class II Restorations: μ-CT, SEM and EDS Evaluations. *Materials* **2021**, *14*, 31. [CrossRef] [PubMed]
60. Putignano, A.; Tosco, V.; Monterubbianesi, R.; Vitiello, A.; Gatto, M.L.; Furlani, M.; Giuliani., A.; Orsini, G. Comparison of three different bulk-filling techniques for restoring class II cavities: μCT, SEM-EDS combined analyses for margins and internal fit assessments. *J. Mech. Behav. Biomed. Mater.* **2021**, *124*, 104812. [CrossRef]

Article

Study of Cytotoxic Properties of an Experimental Preparation with Features of a Dental Infiltrant

Małgorzata Fischer [1,*], Anna Mertas [2], Zenon Paweł Czuba [2] and Małgorzata Skucha-Nowak [1]

[1] Unit of Dental Propedeutics, Department of Conservative Dentistry with Endodontics, Faculty of Medical Sciences in Zabrze, Medical University of Silesia, 40-055 Katowice, Poland; mskuchanowak@sum.edu.pl

[2] Department of Microbiology and Immunology, Faculty of Medical Sciences in Zabrze, Medical University of Silesia, 40-055 Katowice, Poland; amertas@sum.edu.pl (A.M.); zczuba@sum.edu.pl (Z.P.C.)

* Correspondence: skorusmalgorzata@gmail.com or d200698@365.sum.edu.pl; Tel.: +48-322-827-942

Abstract: Microinvasive dentistry is based on the treatment of early carious lesions with the use of dental infiltrants. The commercially available Icon dental infiltrant does not contain any bacteriostatic component. An experimental preparation enriched with the missing component was synthesised. The aim of this study was to evaluate the cytotoxicity of the experimental preparation. Mouse fibroblasts of the L-929 lineage were used for the in vitro study. Cell morphology and viability were assessed. In the cytotoxicity analysis, it was shown that the experimental preparation (42.8 ± 10.3) after 24 h at two-fold dilution showed similar cytotoxicity to Icon (42.7 ± 8.8) ($p > 0.05$), while at four-fold dilution experimental preparation (46.7 ± 3.1), it was less toxic than Icon (34.2 ± 3.1) ($p < 0.05$). The experimental preparation has the potential to provide an alternative to the Icon commercial preparation. Further research is needed to evaluate the cytotoxicity of the experimental preparation over a longer period of time.

Keywords: dental materials; polymers; antimicrobial properties; infiltrants; metronidazole

1. Introduction

Microinvasive dentistry (also called microdentistry) is a modern preventive and therapeutic concept based on an attempt to preserve as much as possible of the patient's healthy tissues and to postpone the onset of invasive (surgical) treatment, that is mechanical removal of carious lesions, in favour of therapeutic (biological) treatment—remineralisation of carious lesions along with modification of the oral cavity environment [1–7]. Currently, the development of technology is moving towards the application of stem cells in numerous fields of medicine. The repair and regeneration of tissue structures can provide an alternative to current treatment protocols. However, there are still no studies showing the reconstruction of enamel tissues and root cement with the participation of stem cells. Due to this fact, non-invasive treatment of early carious lesions on the surfaces of tooth tissues can be treated with the use of dental infiltrants [8]. This trend in conservative dentistry makes it possible to stop the progression of a disease at an early stage and to restore the normal structure and function of dental tissues.

A dental infiltrant is a substance based on low-viscosity polymer resins, capable of deep penetration of demineralised tooth tissues and of filling the inter-crystalline spaces of hard tissues. That process reduces porosity of the decalcified tissues. As a result, the carious spot becomes inaccessible for cariogenic bacteria and their toxins. The result of this process is a halt in the progression of caries' development. Icon (DMG, Hamburg, Germany) is an example of a commercially available dental infiltrant preparation, which was launched in 2009. It is intended for treatment of carious lesions located on smooth and tangential tooth surfaces [9–13]. Icon is recommended for use in early-stage carious lesions to a maximum depth of D1 (Manji radiographic classification of lesion depth) [13,14]. In order to be referred to as a dental infiltrant, the preparation should feature low density, high surface tension, hydrophilicity, no toxic effects on the organism, bacteriostaticity, no interaction with food and drugs, ability to polymerise to a solid state, resistance to chemical and mechanical agents and have good aesthetics after polymerisation [9]. The advantages of the Icon commercial preparation include the ability to polymerise to a solid state, resistance to chemical and mechanical agents and its ability to mimic the tooth colour. Low viscosity and wetting angle, and high surface tension and hydrophilicity allow infiltrants to penetrate pores of decalcified enamel. However, the said preparation is not without its drawbacks. Icon does not contain a component responsible for inhibiting the multiplication of microorganisms on the tooth surface. The bacteriostatic nature of the drug is defined as the effect of the preparation on bacteria by inhibiting their proliferation and preventing the massive breakdown of bacterial cells with the release of antigens (including endotoxins) [15].

The fact that the commercial preparation Icon does not have an ingredient responsible for bacteriostaticity in its composition led the authors to undertake research towards the synthesis of an experimental preparation enriched with the missing component. The developed experimental preparation was enriched with PMMAn-MTZ monomer, which has adhesive properties. The action of metronidazole targets anaerobic and oxygen-deficient organisms. This chemotherapeutic agent is active against protozoa and, at the same time, displays antibacterial activity (against bacteria of the following genera: *Bacteroides*, *Fusobacterium*, *Eubacterium*, *Clostridium*, *Peptococcus*, *Peptostreptococcus*, among others). Metronidazole enters the cells as a prodrug, then it is converted into a short-term nitroso radical. The chemotherapeutic agent in this form is cytotoxic and acts by inhibiting DNA synthesis and damaging it through oxidation reactions. That process results in DNA strand breaks, which leads to degradation and death of the cell [16]. The synthesis of PMMAn with metronidazole provides a component that potentially inhibits microbial proliferation on the surface of dental hard tissue. PMMAn and metronidazole are linked by an ester bond, which is hydrolysed when exposed to water (saliva). In this way, metronidazole can inhibit bacterial growth on the tooth surface in a continuous manner.

One of the basic tests of new materials and therapeutic preparations which are to be used in medicine in the future is to determine the cytotoxic activity of a given substance. Determination of cytotoxicity is the basis for the assessment of the action of the drug and whether its influence on the human body is safe or not. Literature studies show that the development of molecular research enables a better understanding of the aetiology of diseases, complementing clinical practice. The use of profile gene expression can facilitate and replace classical techniques of cytotoxicity analysis. This method can provide much more information than classical methods [17]. Among the available methods for assessing the cytotoxicity of the tested preparations, there are tests that allow direct or indirect measurement of various parameters related to the physiology of cells, against which the activity of a given material is tested, including integrity of the cell membrane, proliferation, activity of cellular metabolic enzymes, as well as the amount of protein or genetic material [18]. There are numerous tests, for example, the test to assess the activity of the lactate dehydrogenase (LDH—lactate dehydrogenase test), the test based on the uptake of neutral red by lysosomes (NR—neutral red test), the test to demonstrate the presence of the lysosomal enzyme N-acetyl-beta-D-glucosaminidase in culture centrum (NAG test), the test determining the number of cells using sulforhodamine (SRB—sulforhodamine test) and one of the most commonly used tests—the MTT test (3-(4,5-dimethylthiazol-2-yl)-2,5-diphenyltetrazolium bromide test)—using the activity of cellular metabolic enzymes [19]. This method allows to estimate the percentage of living cells in cultures contacted with the tested preparation or its extract by measuring mitochondrial dehydrogenase activity, thus making it possible to determine the cytotoxicity of the tested preparation. Based on the recommendations of EN ISO 10993-5 standard: "Biological evaluation of medical devices—Part 5: In vitro cytotoxicity testing", a reduction in the viability of cells contacted with test preparation extracts greater than 30%, when compared to a control cell culture, is considered to have a cytotoxic effect. To obtain a complete cytotoxicity profile for a given substance, tests are performed with different concentrations/dilutions of test extracts [20].

A small sample of literature reports on the subject of cytotoxicity in the area of infiltrants prompted the authors to undertake research in this field [21].

The main aim of our study is to assess the morphology and viability of cells of the model line of mouse fibroblasts (L-929) contacted with extracts of preparations with the characteristics of a dental infiltrant (experimental and commercial preparation).

The first null hypothesis is that the developed experimental preparation with the addition of metronidazole has all the properties of a dental infiltrant. The second null hypothesis is that the developed formulation after 24 h at 2-fold dilution and 4-fold dilution shows at least the same cytotoxicity as the commercial preparation of Icon.

2. Results

Comparing the mean viability of L-929 cells (%) after 24 h of incubation with a 2-fold dilution of the extracts of the tested preparations, it was found that there was no statistically significant difference ($p > 0.05$) between the experimental preparation (42.8 ± 10.3) and the Icon commercial preparation (42.7 ± 7.8), as shown in Table 1.

Table 1. Comparison of the mean viability of L-929 cells (%) after 24 h of incubation with a 2-fold and 4-fold dilution of the extract of the experimental preparation and the Icon preparation.

	Mean Viability of Cell (%)		
Preparation	2-Fold Dilution	4-Fold Dilution	
Experimental preparation	42.8 ± 10.3	46.7 ± 3.1	* $p < 0.05$
Icon	42.7 ± 7.8	34.2 ± 3.1	* $p < 0.05$
	* $p > 0.05$	* $p < 0.05$	

* statistical significance.

Comparing the mean viability of L-929 cells (%) after 24 h of incubation with a 4-fold dilution of the extracts of the tested preparations, it was found that the obtained values were statistically significantly different ($p < 0.05$) between the experimental preparation (46.7 ± 3.1) and the Icon commercial preparation (34.2 ± 3.1), as shown in Table 1 and graphically depicted in Figure 1.

In the next part of the statistical analysis, the mean cell viability of line L-929 (%) after 24 h of incubation with the tested extracts was compared according to the dilution multiplicity of the extracts of the tested preparations. The mean cell viability (%) for each preparation is statistically significantly different in relation to the dilution multiplicity. The experimental preparation in 2-fold dilution (42.8 ± 10.3) and in 4-fold dilution (46.7 ± 3.1) are statistically different ($p < 0.05$). The Icon preparation in 2-fold dilution (42.7 ± 7.8) and in 4-fold dilution (34.2 ± 3.1) are statistically different ($p < 0.05$). That data is shown in Table 1.

The obtained extracts of the tested preparations and the evaluation of L-929 cell viability by MTT assay are shown in Figure 2. Visible intensity of staining after the addition of DMSO is directly proportional to the cell viability.

During microscopic observations, images of L-9329 cells obtained for the experimental preparation were evaluated according to EN ISO 10993-5 standard and compared with control cultures of adhered L-929 cells, as shown in Figure 3.

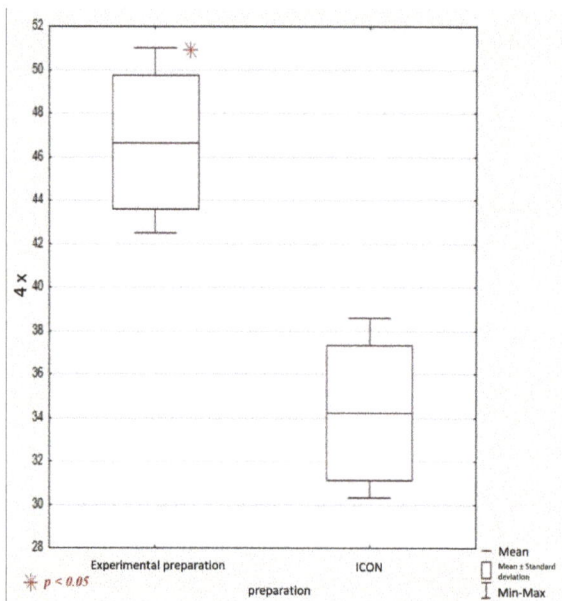

Figure 1. Comparison of mean viability of L-929 cells (%) after 24 h of incubation with a 4-fold dilution of the extract of the experimental preparation and the Icon preparation ($p < 0.05$). * statistical significance.

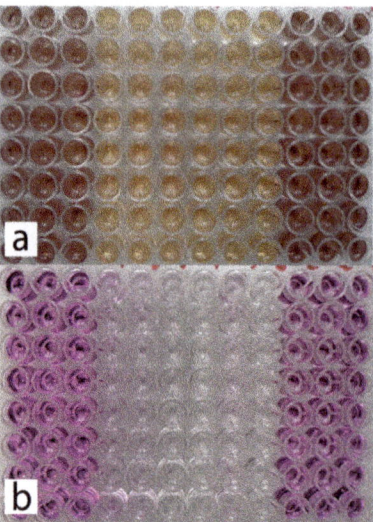

Figure 2. (a) Plate with cultures of L-929 cells immediately after a 24 h incubation period with 2-fold diluted extracts of the test preparations. (b) MTT assay performed for cultures shown in image (a) (after formazan extraction with DMSO).

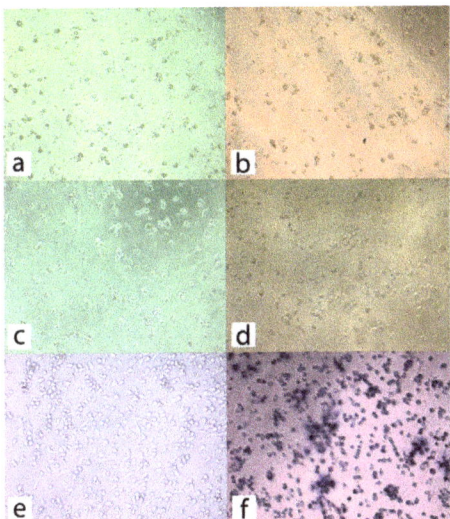

Figure 3. (**a**) L-929 line cells after 24 h incubation with a 2-fold diluted 24 h extract of the experimental preparation, (**b**) L-929 line cells after 24 h incubation with a 2-fold diluted 24 h extract of the experimental preparation after MTT assay, (**c**) L-929 cells after 24 h incubation with 2-fold diluted 24 h Icon extract, (**d**) L-929 cells after 24 h incubation with 2-fold diluted 24 h Icon extract after MTT test. (**e**) Control culture of adhered L-929 cells, (**f**) control culture of adhered L-929 cells after MTT test.

3. Discussion

Microinvasive dentistry presents an innovative model for the treatment of early carious lesions. It is used in numerous branches of dentistry, including in the field of conservative dentistry, dental prosthetics and orthodontics. In the field of prosthetics, thanks to the use of milled and moulded mock-ups, tissue preparation for permanent restorations enables the preservation of more healthy tooth tissues, and in the area of conservative dentistry and orthodontics, e.g., after orthodontic treatment, in areas located in the vicinity of orthodontic brackets, where plaque has been deposited for a long time and tissue demineralisation occurs [22–24]. Mattousch et al. showed in an in vivo study that white spots formed after orthodontic treatment have a limited capacity for spontaneous remineralisation after removal of braces [25]. Infiltration of carious lesions effectively reduces the progression of lesions not only on smooth tooth surfaces but also in the interdental area. Aesthetics is an important aspect of modern dentistry, and infiltration allows for minimally interventional treatment of lesions while maintaining the continuity of the patient's own tissues, which definitely has a positive effect on the final result in terms of aesthetics of the treated area. Studies available in the literature have shown that not only Icon infiltrant, but also the composite sealers Opiguard (Kerr, Orange, CA, USA) and PermaSeal (Ultradent, South Jordan, UT, USA) applied on the surface of early carious lesions proved to be able to infiltrate and, moreover, showed similar improvements in terms of aesthetics, and the colour stability of the infiltrated white spots lasted for at least 2 months [26,27]. That issue was also addressed by Paris et al. and Yuan et al., who found that colour change and masking of early carious lesions can be achieved with use of infiltrants whose RI (refractive index) is close to that of enamel [28,29].

Schmidlin et al. proved in an in vitro study that an infiltrant alone does not prevent enamel demineralisation because it does not have an antimicrobial component [30]. Additionally, Skucha-Nowak et al. observed that the Icon commercial preparation does not have a component that would confer bacteriostatic properties to dental infiltrants, affecting the inhibition of microbial proliferation on the tooth surface [31]. This fact prompted the authors to undertake research in this field.

Metronidazole is a highly active agent with a wide range of action against Gram-negative anaerobic bacteria, as well as Gram-positive anaerobic bacteria [32]. This agent is used, among others, in periodontal practice to reduce anaerobic microflora in pockets of over 5 mm depth. Topical use of metronidazole is beneficial due to its ability to release antimicrobials in concentrations high enough to affect pathogens, even in subgingival biofilms, yet far less than those administered systemically, which prevents a build-up of resistance [33]. Mombelli et al. observed a beneficial clinical and microbiological effect after topical application combined with ornidazole and thymidazole [34]. Other researchers report that topical metronidazole is not fully safe, as there are no studies on adverse effects during long-term metronidazole use [35]. On the other hand, in the study by Kida et al., the release of metronidazole from hydrogel dental dressings follows first-order kinetics, where the reaction rate is proportional to the substrate concentration and its value decreases with time [36]. In their study, Skucha-Nowak et al. demonstrated that the inclusion of a metronidazole component in the composition of an experimental preparation does not affect the process of polymerisation of a preparation by light. They proved that work with a preparation enriched with a bacteriostatic component does not differ in quality from work with a commercially available preparation, Icon [9]. A similar study was performed by Collares et al., who synthesised a preparation with the characteristics of a dental infiltrant by adding polyhexamethylene guanidine hydrochloride (PHMGH) in the amounts of 0.5 and 1 wt%, which is responsible for antibacterial activity against *Streptococcus mutans* and progression of carious lesions. Its mechanism of action is related to electrostatic interaction. The addition of PHMGH reduced the growth and adhesion of carious bacteria on the surface of resin at both tested concentrations, without affecting the physical properties of this material, i.e., conversion rate, wetting angle or surface free energy. However, the researchers did not evaluate the cytotoxic effect of the synthesised formulation, therefore making it impossible to assess its use in vivo and to compare the effect of the applied antimicrobial component on the cells in relation to our antimicrobial component, metronidazole [21]. Zhang et al. synthesised a dental adhesive with an antibacterial component containing dimethylaminododecyl methacrylate (DMADDM), affecting *Streptococcus mutans*, *Streptococcus gordoni* and *Streptococcus sanguis* [37]. In the present study, the authors used an antibacterial component containing metronidazole, which shows broad-spectrum antibacterial activity against *Bacteroides*, *Fusobacterium*, *Eubacterium*, *Clostridium*, *Peptococcus*, *Peptostreptococcus* and protozoa. The supragingival plaque is mainly formed by Gram-positive streptococci, including *Streptococcus mutans*, *Streptococcus salivarius*, *Streptococcus mitis* or *Lactobacillus*, while the subgingival plaque is mainly dominated by Gram-negative anaerobic bacteria, such as *Fusobacterium nucleatum* and *Porphyromonas gingivalis* [38]. The groups of Fu et al. and Wu et al. also studied the influence of the antimicrobial component in the composition of composite resins. According to the researchers, the use of nano-MgO shows a broad antibacterial spectrum against fungi (*Candida albicans*), viruses and bacteria (*Staphylococcus aureus*, *Enterococcus faecalis*, *Escherichia coli*) [39]. The use of an antimicrobial component in the composition of an experimental preparation used for infiltration in our study may potentially inhibit the development of caries in the place where it was applied.

The rapid development of technology means that more and more new preparations for use in dentistry are created.

Mechanical properties are among the key factors when selecting a dental material for use by a dentist.

Chieruzzi et al. undertook an assessment of the mechanical properties of glass-ionomer materials. They also enriched their study with the assessment of the antibacterial effect of a glass-ionomer preparation enriched with the following ingredients: nanohydroxyapatite (Sealent, Miromed Srl, Milano, Italy), ciprofloxacin antibiotic (Ciproxin, Bayer SpA, Milano, Italy) and MDA—zinc L-carnosine (Hepilor, Azienda Farmaceutica Italiana, Ascoli Piceno, Italy) [40]. Khosravani also undertook an evaluation of the mechanical properties of composite materials. In his research, he showed that a greater number of cracks and roughness may result in increased adhesion of bacteria [41].

Modern dentistry mainly uses light-curing composite materials to fill hard tissue defects. These materials include monomers, initiators, retarders and photo-stabilisers, which may potentially cause cytotoxic effects. Most dental composites are based on bis-GMA and TEGDMA. The TEGDMA monomer is used to increase the conversion of vinyl groups during polymerisation. However, an increase in polymerisation shrinkage occurs during the process, causing stress in the composite, which weakens the seal between it and the tooth structure [42]. While evaluating the properties of experimental formulations, other researchers have shown that the addition of hydrophobic monomers and solvents to TEGDMA mixtures does not improve the depth of penetration of the caries-affected enamel surface. However, it affects the degree of conversion and the modulus of elasticity [43]. Inamitsu et al. observed that HEMA monomer and TEGDMA inhibit osteoclast differentiation [44]. Additionally, other researchers have shown that TEGDMA decreases the differentiation capacity of odontoclasts in pulp cells and the differentiation rate of odontoclasts increases proportionally to the amount of TEGDMA monomer [45]. Yoshii in his study demonstrated that bis-GMA is the most cytotoxic compound, followed by UDMA, TEGDMA and finally MMP [46]. The literature describes the possibility of reducing the cytotoxic effect of complex materials by diluting the substance during its penetration through biological in vitro barriers [43]. Moreover, the present research proved that a 4-fold dilution used immediately after the application of the experimental preparation reduces the cytotoxic effect. The results available in the literature have shown that resin-based materials can adversely affect oral tissues, because the release of residual resin monomers can occur up to several hours after application, as a result of incomplete polymerisation. Monomers released from resin-based materials can enter the surrounding tissues, especially when these materials are used in close proximity to gingival mucosa, exerting adverse effects and inducing apoptosis of various types of periodontal cells [47]. The cytotoxicity of infiltrants was studied by Golz et al., who noted that although Icon resin infiltration is a microinvasive technique, it can affect pulp inflammation, as HEMA and TEGDMA, due to their lipophilic nature, have the ability to penetrate the lipid layer of the cell, and in vivo models have demonstrated their presence in the dental pulp. Golz et al. also observed that reduction of the curing time of resin-based composites increases the index of cell apoptosis. In the present study, the experimental preparation with an antimicrobial component in the form of metronidazole also contains HEMA, TEGDMA and CQ in its composition [48]. Our results may suggest that in order to reduce the cytotoxic effect of the experimental preparation on the surrounding soft tissues, it is important to rinse the oral cavity thoroughly with water after its application.

Many practicing dentists pay special attention to the way the preparation is used. This has the effect of interacting with the surrounding tissues in the oral cavity. The contact of the preparation with oral fluids and the lack of the need to isolate the application site makes the procedure much easier for the dentist.

On the basis of their own research, the authors concluded that the developed experimental preparation with metronidazole can be potentially used as a dental infiltrant in dentistry.

The authors also proved that the experimental preparation is characterised by a similar cytotoxicity after 24 h and 2-fold dilution as the commercial preparation Icon, while at 4-fold dilution, it is statistically significantly lower.

The influence of the daily activities performed by the patient in the oral cavity is also important. A question may be asked as to how daily hygienic procedures, such as tooth brushing, the influence of food consumption as the years go on and time may influence the degree of cytotoxicity of the experimental preparation.

Further research should be conducted.

4. Materials and Methods

4.1. Experimental Preparation

An experimental preparation with the characteristics of a dental infiltrant, enriched with a component responsible for bacteriostaticity (metronidazoles), was used in the study, as shown in Table 2 [11]. The experimental preparation formed the research group. A freely available Icon commercial preparation was used in the control group [7].

Table 2. Composition and percentage content of the experimental preparation.

Component	Quantity (g)	Content (%)
TEGDMA	3.75	75
HEMA	1.25	25
PMMAn-MTZ *	0.05	1 *
DMAEMA *	0.05	1 *
CQ *	0.025	0.5 *

* Ratio to total mass of monomers. TEGDMA (triethylene glycol dimethacrylate, Fluka, Buchs, Switzerland); HEMA (2-hydroxyethyl methacrylate, Acros, Belgium, NJ, USA); PMMAn (2-(7-methyl-1,6-dioxo-2,5-dioxa-7-octenyl) trimellitic anhydride); MTZ (metronidazole, Acros, New Jersey, NJ, USA); DMAEMA * (N,N-dimethylaminoethyl methacrylate, Merck, Darmstadt, Germany); CQ * (camphorquinone, Aldrich, St. Louis, MO, USA).

4.2. Polymerisation of Preparations

A wireless LED polymerisation lamp C01-C Premium Plus (Premium Plus International Limited, Bournemouth, UK) was used for the polymerisation process. It emits light with a wavelength between 440 and 480 nm. The length of the optical fibre cable is 8 mm. Full mode with a radiation power of 1200 mW/cm^2 was used for the study (Figure 4).

4.3. Obtaining Extracts of the Tested Preparations

A 10 µL volume of the test preparation was dispensed to the bottom of a 96-well microtiter plate. The test preparations were polymerised, and the microtiter plate was sterilised. Subsequently, 200 µL of RPMI 1640 medium with 10% FBS was dispensed into the wells with the polymerised preparations to obtain extracts of the tested preparations. The culture medium was also placed in the wells without polymerised preparations in order to obtain the medium for further studies, after the 24 h incubation, in the conditions corresponding to those in which extracts of the tested preparations were made. After 24 h of incubation at 37 °C in an atmosphere containing 5% of CO_2 at 100% relative humidity, both extracts and the culture medium used in the next stage of the study to assess cell viability L-929 were ready (Figure 4).

4.4. Evaluation of Cytotoxicity of the Tested Preparation Extracts

Cytotoxicity assessment of the test formulations was performed according to the recommendations of the EN ISO 10993-5 standard, with the use of mouse fibroblast line L-929 (NCTC clone 929) purchased from the American Type Culture Collection (Manassas, VA, USA). The L-929 cell line (ATCC, catalogue number CCL-1) consisted of mouse subcutaneous connective tissue fibroblasts of the C3H/An strain. Under in vitro culture conditions, they were directly contacted with extracts of the test preparations and cell morphology was successively assessed by microscopy, while cell viability was tested with the MTT assay. RPMI 1640 medium with 10% addition of heat-inactivated foetal bovine serum (FBS), penicillin (100 IU/mL) and streptomycin (100 µg/mL) were used to culture L-929 cells (mouse subcutaneous connective tissue fibroblasts). The culture was performed in 25 cm^2 polystyrene adherent cell culture bottles (Nunc EasYFlasksTM NunclonTMDelta by Nunc A/S, Roskilde, Denmark). The cells were continuously cultured in an MCO-17 AIC incubator by Sanyo (Osaka, Japan), which provided stable culture conditions (temperature 37 °C, atmosphere containing 5% CO_2 at 100% relative humidity). The cells were passaged at intervals of 2–3 days. A suspension with a final density of 1×10^5 cells/mL of medium was used for experimental studies. The density of the cell suspension was assessed by microscopy using a Bürker chamber. The cells of the model line L-929 (mouse fibroblasts) under in vitro culture conditions were contacted with previously obtained 24 h extracts of the tested preparations diluted with a 2-fold or 4-fold ratio in the culture medium. The test was conducted with the Norm PN-EN ISO 10993-5, in conditions corresponding to the conditions of the oral cavity. After 24 h of incubation, cell morphology (microscopic method) and cell viability (MTT assay) were evaluated [20].

The stages of the laboratory procedure are presented graphically in Figure 4.

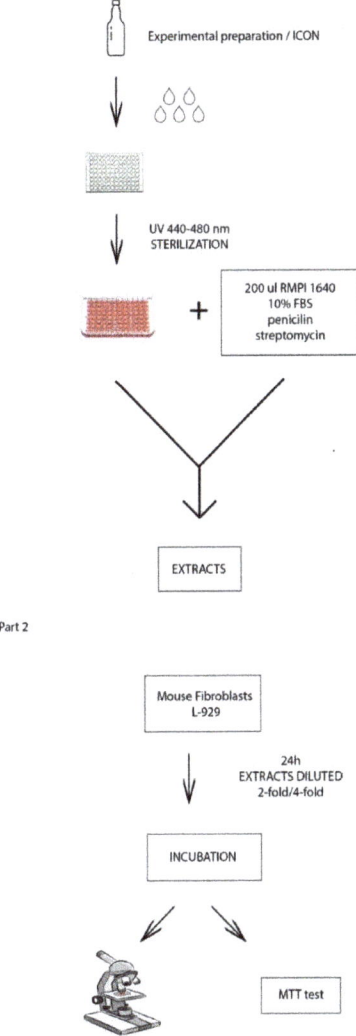

Figure 4. (**Part 1**) Scheme for obtaining 24 h extracts. (**Part 2**) Scheme for assessing cell viability.

4.4.1. Microscopic Observations

An Olympus IX51 inverted fluorescence microscope with a video track with an image recording camera (OLYMPUS, Tokyo, Japan) was used for microscopic observations. During microscopic observation of cells, a magnification of 200 times was used.

4.4.2. Viability Assessment of L-929 Cells Contacted with Extracts of the Tested Preparations

The 3-(4,5-dimethylthiazol-2-yl)-2,5-diphenyltetrazolium bromide test (MTT assay test), conducted by measuring mitochondrial dehydrogenase activity, allowed to assess the percentage of viable cells in cultures contacted with the test extract, thus determining the cytotoxicity of the test preparation. A 200 µL volume of each suspension of the L-929 line cells at a density of 1×10^5 cells/mL in the RPMI 1640 medium with 10% FBS (i.e., 20,000 cells/well) was dispensed into the wells of a 96-well microplate. After 24 h of incubation at 37 °C in the atmosphere containing 5% CO_2 at 100% relative humidity, the supernatant was removed, and 200 µL of the 24 h extract of the specific test preparation, diluted two or four times in the culture medium or the medium with adhered cells after 24 h of incubation, conducted in parallel with obtaining the extracts, were added to the wells, respectively. Control cultures consisted of cells contacted with fresh culture medium.

After 24 h of incubation at 37 °C in an atmosphere containing 5% CO_2 at 100% relative humidity, the viability of L-929 cells was assessed by the MTT assay. For this purpose, a MTT solution with a final concentration of 1.1 mM in the culture medium was dispensed into each well after the culture medium had been removed. After 3 h of incubation at 37 °C in a CO_2-free atmosphere and at constant relative humidity, the filtrate was removed, and 200 µL of each DMSO was added to the test and control cultures to extract MTT formazan. After 20 min, 150 µL of solution was collected from each well and its absorbance was determined at 550 nm using an Eon automated plate reader (BioTek Instruments, Winooski, VT, USA). The intensity of the purple colour of the solution was directly proportional to the amount of formazan formed and thus the number of living cells.

Cell viability, expressed in %, was calculated using the following formula:

$$\text{Cell viability (\%)} = A_b / A_k \times 100\%$$

A_b—absorbance of the test sample,
A_k—absorbance of the control sample.

4.5. Statistical Analysis

The viability of L-929 cells after 24 h of in vitro incubation with 2-fold (or 4-fold) diluted extracts of the experimental preparation and the Icon preparation were examined with the use of statistical analysis methods. At the beginning of the statistical analysis, basic measures of descriptive statistics, i.e., mean, median, minimum, maximum, standard deviation, coefficient of variation, skewness and kurtosis, were calculated. The part of the statistical analysis concerning the significance of the differences began with checking the normality of feature distributions in the tested samples using the Shapiro–Wilk test. Subsequently, the following statistical tests were applied: test of equality of variances (Levene's test), test of equality of feature distributions for two independent samples (Mann–Whitney U test, Kruskal–Wallis H test), test of equality of feature means for two dependent samples (t-test for two dependent samples) and test of equality of feature means for two independent samples (t-test for two independent samples). All hypotheses were verified at the significance level of 0.05. Statistical software was used in the statistical analyses.

5. Conclusions

The experimental preparation has the potential to provide an alternative to the Icon commercial preparation.

Further research should be conducted to evaluate the cytotoxicity of the experimental preparation over a longer period of time.

Author Contributions: Conceptualisation, M.F. and M.S.-N.; Methodology, M.F., A.M., Z.P.C. and M.S.-N.; Software, M.F., A.M. and M.S.-N.; Validation, M.S.-N. and Z.P.C.; Formal analysis, M.F., A.M., Z.P.C. and M.S.-N.; Investigation, M.F., A.M. and Z.P.C.; Resources, M.F., A.M., Z.P.C. and M.S.-N.; Data curation, M.F. and M.S.-N.; Writing—original draft preparation, M.F. and M.S.-N.; Writing—review and editing, visualisation, M.F. and M.S.-N.; Supervision, M.S.-N. and A.M.; Project administration, M.S.-N.; Funding acquisition, M.S.-N. and M.F. All authors have read and agreed to the published version of the manuscript.

Funding: Financing of the work is based on a contract PCN-1-134/N/O/K and KNW-2-144/D/9/N.

Institutional Review Board Statement: In order to conduct this research, an application was filed with the Bioethics Committee of the Medical University of Silesia in Katowice. The Committee issued an approving motion No. KNW/0022/KB/258/1/17 on the day of 14 November 2017.

Informed Consent Statement: Not applicable.

Data Availability Statement: Department of Conservative Dentistry with Endodontics, Unit of Dental Propedeutics, Faculty of Medical Sciences in Zabrze, Medical University of Silesia in Katowice, Plac Akademicki 17, 41-902 Bytom, Poland; Tel.: +48-322-827-942.

Acknowledgments: We would like to thank Anna Korytkowska-Wałach, BEng, from the Faculty of Organic Chemistry, Bioorganic Chemistry and Biotechnology of the Silesian University of Technology in Gliwice for help in preparing and conducting the laboratory part of the research.

Conflicts of Interest: The authors declare no conflict of interest.

References

1. Skucha-Nowak, M.; Skorus, M.; Nowak, M.; Tanasiewicz, M. Stomatologia minimalnie inwazyjna. *Twój Przegląd Stomatol.* 2018, *3*, 23–26.
2. Skucha-Nowak, M.; Fischer, M.; Tanasiewicz, M.; Machorowska-Pieniążek, A.; Skaba, D.; Kielbassa, A. Attempt to modify the chemical model of enamel demineralization used in microinvasive dentistry. *J. Stomatol.* 2019, *72*, 17–22.
3. Showkat, N.; Singh, G.; Singla, K.; Sareen, K.; Chowdhury, C.; Jindal, L. Minimal Invasive Dentistry: Literature Review. *J. CMRO* 2020, *3*, 631–636. [CrossRef]
4. Dawett, B.; Young, S.; Deery, C.; Banerjee, A. Minimally Invasive Selective Caries Removal put into Practice. *Dent. Update* 2020, *47*, 10. [CrossRef]
5. Berczyński, P.; Gmerek, A.; Buczkowska-Radlińska, J. Remineralizing methods in early caries lesions—Review of the literature. *Pom. J. Life Sci.* 2015, *61*, 68–72. [CrossRef]
6. Perkowska, M. Contemporary opinions regarding dental caries treatment—Review of literature. *Nowa Stomatol.* 2010, *2*, 78–81.
7. Skucha-Nowak, M.; Nowa-Wachol, A.; Skaba, D.; Wachol, K.; Korytkowska-Wałach, A. Use of ytterbium trifluoride in the field of microinvasive dentistry—An in vitro preliminary study. *Coatings* 2020, *10*, 915. [CrossRef]
8. Cappare, P.; Tete, G.; Sberna, M.T.; Panine-Bordignon, P. The Emerging Role of Stem Cells in Regenerative Dentistry. *Curr. Gene Ther.* 2020, *20*, 259–268. [CrossRef]
9. Skucha-Nowak, M.; Machorowska-Pieniążek, A.; Tanasiewicz, M. Assesing the Penetrating Abilities of Experimental Preparation with Dental Infiltrant Features Using Optical Microscope: Preliminary Study. *Adv. Clin. Exp. Med.* 2016, *25*, 961–969. [CrossRef]
10. Skucha-Nowak, M.; Fischer, M.; Nowak, M.; Łopaciński, M.; Tanasiewicz, M. Infiltracja odwapnionego szkliwa jako sposób leczenia próchnicy. *Med. Trib. Stomatol.* 2019, *4*, 5–10.
11. Skucha-Nowak, M.; Tanasiewicz, M.; Gibas, M.; Twardawa, H. Analysis of the composition of preparations used as a barrier to protect tissues of the patient against the influence of the environment in the oral cavity. *Pol. J. Environ. Stud.* 2013, *22*, 53–57.
12. Skucha-Nowak, M. Attempt to assess the infiltration of enamel made with experimental preparation using a scanning electron microscope. *J. Med.* 2015, *10*, 238–248. [CrossRef]
13. Zakizade, M.; Davoudi, A.; Akhavan, A.; Shirban, F. Effect of Resin Infiltration Technique on Improving Surface Hardness of Enamel Lesions: A Systematic Review and Meta-analysis. *J. Evid. Based Dent. Pract.* 2020, *2*, 101405. [CrossRef] [PubMed]
14. Kajka-Hawryluk, K.; Furmaniak, K.; Gromak-Zaremba, J.; Szopiński, K. Bitewing radiography in modern pediatric dentistry. *Nowa Stomatol.* 2015, *20*, 73–80. [CrossRef]
15. Zielińska, R.D.; Piątowska, D.; Ostrowska, A.; Bołtacz-Rzepkowska, E. The Evaluation of Caries Lesion Progression After Infiltrantion with a Low-Viscous Resin: In vitro Study. *Dent. Med. Probl.* 2016, *53*, 358–364. [CrossRef]
16. Nowak, Ł.R. Wybrane zagadnienia atybiotykoterapii zakażeń bakteryjnych u chorych na nowotwory złośliwe. *Nowotw. J. Oncol.* 2016, *66*, 222–233. [CrossRef]
17. Crespi, R.; Cappare, P.; Romanos, G.E.; Mariani, E.; Benasciutti, E.; Gherlone, E. Corticocancellous porcine bone in the healing of human extraction sockets: Combining histomorphometry with osteoblast gene expression profiles in vitro. *Int. J. Oral. Maxillofac Implant.* 2021, *26*, 866–872.
18. Brook, I.; Wexler, H.M.; Goldstein, E.J.C. Antianaerobic Antimicrobials: Spectrum and Susceptibility Testing. *Clin. Microbiol. Rev.* 2013, *26*, 526–546. [CrossRef]
19. Krzysztoń-Russjan, J.; Książek, I.; Anuszewska, E. Porównanie użyteczności testów MTT i EZ4U stosowanych do oceny cytotoksyczności ksenobiotyków. *Farm. Pol.* 2009, *65*, 395–402.
20. *European Norm PN-EN ISO 10993-5:2009 Biological Evaluation of Medical Devices—Part 5: Tests for In Vitro Cytotoxicity (ISO 10993-5:2009)*; ISO (the International Organization for Standardization): Warsaw, Poland, 2009.

21. Collares, F.M.; Garcia, I.M.; Bohns, F.R.; Melo, M.A.; Branco Leitune, V.C. Guanidine hydrochloride polymer additive to undertake ultraconservative resin infiltrant against *Streptococcus Mutans*. *Eur. Polym. J.* **2020**, *133*, 109746. [CrossRef]
22. Kraus, A.; Becker, K.; Chrapla, K. Literature demineralization in patients treated with fived orthodontic appliances. *Ortod. W Prakt.* **2013**, *2*, 40–43.
23. Meilnik-Błaszczak, M. Incipient demineralization lesions-causes, signs and therapeutic approach. *Nowa Stomatol.* **2016**, *21*, 74–78.
24. Cattoni, F.; Tete, G.; Calloni, A.M.; Manazza, F.; Gastaldi, G.; Cappare, P. Milled versus moulded mock-ups based on the superimposition of 3D meshes from digital oral impressions: A comparative in vitro study in the aesthetic area. *BMC Oral. Health* **2019**, *19*, 230. [CrossRef] [PubMed]
25. Mattousch, T.J.; Van der Veen, M.H.; Zenter, A. Caries lesions after orthodontic treatment followed by quantitative light-induced fluorescence: A 2-year follow-up. *Eur. J. Orthod* **2007**, *29*, 294–298. [CrossRef] [PubMed]
26. Tamer, G.T. The Esthetic Outcome and the Infiltration Capacity of Three Resin Composite Sealers Compared to ICON (DMG, America). Ph.D. Thesis, University of Iowa, Iowa City, IA, USA, 2018.
27. Prasa, K.L.; Penta, P.K.; Ramya, K.M. Spectrophotometric evaluation of white spot lesion treatment using novel resin infiltration material (ICON). *J. Conserv. Dent.* **2018**, *21*, 531–535.
28. Paris, S.; Schwendicke, F.; Seddig, S.; Muller, W.D.; Dorfer, C.; Meyer-Lueckel, H. Micro-hardness and mineral loss of enamel lesions after infiltration with various resins: Influence of infiltrant composition and application frequency in vitro. *J. Dent.* **2013**, *41*, 543–548. [CrossRef]
29. Yuan, H.; Li, J.; Chen, L.; Cheng, L.; Cannon, R.D.; Mei, L. Esthetic comparison of white-spot lesion treatment modalities using spectrometry and fluorescence. *Angle Orthod.* **2014**, *84*, 343–349. [CrossRef] [PubMed]
30. Schmidlin, P.R.; Sener, B.; Attin, T.; Wiegand, A. Protection of sound enamel and artificial enamel lesions against demineralisation: Caries infiltrant versus adhesive. *J. Dent.* **2012**, *40*, 851–856. [CrossRef]
31. Skucha-Nowak, M.; Mertas, A.; Tanasiewicz, M. Using an Electron Scanning Microscope to Assess the Penetrating Abilities of an Experimental Preparation with Features of a Dental infiltrant: Preliminary Study. *Adv. Clin. Exp. Med.* **2016**, *25*, 1293–1301. [CrossRef]
32. Lofmark, S.; Edlund, C.; Nord, C.E. Metronidazole Is Still the Drug of Choice for Treatment of Anaerobic Infections. *Clin. Infect. Dis.* **2010**, *50*, 16–23. [CrossRef] [PubMed]
33. Toskic-Radojicic, M.; Nonkovic, Z.; Loncar, I.; Varjacic, M. Effects of topical application of metronidazole—Containing mucoadhesive lipogel in periodontal pockets. *Vojn. Pregl.* **2005**, *62*, 565–568. [CrossRef]
34. Mombelli, A.; Samaranayake, L.P. Topical and systemic antibiotics in the management of periodontal diseases. *Int. Dent. J.* **2004**, *54*, 3–14. [CrossRef]
35. Sender-Janeczek, M.; Ziętek, M. Use of Locally Delivered Antiseptics and Antibiotics in the Treatment of Chronic Periodontitis—Review of Literature. *Dent. Med. Probl.* **2007**, *44*, 396–402.
36. Kida, D.; Pluta, J. The effect of selected hydrophilisers on metronidazole release from hydrogels stomatological dressings on the basis of Carbopol 971. *Polim. W Med.* **2010**, *40*, 3–9.
37. Zhang, K.; Wang, S.; Zhou, X. Effect of Antibacterial Dental Adhesive on Multispecies Biofilms Formation. *J. Dent. Res.* **2015**, *94*, 622–629. [CrossRef] [PubMed]
38. Chałas, R.; Wójcik-Chęcińska, I.; Woźniak, M.J.; Grzonka, J.; Święszkowski, W.; Kurzydłowski, K.J. Dental plaque as a biofilm—A risk in oral cavity and methods to prevent. *Postępy Hig. Med. Dosw.* **2015**, *69*, 1140–1148. [CrossRef] [PubMed]
39. Fu, J.; Tonin, B.S.H. Characterization of a new dental resin composite containing nano-MgO. *Dent. Mater.* **2019**, *35*, 22. [CrossRef]
40. Chieruzzi, M.; Pagano, S.; Lombardo, G.; Marinucci, L.; Kenny, J.M.; Torre, L.; Cianetti, S. Effect of nanohydroxyapatite, antibiotic, and mucosal defensive agent on the mechanical and thermal properties of glass ionomer cements for special needs patients. *J. Mater. Res.* **2018**, *33*, 638–649. [CrossRef]
41. Khosravani, M.R. Mechanical behavior of restorative dental composites under various loading conditions. *J. Mech Behav Biomed. Mater.* **2019**, *93*, 151–157. [CrossRef]
42. Radziejewska, M. Cytotoxicity of composite resin and dentine adhesive systems—A literature review. *Nowa Stomatol.* **1999**, *4*, 35–39.
43. Albamonte Araújo, G.S.; Sfalcin, R.A.; Freire Araújo, T.G.; Bruschi Alonso, R.C.; Puppin-Rontanic, M.R. Evaluation of polymerization characteristics and penetration into enamel caries lesions of experimental infiltrants. *J. Dent.* **2013**, *41*, 1014–1019. [CrossRef] [PubMed]
44. Inamitsu, H.; Okamoto, K.; Sakai, E.; Nishishita, K.; Murata, H.; Tsukuba, T. The dental resin monomers HEMA and TEGDMA have inhibitory effects on osteoclast differentiation with low cytotoxicity. *J. Appl. Toxicol.* **2017**, *37*, 817–824. [CrossRef] [PubMed]
45. Oncel Torun, Z.; Torun, D.; Baykal, B.; Oztuna, A.; Yesildal, F.; Avcu, F. Effects of triethylene glycol dimethacrylate (TEGDMA) on the odontoclastic differentiation ability of human dental pulp cells. *J. Appl. Oral. Sci.* **2017**, *25*, 631–640. [CrossRef]
46. Yoshii, E. Cytotoxic effects of acrylates and methacrylates: Relationships of monomer structures and cytotoxicity. *J. Biomed. Mater. Res.* **1997**, *37*, 517–524. [CrossRef]
47. Zingler, S.; Matthei, B.; Diercke, K.; Frese, C.; Ludwig, B.; Kohl, A.; Lux, C.J.; Erber, R. Biological evaluation of enamel sealants in an organotypic model of the human gingiva. *Dent. Mater.* **2014**, *30*, 1039–1050. [CrossRef]
48. Golz, L.; Simonis, R.A.; Reichelt, J.; Stark, H.; Frentzen, M.; Allam, J.P.; Probstmeier, R.; Winter, J.; Kraus, D. In vitro biocompatibility of ICON and TEGDMA on human dental pulp stem cells. *Dent. Mater.* **2016**, *32*, 1052–1064. [CrossRef] [PubMed]

Article

Effect of *Candida albicans* Suspension on the Mechanical Properties of Denture Base Acrylic Resin

Grzegorz Chladek [1,*], Michał Nowak [2], Wojciech Pakieła [1] and Anna Mertas [3]

1. Chair of Engineering Materials and Biomaterials, Faculty of Mechanical Engineering, Silesian University of Technology, 18a Konarskiego Str., 41-100 Gliwice, Poland; wojciech.pakiela@polsl.pl
2. Nova Clinic, 22 Jankego Str., 40-612 Katowice, Poland; michalnowak83@yahoo.pl
3. Department of Microbiology and Immunology, Faculty of Medical Sciences in Zabrze, Medical University of Silesia in Katowice, 19 Jordana Str., 41-808 Zabrze, Poland; amertas@sum.edu.pl
* Correspondence: grzegorz.chladek@polsl.pl

Abstract: Yeast-like fungi such as *Candida albicans* (*C. albicans*) are the primary pathogenic microorganism in the oral cavity of denture wearers. The research available so far, conducted according to a protocol based on the exposure of specimens to a *C. albicans* suspension and their cutting with water cooling, shows that hard polymethyl methacrylate (PMMA) prosthetic materials are not only surface colonized, but also penetrated by microorganisms in a short time. This justifies the hypothesis that exposure to a suspension of the *C. albicans* strain causes the changes in mechanical properties due to surface colonization and/or penetration of the samples. In the current study, the chosen mechanical properties (flexural strength, flexural modulus, tensile strength, impact strength, ball indentation hardness, and surface Vickers hardness at 300 g load) of the PMMA denture base material Vertex RS (Vertex-Dental, The Netherlands) exposed for 30, 60, and 90 days to a suspension of *C. albicans* were investigated. The potential penetration of yeast was examined on the fractured surfaces (interior of specimens) to eliminate the risk of the contamination of samples during cutting. There was no influence on the flexural strength, flexural modulus, tensile strength, impact strength, or ball indentation hardness, but a significant decrease in surface hardness was registered. Microscopic observations did not confirm the penetration of *C. albicans*. On the surface, blastospores and pseudohyphae were observed in crystallized structures and in traces after grinding, which indicates that in clinical conditions, it is not penetration but the deterioration of surface quality, which may lead to the formation of microareas that are difficult to disinfect, causing rapid recolonization.

Keywords: *Candida albicans*; dentures; mechanical properties; colonization; penetration; polymethyl methacrylate

1. Introduction

The functioning of polymethyl methacrylate (PMMA) prosthetic materials in the context of the presence of pathogenic microorganisms in the oral cavity, especially *Candida* yeast strains, is one of the most serious clinical problems. It concerns both the colonization of the tissues under the denture and the properties of the dental materials. The relationship between the pathogenic microflora of the oral cavity and the appearance of stomatitis is well-known [1–3], however, this problem has a much wider context, since the relationship between oral microorganisms and diseases of the lungs, cardiovascular system (including heart), kidneys, and the digestive system has been proven [4,5]. Pathogenic microorganism colonized prostheses are believed to adversely affect the general health of treated patients [6]. This is facilitated by the fact that the microorganisms present in the dentures are both aspirated and swallowed with saliva or chewed food.

It should be emphasized that maintaining the hygiene of dentures and mucosa is a challenge due to the conditions in the oral cavity with 100% humidity, a reduced pH value, increased temperature, no possibility of self-cleaning of the mucosa under the prosthesis by

saliva, reduced access of oxygen, and a significant probability of mucosa injury due to hard denture plates [2,7]. Consequently, yeast-like fungi, especially *Candida albicans* (*C. albicans*), which are a part of the commensal oral microflora in many healthy individuals, for denture wearers, can rapidly become pathogenic and are isolated much more frequently, at least in 50% of patients [1,8], but when stomatitis occurs, the percentage exceeds 90% [9,10]. As a result, the research on denture base materials in terms of their colonization by microorganisms and finding effective methods to counteract this colonization are among the most important in the area. Despite many attempts to provide materials with antifungal properties and the use of cleaning agents, this problem has not been solved due to its negative influence on the mechanical and/or aesthetic properties of the materials [7,11–17], and the limited antimicrobial effectiveness of cleaning agents [18] associated with rapid recolonization (several minutes) after cleaning [19].

Many studies suggest that not only is the surface of the material colonized, but *C. albicans* can penetrate the commonly used acrylic resin [20]. Bulad'a et al. [21] found it inside hard acrylic samples cut with a microtome of *C. albicans* blastopores. Krishnamurthy et al. [22] obtained similar results with a comparable methodology, however, blastopores and hyphae were found inside the material. These studies indicate the probability that the colonization and penetration of materials by *C. albicans* may contribute, especially in the long-term, to the gradual loss of mechanical properties. This seems particularly probable if we take into account that in the above-mentioned studies, the number of cells present inside the samples was counted in the thousands on each of the analyzed planes, and the presence of *C. albicans* inside the materials was demonstrated after a short time [23]. However, this problem has not been investigated thus far. The purpose of the present work was to investigate the influence of exposure to *C. albicans* suspension on the mechanical properties of the polymethyl methacrylate denture base material and to verify previous reports on the penetration of *C. albicans* into the material. Our hypothesis was that exposure to a suspension of *C. albicans* strain causes changes in the mechanical properties due to surface colonization and/or penetration of the samples.

2. Materials and Methods

2.1. Materials and Sample Preparation

2.1.1. Sample Preparation

A transparent (color no. 4) heat-cured acrylic denture base resin Vertex Rapid Simplified (Vertex-Dental, Soesterberg, The Netherlands) was used and all samples were prepared with the standard flasking technique used in prosthetic dentistry, similar to the one described in a previous work [7]. First, the molds were prepared in dental flasks. The wax models of the sample types were flooded with dental stone (type IV gypsum, Zhermack, Badia Polesine, Italy) to prepare the first part of the mold. After the gypsum set, the mold was placed in the first part of the flask and mounted using model plaster (Stodent II, Zhermack, Badia Polesine, Italy). After setting the model, the plaster separation medium (Divosep, Vertex Dental, Soesterberg, The Netherlands) was applied with a brush. When its setting was finished, the second part of the flask was mounted, dental stone was poured to about 1/4 of its volume; after it set, the flask was filled with model plaster and finally closed with a cover. After model plaster setting, the wax models were removed by smelting and the surfaces of both parts of the mold were covered with separate media (Divosep, Vertex Dental, Soesterberg, The Netherlands). The samples were polymerized according to the manufacturer's instructions.

After the whole process, the samples were taken out of the mold and their quality was controlled: if bubbles, discontinuities, or other defects occurred, the sample was discarded. The excess material was cut with a scalpel and the samples were individually wet ground for standarization (Labo-Pol25, Struers, Willich, Germany) with P500-grit abrasive paper (tensile strength, ball hardness, impact strength, and flexural strength) or for surface hardness tests with P500, P800, P1200-grit (Struers, Willich, Germany) and finely with 6 μm diamond paste (Struers, Willich, Germany). After the finishing process, all samples were thoroughly rinsed in distilled water and conditioned in distilled water at 37 ± 1 °C for 48 h to remove most of the residual monomer [24,25].

2.1.2. Incubation in the Suspension of *Candida albicans*

Plasma sterilized samples were incubated for 30, 60, and 90 days in Sabouraud liquid medium (bioMérieux, Craponne, France) diluted five times with PBS (control samples—CO) at 37 °C and in a suspension of the reference strain *C. albicans* ATCC 10231 in liquid Sabouraud medium diluted 5-flod with a 0.9% NaCl solution, the final density of the strain suspension was 3×10^6 CFU/mL (test samples—CA) at 37 °C. The control medium and the *C. albicans* suspension in the medium were changed twice a week.

After the end of each of the incubation times, the samples were removed with tweezers and lightly rinsed with a 4% glutaraldehyde solution in 0.9% NaCl solution, and then placed in the solution as above for 2 h. After this time, the samples were tested.

2.2. Methods

2.2.1. Tensile Strength Test

For dumbbell-shaped tensile strength tests, samples of type 5B specified by EN ISO 527-2 and 1.5-mm thick specimens were mounted with polymeric tweezers in the jaws of a universal testing machine (Zwick Z020, Zwick GmbH & Com, Ulm, Germany). For each of the experiment conditions, 10 samples were tested. The speed was 5 mm/min [26]. The tensile strength was calculated according to the following equation:

$$T_s = \frac{F_{max}}{A} \qquad (1)$$

where T_s is the ultimate tensile strength (MPa); F_{max} is the force at rupture (N); A is the initial cross-sectional area (mm^2).

After testing, the fractured samples were carefully removed in a Petri dish with polymeric tweezers (they were caught near the place of attachment in the jaw) and the parts were placed with a distance that prevents their contact.

2.2.2. Flexural Properties

A three-point bending test based on the ISO 20795-1:2013-07 standard [27] with differences resulting from the aim of the experiment (specimens were tested after exposition). For each of the experimental conditions, 10 specimens measuring 65 mm × 10 mm × 3.3 mm were prepared. The distance between the supports was 50 mm, and the crosshead speed was 5 mm/min. The flexural strength and the flexural modulus were calculated according to the equations.

$$FS = \frac{3Fl}{2bh^2} \qquad (2)$$

$$E = \frac{F_1 l^3}{4bh^3 d} \qquad (3)$$

where *FS* is the flexural strength (MPa); *E* is the flexural modulus (GPa); *l* is the distance between the supports (mm); *b* and *h* are the width and height (mm); *F* is the maximal force (N); F_1 is the load at a chosen point at the elastic region of the stress–strain plot (kN); and *d* is the deflection at F_1 (mm).

During the test, the samples were secured at their ends, protruding over the supports in such a way that they did not fall down at the moment of fracture, and after the test, they were carefully placed on Petri dishes at a distance that prevented contact. If for any reason a broken specimen fell, or another risk of the contamination of the fractured surface occurred, it was removed from further tests.

2.2.3. Charpy Impact Strength

The Charpy impact strength test was based on the ISO 179-1:2010 standard [28] with differences resulting from the aim of the experiment. For each experimental conditions,

10 unnotched samples measuring 50 mm × 6 mm × 4 mm after incubation were placed horizontally on supports (the distance between them was 40 mm) and the test was performed on a pendulum impact tester (HIT 25P, Zwick GmbH & Com, Ulm, Germany) and a pendulum with an energy of 1 J was used. The impact strength was calculated according to the following equation:

$$a_{cU} = \frac{E}{b \times d} \times 10^3 \quad (4)$$

where a_{cU} is the impact strength; E is the energy absorbed by breaking the test specimen (J); b and d are the width and thickness of the specimen, respectively (mm).

2.2.4. Surface Vickers Hardness

For the tests, samples were made with dimensions of 10 × 10 mm and a thickness of 3.0 mm. All samples were made during the same polymerization series to ensure that the results were not affected by this process. Hardness was measured using a Future-Tech FM-700 microhardness tester (Future-Tech Corp, Tokyo, Japan) at a load of 300 g and loading time of 15 s five times for each sample at randomly selected locations with a minimum distance of 2 mm between indentations, and the means of individual specimens were averaged [16,29] on 15 samples per group. The Vickers hardness was calculated automatically by a hardness tester based on the average length of the diagonal left by the indenter.

2.2.5. Ball Hardness

For the tests, samples were made with dimensions of 30 × 40 mm and a thickness of 5.0 mm. All samples were produced during the same polymerization series to ensure that the results were not affected by this process. The ball indentation hardness (H) was determined according to the ISO 2039-1 standard [30] on a Zwick 3106 hardness tester (Zwick GmbH & Com, Ulm, Germany) for samples after a particular incubation time. During the test, a steel ball with a diameter of 5 mm was indented into the material under a load of 358 N for 30 s. During the measurements, the indentation of the depth of the ball was measured, and the ball hardness (H) using the surface of the impression was automatically calculated in N/mm^2. Three samples were made for each condition, and five measurements were made in each of them at randomly selected places with a minimum distance of 10 mm between indentations to calculate the mean hardness value.

2.2.6. Microscopic Evaluation—Colonization and Penetration Evaluation

Scanning electron microscopy (SEM) with a Zeiss SUPRA 35 (Zeiss, Oberkochen, Germany) and an OLYMPUS IX 51 (Olympus, Tokyo, Japan) inverted fluorescence microscope was used for qualitative evaluation to confirm the presence of C. albicans on the surfaces and inside the specimens (the potential penetration was evaluated on fractured surfaces—interior of specimens). During the SEM investigations, we randomly chose five halves of the fractured specimens after the tensile strength and flexural strength tests were investigated (a total of 10 halves of the specimens for each exposition condition). Observations were performed at accelerating voltages of 15 kV, and all samples were gold sputtered [7]. The samples (five for each condition) for fluorescence microscopy were carefully rinsed in PBS, then 1–2 drops of Calcofluor White Stain (Sigma-Aldrich, St. Louis, MO, USA) were placed on the microscopic glass, the investigated surface was placed in it, and after 1–2 min of incubation, it was observed at room temperature under UV light using an inverted fluorescence microscope. Fungal organisms appear fluorescent bright green to blue because the Calcofluor White Stain is a non-specific fluorochrome that binds with the cellulose and chitin contained in the cell walls of fungi and other organisms.

2.2.7. Statistical Analysis

Statistical analysis of the results was conducted using the PQStat ver. Software 1.6.6.204 (PQStat Software, Poznań, Poland). The results of the mechanical properties tests were carried out with one-way ANOVA with F* correction (Brown–Forsythe) when the assumption of the equality of variances was not met ($\alpha = 0.05$). The distributions of the residuals were tested with the Shapiro–Wilk test, and the equality of variances was tested with the Levene test ($\alpha = 0.05$). Tukey HSD post hoc tests were used ($\alpha = 0.05$).

3. Results
3.1. Mechanical Properties

The results of the tensile strength tests are presented in Figure 1a. There were no statistically significant changes in tensile strength after exposure to the control medium and the *C. albicans* suspension, but also in comparison to the baseline (24 h H_2O) ($p = 0.3245$ and $p = 0.2949$, respectively). Furthermore, the tensile strength values did not differ in their statistical significance ($p > 0.05$) for particular times after exposure to the control medium and the suspension of *C. albicans*.

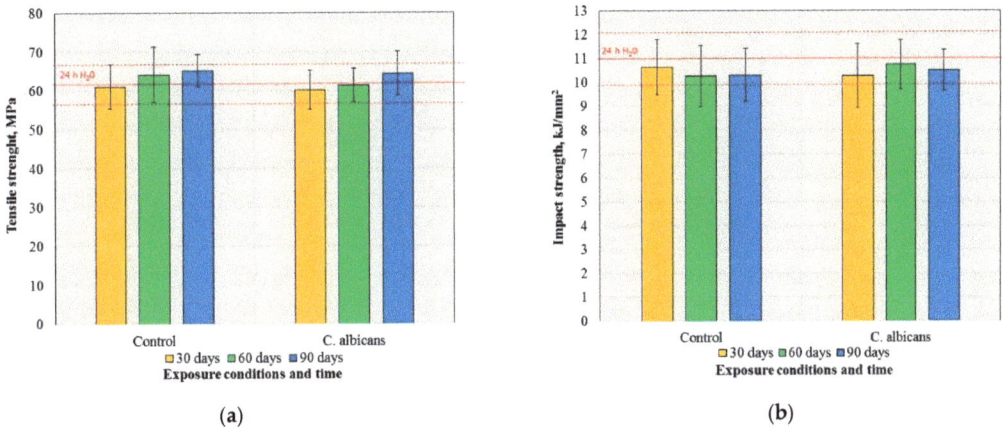

Figure 1. The mean values of the tensile strength (**a**) and impact strength (**b**) with the standard deviations after exposure to the control medium and the *C. albicans* suspension.

The results of the impact strength tests are presented in Figure 1b. There were no statistically significant changes in the impact strength after exposure to the control medium and the suspension of *C. albicans*, but also in comparison to the 24 h H_2O ($p = 0.6457$ and $p = 0.6233$, respectively). Furthermore, the impact strength values did not differ in their statistical significance ($p > 0.05$) for the exposure times after exposure to the control medium and the *C. albicans* suspension.

The results of the flexural strength tests are presented in Figure 2a. There were no statistically significant changes in flexural strength after exposure to the control medium and the *C. albicans* suspension, but also in comparison to the baseline (24 h H_2O) ($p = 0.607$ and $p = 0.7516$, respectively). Furthermore, the flexural strength values did not differ in their statistical significance ($p > 0.05$) for particular exposure times after exposure to the control medium and the *C. albicans* suspension. The results of the flexural modulus are presented in Figure 2b. There were no statistically significant changes in the flexural strength after exposure to the control medium and to the *C. albicans* suspension, but also in comparison to the baseline (24 h H_2O) ($p = 0.6146$ and $p = 0.1848$, respectively). Furthermore, the values of the flexural modulus did not differ in their statistical significance ($p > 0.05$) for individual exposure times after exposure to the control medium and the *C. albicans* suspension.

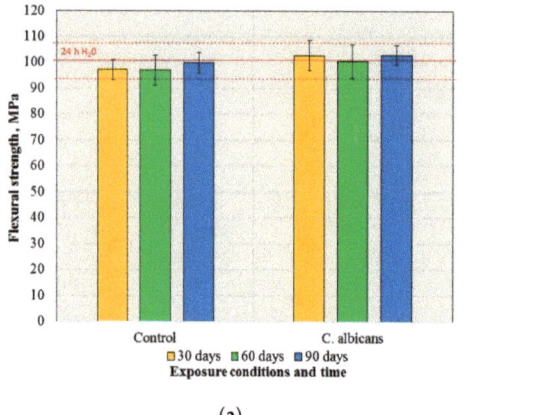
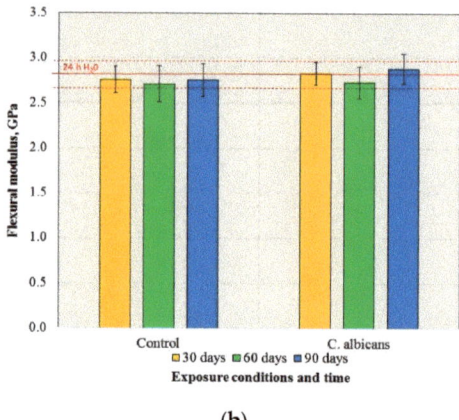

Figure 2. The values of the mean flexural strength (**a**) and flexural modulus (**b**) with the standard deviations after exposure to the control medium and the *C. albicans* suspension.

The results of the ball indentation hardness tests are presented in Figure 3a. There were statistically significant changes in the hardness after exposure to the control medium ($p = 0.0033$) and the suspension of *C. albicans* ($p = 0.0006$), however, the results of the post hoc test (Table 1) showed that statistically significant differences were observed only in comparison to the initial hardness value (24 h H_2O). The ball hardness values did not differ in their statistical significance for individual exposure times after exposure to the control medium and the *C. albicans* suspension.

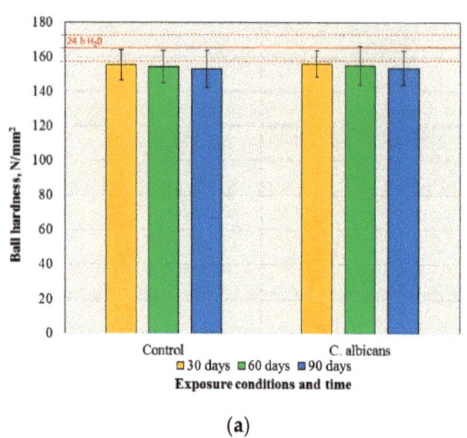
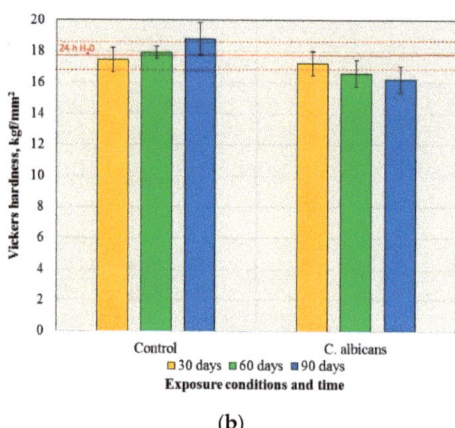

Figure 3. The mean ball hardness (**a**) and Vickers hardness (**b**) values with the standard deviations after exposure to the control medium and the *C. albicans* suspension.

The results of the Vickers hardness tests are presented in Figure 3b. There were statistically significant changes in the hardness after exposure to the control medium ($p = 0.0001$) and the *C. albicans* suspension ($p < 0.0001$). The results of the post hoc test (Table 2) showed that statistically significant differences were observed for the control conditions after 90 days of exposure compared to the results after all other periods including the starting point (increase in hardness). After exposure to the *C. albicans* suspension, successive reduction (also in comparison to starting point) in the hardness (statistically significant differences) with the prolonging of time was registered. The Vickers surface hardness values after exposure to the control medium and the *C. albicans* suspension did not differ in their statistical significance only after 30 days of exposure, but after 60 and 90 days, the differences for the individual exposure times were statistically significant.

Table 1. The results of the Tukey HSD post hoc tests (columns) and the Student *t*-tests (rows) for the ball hardness.

Exposure Time	Control Medium	C. albicans
24 h H$_2$O	A	A
30 days	B; a	B; a
60 days	B; a	B; a
90 days	B; a	B; a

The same uppercase letters (A–B) for each column and the lowercase letters for each row did not show significantly different results at the level of $p < 0.05$.

Table 2. The results of the Tukey's HSD post hoc tests (columns) and *t*-student tests (rows) for the Vickers hardness.

Exposure Time	Control Medium	C. albicans
24 h H$_2$O	A	A
30 days	A; a	A,B; a
60 days	A; a	B,C; b
90 days	B; a	C; b

The same uppercase letters (A–C) for each column and the lowercase letters (a–c) for each row did not show significantly different results at the level of $p < 0.05$.

3.2. Microscopic Evaluation of Surfaces and Fractures

Representative SEM microphotographs that show the surfaces of the control samples are presented in Figure 4. Parallel traces resulting from the grinding of samples with abrasive paper were covered largely with structures related to crystallization in the control medium. With increasing exposure time, a tendency to increase the accumulation of crystalized structures was observed, however, even after 90 days, there were still some microareas where they were hardly present.

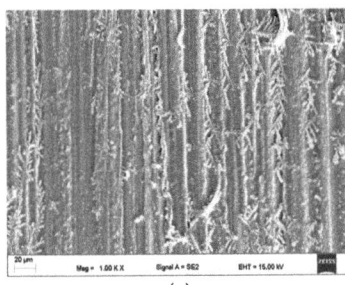
(a)
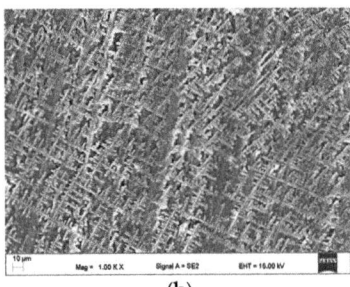
(b)
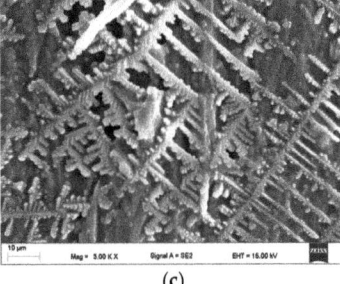
(c)

Figure 4. The SEM microphotographs presenting the surface morphologies after 30 days of exposure (a) and 90 days of exposure in the control medium (b,c).

After incubation in the *C. albicans* suspension, the presence of a microorganism on the surface of the samples was confirmed by SEM and fluorescence microscope observations (Figures 5 and 6). Numerous blastospores as well as pseudohyphae were observed after all incubation times, but the hyphae form was found episodically only after 90 days. Colonies composed of blastospores were the most common, differing in their number and also containing pseudohyphae. There was no obvious trend showing an increase in the number of cells in individual colonies related to an increase in the incubation time. This number usually ranged from a few to over a dozen, and only sporadically more, but there were no colonies characterized by cell numbers visibly exceeding the ones presented in Figures 5 and 6. The SEM observations showed that colonies were formed in crystallized structures and much less frequently inside the scratches after grinding (e.g., Figure 5a, yellow arrow).

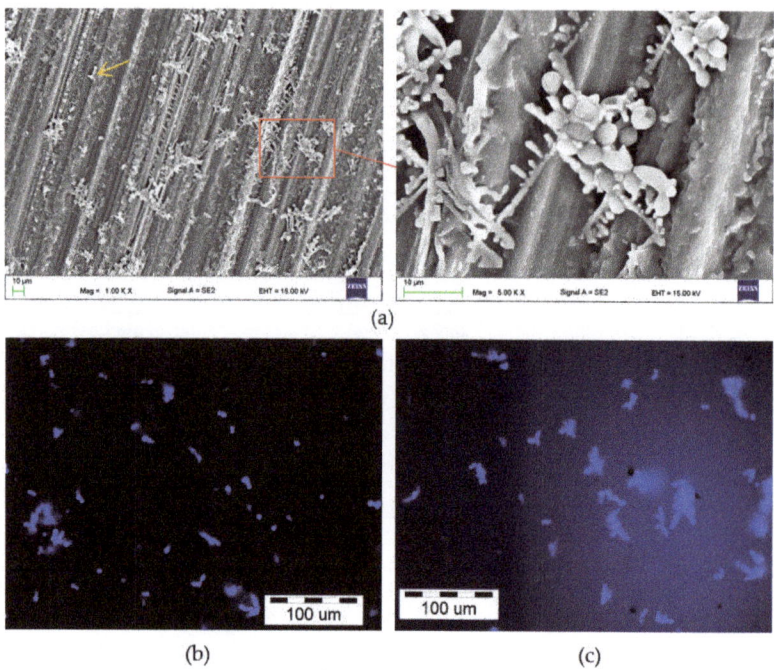

Figure 5. Representative microphotographs of the surface of the samples after 30 days of incubation in the *C. albicans* suspension obtained by SEM (**a**) and fluorescence microscopy (**b**,**c**), yellow arrow—exemplary colony inside the scratch.

For both the control and the samples exposed to *C. albicans*, no signs of surface deterioration (e.g., cracks) were observed.

In Figure 7, exemplar fractures of the PMMA samples after flexural strength tests are shown. They were examined to assess the potential penetration of *C. albicans*. All fractures (after exposure to the control medium and the *C. albicans* suspension) were characterized by a similar morphology. Typical areas representing the brittle fracture mode (compact and smooth surface fields) and an intermediate (brittle to ductile) fracture mode (a jagged and rough appearance) were observed. Cells of *C. albicans* were not found in any of the fractures after 30 days of exposure. After 60 days, one of the fractures analyzed showed the presence of three single blastospores and one colony consisting of two cells (Figure 7b) at a distance of approximately 60–100 μm from the sample surface. In another fracture, two single blastospores were found approximately 200 μm from the sample surface. No blastospores or other forms of *C. albicans* were found in the remaining fractures. In some areas, small particles were visible, which were contaminants after the samples (indicated by a yellow arrow). In the case of samples after 90 days of exposure, one of the fractures analyzed showed the presence of five single blastospores and one colony consisting of two cells at a distance of approximately 400 μm from the sample surface (Figure 7c,d). Blastospores or other forms of *C. albicans* were not found in the remaining failures. In some areas, only single particles were visible from which contaminants were breaking the samples.

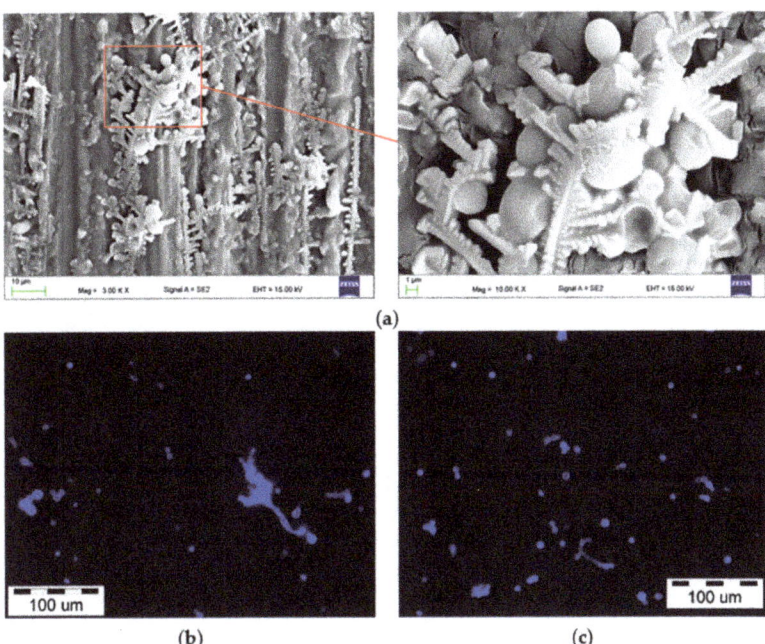

Figure 6. Representative microphotographs of the surface of the samples after 90 days of incubation in the *C. albicans* suspension obtained by SEM (**a**) and fluorescence microscopy (**b**,**c**).

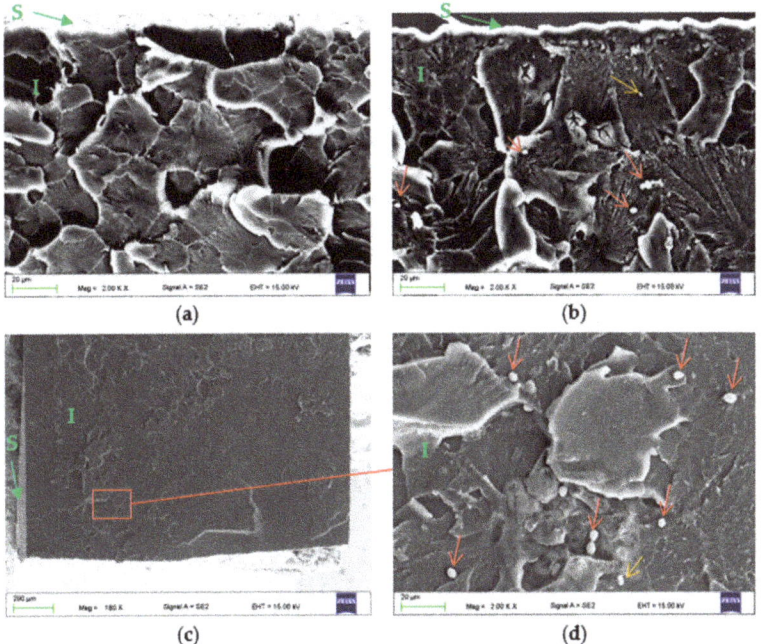

Figure 7. Representative SEM microphotographs of the fractured surfaces (interior of specimens) of the samples after 60 days (**a**,**b**) and 90 days (**c**,**d**) incubation in *C. albicans* suspension. S—surface, I—interior of specimen, red arrows—*C. albicans* blastospores, yellow arrows—contaminations.

4. Discussion

The research carried out thus far indicates that hard acrylic materials may be colonized and penetrated by pathogenic yeasts [21,22]. In these studies, samples of the materials were exposed to a suspension for six weeks, sectioned, and the interior of samples was examined during microscopy observations. Bulad et al. [21] found more than a half thousand blastospores at the deepest level, but a hyphae form was not observed. Krishnamurthy et al. [22] used the same hard acrylic, and an analogous sample preparation method observed blastospores and hyphae forms. These results may suggest that, in the long-term, this phenomenon may have an impact on the mechanical properties, which to a large extent determine the clinical usefulness of materials. In this study, the effect of the presence of the *C. albicans* suspension on hardness, flexural strength, tensile strength, and impact strength was investigated. The Vickers hardness test, due to the applied load and depth of the indentation marks, values should be considered as surface hardness [31,32], while the ball indentation hardness was used to check for possible changes in the macroscale. Flexural strength is considered as a fundamental property of prosthetic materials, which is particularly important when atrophy processes of the alveolar ridges lead to the creation of irregular shapes of the denture bearing area, resulting in uneven support of the prostheses [33,34]. The values of the flexural modulus affect the effectiveness of chewing [35] and the impact strength determines the resistance to dynamic loads (e.g., prosthesis fall) [7].

The conducted experiments did not show statistically significant changes in the flexural strength, flexural modulus, tensile strength, and impact strength, and the values obtained were similar to those reported in other studies for this material [36,37]. The results after exposure to the control medium confirm that changes in the basic strength properties may be insignificant in the case of the materials currently used [38], despite the fact that many previous studies have pointed to a period of even a several dozen days to a significant reduction in these properties [39,40]. The lack of changes in the properties mentioned could also be related to the use of an experimental medium other than distilled water (water is more destructive to acrylates compared to, e.g., artificial saliva) [38,41], and/or conditioning the samples in water prior to the experiment.

Significant hardness changes have been observed after exposure to both the suspension of the control medium and *C. albicans*. In the case of the hardness measured by the ball indentation method, a decrease in hardness was recorded after the first 30 days of the experiment, which should be associated with the plasticization of the material [42,43] caused by the penetration of water molecules between the polymer chains, that weaken their interactions [44,45]. The changes measured by the ball indentation method were approximately 5% and did not translate into changes in the other mechanical properties, probably due to their relatively small influence on the changes in the yield point [46]. For the Vickers hardness, the changes were statistically significant during the 90 day experiment, but the directions of the changes were divergent; in the case of conditioning in the liquid substrate solution, an increase in hardness was recorded, while in the presence of the *C. albicans* suspension, the hardness decreased. With the applied load and the obtained hardness values, the depth of the indentations was ~25 µm, so only the hardness of the material surface was tested. Taking into account that the SEM investigations showed that the surface of the samples after exposure to the control medium showed an increasing tendency to crystallize the components of the medium with time, it can be assumed that their presence caused an increase in the hardness. In the presence of *C. albicans*, the structures formed as a result of crystallization were also visible, but a gradual decrease in the average hardness values was recorded. This process was probably related to the changes taking place in the microenvironment containing *C. albicans*. The liquid Sabouraud medium contains glucose from which *C. albicans* (like other yeasts) is capable of producing, for example, ethanol [47–50], which has a strong plasticizing effect on acrylates [44,51]. Furthermore, it has been shown that *C. albicans* in cultures containing glucose causes a significant decrease in the pH value, even from 7.5 to 3.2 in 48 h [52,53]. The reduction in the pH value in human saliva, with the addition of glucose, is caused by products of glycolysis such as acetic acid, pyruvic acid, and lactic acid [53]. A similar trend of lowering the pH value in the presence of *C. albicans* was also observed in the Sabouraud medium [54]. These data seem to be particularly important in the considered context because Miranda et al. [55] suggested that acid ingredients in liquids with lowered pH values may react with the ester groups of acrylates and lead to the creation of alcohol and carboxylic acid in materials, which further contributes to the accelerated degradation in the polymer properties. Therefore, the products generated during yeast metabolism could have contributed to changes in the properties of the experimental medium, which, in turn, caused the degradation to the surface of the PMMA samples. Due to the low intensity of the process, these changes occurred only at a slight depth in the material; therefore, the hardness test using the ball indentation method (and other mechanical properties) did not show these changes. However, the recorded decrease in hardness, although small in terms of value, should be considered as significant from a clinical point of view, because many works have emphasized the correlation of surface hardness with the tribological wear of materials [56], which is particularly important in the context of dental materials and the penetration of abrasive wear products into the body during the act of swallowing saliva.

The microscopic evaluation of the surface confirmed the presence of blastospores and the pseudohyphae of *C. albicans*, but the occurrence of hyphae was episodic. On the surfaces of the fractured samples (interior of the material), only the presence of up to a few blastospores on three fractures was observed. It can be assumed that their presence was related to accidental contamination related to the movement of the samples immediately after their destruction on the testing machine, rather than the penetration of *C. albicans* into the material. The premise in this regard may be the rare presence of other contaminations (Figure 7, yellow arrows). It should also be noted that the number of *C. albicans* cells was incomparably smaller than that reported in studies carried out in cut samples [21,22]. It should be noted that in the works cited, the sections studied were analyzed after cutting the samples in a water environment (diamond disc/microtome). However, there is no information on how (or if at all) the methodology was checked in terms of whether it was possible to transfer *C. albicans* into the material during the cutting process. In the current study, after the fracture on the testing machine, the samples were not touched in any way on the investigated surfaces, so the only possibility of contamination occurred directly after the break. Therefore, it seems that the possibility of a false positive result has been minimized to a greater extent. Another factor that could influence the results obtained was the composition of the medium in which the samples were incubated. In our study, the Sabouraud substrate was used, similar to the investigations on the colonization and penetration of *C. albicans* into prosthetic materials conducted by Burns et al. [57]. They observed the same morphological forms of *C. albicans* as in the current work, but the presence of microorganisms inside the cut samples of different types of materials was registered. In other works using a hard acrylic material [21,23], the presence of the hyphae form was confirmed, but the experimental media based on artificial saliva were used. The authors of these works do not provide the compositions of the saliva used or any of the other parameters such as the initial pH. Therefore, it is not known what the influence of these features could have been on the results of the observations, taking into account the fact that the filamentation process is favored by, for example, a pH at the level of 8, while inhibiting the hyphae growth and promoting the growth of blastospores, is favored by an acidic or neutral environment with a pH of ~4–7 such as that used in the current research or the work by Burns. However, considering that Burns et al. (based on Sabouraud medium) and the two other indicated works (based on artificial saliva) confirmed the penetration in cut samples regardless of the reported forms of *C. albicans*, it can be assumed that the medium used was not determinative for the results from the aspect of penetration. It should be noted that the works [21,22] did not discuss how the *C. albicans* cells penetrated the materials. Krishnamurthy et al. [22] linked the presence of a large number of blastospores in the interior with its potential porosity resulting from polymerization. However, such a reason seems unlikely considering that SEM studies [58,59] have proven that pores formed during crosslinking have sizes up to a few dozen nanometers, so are several dozen to one hundred times smaller than the size of blastospores, and are closed pores, so no direct migration between them is possible. At the same time, the observed tendency for the presence of *C. albicans* colonies/cells in the crystallized structures or traces after grinding is maintained, according to reports indicating that increased roughness may promote an increased degree of surface colonization [60]. This indicates that problems with the removal of microorganisms from prosthetic materials over time and their rapid recolonization after disinfection [19,61] may not be caused by the penetration of *C. albicans* into materials, but by the formation of surface microcracks or screeches due to daily use including cleaning or the impact of thermal cycles [62–64], which may be difficult to disinfect with cleaning agents. This supposition is supported not only by the currently obtained results, but also by the in vivo investigations by Taylor et al. [65], in which the interior of the prosthetic materials was not penetrated. These results suggest that the problem of the potential penetration of *C. albicans* into dental materials is still a challenge that requires further, intensified research.

5. Conclusions

The presence of the suspension of *C. albicans* did not affect the PMMA properties of the denture material such as the flexural strength, flexural modulus, tensile strength impact strength, or the ball indentation hardness. A decrease in the surface hardness was observed during the experiment, which could have been caused by the presence of yeast metabolism products, which could have caused plasticization and/or surface degradation by lowering the pH of the experimental suspension and the reactions of an acid liquid with the ester groups of acrylates. Only a few blastospores of *C. albicans* that occurred in a small part of the fractures analyzed were recorded. Considering their number and other indications resulting from the analysis of the SEM images, it should be concluded that their presence was not the result of *C. albicans* penetration into the material, but the result of accidental contamination during the mechanical properties tests. Therefore, the investigation did not confirm the penetration of *C. albicans* into the materials. The presence of blastospores and the pseudohyphae of *C. albicans* on the surface, mainly in crystallized structures and traces after grinding, indicates that under clinical conditions, deterioration of the surface quality while using dentures may be an important factor that is conducive to the colonization of materials by microorganisms. This probably led to the formation of difficult to disinfect microareas, which can be a cause of the clinically registered rapid recolonization of dentures by yeasts. The limitation was that the exposure time was not related to the actual oral conditions, but only to the specific laboratory conditions used during the study. Under clinical conditions, there are many additional factors that could influence the results, and these factors should be considered in future tests. The creation of biofilm may be especially important in this context. In the presented work, as in our previous experiments in the field, a biofilm was not formed despite a long exposure time, which may be related to the use of a diluted substrate, so in the future, similar experiments with biofilm formation should be carried out.

Author Contributions: Conceptualization, G.C. and M.N.; Formal analysis, G.C. and M.N.; Investigation, G.C., M.N., W.P. and A.M.; Methodology, G.C., W.P. and A.M.; Supervision, G.C. and A.M; Visualization, G.C. and M.N.; Writing—original draft, G.C. and M.N.; Writing—review & editing, G.C. All authors have read and agreed to the published version of the manuscript.

Funding: This work was financially supported with statutory funds of Faculty of Mechanical Engineering of Silesian University of Technology.

Institutional Review Board Statement: Not applicable.

Informed Consent Statement: Not applicable.

Data Availability Statement: Data supporting reported results are available from the authors.

Conflicts of Interest: The authors declare no conflict of interest.

References

1. Al-Dwairi, Z.N. Isolation of candida species from the oral cavity and fingertips of complete and partial dentures wearers. *J. Dent. Health Oral Disord. Ther.* **2014**, *1*, 420–423. [CrossRef]
2. Spiechowicz, E.; Mierzwińska-Nastalska, E. *Fungal Infections of Oral Cavity*, 1st ed.; Med Tour Press International: Warsaw, Poland, 1998; p. 34.
3. Smith, A.; Williams, D.; Bradshaw, D.; Milward, P.; Kutubi, S.A.; Rowe, W. The effect of residual food stain on candida albicans colonisation of denture acrylics. *Dent. Oral Biol. Craniofacial Res.* **2020**, *3*, 2–5. [CrossRef]
4. Uzel, N.G.; Teles, F.R.; Teles, R.P.; Song, X.Q.; Torresyap, G.; Socransky, S.S.; Haffajee, A.D. Microbial shifts during dental biofilm re-development in the absence of oral hygiene in periodontal health and disease. *J. Clin. Periodontol.* **2011**, *38*, 612–620. [CrossRef] [PubMed]
5. Evren, B.A.; Uludamar, A.; Işeri, U.; Ozkan, Y.K. The association between socioeconomic status, oral hygiene practice, denture stomatitis and oral status in elderly people living different residential homes. *Arch. Gerontol. Geriatr.* **2011**, *53*, 252–257. [CrossRef]
6. Gendreau, L.; Loewy, Z.G. Epidemiology and etiology of denture stomatitis. *J. Prosthodont.* **2011**, *20*, 251–260. [CrossRef]
7. Chladek, G.; Pakieła, K.; Pakieła, W.; Żmudzki, J.; Adamiak, M.; Krawczyk, C. Effect of antibacterial silver-releasing filler on the physicochemical properties of poly(methyl methacrylate) denture base material. *Materials* **2019**, *12*, 4146. [CrossRef]
8. Kinkela Devcic, M.; Simonic-Kocijan, S.; Prpic, J.; Paskovic, I.; Cabov, T.; Kovac, Z.; Glazar, I. Oral candidal colonization in patients with different prosthetic appliances. *J. Fungi* **2021**, *7*, 662. [CrossRef]
9. Scully, C. *Oral and Maxillofacial Medicine—E-Book: The Basis of Diagnosis and Treatment*, 3rd ed.; Churchill Livingstone Elsevier: Edinburgh, UK, 2013; ISBN 978-0-7020-5205-7.
10. Gauch, L.M.R.; Pedrosa, S.S.; Silveira-Gomes, F.; Esteves, R.A.; Marques-da-Silva, S.H. Isolation of candida spp. from denture-related stomatitis in Pará, Brazil. *Braz. J. Microbiol.* **2017**, *49*, 148–151. [CrossRef]
11. De Freitas Oliveira Paranhos, H.; Peracini, A.; Pisani, M.X.; Oliveira, V.D.C.; de Souza, R.F.; Silva-Lovato, C.H. Color stability, surface roughness and flexural strength of an acrylic resin submitted to simulated overnight immersion in denture cleansers. *Braz. Dent. J.* **2013**, *24*, 152–156. [CrossRef]
12. Peracini, A.; Davi, L.R.; de Queiroz Ribeiro, N.; de Souza, R.F.; Lovato da Silva, C.H.; de Freitas Oliveira Paranhos, H. Effect of denture cleansers on physical properties of heat-polymerized acrylic resin. *J. Prosthodont. Res.* **2010**, *54*, 78–83. [CrossRef]
13. Nakahara, T.; Harada, A.; Yamada, Y.; Odashima, Y.; Nakamura, K.; Inagaki, R.; Kanno, T.; Sasaki, K.; Niwano, Y. Influence of a new denture cleaning technique based on photolysis of H_2O_2 the mechanical properties and color change of acrylic denture base resin. *Dent. Mater. J.* **2013**, *32*, 529–536. [CrossRef] [PubMed]
14. Fan, C.; Chu, L.; Rawls, H.R.; Norling, B.K.; Cardenas, H.L.; Whang, K. Development of an antimicrobial resin—A pilot study. *Dent. Mater.* **2011**, *27*, 322–328. [CrossRef] [PubMed]
15. Zidan, S.; Silikas, N.; Alhotan, A.; Haider, J.; Yates, J. Investigating the mechanical properties of ZrO2-impregnated PMMA nanocomposite for denture-based applications. *Materials* **2019**, *12*, 1344. [CrossRef] [PubMed]
16. Alhotan, A.; Yates, J.; Zidan, S.; Haider, J.; Silikas, N. Flexural strength and hardness of filler-reinforced PMMA targeted for denture base application. *Materials* **2021**, *14*, 2659. [CrossRef]
17. Bajunaid, S.O. How effective are antimicrobial agents on preventing the adhesion of candida albicans to denture base acrylic resin materials? A systematic review. *Polymers* **2022**, *14*, 908. [CrossRef]
18. Cruz, P.C.; de Andrade, I.M.; Peracini, A.; de Souza-Gugelmin, M.C.M.; Silva-Lovato, C.H.; de Souza, R.F.; de Freitas Oliveira Paranhos, H. The effectiveness of chemical denture cleansers and ultrasonic device in biofilm removal from complete dentures. *J. Appl. Oral Sci.* **2011**, *19*, 668–673. [CrossRef]
19. Preshaw, P.M.; Walls, A.W.G.; Jakubovics, N.S.; Moynihan, P.J.; Jepson, N.J.A.; Loewy, Z. Association of removable partial denture use with oral and systemic health. *J. Dent.* **2011**, *39*, 711–719. [CrossRef]
20. Van Reenen, J.F. Microbiologic studies on denture stomatitis. *J. Prosthet. Dent.* **1973**, *30*, 493–505.
21. Bulad, K.; Taylor, R.L.; Verran, J.; McCord, J.F. Colonization and penetration of denture soft lining materials by candida albicans. *Dent. Mater.* **2004**, *20*, 167–175. [CrossRef]
22. Krishnamurthy, S.; Hallikerimath, R.B. An in-vitro evaluation of retention, colonization and penetration of commonly used denture lining materials by candida albicans. *J. Clin. Diagn. Res.* **2016**, *10*, ZC84–ZC88. [CrossRef]
23. Rodger, G.; Taylor, R.L.; Pearson, G.J.; Verran, J. In vitro colonization of an experimental silicone by candida albicans. *J. Biomed. Mater. Res. Part B Appl. Biomater.* **2010**, *92*, 226–235. [CrossRef] [PubMed]
24. Kedjarune, U.; Charoenworaluk, N.; Koontongkaew, S. Release of methyl methacrylate from heat-curved and autopolymerized resins: Cytotoxicity testing related to residual monomer. *Aust. Dent. J.* **1999**, *44*, 25–30. [CrossRef] [PubMed]
25. Baker, S.; Brooks, S.C.; Walker, D.M. The release of residual monomeric methyl methacrylate from acrylic appliances in the human mouth: An assay for monomer in saliva. *J. Dent. Res.* **1988**, *67*, 1295–1299. [CrossRef] [PubMed]
26. Shirkavand, S.; Moslehifard, E. Effect of TiO2 nanoparticles on tensile strength of dental acrylic resins. *J. Dent. Res. Dent. Clin. Dent. Prospects* **2014**, *8*, 197–203. [CrossRef]
27. *ISO 20795-1:2013*; Dentistry—Base Polymers—Part 1: Denture Base Polymers. International Organization for Standardization: London, UK, 2013.
28. *ISO 179-1:2010*; Plastics—Determination of Charpy Impact Properties—Part 1: Non-Instrumented Impact Test 2010. International Organization for Standardization: London, UK, 2010.

29. Farina, A.P.; Cecchin, D.; Soares, R.G.; Botelho, A.L.; Takahashi, J.M.F.K.; Mazzetto, M.O.; Mesquita, M.F. Evaluation of vickers hardness of different types of acrylic denture base resins with and without glass fibre reinforcement. *Gerodontology* **2012**, *29*, e155–e160. [CrossRef]
30. *ISO 2039-1:2001*; Plastics—Determination of hardness—Part 1: Ball indentation method. International Organization for Standardization: London, UK, 2001.
31. Ozyilmaz, O.Y.; Akin, C. Effect of cleansers on denture base resins' structural properties. *J. Appl. Biomater. Funct. Mater.* **2019**, *17*, 2280800019827797. [CrossRef]
32. Al-Dwairi, Z.N.; Tahboub, K.Y.; Baba, N.Z.; Goodacre, C.J.; Özcan, M. A comparison of the surface properties of CAD/CAM and conventional polymethylmethacrylate (PMMA). *J. Prosthodont.* **2019**, *28*, 452–457. [CrossRef]
33. Diaz-Arnold, A.M.; Vargas, M.A.; Shaull, K.L.; Laffoon, J.E.; Qian, F. Flexural and fatigue strengths of denture base resin. *J. Prosthet. Dent.* **2008**, *100*, 47–51. [CrossRef]
34. Ajaj-ALKordy, N.M.; Alsaadi, M.H. Elastic modulus and flexural strength comparisons of high-impact and traditional denture base acrylic resins. *Saudi Dent. J.* **2014**, *26*, 15–18. [CrossRef]
35. Wadachi, J.; Sato, M.; Igarashi, Y. Evaluation of the rigidity of dentures made of injection-molded materials. *Dent. Mater. J.* **2013**, *32*, 508–511. [CrossRef]
36. Polychronakis, N.; Sarafianou, A.; Zissis, A.; Papadopoulos, T. The influence of thermocycling on the flexural strength of a polyamide denture base material. *Acta Stomatol. Croat.* **2017**, *51*, 309–315. [CrossRef] [PubMed]
37. Taczała, J.; Fu, C.; Sawicki, J.; Pietrasik, J. Influence different amount of cellulose on the mechanical strength of dental acrylic resin. *IOP Conf. Ser. Mater. Sci. Eng.* **2020**, *743*, 012044. [CrossRef]
38. Jagini, A.S.; Marri, T.; Jayyarapu, D.; Kumari, R. Effect of long-term immersion in water and artificial saliva on the flexural strength of two heat cure denture base resins. *J. Contemp. Dent. Pract.* **2019**, *20*, 341–346. [CrossRef] [PubMed]
39. Takahashi, Y.; Chai, J.; Kawaguchi, M. Equilibrium strengths of denture polymers subjected to long-term water immersion. *Int. J. Prosthodont.* **1999**, *12*, 348–352. [PubMed]
40. Chandrahari, N.; Kumar, C.R.; Salgar, A.R.; Singh, M.; Singh, S. Comparison of fracture resistance of heat cure resins polymerized by conventional and microwave methods after immersion in artificial saliva. *J. Contemp. Dent. Pract.* **2019**, *20*, 71–77.
41. Bacali, C.; Badea, M.; Moldovan, M.; Sarosi, C.; Nastase, V.; Baldea, I.; Chiorean, R.S.; Constantiniuc, M. The influence of graphene in improvement of physico-mechanical properties in PMMA denture base resins. *Materials* **2019**, *12*, 2335. [CrossRef]
42. Sato, S.; Cavalcante, M.R.S.; Orsi, I.A.; de Freitas Oliveira Paranhos, H.; Zaniquelli, O. Assessment of flexural strength and color alteration of heat-polymerized acrylic resins after simulated use of denture cleansers. *Braz. Dent. J.* **2005**, *16*, 124–128. [CrossRef]
43. Ali, I.L.; Yunus, N.; Abu-Hassan, M.I. Hardness, flexural strength, and flexural modulus comparisons of three differently cured denture base systems. *J. Prosthodont.* **2008**, *17*, 545–549. [CrossRef]
44. Ferracane, J.L. Hygroscopic and hydrolytic effects in dental polymer networks. *Dent. Mater.* **2006**, *22*, 211–222. [CrossRef]
45. Al-Dwairi, Z.N.; Tahboub, K.Y.; Baba, N.Z.; Goodacre, C.J. A comparison of the flexural and impact strengths and flexural modulus of CAD/CAM and conventional heat-cured polymethyl methacrylate (PMMA). *J. Prosthodont.* **2020**, *29*, 341–349. [CrossRef]
46. Chladek, G.; Żmudzki, J.; Basa, K.; Pater, A.; Krawczyk, C.; Pakieła, W. Effect of silica filler on properties of PMMA resin. *Arch. Mater. Sci. Eng.* **2015**, *71*, 10.
47. Mishra, J.; Kumar, D.; Samanta, S.; Vishwakarma, M.K. A comparative study of ethanol production from various agro residues by using saccharomyces cerevisiae and candida albicans. *J. Yeast Fungal Res.* **2012**, *3*, 12–17.
48. Aruna, A.; Nagavalli, M.; Girijashankar, V.; Ponamgi, S.P.D.; Swathisree, V.; Rao, L.V. Direct bioethanol production by amylolytic yeast candida albicans. *Lett. Appl. Microbiol.* **2015**, *60*, 229–236. [CrossRef] [PubMed]
49. Mohd Azhar, S.H.; Abdulla, R.; Jambo, S.A.; Marbawi, H.; Gansau, J.A.; Mohd Faik, A.A.; Rodrigues, K.F. Yeasts in sustainable bioethanol production: A review. *Biochem. Biophys. Rep.* **2017**, *10*, 52–61. [CrossRef]
50. Yajima, D.; Motani, H.; Kamei, K.; Sato, Y.; Hayakawa, M.; Iwase, H. Ethanol production by candida albicans in postmortem human blood samples: Effects of blood glucose level and dilution. *Forensic Sci. Int.* **2006**, *164*, 116–121. [CrossRef]
51. Jang, D.-E.; Lee, J.-Y.; Jang, H.-S.; Lee, J.-J.; Son, M.-K. Color stability, water sorption and cytotoxicity of thermoplastic acrylic resin for non metal clasp denture. *J. Adv. Prosthodont.* **2015**, *7*, 278–287. [CrossRef]
52. Samaranayake, L.P.; Geddes, D.A.; Weetman, D.A.; MacFarlane, T.W. Growth and acid production of candida albicans in carbohydrate supplemented media. *Microbios* **1983**, *37*, 105–115.
53. Samaranayake, L.P.; Hughes, A.; Weetman, D.A.; MacFarlane, T.W. Growth and acid production of candida species in human saliva supplemented with glucose. *J. Oral Pathol. Med.* **1986**, *15*, 251–254. [CrossRef]
54. Kapica, L.; Shaw, C.E.; Bartlett, G.W. Inhibition of histoplasma capsulatum by candida albicans and other yeasts on Sabouraud's agar media. *J. Bacteriol.* **1968**, *95*, 2171–2176. [CrossRef] [PubMed]
55. Miranda, D.D.A.; Bertoldo, C.E.D.S.; Aguiar, F.H.B.; Lima, D.A.N.L.; Lovadino, J.R. Effects of mouthwashes on Knoop hardness and surface roughness of dental composites after different immersion times. *Braz. Oral Res.* **2011**, *25*, 168–173. [CrossRef]
56. Dayan, C.; Kiseri, B.; Gencel, B.; Kurt, H.; Tuncer, N. Wear resistance and microhardness of various interim fixed prosthesis materials. *J. Oral Sci.* **2019**, *61*, 447–453. [CrossRef] [PubMed]
57. Burns, D.R.; Burns, D.A.; DiPietro, G.J.; Gregory, R.L. Response of processed resilient denture liners to *Candida albicans*. *J. Prosthet. Dent.* **1987**, *57*, 507–512. [CrossRef]

58. Anusavice, K.; Shen, C.; Rawls, H.R. *Phillips' Science of Dental Materials*, 12th ed.; Saunders: St. Louis, MO, USA, 2013; ISBN 978-0-323-24205-9.
59. Chladek, G.; Basa, K.; Mertas, A.; Pakieła, W.; Żmudzki, J.; Bobela, E.; Król, W. Effect of storage in distilled water for three months on the antimicrobial properties of poly(methyl methacrylate) denture base material doped with inorganic filler. *Materials* **2016**, *9*, 328. [CrossRef]
60. Radford, D.R.; Challacombe, S.J.; Walter, J.D. Denture Plaque and Adherence of Candida Albicans to Denture-Base Materials in Vivo and in Vitro. *Crit. Rev. Oral. Biol. Med.* **1999**, *10*, 99–116. [CrossRef]
61. Dniluk, T.; Fiedoruk, K.; Ściepuk, M. Aerobic bacteria in the oral cavity of patients with removable dentures. *Adv. Med. Sci.* **2006**, *51*, 86–90.
62. Degirmenci, K.; Hayati Atala, M.; Sabak, C.; Degirmenci, K.; Hayati Atala, M.; Sabak, C. Effect of different denture base cleansers on surface roughness of heat polymerised acrylic materials with different curing process. *Odovtos Int. J. Dent. Sci.* **2020**, *22*, 145–153. [CrossRef]
63. Sorgini, D.B.; da Silva-Lovato, C.H.; de Souza, R.F.; Davi, L.R.; de Freitas Oliveira Paranhos, H. Abrasiveness of conventional and specific denture-cleansing dentifrices. *Braz. Dent. J.* **2012**, *23*, 154–159. [CrossRef]
64. Atalay, S.; Çakmak, G.; Fonseca, M.; Schimmel, M.; Yilmaz, B. Effect of thermocycling on the surface properties of CAD-CAM denture base materials after different surface treatments. *J. Mech. Behav. Biomed. Mater.* **2021**, *121*, 104646. [CrossRef]
65. Taylor, R.L.; Bulad, K.; Verran, J.; McCord, J.F. Colonization and deterioration of soft denture lining materials in vivo. *Eur. J. Prosthodont. Restor. Dent.* **2008**, *16*, 50–55. [CrossRef]

Article

Poly(methyl methacrylate) with Oleic Acid as an Efficient *Candida albicans* Biofilm Repellent

Milica Petrović [1], Marina Randjelović [2,3], Marko Igić [4], Milica Randjelović [5], Valentina Arsić Arsenijević [6], Marijana Mionić Ebersold [1,*], Suzana Otašević [2,3] and Irena Milošević [7]

1. Powder Technology Laboratory, Institute of Materials, Ecole Polytechnique Fédérale de Lausanne, 1015 Lausanne, Switzerland; petrovicmilica21@gmail.com
2. Department of Microbiology and Immunology, Faculty of Medicine, University of Niš, Blvd. Dr Zoran Djindjić 81, 18000 Niš, Serbia; marina.randjelovic@medfak.ni.ac.rs (M.R.); otasevicsuzana@gmail.com (S.O.)
3. Public Health Institute Niš, Blvd. Dr Zoran Djindjić 50, 18000 Niš, Serbia
4. Department of Prosthodontics, Faculty of Medicine, University of Niš, Blvd. Dr Zoran Djindjić 81, 18000 Niš, Serbia; marko.igic@medfak.ni.ac.rs
5. Department of Pharmacy, Faculty of Medicine, University of Niš, Blvd. Dr Zoran Djindjić 81, 18000 Niš, Serbia; milica.randjelovic@medfak.ni.ac.rs
6. National Reference Medical Mycology Laboratory, Institute of Microbiology and Immunology, Faculty of Medicine, University of Belgrade, Dr Subotića 1, 11000 Belgarde, Serbia; mikomedlab@yahoo.com
7. University of Applied Sciences of Western Switzerland-Hepia, Hes-So Geneva, 1202 Geneva, Switzerland; irena.milosevic@hesge.ch
* Correspondence: marijanamionic@gmail.com; Tel.: +41-7623-81669

Abstract: Poly(methyl methacrylate) (PMMA), widely used in dentistry, is unfortunately a suitable substrate for *Candida* (C.) *albicans* colonization and biofilm formation. The key step for biofilm formation is *C. albicans* ability to transit from yeast to hypha (filamentation). Since oleic acid (OA), a natural compound, prevents filamentation, we modified PMMA with OA aiming the antifungal PMMA_OA materials. Physico-chemical properties of the novel PMMA_OA composites obtained by incorporation of 3%, 6%, 9%, and 12% OA into PMMA were characterized by Fourier-transform infrared spectroscopy and water contact angle measurement. To test antifungal activity, PMMA_OA composites were incubated with *C. albicans* and the metabolic activity of both biofilm and planktonic cells was measured with a XTT test, 0 and 6 days after composites preparation. The effect of OA on *C. albicans* morphology was observed after 24 h and 48 h incubation in agar loaded with 0.0125% and 0.4% OA. The results show that increase of OA significantly decreased water contact angle. Metabolic activity of both biofilm and planktonic cells were significantly decreased in the both time points. Therefore, modification of PMMA with OA is a promising strategy to reduce *C. albicans* biofilm formation on denture.

Keywords: oleic acid; PMMA; *C. albicans*; filamentation; biofilm; antimicrobial surface

Citation: Petrović, M.; Randjelović, M.; Igić, M.; Randjelović, M.; Arsić Arsenijević, V.; Mionić Ebersold, M.; Otašević, S.; Milošević, I. Poly(methyl methacrylate) with Oleic Acid as an Efficient *Candida albicans* Biofilm Repellent. *Materials* **2022**, *15*, 3750. https://doi.org/10.3390/ma15113750

Academic Editor: Grzegorz Chladek

Received: 1 May 2022
Accepted: 20 May 2022
Published: 24 May 2022

Publisher's Note: MDPI stays neutral with regard to jurisdictional claims in published maps and institutional affiliations.

Copyright: © 2022 by the authors. Licensee MDPI, Basel, Switzerland. This article is an open access article distributed under the terms and conditions of the Creative Commons Attribution (CC BY) license (https://creativecommons.org/licenses/by/4.0/).

1. Introduction

Candida species cause oral candidiasis, the most common fungal infection in the oral cavity, with a high incidence among diabetic, cancer, oral-prosthetic, and immunosuppressed patients and patients on long-term treatment with antibiotic and corticosteroid therapy [1–3]. Especially in the oral cavity of the denture wearer, *Candida* easily colonizes the inner surface of the denture, which is typically made of poly(methyl methacrylate) (PMMA) [4]. This frequently used material is hydrophobic and has a relatively rough surface, facilitating the *Candida* biofilm accumulation [5,6]. However, when *Candida* is in the form of biofilm, it is difficult to treat and remove it since the fungus is encased in an extracellular matrix which protects it from penetration of the antimicrobial agents [7–10]. Thus, the formed *Candida* biofilm on a denture surface acts as a source of infection, continuously reinfecting oral mucosa, [11,12] leading to the development of *Candida*-associated denture stomatitis (CADS) [13–18]. Among *Candida* spp., *Candida albicans* is the most often isolated species that cause infection. However, *C. albicans* biofilm formed is difficult to treat with antifungal drugs. In fact, such treatment is usually unsuccessful due to the persistence of the infection, as a consequence of the biofilm formed on the denture surface, but also due to the fact that the resistance of *C. albicans* within the biofilm [19].

Hence, novel strategies are required to control biofilm formation by pathogens. An alternative approach can be a modification of the current denture materials with antimicrobial compounds, which could prevent *C. albicans* adhesion or biofilm formation on the denture surface such as chlorhexidine, fluconazole, amphotericin B, nystatin, or silver nanoparticles [13,14,17,18,20]. *C. albicans* is often resistant to conventional antimicrobial therapy, which compromises using antimicrobial drugs as fillers. The limiting factor in the use of silver nanoparticles is their cytotoxicity, and their possible releasing from the denture due to the pressure during mastication [21,22]. Therefore, as an antimicrobial agents could be considered molecules which are naturally abundant, such as plant-derived ones. Research showed that incorporation of undecylenic acid, natural compound into PMMA provide antifungal properties of modified PMMA. However, undecylenic acid can be cytotoxic for human cells, in concentration which will completely inhibit biofilm formation on the denture [23]. Interestingly, naturally occurring oleic acid (OA) has been recently reported for its antifungal activity [21,22].

Importantly, OA prevents *C.albicans* transition from yeast to hyphal form [21,22], a crucial step in biofilm formation and invasion of biomaterial.

To date, there has been no reporting carried out on the usage of OA in treating *Candida* infection especially associated with wearing denture., Furthermore, OA has not been yet used for surface modification of dental materials and there has no reported its cytotoxicity neither antimicrobial resistance. In order to overcome this challenge, and keeping in mind that OA inhibits the transition of *C. albicans* yeast to hyphal form and consequently biofilm formation, we proposed to incorporate OA into PMMA in order to obtain a surface that will reduce *C. albicans* biofilm formation. In this study, we have incorporated different amounts of OA into the PMMA matrix, with the aim of developing an antibiofilm composite material. The goal was to create material with OA molecule on the surface of modified PMMA, where OA preserves antifungal properties. Since OA is an unsaturated fatty acid, insoluble in water, it is a challenge to incorporate OA in polymers and create a composite in such way that OA still preserves its antifungal properties. Therefore, we studied the physicochemical properties of the composite's surface, characterized by Fourier- transform infrared spectroscopy (FTIR) and by measuring the water contact angle. Moreover, antimicrobial properties of the PMMA_OA composites were studied with the XTT test by measuring the percentage of metabolically active biofilm (attached to the surface) and planktonic (free-floating) *C. albicans* cells. To better understand how OA affects *C. albicans*, we studied the effect of OA on the morphology and growth of *C. albicans* cells in embedded conditions.

2. Materials and Methods

2.1. Sample Preparation

PMMA-OA composites were made by physical incorporation of suitable amounts of OA into a mixture of cold polymerized acrylic resin-PMMA and MMA (Triplex Cold, Ivoclar Vivadent, Liechtenstein), in a ratio 13 g PMMA with 10 mL MMA, according to the manufacture instruction [4]. Samples of composites containing 0, 3, 6, 9, and 12 wt% OA were made as solid discs (Ø 20 mm) in Teflon molds for all experiments except for the 2,3-bis(2methoxy-4-nitro-5-sulfophenyl)-5-[(phenylamino)carbonyl]-2H-tetrazolium hydroxide (XTT) test. For the XTT test, composites with 0, 3, 6, 9, and 12wt% OA were made in 24-well tissue culture plates (Falcon 353047). Before the experiment, composites were sterilized under a UV-C lamp for 15 min.

2.2. Physico-Chemicals Characterization of PMMA_OA

2.2.1. Fourier Transform Infrared Spectroscopy (FTIR)

Chemical characterization of the surface of the PMMA_OA, PMMA, and OA, was carried out by Fourier- Transform Infrared Spectroscopy, FTIR (Spectrum One spectrometer (series: 69288, Perkin Elmer, Schwerzenbach, Switzerland). All spectra were recorded from 4000–400 cm^{-1}, with 64 scans.

2.2.2. Contact Angle Measurements

The surface wettability of the PMMA_OA composites was tested by sessile drop method by contact angle device (EasyDrop Standard, Krüss, Hamburg, Germany, equipped with a monochrome interline CCD camera) using 20 µL of distilled water at room temperature 21 ± 1 °C. Water was dropped on the PMMA_OA composites' surface using a microliter syringe (Hamilton Typ 1750 TLL). The results were given as an average value of a minimum 3 measurements and ±standard deviation (SD).

2.3. Antifungal Characterization of Composites

Candida Strain and Culture Conditions.

This study used *C. albicans* ATCC90028 reference strain, obtained from the American Type Culture Collection (ATCC, Manassas, VA, USA); the stock was kept at –80 °C; the culture was maintained in Sabouraud 4% Glucose Agar (SGA; Sigma-Aldrich 84088, St. Louis, MO, USA).

2.3.1. Biofilm Formation on PMMA_OA Composites

C. albicans biofilm were formed on the PMMA_OA composites with modification of the previously described method [24–28]. 400 µL of the standardized inoculum 10^6 cell/mL in RPMI medium was added over the composites placed in the 24-well tissue culture plates, previously sterilized under a UV C lamp for 15 min. Plates were covered with a lid, sealed with parafilm, and incubated at 37 °C for 24 h–period of biofilm formation. After incubation, 200 µL of *C. albicans* suspension in RPMI medium incubated with PMMA_OA composites (biofilm supernatant) were transferred to the new 24-well tissue culture plates. The rest of the suspension was discarded, and composites in 24 well plate were left to dry before XTT assay.

2.3.2. XTT Test on PMMA_OA Composites and in Biofilm Supernatant

XTT test was performed to quantify the metabolic activity of *C. albicans* biofilm on the composites as well as metabolic activity of planktonic *C. albicans* cells in medium incubated for 24 h with composites in two time intervals: at the same day and 6 days after composites' preparation (T0 and T6, respectively). Briefly, XTT is prepared as a saturated solution at 0.5 g L^{-1} in sterile PBS (Dulbecco's Phosphate Buffered Saline (DPBS • 1000)). Menadione was added to achieve a final menadione concentration of 1 µM in XTT solution before the experiment. Further, 200 µL of XTT/menadione was added on the PMMA_OA composite placed in the well of 24 well plates, as well as in 200 µL of medium transferred into new well plate. Plates were wrapped in aluminum foil to prevent light penetration and incubated at 37 °C for 3 h. After incubation, 100 µL XTT-menadione solution was transferred into new 96- well tissue culture plates. Changes in color intensity were measured with a micro plate reader (TECAN Infinite M200, Tecan, Männedorf, Switzerland) at 490 nm.

2.4. Antifungal Susceptibility Test

2.4.1. Determination of Minimal Inhibitory Concentration (MIC)

The standard agar dilution method was modified to determine the lowest concentration of OA, which inhibits *C. albicans* growth [29]. Different concentrations of OA ranged from 0.0125, 0.025, 0.05, 0.1, 0.2, 0.4% were dispersed into 40 mL of Yeast Peptone Dextrose (YPD) agar in 50-mL Polypropylene flat tube. Per 1 mL of so-obtained agar containing a corresponding concentration of OA was added to a well of 24 well plate (4 wells per concen-

tration). Control included agar without OA. After agar was solidified, 300 μL 10⁶ cell/mL
C. albicans was added to its surface, and well plates were incubated at 37 °C for 24 h. After
incubation, 100 μL of *C. albicans* suspension was transferred into new 96- well tissue culture
plates and optical density was measured at 600 nm.

2.4.2. Embedded Filamentation Test (EFT)

In parallel with MIC, the effect of OA on the morphological appearance of *C. albicans* was tested in the embedded condition in agar. 50 μL 10⁶ cell/mL *C. albicans* were added in 5 mL of YPD agar containing the lowest and the highest concentration from the previous test (0.0125 wt% and 0.4 wt% of OA), in 50-mL Polystyrene Conical Tube (Sarstedt 352073). This suspension, mixed by inverting tube up and down, slowly and carefully to avoid making bubbles, was poured into the Petri dishes (Ø 30 mm). Plates were incubated at 37 °C for 24 h and 48 h. The morphological appearance of *C. albicans* species through the agar matrix was examined under an optical microscope (Nikon Eclipse Ti-E inverted microscope, Nikon Instruments Europe BV, Amsterdam, The Netherlands) after 24 h and 48 h.

2.4.3. Statistic

Data obtained from the contact angle measurement, XTT assay, and cytotoxicity test were given as means ± SD. The significance of differences between more than two groups was analyzed with ANOVA followed by a post hoc Tukey's test. Pearson's correlation coefficient (r) was used to analyze associations between continuous variables.

3. Results

3.1. Physico-Chemicals Characterization

3.1.1. FTIR (Chemical Characterization of Composite Surface)

FTIR was performed on the composite-discs (PMMA_OA) with different OA concentrations as well as on the native materials (PMMA and OA) (Figure 1). In the spectrum of pure PMMA, bands at 2985 cm^{-1} and 2964 cm^{-1} can be attributed to the –C-H bond stretching vibrations of the –CH3 and –CH2—groups, respectively [30]. The characteristic PMMA band corresponding to the stretching vibrations of the ester group appears at 1727 cm^{-1} (C=O) [31–33]. The bands observed in the PMMA spectra at 1437 cm^{-1} are assigned to the bending vibration of the C-H bond in (CH3) and the stretching of the ester group at 1147 cm^{-1} (C-O-C). The absorption band at 1243 cm^{-1} is due to the C-O-C stretching vibration. The bands at 1388 cm^{-1} and 754 cm^{-1} are due to vibration of the α methyl group. Characteristic absorption vibrations of PMMA can be observed at 1065 cm^{-1}, 987 cm^{-1}, 843 cm^{-1}.

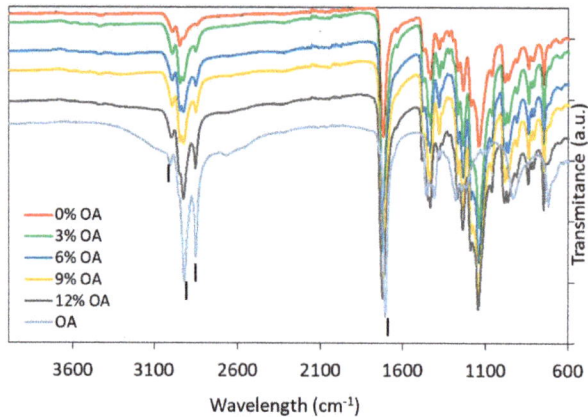

Figure 1. Fourier–transform infrared spectra of pure oleic acid, Poly(methyl methacrylate) and composites with 3%, 6%, 9%, and 12% (*w/w*) OA.

In the spectrum of native OA, the sharp bands at 2920 cm^{-1} and 2850 cm^{-1} can be assigned to C-H stretching in asymmetric and symmetric, respectively, of OA. Moreover, weak absorption bands at 3000 cm^{-1} may be attributed to =CH [34]. The bands that are related to the C=O and C–O stretch of COOH groups are found at 1710 cm^{-1} and 1285 cm^{-1}, respectively [35–37].

The O-H in-plane and out-of-plane bands appeared at 1450 cm^{-1} and 930 cm^{-1}, respectively. However, with the addition of OA in the PMMA, the peak at 930 cm^{-1} disappears.

Vibration peaks of –CH group of OA in region 2920 cm^{-1} and 2850 cm^{-1} overlaps with vibration of –CH group of PMMA spectra in the same region, but shifting of this group is noticed in the spectra of PMMA_OA composites and the peaks are more and more sharp with the increase of OA concentration between 2850 cm^{-1} and 2920 cm^{-1}. Since all of the characteristic peaks of OA are visible in the composites, it can be concluded that OA is present in its native form on the surface of the composite. Also, the interactions between OA and PMMA are physical without new chemical bonding between functional groups of OA and PMMA.

3.1.2. Water Contact Angle Measurement

Water contact angle on PMMA_OA composite-discs decreased from 67.6° for PMMA to 33.8° for PMMA_OA composites with 12% OA (Figure 2). Even for the composites with the lowest OA concentration, 3% OA in PMMA, the water contact angle was significantly decreased to 46.3° ($p < 0.01$), compared to 67.6° (0% OA in PMMA).

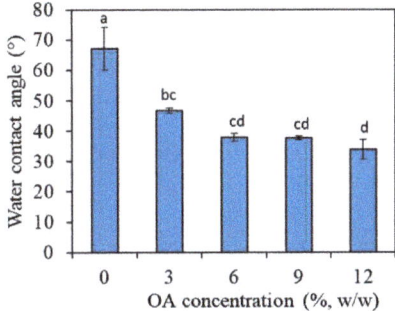

Figure 2. Water contact angle on the surface of composites with 0%, 3%, 6%, 9%, and 12% oleic acid. Bars indicate the mean values of water contact angles and vertical lines standard deviations (SD). Lowercase alphabetical letters above the columns show significant differences among the groups and compared to controls ($p < 0.001$, Tukey's test). Results are presented as a mean ± SD.

The results showed that water contact angle decreased with the increase of OA concentration, indicating that OA changes the surface properties of PMMA in hydrophilic PMMA_OA composites.

3.2. Antifungal Characterization

3.2.1. Antifungal Characterization of PMMA_OA Composites with XTT Test

The metabolic activity of both planktonic and biofilm *C. albicans* cells incubated with PMMA_OA composites compared to the metabolic activity of planktonic and biofilm cells in the control group (incubated with PMMA without OA) is given in Figure 3. In PMMA_OA composites, a biofilm formation was statistically and significantly decreased even at the lowest OA concentration, (3% OA in PMMA), compared to metabolically active biofilm cells on PMMA. Typically, the percentage of metabolically active biofilm cells was 25.80% on composites with 3% OA. The test was performed in the two different time points (T0 and T6) to study the antifungal surface properties of PMMA_OA composites with time. Moreover, on the composites with 3% OA, antifungal activity was lower (46.58% metabolically active biofilm *C. albicans* cells) in the T6 time point than in the T0 time interval

(25.80%). Biofilm formation in T6 decreased on all PMMA_OA composites with ≥6% OA having less than 15% of the metabolically active *C. albicans* cells compared to that on PMMA in T6. In the T0 time point, at the highest OA concentrations of 9% and 12% OA in PMMA, the percentage of metabolically active planktonic *C. albicans* cells was 21.36% and 15.46%, respectively. In the T6 time point, metabolic activity was higher compared to T0, with ~50% metabolically active planktonic *C. albicans* cells.

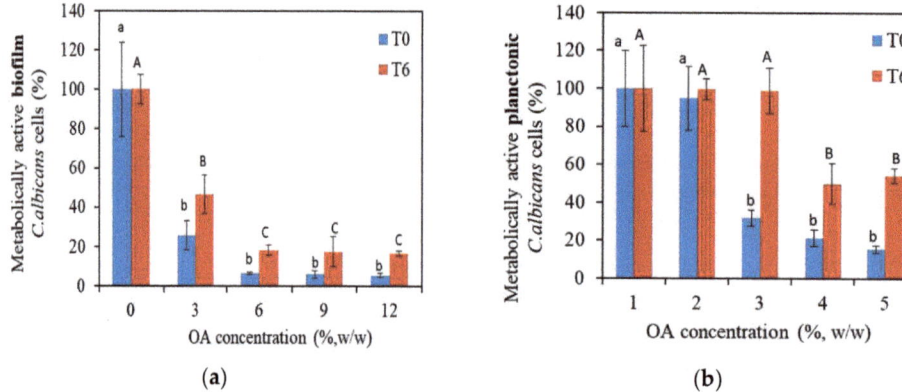

Figure 3. Metabolic activity of *C. albicans* cells on the surface of composites with 0, 3, 6, 9, and 12 wt% of oleic acid (OA) in Poly(Methyl Methacrylate), (**a**) in biofilm and (**b**) in the medium above composites in 2-time points: T0 and T6. Bars indicate the mean values and vertical lines standard deviations (SD). Different lowercase alphabetical letters above the columns show significant differences among the groups at T0 time ($p < 0.001$, Tukey's test). Different uppercase alphabetical letters above the columns show significant differences among the groups at T6 time ($p < 0.001$, Tukey's test). The results were presented as mean ± SD compared to control.

3.2.2. Antifungal Susceptibility Test

In order to test the effect of OA on *C. albicans* growth and morphology, the following two tests were employed:

Determination of Minimal Inhibitory Concentration (MIC)

The susceptibilities of planktonic *C. albicans* cells to OA were examined by measuring of optical density of *C. albicans* suspension on the agar surface containing different OA concentrations. The results show that OA did not inhibit *C. albicans* growth (Figure 4).

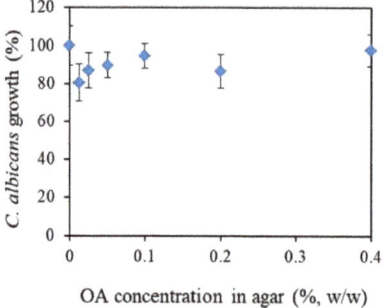

Figure 4. Oleic acid (OA) affection on the *C. albicans* growth 24 h of incubation in agar loaded with different oleic acid (OA) concentrations. Results are given as mean ± standard deviation percentages of A_{620} readings compared to control (0% OA) as a function of OA concentration.

Embedded Filamentation Test (EFT)

This test was carried out to study if OA affects a *C. albicans* cells morphology when embedded in YPD agar and incubated for 24 h and 48 h at 37 °C. The OA concentrations in agar were chosen according to the previous test (i.e., the highest concentration in this test corresponds to the highest concentration in the previous test, 0.4% OA). The morphology of *C. albicans* was assessed under an optical microscope. In embedded condition in YPD agar without OA, after 24 h incubation, *C. albicans* cells formed spindle-shaped colonies, including mainly the yeast cells, while rare hyphae and/or pseudohyphae could be observed peripheral on the colony, as it has been reported previously [23,38,39]. After 48 h of incubation, *C. albicans* cells formed the spindle-shaped yeast colonies with the formation of numerous hyphal branches and lateral yeasts derived from the colonies, which was in agreement with the previous study [23]. However, in agar with both tested OA concentrations, there was no hyphal formation neither after 24 h nor 48 h of incubation (Figure 5).

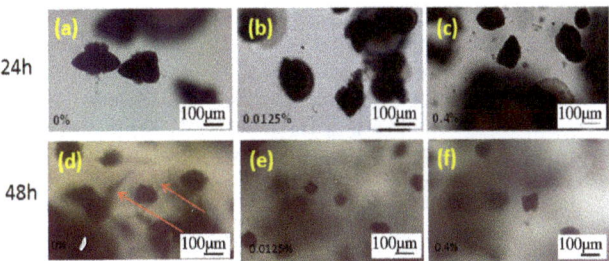

Figure 5. Optical micrographs of the *C. albicans* cells after 24 h (**a,b,c**) and 48 h (**d,e,f**) of incubation, embedded in agar without oleic acid (OA), (control, 0% OA, (**a,d**)), with 0.0125% OA (**b,e**) and 0.4% OA (**c,f**). Arrows mark spindle-shaped colonies with sporadic hyphae and/or pseudohyphae on the margins and lateral yeasts after 48 h incubation.

4. Discussion

The widespread application of PMMA in dentistry has driven biomaterial research to overcome challenges related to biofilm formation on medical devices made from PMMA, such as dentures [40]. PMMA dentures are that it is a suitable substrate for *C. albicans* adhesion and biofilm formation on dentures due to its hydrophobic nature and rough surface texture [19]. Moreover, it is challenging to treat *C. albicans* when it forms a biofilm due to its increasing resistance to conventional antifungal drugs.

Different approaches have been proposed to prevent *C. albicans* adhesion and biofilm formation on dentures by making them antifungal. This can be achieved by incorporating an antifungal compound such as chlorhexidine, fluconazole, amphotericin B, or nystatin into denture PMMA [13,14,17,18,20]. However, increased tolerance and resistance of *Candida* spp. to used antifungal drugs compromise successful treatment [7,25,41,42]. Thus, the advantages of using plant-derived compounds as therapeutic agents include fewer adverse effects, lower chances of antimicrobial resistance, and better efficiency in controlling biofilm-related infections [21].

In this context, the present study examined the inhibitory potential of OA incorporated into PMMA matrix on *C. albicans* biofilm formation on the surface of composites. Results showed that antibiofilm surface properties of PMMA_OA were due to the presence of OA on its surface (confirmed with FTIR analysis) and its inhibitory effect on *C. albicans* filamentation, a key step in biofilm formation.

The goal of this study were to analyze how OA affects surface properties of PMMA and how PMMA_OA composites surface will affect further *C.albicans* attachment and biofilm formation. OA is an unsaturated fatty acid insoluble in water. OA has a hydrophilic, polar head and a hydrophobic tail. OA's polar carboxyl (COOH) group is soluble, while the tail is insoluble in water. The possible reason for changing of PMMA_OA surface properties in hydrophilic one could be the orientation of OA molecule in the matrix, wherein the polar head would be orientated on the surface while the tail would be orientated out of the surface [23,43]. However, according to Garland et al. vapor-deposited OA on both polar (silica) and nonpolar (polystyrene) substrates resulted in the hydrophobic surface at high coverages of OA which suggests that the hydrocarbon chain on the OA molecule is facing away from the surface [43]. Hence, the challenge was to test if hydrophilic surface of PMMA_OA composites affects further *C. albicans* attachment and biofilm formation.

It has been reported that hydrophilic surfaces reduce fungal adhesion and the consequent biofilm formation to polymeric biomaterials [19,40,44]. In this regard, the correlation between the water contact angle of the PMMA_OA and the percentage of metabolically active biofilm cells attached to the composite surface was determined [44]. We showed a positive correlation (r = 0.987) between the value of water contact angle and the percentage of metabolically active biofilm cells on PMMA_OA composites. It means that the addition of OA into PMMA affects the changes of wetting properties of the surface (in hydrophilic one) and significantly decreases the percentage of metabolically active biofilm cells in all samples of PMMA modified with all tested OA concentrations. Therefore, the decreased biofilm formation on the PMMA_OA composite surface could be a consequence of the combined action of the antibiofilm surface properties of the PMMA_OA and the increase in surface hydrophilicity.

Furthermore, we showed that PMMA_OA composites decrease metabolic activity of planktonic cells, suggesting that OA may be released from composites and influences planktonic cells as well.

In contrast to our study that incorporation of OA into PMMA affects metabolic activity of both biofilm and planktonic cells, Muthamil et al. have recently reported the non-fungicidal effect of OA against *Candida* spp. by XTT assay. According to their study, incubation of *Candida* in spider broth in the presence of OA (at different concentrations 5, 10, 20, 40, 80, 160 and 320 µg mL^{-1}), resulted in a thinner biofilm formed on glass slides with reduced biomass and architecture of mature biofilm compared to the controls (in the absence of OA) [21].

Therefore, to understand the mode of action of OA when it is not being incorporated into PMMA, we studied the effect of OA on the growth and morphology of *C. albicans* cells in agar. In our study, OA was not dissolved in any non-aqueous solvent to avoid the possible influence of non-aqueous solvents on *C. albicans*, but it has been dispersed in agar. Thus, we studied the effect of, pure" OA on *C. albicans* cells after incubation on an agar surface containing different OA concentrations. Results showed that OA did not inhibit *C. albicans* growth, which was in agreement with Muthamil et al. (2020) [21]. However, Lee et al. (2020) have reported that OA inhibits the growth of *C. albicans*, but at higher concentration, such as >500 µg mL^{-1} [22].

In this study, the filamentation test showed that even the highest OA concentration, previously tested for MIC (0.4%), did not inhibit *C. albicans* growth, but it prevented hyphae formation (filamentation) after 24 h and 48 h of incubation. Similarly to reports of proved antibiofilm effects of some fatty acid at concentrations lower of their MICs, suppression of *C. albicans* biofilm formation occurs by inhibiting hyphal growth and cell aggregation [22]. Additionally, it has been shown that OA treatment could significantly reduce the extracellular polymeric (EPS) matrix's carbohydrates, lipids, and eDNA content of the EPS matrix [21], which protects biofilm cells from the host immune system and the antifungal agents. Given that OA changes the ergosterol content of *Candida* spp., it qualifies it as a more potent drug than standard antifungal agents [21].

Within the limitations of this in vitro study, the first results of here developed PMMA_OA composites focusing on its physico-chemicals characterization and antifungal properties have been reported. However, the morphological analyses of the samples with SEM should be performed in future research to provide more information how OA incorporation into PMMA affect surface properties of PMMA_OA composites. Furthermore, OA is non-toxic compound [45,46] and a potential of PMMA_OA composites for biomedical applications could be considered.

5. Conclusions

This study demonstrates, for the first time, incorporation of OA into PMMA and development of PMMA_OA composites with antibiofilm surface properties. This new PMMA_OA composite with ≥3% OA significantly reduces metabolic activity of biofilm cells even six days after PMMA_OA preparation. Moreover, OA present on the composites surface, results in increased hydrophilic surface properties of this developed composites. This study confirmed that OA prevents filamentation and, consequently, the early stage of *C. albicans* biofilm formation on PMMA_OA composites surface. Since OA is naturally occurring non-toxic molecule and has no antimicrobial resistance, it could be a promising agent for modifying dental material such as denture and preventing the *Candida* associated denture stomatitits. Within the limitations of this study, it can be concluded that PMMA_OA may be used as a dental polymer to reline inner surface of denture having the potential to prevent and to treat *Candida* associated infection in denture wearers. For this purpose, further research is required to evaluate additional biological and mechanical parameters of PMMA_OA for clinical applications.

Author Contributions: Conceptualization, M.M.E. and I.M.; methodology, M.M.E.; validation, M.M.E., V.A.A., I.M. and S.O.; formal analysis, M.P., M.R. (Marina Randjelović), M.I., M.R. (Milica Randjelović) and V.A.A.; investigation, M.P.; writing—original draft preparation, M.P. and M.R. (Milica Randjelović); writing—review and editing, M.M.E., S.O. and I.M.; visualization, M.M.E.; supervision, M.M.E. All authors have read and agreed to the published version of the manuscript.

Funding: This research was funded by the Swiss Government Excellence Scholarship No. 2014.0040/Serbia/OP; Swiss Confederation, Federal Department of Economic Affairs, Education and Research EAER.

Institutional Review Board Statement: Not applicable.

Informed Consent Statement: Not applicable.

Acknowledgments: All authors express gratitude to Heinrich Hofmann, Powder Technology Laboratory, Institute of Materials, Ecole Polytechnique Fédérale de Lausanne, Switzerland for his great support of this work and his hospitality. The authors acknowledge inSTI-HEPIA for financial support. The authors thank the Ministry of Education, Science and Technological Development of the Republic of Serbia (Project No. 451-03-68/2022-14/200113) and Science Fund of the Republic of Serbia, Program Ideas The project FungalCaseFinder No 7754282.

Conflicts of Interest: The authors declare no conflict of interest. The funders had no role in the design of the study; in the collection, analyses, or interpretation of data; in the writing of the manuscript, or in the decision to publish the results.

References

1. Beltrán-Partida, E.; Valdez-Salas, B.; Curiel-Álvarez, M.; Castillo-Uribe, S.; Escamilla, A.; Nedev, N. Enhanced Antifungal Activity by Disinfected Titanium Dioxide Nanotubes via Reduced Nano-Adhesion Bonds. *Mater. Sci. Eng. C* **2017**, *76*, 59–65. [CrossRef] [PubMed]
2. Nobile, C.J.; Johnson, A.D. Candida Albicans Biofilms and Human Disease. *Annu. Rev. Microbiol.* **2015**, *69*, 71–92. [CrossRef] [PubMed]
3. Tsui, C.; Kong, E.F.; Jabra-Rizk, M.A. Pathogenesis of Candida Albicans Biofilm. *Pathog. Dis.* **2016**, *74*, ftw018. [CrossRef]
4. Petrović, M.; Hofmann, H.; Mionić Ebersold, M. *PMMA-Oleic Acid Composites as Candida Biofilm Repellent*; eCM Meeting Abstracts 2017, Collection 3; SSB+RM; European Cells and Materials/ECM, Empa: St.Gallen, Switzerland, 2017; p. 51.
5. Cierech, M.; Szerszeń, M.; Wojnarowicz, J.; Łojkowski, W.; Kostrzewa-Janicka, J.; Mierzwińska-Nastalska, E. Preparation and Characterisation of Poly(Methyl Metacrylate)-Titanium Dioxide Nanocomposites for Denture Bases. *Polymers* **2020**, *12*, 2655. [CrossRef]
6. Lee, M.-J.; Kim, M.-J.; Oh, S.-H.; Kwon, J.-S. Novel Dental Poly (Methyl Methacrylate) Containing Phytoncide for Antifungal Effect and Inhibition of Oral Multispecies Biofilm. *Materials* **2020**, *13*, 371. [CrossRef]
7. Davies, D. Understanding Biofilm Resistance to Antibacterial Agents. *Nat. Rev. Drug Discov.* **2003**, *2*, 114–122. [CrossRef] [PubMed]
8. Nobile, C.J.; Fox, E.P.; Nett, J.E.; Sorrells, T.R.; Mitrovich, Q.M.; Hernday, A.D.; Tuch, B.B.; Andes, D.R.; Johnson, A.D. A Recently Evolved Transcriptional Network Controls Biofilm Development in Candida Albicans. *Cell* **2012**, *148*, 126–138. [CrossRef] [PubMed]

9. Tobudic, S.; Kratzer, C.; Lassnigg, A.; Presterl, E. Antifungal Susceptibility of Candida Albicans in Biofilms. *Mycoses* **2012**, *55*, 199–204. [CrossRef]
10. Shehabeldine, A.; El-Hamshary, H.; Hasanin, M.; El-Faham, A.; Al-Sahly, M. Enhancing the Antifungal Activity of Griseofulvin by Incorporation a Green Biopolymer-Based Nanocomposite. *Polymers* **2021**, *13*, 542. [CrossRef]
11. Jose, A.; Coco, B.J.; Milligan, S.; Young, B.; Lappin, D.F.; Bagg, J.; Murray, C.; Ramage, G. Reducing the Incidence of Denture Stomatitis: Are Denture Cleansers Sufficient? *J. Prosthodont.* **2010**, *19*, 252–257. [CrossRef]
12. Chladek, G.; Pakieła, K.; Pakieła, W.; Żmudzki, J.; Adamiak, M.; Krawczyk, C. Effect of Antibacterial Silver-Releasing Filler on the Physicochemical Properties of Poly(Methyl Methacrylate) Denture Base Material. *Materials* **2019**, *12*, 4146. [CrossRef] [PubMed]
13. Da Silva, P.M.; Acosta, E.J.; de Rezende Pinto, L.; Graeff, M.; Spolidorio, D.M.; Almeida, R.S.; Porto, V.C. Microscopical Analysis of Candida Albicans Biofilms on Heat-Polymerised Acrylic Resin after Chlorhexidine Gluconate and Sodium Hypochlorite Treatments. *Mycoses* **2011**, *54*, e712–e717. [CrossRef] [PubMed]
14. Monteiro, D.R.; Silva, S.; Negri, M.; Gorup, L.F.; De Camargo, E.R.; Oliveira, R.; Barbosa, D.D.; Henriques, M. Silver Nanoparticles: Influence of Stabilizing Agent and Diameter on Antifungal Activity against Candida Albicans and Candida Glabrata Biofilms. *Lett. Appl. Microbiol.* **2012**, *54*, 383–391. [CrossRef] [PubMed]
15. Salerno, C.; Pascale, M.; Contaldo, M.; Esposito, V.; Busciolano, M.; Milillo, L.; Guida, A.; Petruzzi, M.; Serpico, R. Candida-Associated Denture Stomatitis. *Med. Oral Patol. Oral Cir. Bucal* **2011**, *16*, e139–e143. [CrossRef]
16. Sun, X.; Cao, Z.; Yeh, C.-K.; Sun, Y. Antifungal Activity, Biofilm-Controlling Effect, and Biocompatibility of Poly(N-Vinyl-2-Pyrrolidinone)-Grafted Denture Materials. *Colloids Surf. B Biointerfaces* **2013**, *110*, 96–104. [CrossRef]
17. Wady, A.F.; Machado, A.L.; Zucolotto, V.; Zamperini, C.A.; Berni, E.; Vergani, C.E. Evaluation of Candida Albicans Adhesion and Biofilm Formation on a Denture Base Acrylic Resin Containing Silver Nanoparticles. *J. Appl. Microbiol.* **2012**, *112*, 1163–1172. [CrossRef]
18. Wen, J.; Jiang, F.; Yeh, C.-K.; Sun, Y. Controlling Fungal Biofilms with Functional Drug Delivery Denture Biomaterials. *Colloids Surf. B Biointerfaces* **2016**, *140*, 19–27. [CrossRef]
19. Bajunaid, S.O.; Baras, B.H.; Balhaddad, A.A.; Weir, M.D.; Xu, H.H. Antibiofilm and Protein-Repellent Polymethylmethacrylate Denture Base Acrylic Resin for Treatment of Denture Stomatitis. *Materials* **2021**, *14*, 1067. [CrossRef]
20. Salim, N.; Silikas, N.; Satterthwaite, J.D.; Moore, C.; Ramage, G.; Rautemaa, R. Chlorhexidine-Impregnated PEM/THFM Polymer Exhibits Superior Activity to Fluconazole-Impregnated Polymer against Candida Albicans Biofilm Formation. *Int. J. Antimicrob. Agents* **2013**, *41*, 193–196. [CrossRef]
21. Muthamil, S.; Prasath, K.G.; Priya, A.; Precilla, P.; Pandian, S.K. Global Proteomic Analysis Deciphers the Mechanism of Action of Plant Derived Oleic Acid against Candida Albicans Virulence and Biofilm Formation. *Sci. Rep.* **2020**, *10*, 5113. [CrossRef]
22. Lee, J.-H.; Kim, Y.-G.; Khadke, S.K.; Lee, J. Antibiofilm and Antifungal Activities of Medium-Chain Fatty Acids against Candida Albicans via Mimicking of the Quorum-Sensing Molecule Farnesol. *Microb. Biotechnol.* **2021**, *14*, 1353–1366. [CrossRef] [PubMed]
23. Petrović, M.; Bonvin, D.; Hofmann, H.; Mionić Ebersold, M. Fungicidal PMMA-Undecylenic Acid Composites. *Int. J. Mol. Sci.* **2018**, *19*, 184. [CrossRef] [PubMed]
24. Chandra, J.; Mukherjee, P.K.; Leidich, S.D.; Faddoul, F.F.; Hoyer, L.L.; Douglas, L.J.; Ghannoum, M.A. Antifungal Resistance of Candidal Biofilms Formed on Denture Acrylic in Vitro. *J. Dent. Res.* **2001**, *80*, 903–908. [CrossRef] [PubMed]
25. Chandra, J.; Kuhn, D.M.; Mukherjee, P.K.; Hoyer, L.L.; McCormick, T.; Ghannoum, M.A. Biofilm Formation by the Fungal Pathogen Candida Albicans: Development, Architecture, and Drug Resistance. *J. Bacteriol.* **2001**, *183*, 5385–5394. [CrossRef]
26. Pierce, C.G.; Uppuluri, P.; Tristan, A.R.; Wormley, F.L.; Mowat, E.; Ramage, G.; Lopez-Ribot, J.L. A Simple and Reproducible 96-Well Plate-Based Method for the Formation of Fungal Biofilms and Its Application to Antifungal Susceptibility Testing. *Nat. Protoc.* **2008**, *3*, 1494–1500. [CrossRef]
27. Ferreira, J.A.G.; Carr, J.H.; Starling, C.E.F.; Resende, M.A.d.; Donlan, R.M. Biofilm Formation and Effect of Caspofungin on Biofilm Structure of Candida Species Bloodstream Isolates. *Antimicrob. Agents Chemother.* **2009**, *53*, 4377–4384. [CrossRef]
28. Ahmad, A.; Khan, A.; Akhtar, F.; Yousuf, S.; Xess, I.; Khan, L.A.; Manzoor, N. Fungicidal Activity of Thymol and Carvacrol by Disrupting Ergosterol Biosynthesis and Membrane Integrity against Candida. *Eur. J. Clin. Microbiol. Infect. Dis.* **2011**, *30*, 41–50. [CrossRef]
29. Wiegand, I.; Hilpert, K.; Hancock, R.E. Agar and Broth Dilution Methods to Determine the Minimal Inhibitory Concentration (MIC) of Antimicrobial Substances. *Nat. Protoc.* **2008**, *3*, 163–175. [CrossRef]
30. Duan, G.; Zhang, C.; Li, A.; Yang, X.; Lu, L.; Wang, X. Preparation and Characterization of Mesoporous Zirconia Made by Using a Poly (Methyl Methacrylate) Template. *Nanoscale Res. Lett.* **2008**, *3*, 118–122. [CrossRef]
31. Totu, E.E.; Nechifor, A.C.; Nechifor, G.; Aboul-Enein, H.Y.; Cristache, C.M. Poly(Methyl Methacrylate) with TiO$_2$ Nanoparticles Inclusion for Stereolitographic Complete Denture Manufacturing—The Fututre in Dental Care for Elderly Edentulous Patients? *J. Dent.* **2017**, *59*, 68–77. [CrossRef]
32. Moreno, K.; García-Miranda, J.; Hernández-Navarro, C.; Ruiz-Guillén, F.; Aguilera-Camacho, L.; Lesso, R.; Arizmendi-Morquecho, A. Preparation and Performance Evaluation of PMMA/HA Nanocomposite as Bulk Material. *J. Compos. Mater.* **2015**, *49*, 1345–1353. [CrossRef]
33. Elshereksi, N.W.; Ghazali, M.J.; Muchtar, A.; Azhari, C.H. Studies on the Effects of Titanate and Silane Coupling Agents on the Performance of Poly (Methyl Methacrylate)/Barium Titanate Denture Base Nanocomposites. *J. Dent.* **2017**, *56*, 121–132. [CrossRef]

34. Hong, J.; Yamaoka-Koseki, S.; Yasumoto, K. Determination of Palmitic Acid, Oleic Acid and Linoleic Acid by Near-Infrared Transflectance Spectroscopy in Edible Oils. *Food Sci. Technol. Int.* **1996**, *2*, 146–149. [CrossRef]
35. Lee, D.H.; Condrate, R.A. FTIR Spectral Characterization of Thin Film Coatings of Oleic Acid on Glasses: I. Coatings on Glasses from Ethyl Alcohol. *J. Mater. Sci.* **1999**, *34*, 139–146. [CrossRef]
36. Wu, N.; Fu, L.; Su, M.; Aslam, M.; Wong, K.C.; Dravid, V.P. Interaction of Fatty Acid Monolayers with Cobalt Nanoparticles. *Nano Lett.* **2004**, *4*, 383–386. [CrossRef]
37. Nájera, J.J.; Horn, A.B. Infrared Spectroscopic Study of the Effect of Oleic Acid on the Deliquescence Behaviour of Ammonium Sulfate Aerosol Particles. *Phys. Chem. Chem. Phys. PCCP* **2009**, *11*, 483–494. [CrossRef]
38. Lindsay, A.K.; Deveau, A.; Piispanen, A.E.; Hogan, D.A. Farnesol and Cyclic AMP Signaling Effects on the Hypha-to-Yeast Transition in Candida Albicans. *Eukaryot. Cell* **2012**, *11*, 1219–1225. [CrossRef]
39. Pendrak, M.L.; Roberts, D.D. Hbr1 Activates and Represses Hyphal Growth in Candida Albicans and Regulates Fungal Morphogenesis under Embedded Conditions. *PLoS ONE* **2015**, *10*, e0126919. [CrossRef]
40. Mangal, U.; Kim, J.-Y.; Seo, J.-Y.; Kwon, J.-S.; Choi, S.-H. Novel Poly(Methyl Methacrylate) Containing Nanodiamond to Improve the Mechanical Properties and Fungal Resistance. *Materials* **2019**, *12*, 3438. [CrossRef]
41. Bennett, J.E.; Izumikawa, K.; Marr, K.A. Mechanism of Increased Fluconazole Resistance in Candida Glabrata during Prophylaxis. *Antimicrob. Agents Chemother.* **2004**, *48*, 1773–1777. [CrossRef]
42. Mathé, L.; Dijck, P.V. Recent Insights into Candida Albicans Biofilm Resistance Mechanisms. *Curr. Genet.* **2013**, *59*, 251–264. [CrossRef]
43. Garland, E.R.; Rosen, E.P.; Clarke, L.I.; Baer, T. Structure of Submonolayer Oleic Acid Coverages on Inorganic Aerosol Particles: Evidence of Island Formation. *Phys. Chem. Chem. Phys.* **2008**, *10*, 3156–3161. [CrossRef] [PubMed]
44. Chandra, J.; Patel, J.D.; Li, J.; Zhou, G.; Mukherjee, P.K.; McCormick, T.S.; Anderson, J.M.; Ghannoum, M.A. Modification of Surface Properties of Biomaterials Influences the Ability of Candida Albicans to Form Biofilms. *Appl. Environ. Microbiol.* **2005**, *71*, 8795–8801. [CrossRef] [PubMed]
45. Carrillo Pérez, C.; Cavia Camarero, M.D.; Alonso de la Torre, S. Role of Oleic Acid in Immune System; Mechanism of Action: A Review. *Nutr. Hosp.* **2012**, *27*, 978–990. [CrossRef]
46. Sales-Campos, H.; Reis de Souza, P.; Crema Peghini, B.; Santana da Silva, J.; Ribeiro Cardoso, C. An Overview of the Modulatory Effects of Oleic Acid in Health and Disease. *Mini Rev. Med. Chem.* **2013**, *13*, 201–210. [CrossRef]

Article

Effect of Antibacterial Silver-Releasing Filler on the Physicochemical Properties of Poly(Methyl Methacrylate) Denture Base Material

Grzegorz Chladek [1,*], Katarzyna Pakieła [1], Wojciech Pakieła [1], Jarosław Żmudzki [1], Marcin Adamiak [1] and Cezary Krawczyk [2]

1. Faculty of Mechanical Engineering, Silesian University of Technology, ul. Konarskiego 18a, 44-100 Gliwice, Poland; katarzyna.pakiela@polsl.pl (K.P.); wojciech.pakiela@polsl.pl (W.P.); Jaroslaw.Zmudzki@polsl.pl (J.Ż.); marcin.adamiak@polsl.pl (M.A.)
2. Department of Dental Technology, Medical College, ul. 3 Maja 63, 41-800 Zabrze, Poland; crkrawczyk@interia.pl
* Correspondence: grzegorz.chladek@polsl.pl; Tel.: +48-32-237-29-07

Received: 23 October 2019; Accepted: 9 December 2019; Published: 11 December 2019

Abstract: Colonization of polymeric dental prosthetic materials by yeast-like fungi and the association of these microorganisms with complications occurring during prosthetic treatment are important clinical problems. In previously presented research, submicron inorganic particles of silver sodium hydrogen zirconium phosphate (S–P) were introduced into poly(methyl methacrylate) (PMMA) denture base material which allowed for obtaining the antimicrobial effect during a 90 day experiment. The aim of the present study was to investigate the flexural strength, impact strength, hardness, wear resistance, sorption, and solubility during three months of storage in distilled water. With increasing S–P concentration after 2 days of conditioning in distilled water, reduced values of flexural strength (107–72 MPa), impact strength (18.4–5.5 MPa) as well as enhanced solubility (0.95–1.49 µg/mm^3) were registered, but they were at acceptable levels, and the sorption was stable. Favorable changes included increased hardness (198–238 MPa), flexural modulus (2.9–3.3 GPa), and decreased volume loss during wear test (2.9–0.2 mm^3). The percentage changes of the analyzed properties during the 90 days of storage in distilled water were similar for all materials.

Keywords: polymethyl methacrylate; denture; antibacterial properties; silver; mechanical properties; sorption; solubility; wear resistance

1. Introduction

In the United States, only 34% of adults aged 40–64 have retained all of their permanent teeth, and nearly 19% of patients aged 65 and over suffer from edentulism [1]. Similarly, 16.3% of Indians and almost 22% of Mexicans aged 50 and above are edentulous [2]. These data show how common the problem of missing teeth is in modern society, regardless of race or region of the world. Many patients are users of conventional removable dentures, mainly made of polymethyl methacrylate (PMMA) due to the fact that they allow for obtaining products at an affordable price, in comparison to, for example, implant-fixed dentures [3]. In addition, they are characterized by acceptable quality in terms of improving oral function, enhancing phonetics, facilitating social engagement, and aesthetics [4].

Colonization of polymeric dental prosthetic materials by *Candida* species and the association of these microorganisms with complications, such as denture stomatitis occurring during prosthetic treatment, is an important clinical problem which has been widely described in the literature [5–7]. Fungi and bacteria occurring in the mouth are the sources of many ailments and systemic diseases including heart, circulatory system, kidney, stomach or esophagus problems [8,9]; therefore, poor microbiological status of dentures can contribute to the deterioration of overall health [10]. Humid microenvironment under prostheses and decreased possibilities of mucosal self-cleaning by saliva promote growth of microorganisms [6,11], and only a few dozen minutes after cleaning, the denture surface begins to be re-colonized by bacteria and fungi [12,13].

In order to reduce the indicated problems, various strategies are proposed. Antifungal drugs, such as nystatin or amphotericin B, can eliminate pathogenic microorganisms from the surface of tissues [14] and have also been added experimentally to PMMA material [15]. However, *Candida albicans* show increasing resistance during treatment of oral fungal infections [16].

Numerous studies have proved the varied effectiveness of removing microorganisms from the surface of prosthetic materials, for example, by using chlorhexidine gluconate, guanidine solution, peroxides, irradiation microwaves, or buy brushing with toothpaste, but these methods result in the loss of various functional properties, including increased roughness [17–22], which can facilitate the recolonization of prosthetic materials [23]. It should be emphasized that there are studies questioning the possibility of fully effective removal of microorganisms from the denture using mechanical or chemical cleaning methods [24], suggesting the possibility of penetration of *C. albicans* into the interior of acrylic materials [25–27] which indicates limited disinfection possibilities. Due to the problems associated with colonization by *Candida* species, investigations related with the development of new materials are being conducted in two directions: the introduction of additional monomers with antimicrobial properties and the manufacturing of composites by introducing the fillers with antimicrobial properties [28]. Such materials would be characterized by both increased resistance to microbial colonization and support for the treatment of, for example, *Candida*-infected mucosa. Antimicrobial efficacy in vitro has been confirmed so far in laboratory experiments conducted with PMMA denture base materials modified with numerous metal and metal oxides nanoparticles such as ZrO_2 [29], TiO_2 [30,31], ZnO [32], platinum [33], silver [34,35] or silver microparticles [36]. Silver nanoparticles have particularly strong antimicrobial properties; however, studies have simultaneously shown that the introduction of this type of additive to prosthetic materials causes an intense brown color of PMMA resin due to the plasmon effect [35] which is unacceptable for aesthetic reasons. Despite these problems, materials containing silver are still considered an attractive antimicrobial additive; therefore, in the previously presented research, it was proposed to introduce submicron inorganic particles of silver sodium hydrogen zirconium phosphate (S–P) as an antimicrobial additive to PMMA [37]. This white filler does not cause the initial dark coloring of modified materials.

In the published first part of our investigations, the morphology and antimicrobial properties of the developed composites were investigated [37]. Most of the experimental materials presented efficacy against *C. albicans*, even when samples were stored in distilled water for a three-month period. However, the morphologies of the composites were not homogenous. This indicated a risk of unfavorable changes in physicochemical properties.

Appropriate mechanical properties and their stability are particularly important for the functioning of complete and partial dentures. Denture-based materials must show strength, ensuring long-term functioning of the prosthesis loaded with functional and parafunctional masticatory forces [38]. Even 68% of the mentioned types of prostheses are damaged during the first few years after manufacturing which clearly shows the scale of this clinical problem and the importance of mechanical properties [39]. The denture fracture may result from flexural forces due to, for example, the improper fabrication, poor fit or lack of balanced occlusion [40]. Moreover, most PMMA partial or complete dentures are removable; thus, their resistance to shock-induced fractures, represented by impact strength, is no less important due to the possibility of their falling or being damaged during the action of violent forces caused by other events [41]. Between 39.5% and 56% of fractures occur as a result of the fall of dentures [41,42]. Other material properties that affect the durability of PMMA dentures are their hardness and wear resistance which determine the surface conditions of prosthetic materials [43,44]. The significance of the abovementioned properties is demonstrated by a significant number of studies on their changes in the context of various aspects of the functioning of materials such as the use of cleaning agents or the consumption of hot/cold foods and drinks (thermocycling) [22,45]. Considering that these materials function in an environment with 100% humidity, their behavior under these conditions is equally important, not only in the context of mechanical/tribological properties, but also due to the fact of water absorption and release of material components into the environment. Therefore, the aim of this paper was to investigate the effect of the introduction of S–P on the mechanical properties, wear resistance, and the sorption and solubility of the modified PMMA denture base material during three months of storage in distilled water. Our hypothesis was that composites filled with silver sodium hydrogen zirconium phosphate would show physicochemical properties relevant to the application being considered.

2. Materials and Methods

2.1. Material and Sample Preparation

2.1.1. Material Preparation

The method of the composites' preparation was described in a previously published work [37]. The PMMA heat-cured denture base resin, Meliodent Heat Cure (Heraeus Kulzer, Hanau, Germany), was used as a matrix and silver sodium hydrogen zirconium phosphate (Milliken Chemical, Spartanburg, SC, USA) as filler. To exclude sedimentation during material storage, the filler was added only to the pre-polymerized "powder" component of the *"powder–liquid"* system. The components were mixed using a planetary ball mill (Pulverisette 5, Fritsch, Idar–Oberstein, Germany) with 50 ZrO_2 balls with a diameter of 10 mm. A milling time of 5 min with a frequency of rotation of 400 rpm was used. The mass of modified components (PMMA with S-P) obtained during one milling process was from 10–10.4 g. During the materials' preparation, we used the AS 110/C/2 analytic scale (Radwag, Radom, Poland) with a measurement accuracy of 0.1 mg, but the real mass of the S–P mixed with PMMA powder was determined by the possibility of manual dosing, and the error did not exceed 0.01%. Moreover, the concentrations of S–P in liquid, listed in Table 1, are only theoretical. Dosing of the liquid component was possible with an accuracy of one drop (0.01–0.015 g), so the error mainly depended on the mass of the used liquid necessary for the particular samples' preparation in flasks and was not higher than 0.05%. However, during cross-linking, many necessary activities must be performed (mixing, packing into flask, etc.) and some mass of the monomer evaporates and, in fact, those considerations have only limited practical value. The list of the obtained materials with filler concentrations is presented in Table 1. It should be noted that the filler was mixed with PMMA powder, but after polymerization of the samples, the filler was present in areas of materials formed during polymerization from the "liquid" component (mainly methyl methacrylate). Samples were polymerized in accordance with the instructions of the resin manufacturer.

Table 1. Mass concentrations of the antimicrobial filler in relation to the individual components of the "powder–liquid" system.

Material Code	Concentration of Filler after Milling with Powder, %	Concentration of Filler in the Weight of The Liquid, %
A0 (Control)	0	0
A1	0.25	0.7
A2	0.5	1.3
A3	1	2.6
A4	2	5.1
A5	4	9.9
A6	8	18.6

2.1.2. Sample Preparation

The samples for most tests (i.e., flexural properties, impact strength, hardness, wear resistance) were prepared using a standard flasking technique used in dental prosthetics to maintain the typical polymerization conditions provided by the manufacturer of the PMMA resin. The most important stages of the procedure are related with mold preparation and are shown in Figure 1. First, the prepared models of the samples were filled with dental stone (gypsum type III, Stodent III, Zhermack, Badia Polesine, Italy) (Figure 1a–b) to create the mold. After setting of the dental stone, the mold was trimmed to obtain the required external dimension, determined by the internal dimensions of the flask (Figure 1c). Next, each mold was placed in the first part of flask and mounted using model plaster (gypsum type II, Stodent II, Zhermack, Badia Polesine, Italy) (Figure 1d). After setting, the dental plaster separating medium (Isofix 2000, Renfert GmbH, Hilzingen, Germany) was applied. Next, the second part of the flask was mounted and dental stone was poured to about 1/3 of its height. After setting, the dental plaster was used to fill the flask and flask was closed with a cover. After setting of the dental plaster, the models of the samples were removed and the molds (Figure 1e) were covered with separating media (Izolit SL, Chema-Elektromet, Rzeszów, Poland) before PMMA material and composite packing (Figure 1f). The samples of all materials were cured in accordance with the instructions of the manufacturer of the used resin.

Figure 1. The stages of mold manufacturing for sample preparation: models of the samples before (**a**) and after (**b**) filling with dental stone; mold with models of samples (**c**); mold mounted in the first part of flask (**d**); ready mold in the flask (**e**), and "packing" of the material during the curing process (**f**).

After curing, the samples were taken out of the mold, the excess of material was cut off, and the specimens were then wet-ground (Labo-Pol25, Struers, Willich, Germany) with P220 and, finally, P500-grit abrasive paper and thoroughly rinsed in water. Samples were divided into five groups (storing conditions), and one sample per group was prepared from one powder + liquid mixture. The groups of samples were stored in distilled water at 37 ± 1 °C for 2 days ± 2 h, 7 days ± 2 h, 30 days ± 2 h, 60 days ± 2 h, and 90 days ± 2 h. Potential detailed procedures for sample preparation and sample dimensions for specific tests are provided in the descriptions of the test methods.

Samples for sorption and solubility tests were cured in stainless steel molds in accordance with the ISO standard [46]. The use of this technique was justified by the restrictive requirements of the sample dimensions and sample purity (elimination of separating medium and gypsum). A 23 μm thick polyester film (DuPont Teijin Films, Chester, USA) was used as a spacer to prevent sticking of parts of molds/adhesion of materials during the curing process [46].

2.2. Methods

2.2.1. Flexural Properties

A three-point bending test was carried out using a universal testing machine (Zwick Z020, Zwick GmbH & Com, Ulm, Germany) based on the ISO 20795-1:2013-07 standard [46] with some modifications, i.e., additional storing times in the water were used. Specimens measuring 65 mm × 10 mm × 3.3 mm were prepared using a method described in Section 2.1.2. Twenty-five samples were prepared from each material. After conditioning, the specimen was removed from the water, placed on supports, and the test was performed at a cross-head speed of 5 mm/min. The distance between the supports was 50 mm. Flexural strength and flexural modulus were calculated according to the equations:

$$\sigma = \frac{3Fl}{2bh^2} \tag{1}$$

$$E = \frac{F_1 l^3}{4bh^3 d} \tag{2}$$

where σ is the flexural strength (MPa); E is the flexural modulus (GPa); l is the distance among the supports (mm); b and h are the specimen's width and height (mm); F is the maximal force (N); F_1 is the load at a chosen point at the elastic region of the stress–strain plot (kN); and d is the deflection at F_1 (mm).

2.2.2. Impact Strength

The Charpy impact strength test was conducted in accordance with the ISO 179-1:2010 standard [47] on a pendulum impact tester (HIT 25, Zwick GmbH & Com, Ulm, Germany). The unnotched specimens measuring 80 mm × 10 mm × 4 mm were prepared as described in Section 2.1.2, and, after conditioning, the specimen was removed from water, placed on supports, and the test was performed. The distance among the supports was 62 mm. The following formula was applied to calculate impact strength:

$$a_{cU} = \frac{E}{b \times d} \times 10^3 \tag{3}$$

where a_{cU} is the energy absorbed by breaking the test specimen (J); b and d are the width and thickness of the specimen, respectively (mm).

2.2.3. Fracture Analysis

Scanning electron microscopy, on a Zeiss SUPRA 35 (Zeiss, Oberkochen, Germany), was used to characterize the fractures of the samples broken by the impact and the three-point bending tests. All specimens were sputtered with gold before observations. Compact and smooth surface fields present brittle fracture modes, while a rough and jagged appearance presents intermediate (brittle to ductile) fracture modes [48,49]. Particular attention was paid to the impact of filler presence and aggregation on the appearance of fractures. If the fragments of fractured specimens could be repositioned at the fractured line presenting a smooth surface, the fractures were classified as brittle. Conversely, those presenting plastic deformation, exhibiting rough and jagged surfaces, were recorded as ductile [12]. Observations were performed at accelerating voltages from 5 to 10 kV.

2.2.4. Hardness

The ball indentation hardness (H) was determined according to the ISO 2039-1 standard [50] on the Zwick 3106 hardness tester (Zwick GmbH & Com, Ulm, Germany). The specimens measuring 65 mm

× 65 mm × 4 mm were prepared as described in Section 2.1.2. Four samples for each material/storing time were prepared, and three indentations were made on each sample. A steel ball, 5 mm in diameter, was indented into the materials. A test load of 358 N was applied for 30 seconds. The ball indentation hardness (*H*) was calculated according to the following equation:

$$H = F_m \times \frac{\left(\frac{0.21}{h-0.25+0.21}\right)}{\pi d h_r} \quad (4)$$

where, H is the ball indentation hardness (MPa); h is the depth of impression after correcting for the deformation of the frame (mm); d is the diameter of the ball indenter (mm); F_m is the test load on the indenter (N).

2.2.5. Wear Resistance

The samples for tribological tests measuring 30 mm × 30 mm × 10 mm were prepared as described in 2.1.2. The experiment was performed on a CSM Tribometer (CSM Instruments, Peseux, Switzerland) using the "ball on disc" method based on the methodology presented in the standards ASTM G99-95A [51] and ISO/TS 14569-2:2001 [52]. During wear testing the specimens were subjected to rotational motion and were kept in permanent contact with the spherical antagonist – a Al_2O_3 ball, 6 mm in diameter (Gewa, Zabrze, Poland). The experiment was conducted in distilled water at 37 ± 1 °C. The normal force applied by weight was 10 N, the relative sliding velocity was kept constant at 10 mm/s, the diameter at which the track was made was 4 mm, the total sliding distance was 100 m. After tests for each sample the profile of the cross-sectional area of wear track was measured at 6 points spaced about 60° apart (TalyProfile Lite, Tylor-Hubson, Leicester, U K) the mean area of profile was determined and finally the volume loss (mm^3) was calculated. Scanning electron microscopy, on a Zeiss SUPRA 35 (Zeiss, Oberkochen, Germany), was used to characterize traces left after wear tests. Observations were performed at accelerating voltages 5 kV.

2.2.6. Sorption and Solubility

Sorption and solubility were tested using the method presented in the ISO standard [46]. Five test samples of each material, measuring 50 mm in diameter and 0.5 mm in thickness, were cured in stainless steel molds. The samples were dried inside desiccators with freshly dried silica gel at 37 ± 1 °C and weighed daily (Analytic Scale AS/X, Radwag, Radom, Poland) until changes in mass were no higher than 0.2 mg (mass recorded as m_1). The thickness and diameter of the samples were measured with a digital caliper with an accuracy of 0.01 mm. Samples were placed in distilled water at 37 ± 1 °C for 7 days and weighed after storing (mass m_2). After that, the drying process, as described above, was repeated, and the stable mass was denoted as m_3. Sorption and solubility were calculated using the following equations:

$$w_{sp} = \frac{m_2 - m_3}{V} \quad (5)$$

$$w_{sl} = \frac{m_1 - m_3}{V} \quad (6)$$

where w_{sp} is the sorption, w_{sl} is the solubility, m_1 is the initial mass of the dried sample (µg); m_2 is the mass after storing (µg), m_3 is the mass after the second drying (µg), and V is the volume of the sample, (mm^3).

2.2.7. Statistical Analyses

Statistical analysis of the results was performed with the use of the Statistica 13.1 software (TIBCO Software Inc., Palo Alto, CA, USA). The distributions of the residuals were tested with the Shapiro–Wilk test, and the equality of variances was tested with Bartlett test. When the distribution of the residuals was normal and the variances were equal, the one-way or two-way ANOVA with Tukey's HSD post-hoc tests were used ($\alpha = 0.05$), otherwise the non-parametric Kruskal–Wallis test ($\alpha = 0.05$) was used.

3. Results

3.1. Flexural Properties

The mean flexural strength values are presented in Figure 2. The S–P concentration significantly decreased the flexural strength values of the composites (Table 2). The mean values after 2 days of conditioning in distilled water were 107.2 MPa for the A0 material and 72.4 MPa for the A6 composite (a reduction of 32%). Storing time also had a significant influence ($p < 0.0001$) on the hardness values. The mean values decreased after 90 days and were 96.9 MPa for A0 and 62.6 MPa for A6. For all materials, similar reductions in the flexural strength values were noted (from 9% to 12%).

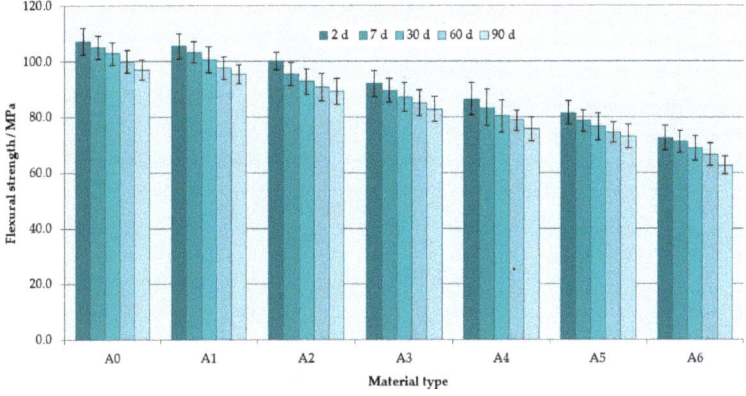

Figure 2. Mean flexural strength values with standard deviations.

Table 2. The results of one-way ANOVA and Tukey's HSD post-hoc tests for flexural strength. *

Material Code	Storing Time /Days				
	2	7	30	60	90
	($p < 0.0001$)	($p < 0.0001$)	($p < 0.0001$)	($p < 0.0001$)	($p < 0.0001$)
A0 ($p = 0.0076$)	A; a	A; a	A; a,b	A; a,b	A; b
A1 ($p = 0.0066$)	A; a	A,B; a	A,B; a,b	A,B; a,b	A; b
A2 ($p = 0.0062$)	A,B; a	B,C; a,b	B,C; a,b	B,C; b	A,B; b
A3 ($p = 0.0459$)	B,C; a	C,D; a,b	C,D; a,b	C,D; a,b	B,C; b
A4 ($p = 0.0427$)	C,D; a	D,E; a,b	D,E; a,b	D,E; a,b	C,D; b
A5 ($p = 0.0342$)	D,E; a	E,F; a,b	E,F; a,b	E,F; a,b	D; b
A6 ($p = 0.0084$)	E; a	F; a	F; a,b	F; a,b	E; b

* The different uppercase letters (A–F) for each column and lowercase letters (a–b) for each row show significantly different results at the $p < 0.05$ level.

The flexural modulus (Figure 3) increased with increasing S–P concentration; however, the changes were statistically significant (Table 3) only for three longer storing times. The highest increase in flexural modulus values between A0 and A6 materials was 0.5 GPa (after 90 days). The storing time had no statistically significant influence ($p > 0.05$) on the flexural modulus values, but they were noticeably reduced from 0.18 GPa to 0.35 GPa.

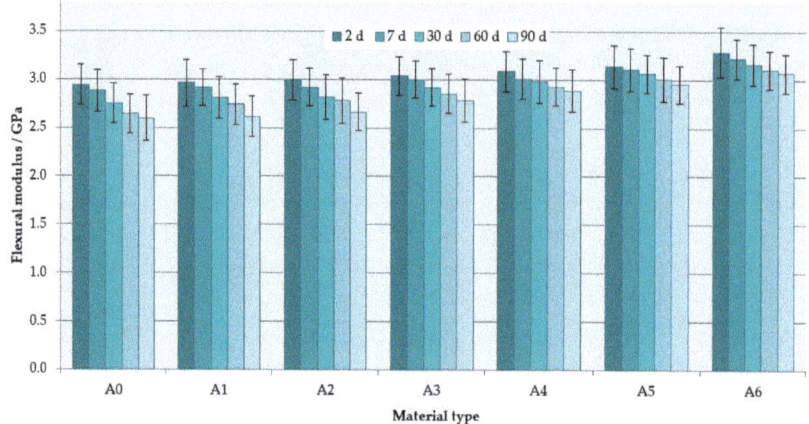

Figure 3. Mean flexural modulus values with standard deviations; the different uppercase letters (A–D) for each storing time show significantly different results at the $p < 0.05$ level.

Table 3. The results of one-way ANOVA and Tukey's HSD post-hoc tests for flexural modulus.*

Material Code	Storing Time /Days				
	2	7	30	60	90
	($p = 0.2185$)	($p = 0.1563$)	($p = 0.0447$)	($p = 0.0237$)	($p = 0.0076$)
A0 ($p = 0.0848$)	-	-	A	A	A
A1 ($p = 0.1101$)	-	-	A,B	A,B	A,B
A2 ($p = 0.1662$)	-	-	A,B	A,B	A,B
A3 ($p = 0.3100$)	-	-	A,B,C	A,B,C	B,C
A4 ($p = 0.6143$)	-	-	A,B,C	B,C	B,C,D
A5 ($p = 0.6612$)	-	-	B,C	C	C
A6 ($p = 0.5290$)	-	-	C	C	D

* The different uppercase letters (A–D) for each column show significantly different results at the $p < 0.05$ level.

The SEM observations of fractures showed that the morphology of the deformed regions changed with increasing S–P concentration. For the control material (Figure 4a), areas representing a brittle fracture mode (i.e., smooth and compact surface) and an intermediate form of fracture (i.e., jagged and rough appearance) were observed. For the obtained composites, a brittle fracture mode (smooth surface) was clearly visible in the central part of the fractures, but near the lower and upper edge of the samples, there were characteristic areas showing the presence of spherical shapes determined by the shape of PMMA pre-polymerized particles of the "powder" component (Figure 4b–e), the examples of which are indicated by black and white arrows. The area of this type of morphology increased with the increasing concentration of the filler (please compare Figure 4b,d). In Figure 4f, it is shown the presence in spherical structures of a large number of cubic filler particles and the spaces formed after pulling out S–P particles from the matrix.

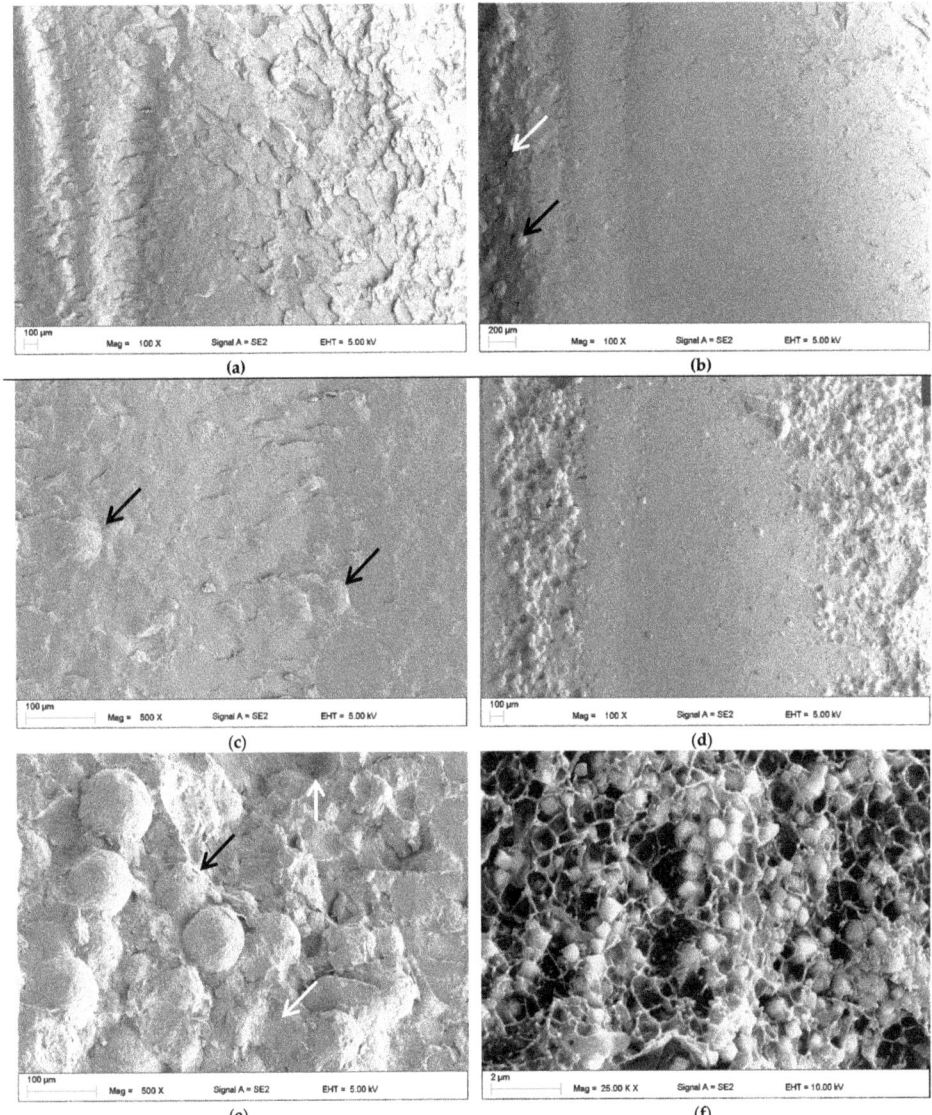

Figure 4. Representative SEM images presenting the fracture surfaces of flexural test specimens. Control (A0) (**a**). composites with antimicrobial filler concentrations of 1% (**b**,**c**) and 4% (**d**–**f**). Black arrows indicate the areas determined by the shape of poly(methyl methacrylate) (PMMA) pre-polymerized spheres; white arrows indicate the spherical niches remaining after them.

3.2. Impact Strength

The mean impact strength values are presented in Figure 5. The S–P concentration significantly decreased the impact strength values of the composites (Table 4). The mean values after 2 days of conditioning in distilled water were 18.4 kJ/mm2 for A0 material and 5.6 kJ/mm^2 for A6 composite (a reduction of 70%). Storing in distilled water also caused a reduction of mean values of impact strength, and, for four out of six materials, it was statistically significant ($p < 0.05$). After 90 days, the reduction was 17% for control material, whereas for the A6 composite, it was 42%.

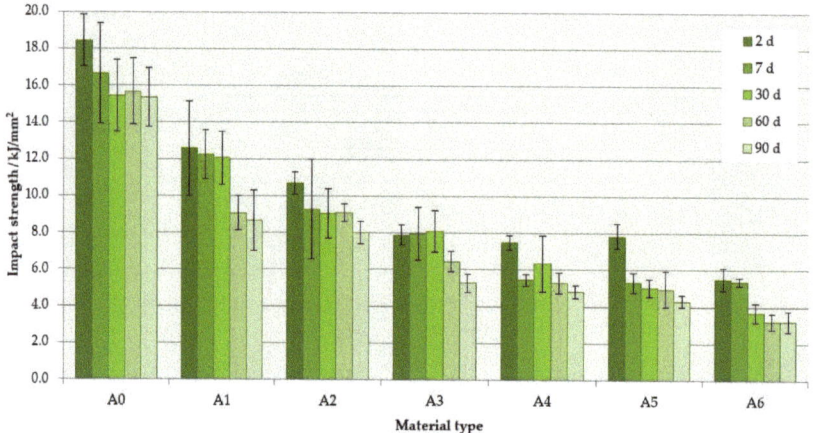

Figure 5. Mean impact strength values with standard deviations. The different uppercase letters (A–D) for each storing time and lowercase letters (a–c) for each material show significantly different results at the $p < 0.05$ level.

Table 4. The results of one-way ANOVA and Tukey's HSD post-hoc tests for impact strength. *

Material Code	Storing Time /Days				
	2	7	30	60	90
	($p < 0.0001$)	($p < 0.0001$)	($p < 0.0001$)	($p < 0.0001$)	($p < 0.0001$)
A0 ($p = 0.1882$)	A	A	A	A	A
A1 ($p = 0.0084$)	B; a	B; a,b	B; a,b	B; a,b	B; b
A2 ($p = 0.1760$)	B	B,C	C	B	B
A3 ($p = 0.0017$)	C,D; a	C,D; a	C; a	C; a,b	C,D; b
A4 ($p = 0.0017$)	D; a	D; a	C,D; a,b	C,D; b	D; b
A5 ($p < 0.0001$)	D; a	D; b	D; b	C,D; b	D; b
A6 ($p < 0.0001$)	D; a	D; a	D; b	D; b	D; b

* The different uppercase letters (A–D) for each column and lowercase letters (a–b) for each row show significantly different results at the $p < 0.05$ level.

The SEM observations (Figure 6) of impact fractures showed that the morphology where a rough and jagged appearance presents an intermediate fracture mode (brittle to ductile) was, in general, similar for all materials. However, the areas showing the presence of spherical shapes, determined by the shape of PMMA pre-polymerized particles (Figure 6b) were present, and their number increased with the increasing concentration of the filler. In comparison to the morphologies of fracture after flexural tests, these structures were observed much less frequently for the same filler concentrations, and the tendency to create areas of this type near the upper/lower surface of the samples was not clearly demonstrated.

Figure 6. Representative SEM images presenting the fracture surfaces of impact strength test specimens. Control (A0) (**a**). Composites with an antimicrobial filler concentration of 4% (**b**). Black arrows indicate the areas determined by the shape of PMMA pre-polymerized spheres; white arrows indicate the spherical niches remaining after them.

3.3. Hardness

The mean hardness values are presented in Figure 7. The hardness values significantly (Table 5) increased with the antimicrobial filler concentration. The mean hardness values after 2 days of conditioning in distilled water were 192 N/mm^2 for A0 and 232 N/mm^2 for A6 composite. Storing time also had a significant influence ($p < 0.0001$) on the hardness values. The hardness values decreased with storing time. After 90 days, the mean values for the A0 and A6 materials were 177 N/mm^2 and 211 N/mm^2, respectively. The reduction of hardness values was from 9% to 12%; thus, it was similar for all materials.

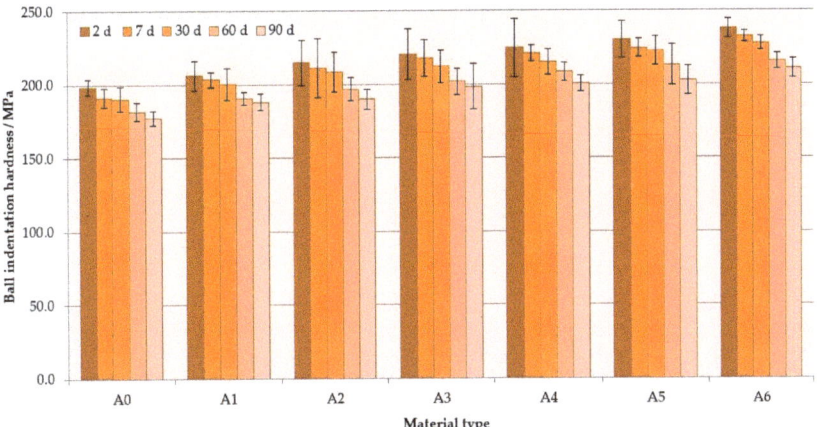

Figure 7. Mean ball indentation hardness values with standard deviations. The different uppercase letters (A–E) for each storing time and lowercase letters (a–c) for each material show significantly different results at the $p < 0.05$ level.

Table 5. The results of one-way ANOVA and Tukey's HSD post-hoc tests for ball indentation hardness. *

Material Code	Storing Time /Days				
	2	7	30	60	90
	($p < 0.0001$)	($p < 0.0001$)	($p < 0.0001$)	($p < 0.0001$)	($p < 0.0001$)
A0 ($p < 0.0001$)	A; a	A; b	A; b	A; c	A; c
A1 ($p < 0.0001$)	A,B; a	B; a	A,B; a	A,B; b	A,B; b
A2 ($p < 0.0001$)	B,C; a	B,C; a	B,C; a,b	B,C; b,c	B,C; c
A3 ($p < 0.0001$)	B,C; a	C,D; a	C,D; a,b	C,D; b,c	C,D; c
A4 ($p < 0.0001$)	C,D; a	C,D; a	C,D; a,b	D,E; b,c	D; c
A5 ($p < 0.0001$)	C,D; a	D,E; a	D,E; a,b	E; b,c	D,E; c
A6 ($p < 0.0001$)	D; a	E; a,b	E; b	E; c	E; c

* The different uppercase letters (A–E) for each column and lowercase letters (a–c) for each row show significantly different results at the p < 0.05 level.

3.4. Wear Resistance

The mean values of volume loss after wear tests are presented in Figure 8. For higher S–P concentrations, significantly decreased volume losses were recorded (Table 6), and increased storing time caused a significant increase of volume loss values ($p < 0.05$). The mean values after 2 days of conditioning in distilled water were 2.98 mm^3 for the A0 material and 0.15 mm^3 for the A6 composite (a reduction of 95%). After 90 days, volume losses were 5.6 mm^3 for A0 and 0.35 mm^3 for the A6 composite.

The representative SEM images of the traces left after the wear test showed that the intensity of the occurrence of wear mechanisms was associated with the mass concentration of the S–P and the conditioning time. The dominant feature was abrasive wear as a result of plowing. There were numerous scratches and micro-craters parallel to the direction of counter-sample movement on the surface of the traces, and their number and depth decreased with the increase of the filler content (Figure 9a–d). For the materials from A0 to A3, the areas of surface delamination were noted. On the wear traces for materials from A0 to A4, the areas indicating plastic deformation and fatigue wear mechanism were observed. For samples of the A5 composite stored up to 30 days, only uniform abrasion was observed with slight surface scratches in the direction of the movements of the antagonist which indicated the appearance of plowing (Figure 9e). A similar situation was observed for the A6 composite (all storing times). For the A5 composites conditioned from 60 to 90 days, traces of fatigue wear and microcracks were observed over the entire surface of the wear track (Figure 9f).

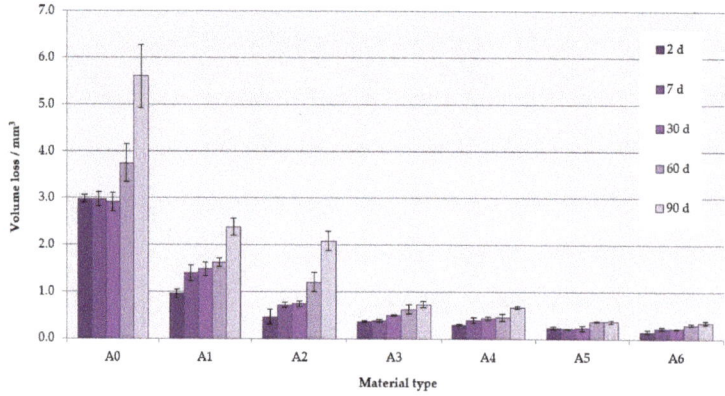

Figure 8. Mean volume loss values with standard deviations. The different uppercase letters (A–E) for each storing time and lowercase letters (a–c) for each material show significantly different results at the $p < 0.05$ level.

Table 6. The results of one-way ANOVA and Tukey's HSD post-hoc tests for volume loss. *

Material Code	Storing Time /Days				
	2	7	30	60	90
	($p < 0.0001$)	($p < 0.0001$)	($p < 0.0001$)	($p < 0.0001$)	($p < 0.0001$)
A0 ($p < 0.0425$)	A; a,b	A; a,b	A; a	A; b	A; c
A1 ($p < 0.0001$)	B; a	B; b	B; b	B; b	B; c
A2 ($p < 0.0001$)	C; a	C; a	C; a	B; b	B; c
A3 ($p < 0.0001$)	C,D; a	D; a	C,D; b	C; c	C; c
A4 ($p < 0.0001$)	D,E; a	D; a,b	D; b	C; b	C; c
A5 ($p < 0.0001$)	D,E; a	D; a	E; a	C; b	C; b
A6 ($p < 0.0001$)	E; a	D; b	E; b	C; c	C; c

* The different uppercase letters (A–E) for each column and lowercase letters (a–c) for each row show significantly different results at the p < 0.05 level.

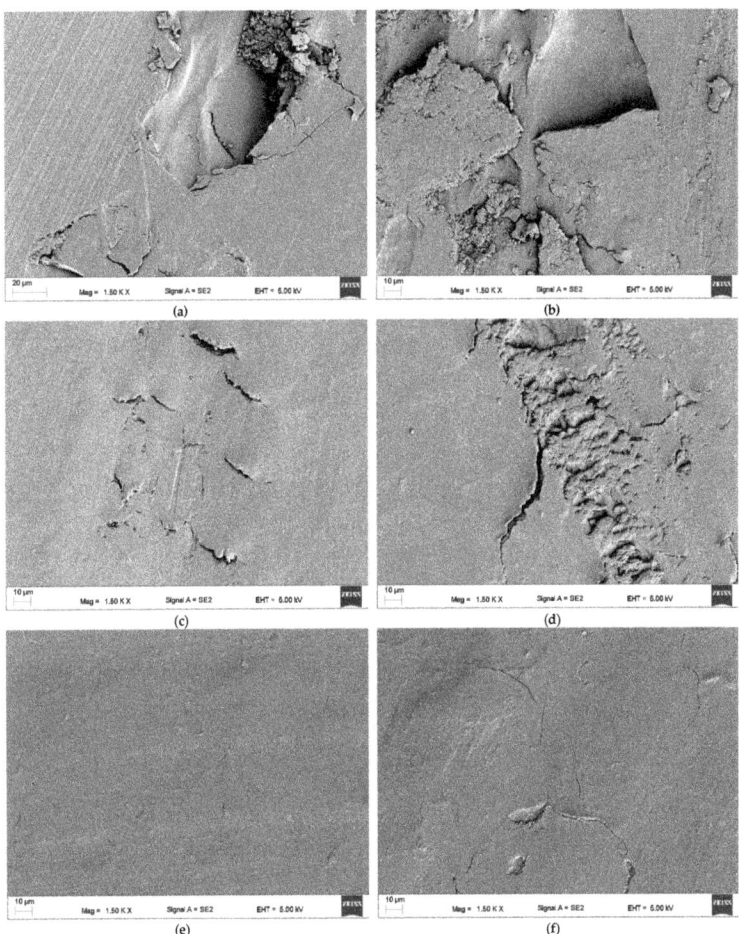

Figure 9. Representative SEM images presenting the traces left after wear tests on the samples of materials A0 (**a,b**), A3 (**c,d**), and A5 (**e,f**) stored in distilled water for 2 days (**a,c,e**) and 90 days (**b,d,f**).

3.5. Sorption and Solubility

There were no statistically significant differences between mean sorption values ($p = 0.9248$), presented in Figure 10a, which fell within the range of 23.20 µg/mm^3 (A4) to 24.49 µg/mm^3 (A6). The obtained values for all samples of each tested material were below the maximum limit of 32 µg/mm^3 allowed by the ISO standard [46].

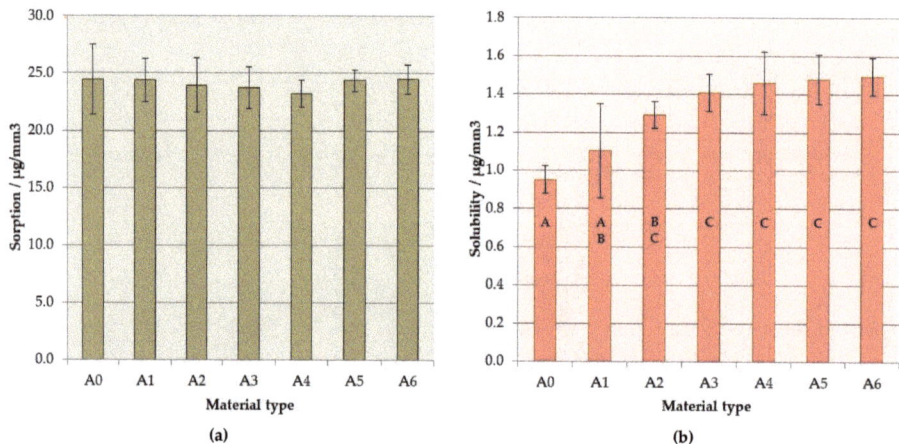

Figure 10. Mean and standard deviations of (**a**) sorption and (**b**) solubility. Different lowercase letters (a–c) show significantly different results at the $p < 0.05$ level.

The solubility values (Figure 10b) increased with increasing S–P concentration ($p < 0.0001$); however, the changes were not statistically significant ($p > 0.05$) for materials from A2 to A6. Excluding one sample of A6 composite, all the obtained sorption values were below the maximum limit of 1.6 µg/mm^3, so all materials were within the limit [46].

4. Discussion

The paper presents the results of the second stage of research on the influence of S–P introduction as an antimicrobial filler on the properties of PMMA denture base material. In the previously published part [37], the antimicrobial properties were confirmed via three-month in vitro experiments; thus, further tests were needed to investigate the other properties of the obtained composites related to their application. Additionally, S–P has also been investigated in our other works as an additive into silicone soft lining material [53] and direct restorative photopolymerizable resin-based composites [54], where only a slight influence of the filler on some physicochemical properties was noted. However, for the currently tested composites, the matrix and the method of introducing the filler into it were different which determined the inhomogeneous morphology of polymerized composites related with the used components (PMMA pre-polymerized particles) [37]. This could affect the physicochemical properties, thus investigations were conducted.

During most tests (excluding sorption and solubility), the conditioning of the samples for 90 days was conducted. The period of the samples' storing was based on the literature. Although studies involving tests of mechanical properties of materials aged in clinical conditions are rare, it is proven that the use of dentures in the oral cavity by patients for a period of 2 to 10 years is the cause of the deterioration of acrylate materials' hardness [55]. However, the dynamics of this process under clinical conditions was not tracked, because it requires obtaining the samples and conducting material tests at specific time intervals. This is one of the reasons why laboratory tests are performed much more frequently in this regard. These tests are usually conducted for 60 to 120 days, and the results have shown that mechanical properties of denture polymers reach equilibrium after up to 4 months [56], but most often for modern materials, this period does not exceed 60 days [57,58]. For this reason, the duration of the experiment in this study was limited to 90 days, and, after 60 days, no statistically significant changes in the mechanical properties of the materials were found, although insignificant changes were still visible.

During conditioning, the choice of medium may also influence the results. Typical media used are water and artificial saliva; however, using water is recommended by the ISO 20795-1:2013-07 standard for testing the mechanical properties of denture base materials [46]. Moreover, investigations indicate that when distilled water is used, the values of mechanical properties are lower and liquid absorption is higher than after storing in artificial saliva, even if these differences are not statistically significant [58,59]. This shows that using distilled water in these types of experiments was a rational choice, because water has, at least, the same degrading effect in comparison to artificial saliva. In addition, the use of water is justified in the first stages of research for practical reasons, because it allows creating very repetitive conditions. However, it should be noted that both mentioned liquids (i.e., water and artificial saliva) do not fully reflect real conditions, because salivary enzymes may also be the cause of polymer degradation and, as a consequence, lead to a reduction in surface hardness or wear resistance [60]. Moreover, Miranda et al. [61] suggested that liquids with lowered pH values (higher acidity) may influence the polymeric matrix of the resin by reacting with ester groups from acrylates which can create molecules of alcohol and carboxylic acid. This may lower the pH value inside the resin matrix and accelerate the degradation of the materials which should also be considered in the context of the use of dental prostheses.

The flexural strength and modulus using three-point bending tests are mechanical properties with limits that are specified by the ISO 20795-1:2013-07 standard for denture base materials, and they are related with the behavior of materials under clinical conditions. The PMMA resins during service in the mouth are subject to flexural fatigue as the denture base undergoes repeated masticatory loading [62]; thus, the high flexural strength is considered essential to denture durability, especially when gradual and irregular alveolar absorption processes cause tissue-borne dentures to be unevenly supported [63,64] but also when a perfect fit of a denture to the well-developed convex residual ridges occurs, because the denture is lifted at the non-working-side [65,66]. Flexural modulus influences denture stiffness. Their lower values are favorable in increasing the absorbed energy before fracture of the denture base, but a higher flexural modulus is recognized as clinically advantageous [67]. Partial dentures made of materials with a lower modulus of elasticity are more easily deformed during chewing and, as a result of which, locally higher loads can be transferred to the mucosa under the prosthesis [68]. Therefore, the use of materials with a lower flexural modulus may be the reason for increased pain associated with the increase in mobility of the dentures and their worse stabilization which can be the cause of the recorded decrease in chewing efficiency [69]. The increase in the antimicrobial filler mass concentration resulted in reduced flexural strength but, on the other hand, caused an increase in flexural modulus values. For all composites, the obtained values were higher than the indicated minimum (flexural strength –65 MPa, flexural modulus –2 GPa). Moreover, the registered flexural strength and modulus values for commercially available hot polymerized prosthetic PMMA resins ranged from 60 to 120 MPa [48,70–73], so it is the range analogous to that obtained for all the considered experimental composites. The flexural modulus values for heat-cured acrylic denture base resins reported in literature ranged from 2.1 to 3.1 GPa [48,74] which means that the analyzed materials are comparable in this respect with the best commercially available resins. The decrease in flexural strength and modulus values during conditioning in distilled water is a typical process noted for acrylates and related to the interaction of the liquid with the polymer matrix [75,76]. This is usually linked with the plasticizing effect exerted by water molecules penetrating into the materials [48], and this process is partially reversible [77]; however, other changes related to the release of components into the environment and the degradation of acrylates due to the hydrolysis are irreversible [78].

The percentage of reduction was the same regardless of filler content which indicates that this process was not determined by the filler content but by the properties of the polymer matrix.

Another important mechanical property of prosthetic materials is their impact strength which represents the resistance of materials to dynamic loads occurring in practice, e.g., during a prosthesis' fall. The amount of energy absorbed by materials before they are fractured is evaluated using the Charpy or Izod tests. The values recorded with each of these two methods differed for the same materials; however, good correlations among them have been found [79,80]. In this study, the Charpy method was used with the un-notched specimens, because the notching processes is time consuming, criticized for creating stresses in the PMMA specimens, and leads to problems with reproducibility [48,81]. A significant reduction in impact strength was obtained with increasing filler concentration. The impact values obtained for the A0 material were similar or higher than those recorded for PMMA denture base materials tested using an analogous methodology [48]. Impact strength decreased with conditioning time which corresponds well to the other results [82].

This reduction of flexural and impact strength values was related to the obtained inhomogeneous morphology and presence of aggregation, described in the previously published paper focused on antimicrobial properties and their stability [37]. The strong tendency of the inorganic submicron or nanofillers to aggregate is typical and related with their large surface area that provides high surface energy [83]. This problem may lead to decreased chemical interaction between the particles and the polymeric matrix [84]. The inhomogeneities in materials act as structural defects causing stress concentrations and strength reduction [85,86]. This corresponds well with the results of SEM observations (Figures 4 and 6) which showed the changes of morphologies of fractured samples and more brittle behavior of the composites. Observed areas showing the presence of spherical shapes, determined by the shape of PMMA pre-polymerized particles of the "powder" component, indicated a significant local reduction in material strength as a consequence of filler aggregations, leading to uprooting of these particles from the cured material and probably accelerated fracture of samples. A similar decrease in the mechanical properties along with the increase in mass concentrations of fillers has been noted after the introduction of particles, e.g., ZrO_2, nanodiamonds, Al_2O_3, hydroxyapatite, titanium oxide, ground fillers of natural origin or even glass fibers [83,87–89]. On the other hand, potential reinforcement in many cases can be achieved by using the synergistic effect of various additives (e.g., different particles, glass meshes, and glass fibers) [90].

Hardness is another important mechanical property of denture base materials. There are several methods for testing the hardness of polymer materials, but, frequently, the experiments conducted with denture base materials use Vickers microhardness test [91,92]. However, in this study, the ball indentation hardness test was used because of the morphology of materials determined by the used powder–liquid system. If one considers that the mass of the used monomer is about 27% of the mass of components, it becomes obvious that only part of the surface of the samples shows the presence of S–P particles. Moreover, the size of the used pre-polymerized PMMA particles was similar or larger than the expected size of indentation left during the Vickers microhardness test, so the risk of indentation on unmodified areas of materials can be assumed. For this reason, it was considered that making larger-sized indentations would provide more representative results. The increase in hardness after the introduction of the antimicrobial inorganic filler into the material was consistent with the results of other works, where a similar effect was reported as a result of the addition of metal oxides, mica, and glass particles [71,83,84,93]. A reduction in hardness during 90 days of storing was expected because water, like many other liquids, acts as a solvent to the acrylates which has been described as the plasticizing effect [94]. Water molecules penetrate into the material which leads to the separation of polymer chains, because molecules do not form basic chemical bonds with it but only occupy spaces and reduce interactions among chains such as secondary bonding and entanglements [95]. This process is typical, and a reduction in hardness values for acrylates or acrylate-based composites is frequently reported [95–98].

Higher hardness values are usually correlated with wear resistance [83,99,100]. However, for inhomogeneous materials, selective damage during wear may occur, so the hardness measurements are not sufficient to determine how the material will behave in this respect. The wear resistance of acrylate-based materials is usually considered in the context of using denture teeth [101–103] or restorative composites [104–106]. However, denture base materials should also present sufficient abrasion resistance to avoid wear by food, abrasive denture cleansers [76] or other functional forces created, for example, by the tongue [107,108]. The introduction of inorganic filler allowed to reduce abrasion with increasing S–P concentration, even by 95%, and allowed to gradually change the intensity of the occurrence of scratches, areas of surface delamination, areas indicating plastic deformation and fatigue wear mechanism. For samples of A5 composite stored up to 30 days, only uniform abrasion was observed with slight surface scratches in the direction of the movements of the antagonist. These changes are beneficial because they lead to a reduced risk of abrasion products such as polymer particles or fillers getting into the body, for example, with saliva or foods. Interestingly, this problem and its potential long-term consequences for patients' health have not yet been studied. The reduction in the number and size of scratches, craters, and other damage that occurs on the surface of materials during abrasion is also important due to the fact that they are potential areas where increased adherence of yeast-like fungi to the surface of materials can occur [21,23,109].

After seven days of storing in distilled water, all experimental materials showed values of sorption and solubility below the maximum limit of 32 $\mu g/mm^3$ and 1.6 $\mu g/mm^3$, respectively, allowed by the EN ISO 20795-1:2013-07 standard. Convergent results for PMMA denture base materials were obtained in other works for commercial and experimental materials [110–113]. Furthermore, Ergun et al. [85] reported a two-fold increase in sorption and more than three-fold increase in solubility with increasing concentration of zirconium oxide nanoparticles introduced to PMMA denture base resin. The penetration of water or aqueous solutions into the material also has an impact on its properties, because acrylates can undergo slow degradation due to the fact of hydrolysis as well as enzymatic reactions [78] which affects cytotoxicity and tribological and mechanical properties [114,115]. In this background, the lack of differences in the materials' sorption values is favorable. The low solubility value is particularly important, because the leaching of residuals of monomers and other additives used in prosthetic materials, as well as their penetration into the organism, are considered unfavorable [116,117]. In this context, the enhanced solubility justifies future research with complimentary techniques to understand the release of specific ions and chemical compounds from the materials. Research using techniques such as ICP-OES (optical emission spectrometry in inductively coupled plasma) or ICP-MS (inductively coupled plasma mass spectrometry) may provide answers to questions related to the release of, for example, silver or zirconium ions from composites containing S–P [118–120]. Chromatography techniques can be used to determine the release of compounds such as the residual monomer or dibenzoyl peroxide [121–123]. These investigations can be considered for selected composites together with analyses for other materials filled with S–P and based on different matrixes [53,54].

To sum up, on the basis of previously published and current laboratory studies, it should be stated that promising compilations of different properties were obtained for materials from A4 to A6. Those composites showed a strong effect against *C. albicans*, over 90 days of in vitro investigation, and acceptable physicochemical properties. However, the disadvantages of introducing these concentrations of S–P were a significant reduction of flexural strength and impact strength which was caused by the presence of structured defects and the more brittle behavior of tested composites. The reduction in impact strength was particularly negative. The increase in solubility recorded for these materials was also unfavorable. Determining its exact causes requires further research. Increased wear resistance, elimination of surface delamination, plastic deformation, and fatigue wear mechanisms observed during wear tests were favorable and can partially prevent entering abrasion products into the organism. Changes in hardness were also beneficial, because they indicated increased resistance to plastic deformation under localized mechanical loads. For materials with lower filler concentrations

(A1–A3), changes in physicochemical properties were smaller, but they were not outweighed by other benefits, because the antimicrobial effect was short-lived.

5. Conclusions

Within the limits of this study, it can be concluded that the experimental composites showed a satisfactory combination of physicochemical properties. With increasing S–P concentration after 2 day conditioning in distilled water, reduced values of flexural strength (from 107 to 72 MPa), impact strength (from 18.4 to 5.5 MPa) as well as enhanced solubility (from 0.95 to 1.49 µg/mm^3) were reported. These changes were unfavorable, but the recorded values were at acceptable levels and also within the context of the requirements of the ISO 20795-1:2013 standard. Favorable changes included increased hardness (from 198 to 238 MPa), flexural modulus (from 2.9 to 3.3 GPa), and decreased volume loss during wear test (from 2.9 to 0.2 mm^3). The sorption values were stable. The percentage changes of the analyzed properties during storage in distilled water were similar for all materials. Cytotoxic tests need to be performed in future experiments as well as long-term studies on the release of ions and other components, such as MMA monomer or dibenzoyl peroxide, from the materials.

Author Contributions: Conceptualization, G.C.; Formal analysis, G.C., K.P., J.Ż. and M.A.; Investigation, K.P. and W.P.; Methodology, G.C., K.P., W.P. and C.K.; Resources, K.P. and C.K.; Supervision, G.C.; Visualization, G.C. and J.Ż.; Writing—original draft, G.C., K.P. and W.P.; Writing—review and editing, G.C., J.Ż. and M.A.

Acknowledgments: This research was funded by a statutory grant from the Faculty of Mechanical Engineering of the Silesian University of Technology.

Conflicts of Interest: The authors declare no conflict of interest.

References

1. Dye, B.; Thornton-Evans, G.; Li, X.; Iafolla, T. Dental caries and tooth loss in adults in the United States, 2011–2012. *NCHS Data Brief* **2015**, *197*, 1–7.
2. Peltzer, K.; Hewlett, S.; Yawson, A.E.; Moynihan, P.; Preet, R.; Wu, F.; Guo, G.; Arokiasamy, P.; Snodgrass, J.J.; Chatterji, S.; et al. Prevalence of loss of all teeth (edentulism) and associated factors in older adults in China, Ghana, India, Mexico, Russia and South Africa. *Int. J. Environ. Res. Public Health* **2014**, *11*, 11308–11324. [CrossRef] [PubMed]
3. Kosuru, K.R.V.; Devi, G.; Grandhi, V.; Prasan, K.K.; Yasangi, M.K.; Dhanalakshmi, M. Denture care practices and perceived denture status among complete denture wearers. *J. Int. Soc. Prev. Community Dent* **2017**, *7*, 41–45. [CrossRef] [PubMed]
4. Piampring, P. Problems with complete dentures and related factors in patients in rajavithi hospital from 2007 to 2012. *J. Med. Assoc. Thai.* **2016**, *99*, 182–187.
5. Webb, B.C.; Thomas, C.J.; Willcox, M.D.P.; Harty, D.W.S.; Knox, K.W. Candida-associated denture stomatitis. Aetiology and management: A review. Part 2. Oral diseases caused by candida species. *Aust. Dent. J.* **1998**, *43*, 160–166. [CrossRef] [PubMed]
6. Akpan, A. Oral candidiasis. *Postgrad. Med. J.* **2002**, *78*, 455–459. [CrossRef]
7. Al-Dwairi, Z.N. Isolation of candida species from the oral cavity and fingertips of complete and partial dentures wearers. *J. Dent. Health Oral Disord. Ther.* **2014**, *1*, 00015. [CrossRef]
8. Uzel, N.G.; Teles, F.R.; Teles, R.P.; Song, X.Q.; Torresyap, G.; Socransky, S.S.; Haffajee, A.D. Microbial shifts during dental biofilm re-development in the absence of oral hygiene in periodontal health and disease. *J. Clin. Periodontol.* **2011**, *38*, 612–620. [CrossRef]
9. Evren, B.A.; Uludamar, A.; Işeri, U.; Ozkan, Y.K. The association between socioeconomic status, oral hygiene practice, denture stomatitis and oral status in elderly people living different residential homes. *Arch. Gerontol. Geriatr.* **2011**, *53*, 252–257. [CrossRef]
10. Gendreau, L.; Loewy, Z.G. Epidemiology and etiology of denture stomatitis. *J. Prosthodont.* **2011**, *20*, 251–260. [CrossRef]
11. Spiechowicz, E.; Mierzwińska-Nastalska, E. *Fungal Infections of Oral Cavity*, 1st ed.; Med Tour Press International: Warszawa, Poland, 1998.

12. Dniluk, T.; Fiedoruk, K.; Ściepuk, M. Aerobic bacteria in the oral cavity of patients with removable dentures. *Adv. Med Sci.* **2006**, *51*, 86–90.
13. Preshaw, P.M.; Walls, A.W.G.; Jakubovics, N.S.; Moynihan, P.J.; Jepson, N.J.A.; Loewy, Z. Association of removable partial denture use with oral and systemic health. *J. Dent.* **2011**, *39*, 711–719. [CrossRef] [PubMed]
14. Niimi, M.; Firth, N.A.; Cannon, R.D. Antifungal drug resistance of oral fungi. *Odontology* **2010**, *98*, 15–25. [CrossRef] [PubMed]
15. Amin, W.M.; Al-Ali, M.H.; Salim, N.A.; Al-Tarawneh, S.K. A new form of intraoral delivery of antifungal drugs for the treatment of denture-induced oral candidosis. *Eur. J. Dent.* **2009**, *3*, 257–266. [CrossRef]
16. Chandra, J.; Mukherjee, P.K.; Leidich, S.D.; Faddoul, F.F.; Hoyer, L.L.; Douglas, L.J.; Ghannoum, M.A. Antifungal resistance of candidal biofilms formed on denture acrylic in vitro. *J. Dent. Res.* **2001**, *80*, 903–908. [CrossRef]
17. Paranhos, H.D.; Peracini, A.; Pisani, M.X.; Oliveira, V.D.; Souza, R.F.; Silva-Lovato, C.H. Color stability, surface roughness and flexural strength of an acrylic resin submitted to simulated overnight immersion in denture cleansers. *Braz. Dent. J.* **2013**, *24*, 152–156. [CrossRef]
18. Koray, M.; Ak, G.; Kurklu, E.; Issever, H.; Tanyeri, H.; Kulekci, G.; Guc, U. Fluconazole and/or hexetidine for management of oral candidiasis associated with denture-induced stomatitis. *Oral Dis.* **2005**, *11*, 309–313. [CrossRef]
19. Carmen, S.; Michelangelo, P.; María, C.; Vincenzo, E.; Maurizio, B.; Lucio, M.; Agostino, G.; Massimo, P.; Rosario Serpico, S. Candida-associated denture stomatitis. *Oral Med. Pathol.* **2011**, *16*, 139–143.
20. Richmond, R.; Macfarlane, T.V.; McCord, J.F. An evaluation of the surface changes in PMMA biomaterial formulations as a result of toothbrush/dentifrice abrasion. *Dent. Mater.* **2004**, *20*, 124–132. [CrossRef]
21. Charman, K.M.; Fernandez, P.; Loewy, Z.; Middleton, A.M. Attachment of streptococcus oralis on acrylic substrates of varying roughness. *Lett. Appl. Microbiol.* **2009**, *48*, 472–477. [CrossRef]
22. Kurt, A.; Erkose-Genc, G.; Uzun, M.; Sarı, T.; Isik-Ozkol, G. The effect of cleaning solutions on a denture base material: elimination of candida albicans and alteration of physical properties. *J. Prosthodont.* **2018**, *27*, 577–583. [CrossRef] [PubMed]
23. Köroğlu, A.; Şahin, O.; Dede, D.Ö.; Deniz, Ş.T.; Sever, N.K.; Özkan, S. Efficacy of denture cleaners on the surface roughness and candida albicans adherence of sealant agent coupled denture base materials. *Dent. Mater. J.* **2016**, *35*, 810–816. [CrossRef] [PubMed]
24. Shibata, T.; Hamada, N.; Kimoto, K.; Sawada, T.; Sawada, T.; Kumada, H.; Umemoto, T.; Toyoda, M. Antifungal Effect of acrylic resin containing apatite-coated TiO_2 photocatalyst. *Dent. Mater. J.* **2007**, *26*, 437–444. [CrossRef] [PubMed]
25. Bulad, K.; Taylor, R.L.; Verran, J.; McCord, J.F. Colonization and penetration of denture soft lining materials by candida albicans. *Dent. Mater.* **2004**, *20*, 167–175. [CrossRef]
26. Rodger, G.; Taylor, R.L.; Pearson, G.J.; Verran, J. In vitro colonization of an experimental silicone by candida albicans. *J. Biomed. Mater. Res. Part B Appl. Biomater.* **2010**, *92*, 226–235. [CrossRef]
27. Krishnamurthy, S.; Hallikerimath, R.B. An in-vitro evaluation of retention, colonization and penetration of commonly used denture lining materials by candida albicans. *J. Clin. Diagn. Res.* **2016**, *10*, 84–88. [CrossRef]
28. Zhang, K.; Baras, B.; Lynch, C.D.; Weir, M.D.; Melo, M.A.S.; Li, Y.; Reynolds, M.A.; Bai, Y.; Wang, L.; Wang, S.; et al. Developing a new generation of therapeutic dental polymers to inhibit oral biofilms and protect teeth. *Materials* **2018**, *11*, 1747. [CrossRef]
29. Gad, M.M.; Al-Thobity, A.M.; Shahin, S.Y.; Alsaqer, B.T.; Ali, A.A. Inhibitory effect of zirconium oxide nanoparticles on candida albicans adhesion to repaired polymethyl methacrylate denture bases and interim removable prostheses: a new approach for denture stomatitis prevention. *Int. J. Nanomed.* **2017**, *12*, 5409–5419. [CrossRef]
30. Venkatesh Anehosur, G.; Kulkarni, R.D. Synthesis and determination of antimicrobial activity of visible light activated TiO_2 nanoparticles with polymethyl methacrylate denture base resin against staphylococcus aureus. *J. Gerontol. Geriatr. Res.* **2012**, *1*, 1–8. [CrossRef]
31. Totu, E.E.; Nechifor, A.C.; Nechifor, G.; Aboul-Enein, H.Y.; Cristache, C.M. Poly (methyl methacrylate) with TiO_2 nanoparticles inclusion for stereolitographic complete denture manufacturing—The fututre in dental care for elderly edentulous patients? *J. Dent.* **2017**, *59*, 68–77. [CrossRef]

32. Cierech, M.; Kolenda, A.; Grudniak, A.M.; Wojnarowicz, J.; Woźniak, B.; Gołaś, M.; Swoboda-Kopeć, E.; Łojkowski, W.; Mierzwińska-Nastalska, E. Significance of polymethylmethacrylate (PMMA) modification by zinc oxide nanoparticles for fungal biofilm formation. *Int. J. Pharm.* **2016**, *510*, 323–335. [CrossRef] [PubMed]
33. Nam, K.Y. Characterization and bacterial anti-adherent effect on modified PMMA denture acrylic resin containing platinum nanoparticles. *J. Adv. Prosthodont.* **2014**, *6*, 207–214. [CrossRef] [PubMed]
34. Acosta-Torres, L.S.; Mendieta, I.; Nuñez-Anita, R.E.; Cajero-Juárez, M.; Castaño, V.M. Cytocompatible antifungal acrylic resin containing silver nanoparticles for dentures. *Int. J. Nanomed.* **2012**, *7*, 4777–4786.
35. Fan, C.; Chu, L.; Rawls, H.R.; Norling, B.K.; Cardenas, H.L.; Whang, K. Development of an antimicrobial resin—A pilot study. *Dent. Mater.* **2011**, *27*, 322–328. [CrossRef] [PubMed]
36. Kurt, A.; Erkose-Genc, G.; Uzun, M.; Emrence, Z.; Ustek, D.; Isik-Ozkol, G. The antifungal activity and cytotoxicity of silver containing denture base material. *Niger. J. Clin. Pract.* **2017**, *20*, 290–295. [CrossRef]
37. Chladek, G.; Basa, K.; Mertas, A.; Pakieła, W.; Żmudzki, J.; Bobela, E.; Król, W. Effect of storage in distilled water for three months on the antimicrobial properties of poly (methyl methacrylate) denture base material doped with inorganic filler. *Materials* **2016**, *9*, 328. [CrossRef]
38. Meng, T.R.; Latta, M.A. Physical properties of four acrylic denture base resins. *J. Contemp. Dent. Pract.* **2005**, *6*, 93–100. [CrossRef]
39. Oleiwi, J.K.; Hamad, Q.A. Studying the mechanical properties of denture base materials fabricated from polymer composite materials. *Al-Khwarizmi Eng. J.* **2018**, *14*, 100–111. [CrossRef]
40. Choudhary, S. Complete denture fracture—a proposed classification system and its incidence in national capital region population: A survey. *J. Indian Prosthodont. Soc.* **2019**, *19*, 307–312. [CrossRef]
41. Takamiya, A.S.; Monteiro, D.R.; Marra, J.; Compagnoni, M.A.; Barbosa, D.B. Complete denture wearing and fractures among edentulous patients treated in university clinics. *Gerodontology* **2012**, *29*, 728–734. [CrossRef]
42. Shakir, S.; Jalil, H.; Khan, M.A.; Qayum, B.; Qadeer, A. Causes and types of denture fractures—A study. *Pak. Oral Dent. J.* **2017**, *37*, 634–637.
43. de Freitas Pontes, K.M.; de Holanda, J.C.; Fonteles, C.S.R.; de Barros Pontes, C.; da Silva, C.H.L.; Paranhos, H.D.F.O. Effect of toothbrushes and denture brushes on heat-polymerized acrylic resins. *Gen. Dent.* **2016**, *64*, 49–53. [PubMed]
44. Hamanaka, I.; Iwamoto, M.; Lassila, L.V.J.; Vallittu, P.K.; Takahashi, Y. Wear resistance of injection-molded thermoplastic denture base resins. *Acta Biomater. Odontol. Scand.* **2016**, *2*, 31–37. [CrossRef] [PubMed]
45. Kumar, V.; Kumar, L.; Sehgal, K.; Datta, K.; Pal, B. A Comparative evaluation of effect of reinforced autopolymerizing resin on the flexural strength of repaired heat-polymerized denture base resin before and after thermocycling. *J. Int. Soc. Prev. Community Dent.* **2017**, *7*, 99–106. [CrossRef]
46. ISO 20795-1:2013. *Dentistry—Base Polymers -Part 1: Denture Base Polymers*; ISO: Geneva, Switzerland, 2013.
47. ISO 179-1:2010. *Plastics—Determination of Charpy Impact Properties—Part 1: Non-instrumented Impact Test*; ISO: Geneva, Switzerland, 2010.
48. Al-Dwairi, Z.N.; Tahboub, K.Y.; Baba, N.Z.; Goodacre, C.J. A comparison of the flexural and impact strengths and flexural modulus of CAD/CAM and conventional heat-cured polymethyl methacrylate (PMMA). *J. Prosthodont.* **2018**, 1–9. [CrossRef]
49. Faot, F.; Panza, L.H.V.; Garcia, R.C.R.; Cury, A.D.B. Impact and flexural strength, and fracture morphology of acrylic resins with impact modifiers. *In Proceedings of the Open Dentistry Journal, Open Dent. J.* **2019**, *3*, 137. [CrossRef]
50. ISO 2039-1:2001. *Plastics—Determination of Hardness—Part 1: Ball Indentation Method*; ISO: Geneva, Switzerland, 2001.
51. ASTM G99—95a (2000). *Test Method for Wear Testing with a Pin-on-Disk Apparatus*; ASTM International: West Conshohocken, PA, USA, 2000.
52. ISO/TS 14569-2:2001. *Dental materials—Guidance on Testing of Wear—Part 2: Wear by Two—and/or Three Body Contact*; ISO: Geneva, Switzerland, 2001.
53. Jabłońska-Stencel, E.; Pakieła, W.; Mertas, A.; Bobela, E.; Kasperski, J.; Chladek, G. Effect of silver-emitting filler on antimicrobial and mechanical properties of soft denture lining material. *Materials* **2018**, *11*, 318. [CrossRef]
54. Stencel, R.; Kasperski, J.; Pakieła, W.; Mertas, A.; Bobela, E.; Barszczewska-Rybarek, I.; Chladek, G. Properties of experimental dental composites containing antibacterial silver-releasing filler. *Materials* **2018**, *11*, 1031. [CrossRef]

55. Matsuo, H.; Suenaga, H.; Takahashi, M.; Suzuki, O.; Sasaki, K.; Takahashi, N. Deterioration of polymethyl methacrylate dentures in the oral cavity. *Dent. Mater. J.* **2015**, *34*, 234–239. [CrossRef]
56. Takahashi, Y.; Chai, J.; Kawaguchi, M. Equilibrium strengths of denture polymers subjected to long-term water immersion. *Int. J. Prosthodont.* **1999**, *12*, 348–352.
57. Chandrahari, N.; Kumar, C.R.; Singh, M.; Singh, S. Comparison of fracture resistance of heat cure resins polymerized by conventional and microwave methods after immersion in artificial saliva. *J. Contemp. Dent. Pract.* **2019**, *20*, 71–77. [PubMed]
58. Jagini, A.S.; Marri, T.; Jayyarapu, D.; Kumari, R.D. Effect of long-term immersion in water and artificial saliva on the flexural strength of two heat cure denture base resins. *J. Contemp. Dent. Pract.* **2019**, *20*, 341–346. [CrossRef] [PubMed]
59. Bacali, C.; Badea, M.; Moldovan, M.; Sarosi, C.; Nastase, V.; Baldea, I.; Chiorean, R.S.; Constantiniuc, M. The influence of graphene in improvement of physico-mechanical properties in PMMA Denture Base Resins. *Materials* **2019**, *12*, 2335. [CrossRef] [PubMed]
60. Bettencourt, A.F.; Neves, C.B.; de Almeida, M.S.; Pinheiro, L.M.; Oliveira, S.A.E.; Lopes, L.P.; Castro, M.F. Biodegradation of acrylic based resins: A review. *Dent. Mater.* **2010**, *26*, 171–180. [CrossRef] [PubMed]
61. Miranda, D.D.A.; Bertoldo, C.E.D.S.; Aguiar, F.H.B.; Lima, D.A.N.L.; Lovadino, J.R. Effects of mouthwashes on Knoop hardness and surface roughness of dental composites after different immersion times. *Braz. Oral Res.* **2011**, *25*, 168–173. [CrossRef] [PubMed]
62. Hirajima, Y.; Takahashi, H.; Minakuchi, S. Influence of a denture strengthener on the deformation of a maxillary complete denture. *Dent. Mater. J.* **2009**, *28*, 507–512. [CrossRef]
63. Diaz-Arnold, A.M.; Vargas, M.A.; Shaull, K.L.; Laffoon, J.E.; Qian, F. Flexural and fatigue strengths of denture base resin. *J. Prosthet. Dent.* **2008**, *100*, 47–51. [CrossRef]
64. Ajaj-ALKordy, N.M.; Alsaadi, M.H. Elastic modulus and flexural strength comparisons of high-impact and traditional denture base acrylic resins. *Saudi Dent. J.* **2014**, *26*, 15–18. [CrossRef]
65. Żmudzki, J.; Chladek, G.; Malara, P. Use of finite element analysis for the assessment of biomechanical factors related to pain sensation beneath complete dentures during mastication. *J. Prosthet. Dent.* **2018**, *120*, 934–941. [CrossRef]
66. Żmudzki, J.; Chladek, G.; Kasperski, J. Biomechanical factors related to occlusal load transfer in removable complete dentures. *Biomech. Model. Mechanobiol.* **2015**, *14*, 679–691. [CrossRef]
67. Ucar, Y.; Akova, T.; Aysan, I. Mechanical properties of polyamide versus different PMMA denture base materials. *J. Prosthodont.* **2012**, *21*, 173–176. [CrossRef] [PubMed]
68. Wadachi, J.; Sato, M.; Igarashi, Y. Evaluation of the rigidity of dentures made of injection-molded materials. *Dent. Mater. J.* **2013**, *32*, 508–511. [CrossRef] [PubMed]
69. Macura-Karbownik, A.; Chladek, G.; Żmudzki, J.; Kasperski, J. Chewing efficiency and occlusal forces in PMMA, acetal and polyamide removable partial denture wearers. *Acta Bioeng. Biomech.* **2016**, *18*, 137–144. [PubMed]
70. Gunduz, O.; Unalan, F.; Dikbas, I. Comparison of the transverse strength of six acrylic denture resins. *OHDMBSC* **2010**, *9*, 21–24.
71. Vojdani, M.; Bagheri, R.; Khaledi, A.A.R. Effects of aluminum oxide addition on the flexural strength, surface hardness, and roughness of heat-polymerized acrylic resin. *J. Dent. Sci.* **2012**, *7*, 238–244. [CrossRef]
72. Chander, N.G.; Jayaraman, V.; Sriram, V. Comparison of ISO and ASTM standards in determining the flexural strength of denture base resin. *Eur. Oral Res.* **2019**, *53*, 137–140. [CrossRef]
73. Barbosa, D.B.; de Souza, R.F.; Pero, A.C.; Marra, J.; Compagnoni, M.A. Flexural strength of acrylic resins polymerized by different cycles. *J. Appl. Oral Sci.* **2007**, *15*, 424–428. [CrossRef]
74. Yunus, N.; Rashid, A.A.; Azmi, L.L.; Hassan, M.I.A. Some flexural properties of a nylon denture base polymer. *J. Oral Rehabil.* **2005**, *32*, 65–71. [CrossRef]
75. Sato, S.; Cavalcante, M.R.S.; Orsi, I.A.; deParanhos, H.F.O.; Zaniquelli, O. Assessment of flexural strength and color alteration of heat-polymerized acrylic resins after simulated use of denture cleansers. *Braz. Dent. J.* **2005**, *16*, 124–128. [CrossRef]
76. Ali, I.L.; Yunus, N.; Abu-Hassan, M.I. Hardness, flexural strength, and flexural modulus comparisons of three differently cured denture base systems. *J. Prosthodont.* **2008**, *17*, 545–549. [CrossRef]
77. Takahashi, Y.; Yoshida, K.; Shimizu, H. Fracture resistance of maxillary complete dentures subjected to long-term water immersion. *Gerodontology* **2012**, *29*, 1086–1091. [CrossRef] [PubMed]

78. Santerre, J.P.; Shajii, L.; Leung, B.W. Relation of dental composite formulations to their degradation and the release of hydrolyzed polymeric-resin-derived products. *Crit. Rev. Oral Biol. Med.* **2001**, *12*, 136–151. [CrossRef] [PubMed]
79. Zappini, G.; Kammann, A.; Wachter, W. Comparison of fracture tests of denture base materials. *J. Prosthet. Dent.* **2003**, *90*, 578–585. [CrossRef]
80. Faot, F.; Costa, M.A.; Del Bel Cury, A.A.; Rodrigues Garcia, R.C.M. Impact strength and fracture morphology of denture acrylic resins. *J. Prosthet. Dent.* **2006**, *96*, 367–373. [CrossRef] [PubMed]
81. Dikbas, I.; Gurbuz, O.; Unalan, F.; Koksal, T. Impact strength of denture polymethyl methacrylate reinforced with different forms of E-glass fibers. *Acta Odontol. Scand.* **2013**, *71*, 727–732. [CrossRef] [PubMed]
82. Arundati, R.; Patil, N.P. An investigation into the transverse and impact strength of a new indigenous high-impact denture base resin, DPI- tuff and its comparison with most commonly used two denture base resins. *J. Indian Prosthodont. Soc.* **2006**, *6*, 133–138. [CrossRef]
83. Zidan, S.; Silikas, N.; Alhotan, A.; Haider, J.; Yates, J. Investigating the mechanical properties of ZrO2-impregnated PMMA nanocomposite for denture-based applications. *Materials* **2019**, *12*, 1344. [CrossRef]
84. Zhang, X.-Y.; Zhang, X.-J.; Huang, Z.-L.; Zhu, B.-S.; Chen, R.-R. Hybrid effects of zirconia nanoparticles with aluminum borate whiskers on mechanical properties of denture base resin PMMA. *Dent. Mater. J.* **2014**, *33*, 141–146. [CrossRef]
85. Ergun, G.; Sahin, Z.; Ataol, A.S. The effects of adding various ratios of zirconium oxide nanoparticles to poly(methyl methacrylate) on physical and mechanical properties. *J. Oral Sci.* **2018**, *60*, 304–315. [CrossRef]
86. Ali, M.; Al Mgoter, I.N. Evaluation the effect of modified nano-fillers addition on some properties of heat cured acrylic denture base material. *J. Baghdad Coll. Dent.* **2011**, *23*, 23–29.
87. Al-Harbi, F.A.; Abdel-Halim, M.S.; Gad, M.M.; Fouda, S.M.; Baba, N.Z.; AlRumaih, H.S.; Akhtar, S. Effect of nanodiamond addition on flexural strength, impact strength, and surface roughness of PMMA denture base: nanodiamond effect on PMMA denture base. *J. Prosthodont.* **2019**, *28*, 417–425. [CrossRef] [PubMed]
88. Sabbatini, A.; Lanari, S.; Santulli, C.; Pettinari, C. Use of almond shells and rice husk as fillers of poly(methyl methacrylate) (PMMA) composites. *Materials* **2017**, *10*, 872. [CrossRef] [PubMed]
89. Somani, M.V.; Khandelwal, M.; Punia, V.; Sharma, V. The effect of incorporating various reinforcement materials on flexural strength and impact strength of polymethylmethacrylate: A meta-analysis. *J. Indian Prosthodont. Soc.* **2019**, *19*, 101–112. [CrossRef] [PubMed]
90. Gad, M.M.; Al-Thobity, A.M.; Rahoma, A.; Abualsaud, R.; Al-Harbi, F.A.; Akhtar, S. Reinforcement of PMMA denture base material with a mixture of ZrO2 nanoparticles and glass fibers. Available online: https://www.hindawi.com/journals/ijd/2019/2489393/abs/ (accessed on 21 June 2019).
91. Ozyilmaz, O.Y.; Akin, C. Effect of cleansers on denture base resins' structural properties. *J. Appl. Biomater. Funct. Mater.* **2019**, *17*. [CrossRef] [PubMed]
92. Al-Dwairi, Z.N.; Tahboub, K.Y.; Baba, N.Z.; Goodacre, C.J.; Özcan, M. A Comparison of the surface properties of CAD/CAM and conventional polymethylmethacrylate (PMMA). *J. Prosthodont.* **2019**, *28*, 452–457. [CrossRef]
93. Unalan, F.; Dikbas, I. Effects of mica and glass on surface hardness of acrylic tooth material. *Dent. Mater. J.* **2007**, *26*, 545–548. [CrossRef]
94. Wu, W.; Mckinney, J.E. Influence of chemicals on wear of dental composites. *J. Dent. Res.* **1982**, 1180–1183. [CrossRef]
95. Ferracane, J.L. Hygroscopic and hydrolytic effects in dental polymer networks. *Dent. Mater.* **2006**, *22*, 211–222. [CrossRef]
96. Chladek, G.; Basa, K.; Żmudzki, J.; Malara, P.; Nowak, A.J.; Kasperski, J. Influence of aging solutions on wear resistance and hardness of selected resin-based dental composites. *Acta Bioeng. Biomech.* **2016**, *18*, 43–52.
97. Chladek, G.; Żmudzki, J.; Basa, K.; Pater, A.; Krawczyk, C.; Pakieła, W. Effect of silica filler on properties of PMMA resin. *Arch. Mater. Sci. Eng.* **2015**, *71*, 10.
98. Jang, D.-E.; Lee, J.-Y.; Jang, H.-S.; Lee, J.-J.; Son, M.-K. Color stability, water sorption and cytotoxicity of thermoplastic acrylic resin for non metal clasp denture. *J. Adv. Prosthodont.* **2015**, *7*, 278–287. [CrossRef] [PubMed]

99. Poggio, C.; Lombardini, M.; Gaviati, S.; Chiesa, M. Evaluation of Vickers hardness and depth of cure of six composite resins photo-activated with different polymerization modes. *J. Conserv. Dent.* **2012**, *15*, 237–241. [CrossRef] [PubMed]
100. Dayan, C.; Kiseri, B.; Gencel, B.; Kurt, H.; Tuncer, N. Wear resistance and microhardness of various interim fixed prosthesis materials. *J. Oral Sci.* **2019**, *61*, 447–453. [CrossRef] [PubMed]
101. Suwannaroop, P.; Chaijareenont, P.; Koottathape, N.; Takahashi, H.; Arksornnukit, M. In vitro wear resistance, hardness and elastic modulus of artificial denture teeth. *Dent. Mater. J.* **2011**, *30*, 461–468. [CrossRef]
102. Munshi, N.; Rosenblum, M.; Jiang, S.; Flinton, R. In vitro wear resistance of nano-hybrid composite denture teeth. *J. Prosthodont.* **2017**, *26*, 224–229. [CrossRef]
103. Kamonwanon, P.; Yodmongkol, S.; Chantarachindawong, R.; Thaweeboon, S.; Thaweeboon, B.; Srikhirin, T. Wear resistance of a modified polymethyl methacrylate artificial tooth compared to five commercially available artificial tooth materials. *J. Prosthet. Dent.* **2015**, *114*, 286–292. [CrossRef]
104. Yu, H.; Wegehaupt, F.J.; Wiegand, A.; Roos, M.; Attin, T.; Buchalla, W. Erosion and abrasion of tooth-colored restorative materials and human enamel. *J. Dent.* **2009**, *37*, 913–922. [CrossRef]
105. Tsujimoto, A.; Jurado, C.; Villalobos-Tinoco, J.; Barkmeier, W.; Fischer, N.; Takamizawa, T.; Latta, M.; Miyazaki, M. Wear resistance of indirect composite resins used for provisional restorations supported by implants. *J. Adv. Prosthodont.* **2019**, *11*, 232–238. [CrossRef]
106. Tanoue, N.; Matsumura, H.; Atsuta, M. Wear and surface roughness of current prosthetic composites after toothbrush/dentifrice abrasion. *J. Prosthet. Dent.* **2000**, *84*, 93–97. [CrossRef]
107. Oncescu Moraru, A.M.; Preoteasa, C.T.; Preoteasa, E. Masticatory function parameters in patients with removable dental prosthesis. *J. Med. Life* **2019**, *12*, 43–48. [CrossRef]
108. Żmudzki, J.; Chladek, G.; Krawczyk, C. Relevance of tongue force on mandibular denture stabilization during mastication. *J. Prosthodont.* **2017**, *28*, 27–33. [CrossRef] [PubMed]
109. Morgan, T.D.; Wilson, M. The effects of surface roughness and type of denture acrylic on biofilm formation by Streptococcus oralis in a constant depth film fermentor. *J. Appl. Microbiol.* **2001**, *91*, 47–53. [CrossRef] [PubMed]
110. Asar, N.V.; Albayrak, H.; Korkmaz, T.; Turkyilmaz, I. Influence of various metal oxides on mechanical and physical properties of heat-cured polymethyl methacrylate denture base resins. *J. Adv. Prosthodont.* **2013**, *5*, 241–247. [CrossRef] [PubMed]
111. Duraid, M.; Mudhaffar, M. Effect of modified zirconium oxide nano-fillers addition on some properties of heat cure acrylic denture base material. *Restor. Dent.* **2012**, *24*, 7.
112. Pfeiffer, P.; Rosenbauer, E.-U. Residual methyl methacrylate monomer, water sorption, and water solubility of hypoallergenic denture base materials. *J. Prosthet. Dent.* **2004**, *92*, 72–78. [CrossRef]
113. Miettinen, V.M.; Vallittu, P.K. Water sorption and solubility of glass fiber-reinforced denture polymethyl methacrylate resin. *J. Prosthet. Dent.* **1997**, *77*, 531–534. [CrossRef]
114. Lin, B.A.; Jaffer, F.; Duff, M.D.; Tang, Y.W.; Santerre, J.P. Identifying enzyme activities within human saliva which are relevant to dental resin composite biodegradation. *Biomaterials* **2005**, *26*, 4259–4264. [CrossRef]
115. Finer, Y.; Santerre, J.P. Salivary esterase activity and its association with the biodegradation of dental composites. *J. Dent. Res.* **2004**, *83*, 22–26. [CrossRef]
116. Rashid, H.; Sheikh, Z.; Vohra, F. Allergic effects of the residual monomer used in denture base acrylic resins. *Eur. J. Dent.* **2015**, *9*, 614–619. [CrossRef]
117. Goiato, M.C.; Freitas, E.; dos Santos, D.; de Medeiros, R.; Sonego, M. Acrylic resin cytotoxicity for denture base-literature review. *Adv. Clin. Exp. Med.* **2015**, *24*, 679–686. [CrossRef]
118. Cierech, M.; Wojnarowicz, J.; Kolenda, A.; Krawczyk-Balska, A.; Prochwicz, E.; Woźniak, B.; Łojkowski, W.; Mierzwińska-Nastalska, E. Zinc Oxide Nanoparticles cytotoxicity and release from newly formed PMMA–ZnO nanocomposites designed for denture bases. *Nanomaterials* **2019**, *9*, 1318. [CrossRef] [PubMed]
119. Sokołowski, K.; Szynkowska, M.I.; Pawlczyk, A.; Łukomska-Szymańska, M.; Sokołowski, J. The impact of nanosilver addition on element ions release form light-cured dental composite and compomer into 0.9% NaCl. *Acta Biochim. Pol.* **2014**, *61*, 317–323. [CrossRef] [PubMed]
120. Sleibi, A.; Tappuni, A.R.; Karpukhina, N.G.; Hill, R.G.; Baysan, A. A comparative evaluation of ion release characteristics of three different dental varnishes containing fluoride either with CPP-ACP or bioactive glass. *Dent. Mater.* **2019**, *35*, 1695–1705. [CrossRef] [PubMed]

121. Boeckler, A.F.; Morton, D.; Poser, S.; Dette, K.-E. Release of dibenzoyl peroxide from polymethyl methacrylate denture base resins: an in vitro evaluation. *Dent. Mater.* **2008**, *24*, 1602–1607. [CrossRef] [PubMed]
122. Ayman, A.-D. The residual monomer content and mechanical properties of CAD\CAM resins used in the fabrication of complete dentures as compared to heat cured resins. *Electron. Physician* **2017**, *9*, 4766–4772. [CrossRef]
123. Nguyen, L.G.; Kopperud, H.M.; Øilo, M. Water sorption and solubility of polyamide denture base materials. *Acta Biomater. Odontol. Scand.* **2017**, *3*, 47–52. [CrossRef]

© 2019 by the authors. Licensee MDPI, Basel, Switzerland. This article is an open access article distributed under the terms and conditions of the Creative Commons Attribution (CC BY) license (http://creativecommons.org/licenses/by/4.0/).

Article

Changes in the Surface Texture of Thermoplastic (Monomer-Free) Dental Materials Due to Some Minor Alterations in the Laboratory Protocol—Preliminary Study

Bozhana Chuchulska [1], Ilian Hristov [1,*], Boyan Dochev [2] and Raycho Raychev [2]

[1] Department of Prosthetic Dental Medicine, Faculty of Dental Medicine, Medical University of Plovdiv, 4000 Plovdiv, Bulgaria
[2] Department of Mechanics, Faculty of Mechanical Engineering, Technical University of Sofia, Branch Plovdiv, 4000 Plovdiv, Bulgaria
* Correspondence: ilian.hristov@mail.bg

Abstract: Contemporary thermoplastic monomer-free prosthetic materials are widely used nowadays, and there are a great variety available on the market. These materials are of interest in terms of the improvement of the quality features of the removable dentures. The aim of this study is to establish how minimal changes in the laboratory protocol of polyamide prosthetic base materials influence the surface texture. Two polyamide materials intended for the fabrication of removable dentures bases were used—Perflex Biosens (BS) and VertexTM ThermoSens (TS). A total number of 20 coin-shaped samples were prepared. They were injected under two different modes—regular, as provided by the manufacturer, and modified, proposed by the authors of this study. Scanning electronic microscopy (SEM) under four magnifications—×1000, ×3000, ×5000, and ×10,000—was conducted. With minimal alterations to the melting temperature (5 °C) and the pressure (0.5 Bar), in Biosens, no changes in terms of surface improvement were found, whereas in ThermoSens, the surface roughness of the material significantly changed in terms of roughness reduction. By modifying the technological mode during injection molding, a smoother surface was achieved in one of the studied materials.

Keywords: thermoplastic materials; laboratory protocol; dentures; texture; roughness

1. Introduction

The quality and efficiency of prosthetic treatments depend on the properties of the base prosthetic materials. It is often the case that with removable dentures, complications occur, e.g., denture stomatitis, caused by microflora with various degrees of virulence. Dental prostheses are potential sites of adsorption and colonization of various microorganisms. One of the conditions determining the degree of bacterial adhesion and colonization resistance is the surface structure of the base material.

Denture surface can be affected by various mechanisms. Such mechanisms may include aging and wear and tear [1]. Professional hygienic and cleaning procedures, as well as the instruments used during these procedures, increase the roughness of the material and the risk of future bacterial or fungal contamination [2]. Substantial changes in the surface morphology, increased hydrophilicity and higher optical density of the adhered microorganisms are observed when various chemical agents are used for denture cleaning [3].

Studying at a molecular level the correlation between the surface of the restorative material and the microorganisms in the oral cavity, G. Allias concluded that a conditio sine qua non for micro-floral contamination is related to the material's texture and depends on the surface tension [4]. The higher the surface tension, the higher the probability for pathogenic microbial contamination is. The surface tension of a given material depends on the material's technology and processing algorithm, as well as the inclusion of other materials over the prosthetic material's surface that alter the surface tension [5].

The most common pathogen causing denture stomatitis is *C. albicans* [6]. *C. albicans*, as a conditionally pathogenic species, can asymptomatically colonize both the surfaces of the denture and the mucosa [7]. Al-Dwairi emphasized the significance of *Candida* spp. isolated from the fingertips of removable denture wearers as a source of re-infection of the oral cavity [8]. L. Gendreau identified the spread of denture stomatitis in approximately 70% of the removable denture wearers, and the frequency is higher in elder patients of the female gender [9].

Conventional acrylic resin exhibits highly hydrophilic properties and solubility [10], as well as heterogeneity of the surface texture, further causing internal and surface tension and the formation of cavities where microorganisms infiltrate and propagate. This leads to the disturbance of the micro-biocenosis in the oral cavity, inflammation of the mucosa beneath the denture and the development of denture stomatitis of various etiologies [11].

Nowadays, a great variety of prosthetic materials are available on the market. However, the issue of their interaction with the oral microflora, as well as how the microflora affects these materials, is still understudied. Therefore, the correlation between the microfloral adhesion to the various prosthetic materials and their texture remains a topical question, as does the search for solutions for the improvement of the microstructure and degree of roughness of these materials.

Contemporary thermoplastic monomer-free prosthetic materials are of interest in terms of the improvement of the quality features of removable dentures. However, they are still not sufficiently explored regarding their microbial contamination and colonization. Reliable information can be obtained by performing microbiological and high-magnification microscopic studies in parallel. This would allow for exploring the structures at a nano level. The purpose of this study is to establish how minor alterations in the laboratory protocol of polyamide prosthetic base materials influence the surface texture of these materials.

2. Materials and Methods

2.1. Materials and Samples

In this study, two polyamide materials intended for the fabrication of removable dentures were used—Perflex Biosens (BS) and VertexTM ThermoSens (TS). A total number of 20 coin-shaped samples were prepared with a diameter of 5 mm and 1 mm thickness (Figure 1).

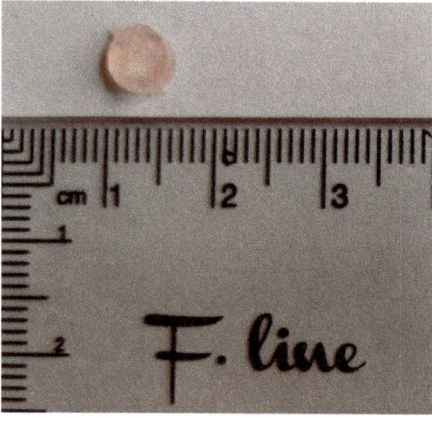

Figure 1. Test samples for observation.

2.2. Methods

2.2.1. Technological Mode

The samples were injected under two different modes—regular, as provided by the manufacturer (Table 1), and modified, proposed by the authors of this study (Table 2). Ten samples of the two tested materials were injected under the regular technological mode, and the other ten samples, five of each material, were injected under a modified mode.

Table 1. Materials, technological parameters, and manufacturer.

Material	Type	Time	Temperature	Pressure	System	Manufacturer
Perflex Biosens (BS)	Polyamide (MSDS: no declaration)	18 min	300 °C	8–9 Bar	Thermopress 400	Perflex, Israel
VertexTM ThermoSens (TS)	Polyamide (MSDS: no declaration)	18 min	290 °C	6 Bar	Vertex Thermoject 22	Vertex Dental B.V., The Netherlands

Table 2. Materials, modified technological parameters, and manufacturer.

Material	Type	Time	Temperature	Pressure	System	Manufacturer
Perflex Biosens (BS)	Polyamide (MSDS: no declaration)	18 min	305 °C	9.5 Bar	Thermopress 400	Perflex, Israel
VertexTM ThermoSens (TS)	Polyamide (MSDS: no declaration)	18 min	295 °C	6.5 Bar	Vertex Thermoject 22	Vertex Dental B.V., The Netherlands

2.2.2. Scanning Electronic Microscopy (SEM)

The test samples from the two polyamide materials, under the two different technological modes, were plated in 24-carat gold powder (Figure 2) and were scanned using SEM in four different magnifications: ×1000, ×3000, ×5000, and ×10,000.

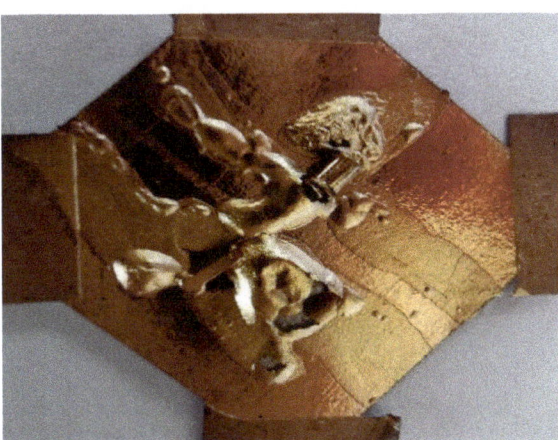

Figure 2. Samples ready for scanning.

2.2.3. Microbiological Evaluation

Microbiological evaluation of mucosal and denture surface samples was performed. Samples were collected by swabbing and transported to the laboratory of microbiology within the same day. Swabs were cultured on Sabouraud-dextrose agar (SDA) and incubated for up to 48 h at 30 °C. Colony identification was performed by using matrix-assisted laser desorption time-of-flight mass spectrometry (MALDI-TOF MS, Vitek MS, bioMerieux, Craponne, France). Samples were stained with Löffler methylene blue and observed using ×100 immersion oil microscopy.

3. Results

3.1. Samples under Regular Technological Mode

The investigation with SEM methods of the samples injected under the regular technological mode showed different types of defects and numerous spots of unevenness on the surface of both materials under all magnifications. (Figures 3a,b and 4a,b).

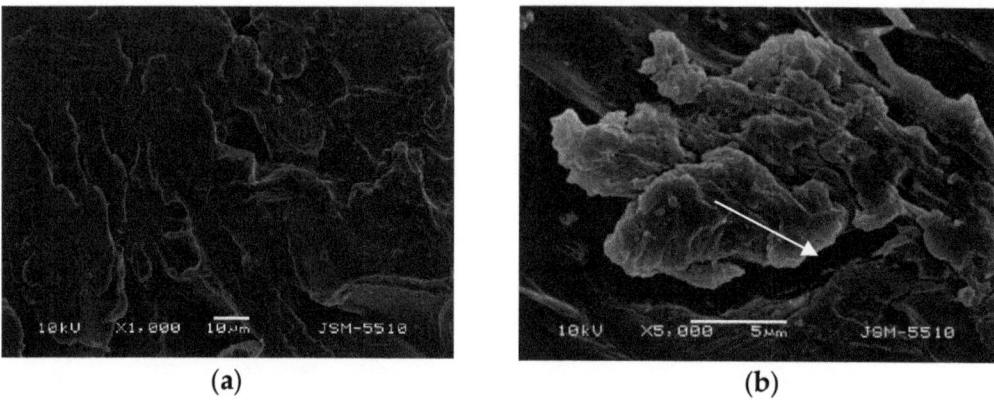

Figure 3. (a) BS under magnification ×1000. (b) BS under magnification ×5000.

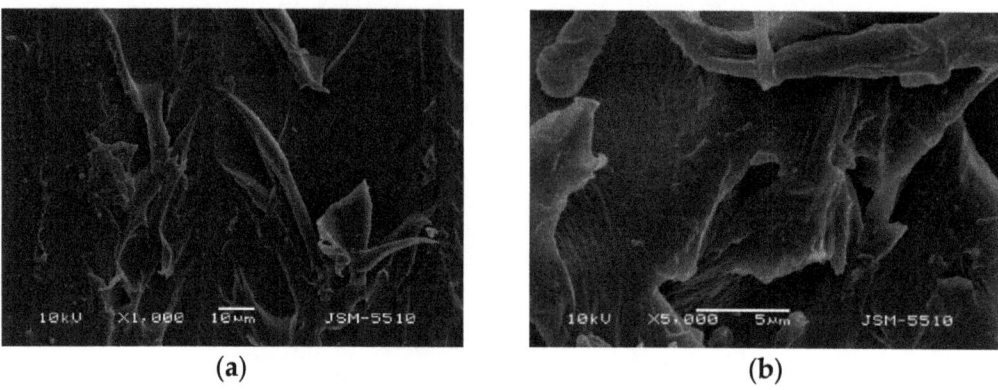

Figure 4. (a) TS under magnification ×1000. (b) TS under magnification ×5000.

3.1.1. Samples Made of Biosens

On the surface of the BS test samples, holes, openings, deep grooves, caverns, and some areas of a rough surface resembling orange peel can be observed. In the ×5000 magnification photo, the dimensions of these surface defects can be measured, and they vary a lot. Portions of the surface display a mica-like texture.

3.1.2. Samples Made of Thermosens

On the TS surface, expressed unevenness with openings, grooves, caverns and a mica-like surface can be observed, along with some bulging formations and deep and undermined areas, and at some points surface destruction can be observed. In the ×5000 magnification photo, it can be observed that these defects form undermined and predilection zones for the retention of different microorganisms.

3.2. Samples under Modified Technological Mode

3.2.1. Samples Made of Biosens

Observations of the surface of BS samples prepared under the modified technological mode do not show any significant differences in the defects compared to the test samples injected under optimal fabrication parameters. Under ×1000 magnification, slight smoothing of the texture is observed; however, the mica-like surface remains unchanged, and the presence of openings and canals is clearly visible (Figure 5a,b). A magnification of ×5000 reveals that these openings grow into deep caverns more than 20 microns in size.

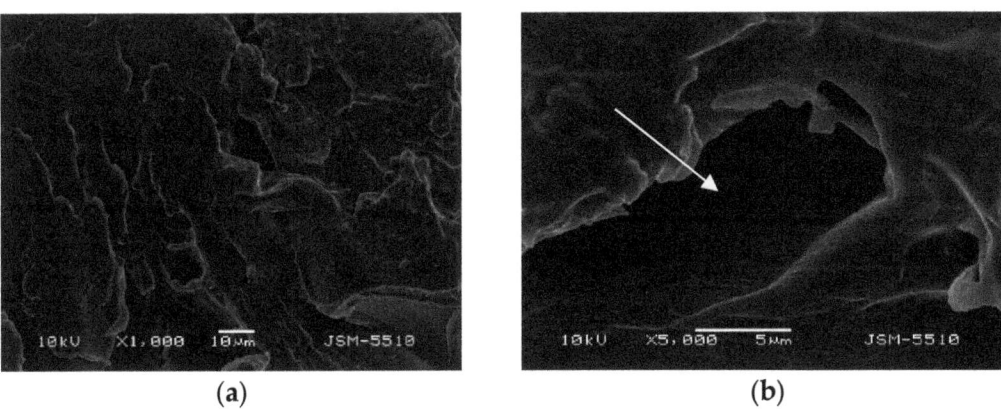

Figure 5. (a) BS under magnification ×1000. (b) BS under magnification ×5000.

3.2.2. Samples Made of Thermosens

Observations of the surface of TS test samples prepared under the modified technological mode show significant differences in the surface characteristics compared to the test samples injected under the optimal technological mode. Under ×1000 magnification, smoothening of the texture is observed, where shallow grooves and unevenness with a bubble-like shape can be seen; the structure is slightly wavy (Figure 6a,b). Under a magnification of ×5000, the surface is orange peel-textured; however, the uneven areas and deep defects do not exceed 1–3 microns. It should be noted that the refinement of the surface texture of this material is a direct result of the technological mode modification, but the effect on the mechanical properties has yet not been investigated.

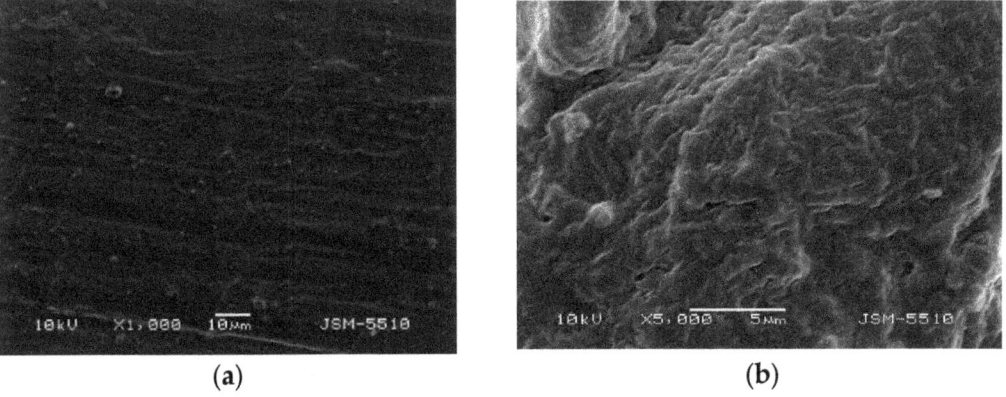

Figure 6. (a) TS under magnification ×1000. (b) TS under magnification ×5000.

Although the form of the defects is too complex to be measured precisely, some dimensions are given in the following table (Table 3).

Table 3. Materials, mode, and dimensions of the defects.

Material	Mode	Sample No.1 Length/Width (Microns)	Sample No.2 Length/Width (Microns)	Sample No.3 Length/Width (Microns)	Sample No.4 Length/Width (Microns)	Sample No.5 Length/Width (Microns)	Mean Value Length/Width (Microns)
Thermosens	Regular mode	15/18	12/12	20/14	15/12	16/15	15.6/14.2
	Modified mode	1/1	1.5/1	3/1	1.2/1	1/1	1.54/1
Biosens	Regular mode	20/25	23/21	15/14	21/20	28/25	21.4/21
	Modified mode	12/10	14/10	15/15	10/10	12/10	12.6/11

Ten patients were included in this pilot study. Five Thermosens dentures and five Biosens dentures were created. The patients were examined during regular (every two weeks) follow-ups. Two of them (one male, 72 years old and one female patient, 69 years old) showed clinical symptoms of denture stomatitis (Figure 7).

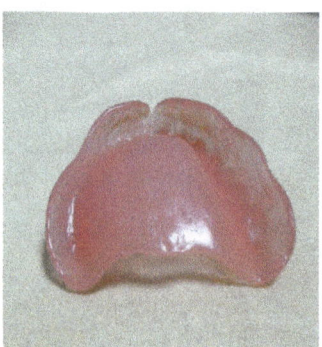

Figure 7. Patient's denture made from ThermoSens using the conventional laboratory method.

The method of direct fluorescence visualization with the help of the VELscope® (LED Dental, Inc., White Rock, BC, Canada) device was applied. Contamination not only of the mucosa (Figure 8a,b) beneath the denture but also on the denture surface itself was ascertained (Figure 9).

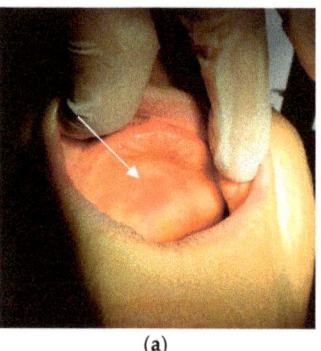 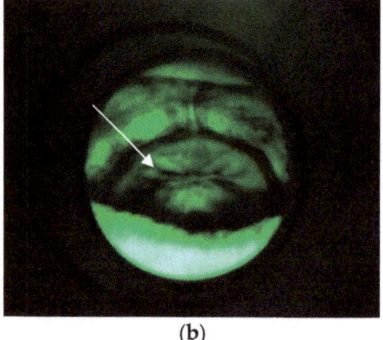

(a)   (b)

Figure 8. (a) Mucosa with lesions. (b) Fungal colonies over the mucosa.

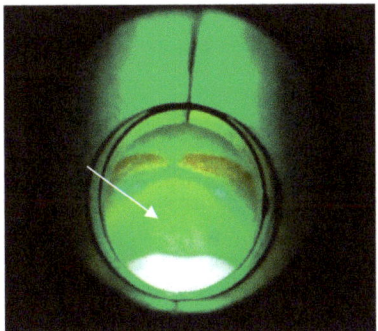

Figure 9. Colonized ThermoSens denture.

The symptoms started at the end of the sixth week for the female patient and at the beginning of the tenth week for the male patient. Neither of them suffered any general disease (except high blood pressure for the male patient and osteoporosis for the female patient). Both were treated with dentures made from Thermosens (under regular laboratory mode).

After culturing of the samples on Sabouraud-dextrose agar (SDA), the present colonies were subsequently identified by MALDI-TOF MS as *Candida albicans*. (Figure 10a,b).

 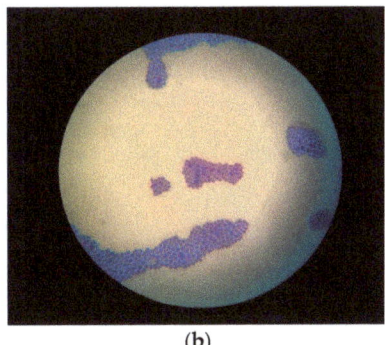

(a)   (b)

Figure 10. (a) SDA agar. Fungal growth on 3 (sample taken from the palate) and 4 (sample taken from the denture). No fungal growth observed on the negative control (n. c.). 2—sample from the vibrating line area, 5—sample from tuber maxillae sin., 6—sample from tuber maxillae dex. (b) Fungi, methylene blue stain, ×100 immersion oil microscopy.

4. Discussion

The oral cavity is a habitat for microorganisms in large quantities and numerous varieties—pathogens, conditional pathogens, and saprophytes. The coarse and rough surface of dental prosthesis, the retention of food, and the constant humidity and temperature present suitable conditions for microbial contamination, colonization and propagation.

The surface characteristics of thermoplastic polymers exhibit numerous defects and a high level of roughness [12] that allow for the microbial colonization of their surface. Thermoplastic materials are challenging in terms of mechanical processing, making it difficult to produce a smooth and glossy surface [13]. The lack of this smoothness represents the optimal conditions [14] for the adhesion of microbial cells [15]. Although polyamide materials are characterized by a high level of mechanical properties, a modification [16] of the technological parameters [17,18] of their injection could be attempted to achieve an optimal texture. This modified surface needs to be resistant to impacts that would increase roughness [19,20] or deteriorate the quality of the material [20,21].

Surface modification could be a possible approach to identify surfaces that possess anti-biofilm properties [22]. The injection mode is precise and too short in duration, yet it depends on conditions and factors that could be manipulated, and the injection molding devices allow for it.

Attempting to improve the polyamide materials' surface characteristics so that a surface with better anti-microbial [23] and bacterial attack inhibition effects is obtained, as well as a reduction in microbial activity [24–30], the authors altered some of the factors in the injection molding process. These factors and conditions are interdependent and mutually affecting.

With minimal alteration of the melting temperature (5 °C) and the pressure (0.5 Bar), no changes in terms of surface improvement were found in Biosens. What led us to apply variation of the temperature was the expectation that this would result in more even and more thorough melting of the material inside the machine tumbler. On the other hand, with a rise in temperature, the melt flow speed in the sprues changes as well, leading to a quicker filling of the mold, preventing uneven cooling of the material.

With minimal alteration to both the temperature (5 °C) and the pressure (0.5 Bar) in ThermoSens, the surface roughness of the material is significantly changed [31] in terms of roughness reduction [32]. This positive change in the surface texture is likely to result in: an improvement [33] in the mechanical strength and physical properties, a lack of microflora [34] or minimal changes [21], as well as a reduction in the conditions for colony formation [35,36]. To ascertain the presence or absence of such changes, further studies are necessary, including not only in vitro, but in vivo tests as well. A few volunteers are planned to be examined, treated with dentures manufactured using the modified laboratory protocol in a future study.

Ayaz et al. stated that striving for improvements in the texture of injection-molded materials is based on the fact that surface imperfection affects the adhesion and colonization of pathogenic microorganisms. Biofilm accumulation is the main factor in the etiology of denture stomatitis, emerging due to surface irregularities [37].

Verran and Maryan [38], Quirynen et al. [39], and Radford et al. [40] reported that dental materials on polyamide bases are rougher than PMMA materials. This statement is in agreement with Yunus et al. [41], Ucar et al. [42], and Kurkcuoglu et al. [43]. In their studies, they found a direct correlation between the surface roughness and adhesion of microorganisms. These findings correspond with some previous investigations of the authors of this article.

Kohli and Bhatia stated that the hydrophilic behavior of polyamide materials is due to the amide groups in their polymeric chain. Nylon, being hygroscopic, swells when immersed in a humid medium, increasing its irregularities [44].

Some substances, including saliva, alcohol, and acids produced by bacteria, may affect the structure and surface features of the restorations [45]. Arslan et al. assumed that material aging increases roughness and hydrophilicity [46]. Atalaya et al., in their study, declared that a smoother surface guarantees higher hydrophobicity and lower surface tension [47]. Liebermann et al. concluded that increased temperature and pressure of injection may change the polarity of the molecules, and that this can consequently cause alterations in the surface structure and wetting [48].

It is assumed that raised temperature leads to better and more even melting of the material, while increasing pressure leads to quicker and more uniform filling-up of the mold and therefore to its more uniform cooling down, both on the surface and internally. Both factors can reduce the cooling-induced tension on the surface and within the mold, and finally, this can cause the smoothing of the surface texture of the injected material.

5. Conclusions

By modifying the technological mode during injection molding, a smoother surface was achieved in one of the studied materials, and this variation could affect other factors and conditions during the process. Further studies should be conducted to find out whether such changes in the laboratory protocol affect the mechanical properties of these materials, and if so, in what range.

Author Contributions: Conceptualization: B.C., Methodology: B.D., R.R., Writing—review and editing: I.H. All authors have read and agreed to the published version of the manuscript.

Funding: This research received no external funding.

Institutional Review Board Statement: Not applicable.

Informed Consent Statement: Written informed consent was obtained from the patient(s) to publish this paper.

Acknowledgments: We thank Yordan Kalchev for his valuable help in terms of the microbiological evaluation of the test samples.

Conflicts of Interest: The authors declare no conflict of interest.

References

1. McKellop, H.A. The lexicon of polyethylene wear in artificial joints. *Biomaterials* **2007**, *28*, 5049–5057. [CrossRef] [PubMed]
2. Grande, F.; Zamperoli, E.M.; Pozzan, M.K.; Tesini, F.; Catapano, S. Qualitative Evaluation of the Effects of Professional Oral Hygiene Instruments on Prosthetic Ceramic Surfaces. *Materials* **2022**, *15*, 21. [CrossRef] [PubMed]
3. Yu-Shana, H.; Cheng-Yuana, H.; Her-Hsiung, H. Surface changes and bacterial adhesion on implant abutment materials after various clinical cleaning procedures. *J. Chin. Med. Assoc.* **2019**, *82*, 643–650. [CrossRef]
4. Qurynen, M.; Listgarten, M. oa oo a. *Clin. Oral. Implant. Res.* **1990**, *1*, 13.
5. Allais, G. Biofilm of the oral cavity. *New Dent.* **2005**, *4*, 4–14.
6. Kinkela-Devcic, M.; Simonic-Kocijan, S.; Prpic, J.; Paskovic, I.; Cabov, T.; Kovac, Z.; Glazar, I. Oral candidal colonization in patients with different prosthetic appliances. *J. Fungi* **2021**, *7*, 662. [CrossRef]
7. Tsui, C.; Kong, E.F.; Jabra-Rizk, M.A. Pathogenesis of Candida Albicans Biofilm. *Pathog. Dis.* **2016**, *74*, ftw018. [CrossRef]
8. Al-Dwairi, Z.N. Isolation of candida species from the oral cavity and fingertips of complete and partial dentures wearers. *J. Dent. Health Oral Disord. Ther.* **2014**, *1*, 420–423. [CrossRef]
9. Gendreau, L.; Loewy, Z.G. Epidemiology and etiology of denture stomatitis. *J. Prosthodont.* **2011**, *20*, 251–260. [CrossRef]
10. Shah, J.; Bulbule, N.; Kulkarni, S. Comparative evaluation of sorption, solubility and microhardness of heat cure polymethylmethacrylate denture base resin and flexible denture base resin. *J. Clin. Diagn. Res.* **2014**, *8*, ZF01–ZF04. [CrossRef]
11. Gozhaya, L.D. Oral Mucosa Diseases Caused by Denture Materials. In *Abstract of a Doctoral Thesis of Medical Sciences*; 2001; p. 20.
12. Durkan, R.; Ayaz, E.A.; Bagis, B. Comparative effects of denture cleansers on physical properties of polyamide and polymethyl methacrylate base polymers. *Dent. Mater. J.* **2013**, *32*, 367–375. [CrossRef] [PubMed]
13. O' Brien, W.J. Chicago: Quintessence. In *Dental Materials and Their Selection*, 4th ed.; Pub. Co., Inc.: Singapore, 2009; pp. 78–79. Available online: http://www.quintpub.com/PDFs/book_preview/B4375.pdf (accessed on 11 July 2022).
14. Rubtsova, .A.; Chirkova, N.V.; Polushkina, N.A.; Kartavtseva, N.G.; Vecherkina, Z.V.; Popova, T.A. Evaluation Of the Microbiological Examination of Removable Dentures of Thermoplastic Material. *J. New Med. Technol.* **2017**, *2*, 314–322.
15. Bulad, K.; Taylor, R.L.; Verran, J.; McCord, J.F. Colonization and penetration of denture soft lining materials by candida albicans. *Dent. Mater.* **2004**, *20*, 167–175. [CrossRef]
16. Vojdani, M.; Giti, R. Polyamide as a denture base material: A literature review. *J. Dent. Shiraz.* **2015**, *16*, 1–9. [PubMed]
17. Srinivasan, M.; Kamnoedboon, P.; McKenna, G.; Angst, L.; Schimmel, M.; Ozcan, M.; Müller, F. CAD-CAM removable complete dentures: A systematic review and meta-analysis of trueness of fit, biocompatibility, mechanical properties, surface characteristics, color stability, time-cost analysis, clinical and patient-reported outcomes. *J. Dent.* **2021**, *113*, 103777. [CrossRef]
18. Fouda, S.M.; Gad, M.M.; Abualsaud, R.; Ellakany, P.; AlRumaih, H.S.; Khan, S.Q.; Akhtar, S.D.; Al-Qarn, F.; Al-Harbi, F.A. Flexural Properties and Hardness of CAD-CAM Denture Base Materials. *J. Prosthodont.* **2022**, *2*, 1–7. [CrossRef]
19. Tripathi, P.; Phukela, S.S.; Yadav, B.; Malhotra, P. An in vitro study to evaluate and compare the surface roughness in heat-cured denture-based resin and injection-molded resin system as affected by two commercially available denture cleansers. *J. Indian Prosthodont. Soc.* **2018**, *18*, 291–298. [CrossRef]
20. Szczesio-Wlodarczyk, A.; Domarecka, M.; Kopacz, K.; Sokolowski, G.; Bociong, K. Evaluation of the Properties of Urethane Dimethacrylate-Based Dental Resins. *Materials* **2021**, *14*, 2727. [CrossRef]
21. Chladek, G.; Nowak, M.; Pakieła, W.; Mertas, A. Effect of Candida albicans Suspension on the Mechanical Properties of Denture Base Acrylic Resin. *Materials* **2022**, *15*, 3841. [CrossRef]
22. D'Ercole, S.; De Angelis, F.; Biferi, V.; Noviello, C.; Tripodi, D.; Di Lodovico, S.; Cellini, L.; D'Arcangelo, C. Antibacterial and Antibiofilm Properties of Three Resin-Based Dental Composites against Streptococcus mutans. *Materials* **2022**, *15*, 1891. [CrossRef]
23. Chladek, G.; Pakieła, K.; Pakieła, W.; Żmudzki, J.; Adamiak, M.; Krawczyk, C. Effect of antibacterial silver-releasing filler on the physicochemical properties of poly (methyl methacrylate) denture base material. *Materials* **2019**, *12*, 4146. [CrossRef] [PubMed]

24. Fan, C.; Chu, L.; Rawls, H.R.; Norling, B.K.; Cardenas, H.L.; Whang, K. Development of an antimicrobial resin—A pilot study. *Dent. Mater.* **2011**, *27*, 322–328. [CrossRef] [PubMed]
25. Bajunaid, S.O. How effective are antimicrobial agents on preventing the adhesion of candida albicans to denture base acrylic resin materials? A systematic review. *Polymers* **2022**, *14*, 908. [CrossRef]
26. Paranhos, H.F.; Davi, L.R.; Peracini, A. Comparison of physical and mechanical properties of microwave-polymerized acrylic resin after disinfection in sodium hypochlorite solutions. *Braz. Dent. J.* **2009**, *20*, 331–335. [CrossRef] [PubMed]
27. Totu, E.E.; Nechifor, A.C.; Nechifor, G.; Aboul-Enein, H.Y.; Cristache, C.M. Poly (Methyl Methacrylate) with TiO2 Nanoparticles Inclusion for Stereolithographic Complete Denture Manufacturing—The Future in Dental Care for Elderly Edentulous Patients? *J. Dent.* **2017**, *59*, 68–77. [CrossRef]
28. Wady, A.F.; Machado, A.L.; Zucolotto, V.; Zamperini, C.A.; Berni, E.; Vergani, C.E. Evaluation of Candida Albicans Adhesion and Biofilm Formation on a Denture Base Acrylic Resin Containing Silver Nanoparticles. *J. Appl. Microbiol.* **2012**, *112*, 1163–1172. [CrossRef]
29. Lee, M.-J.; Kim, M.-J.; Oh, S.-H.; Kwon, J.-S. Novel Dental Poly (Methyl Methacrylate) Containing Phytoncide for Antifungal Effect and Inhibition of Oral Multispecies Biofilm. *Materials* **2020**, *13*, 371. [CrossRef]
30. Chladek, G.; Basa, K.; Mertas, A.; Pakieła, W.; Żmudzki, J.; Bobela, E.; Król, W. Effect of storage in distilled water for three months on the antimicrobial properties of poly (methyl methacrylate) denture base material doped with inorganic filler. *Materials* **2016**, *9*, 328. [CrossRef]
31. Romanov, B.G. Design of Complex Plastic Elements Based on Modeling and Study of the Filling Process Through Virtual Prototyping. Ph.D. Thesis, Technical University, Sofia, Bulgaria, 2015.
32. Ozyilmaz, O.Y.; Akin, C. Effect of cleansers on denture base resins' structural properties. *J. Appl. Biomater. Funct. Mater.* **2019**, *17*, 2280800019827797. [CrossRef]
33. Beltrán-Partida, E.; Valdez-Salas, B.; Curiel-Álvarez, M.; Castillo-Uribe, S.; Escamilla, A.; Nedev, N. Enhanced Antifungal Activity by Disinfected Titanium Dioxide Nanotubes via Reduced Nano-Adhesion Bonds. *Mater. Sci. Eng. C* **2017**, *76*, 59–65. [CrossRef]
34. Mangal, U.; Kim, J.-Y.; Seo, J.-Y.; Kwon, J.-S.; Choi, S.-H. Novel Poly (Methyl Methacrylate) Containing Nanodiamond to Improve the Mechanical Properties and Fungal Resistance. *Materials* **2019**, *12*, 3438. [CrossRef] [PubMed]
35. Takabayashi, Y. Characteristics of denture thermoplastic resins for non-metal clasp dentures. *Dent. Mater. J.* **2010**, *29*, 353–361. [CrossRef] [PubMed]
36. Bajunaid, S.O.; Baras, B.H.; Balhaddad, A.A.; Weir, M.D.; Xu, H.H. Antibiofilm and Protein-Repellent Polymethylmethacrylate Denture Base Acrylic Resin for Treatment of Denture Stomatitis. *Materials* **2021**, *14*, 1067. [CrossRef] [PubMed]
37. Ayaz, .A.; Bagis, B.; Turgut, S. Effects of thermal cycling on surface roughness, hardness and flexural strength of polymethylmethacrylate and polyamide denture base resins. *J. Appl. Biomater. Funct. Mater.* **2015**, *13*, e280–e286. [CrossRef]
38. Verran, J.; Maryan, C.J. Retention of Candida albicans on acrylic resin and silicone of different surface topography. *J. Prosthet. Dent.* **1997**, *77*, 535–539. [CrossRef]
39. Quirynen, M.; Marechal, M.; Busscher, H.J.; Weerkamp, A.H.; Darius, P.L.; van Steenberghe, D. The influence of surface free energy and surface roughness on early plaque formation: An in vivo study in man. *J. Clin. Periodontol.* **1990**, *17*, 138–144. [CrossRef]
40. Radford, D.R.; Sweet, S.P.; Challacombe, S.J.; Walter, J.D. Adherence of Candida albicans to denture-base materials with different surface finishes. *J. Dent.* **1998**, *26*, 577–583. [CrossRef]
41. Yunus, N.; Rashid, A.A.; Azmi, L.L.; Abu-Hassan, M.I. Some flexural properties of a nylon denture base polymer. *J. Oral. Rehabil.* **2005**, *32*, 65–71. [CrossRef]
42. Ucar, Y.; Akova, T.; Aysan, I. Mechanical properties of polyamide versus different PMMA denture base materials. *J. Prosthodont.* **2012**, *21*, 173–176. [CrossRef]
43. Kurkcuoglu, I.; Koroglu, A.; Ozkır, S.; Ozdemir, T.A. Comparative study of polyamide and PMMA denture base biomaterials: I. thermal, mechanical, and dynamic mechanical properties. *Int. J. Polym. Mater.* **2012**, *61*, 768–777. [CrossRef]
44. Kohli, S.; Bhatia, S. Polyamides in dentistry. *Int. J. Sci. Study* **2013**, *1*, 120–125.
45. Munchow, E.A.; Ferreira, A.C.; Machado, R.M.; Ramos, T.S.; Rodrigues-Junior, S.A.; Zanchi, C.H. Effect of acidic solutions on the surface degradation of a micro-hybrid composite resin. *Braz. Dent. J.* **2014**, *25*, 321–326. [CrossRef] [PubMed]
46. Arslan, .; Murat, S.; Alp, G.; Zaimoglu, A. Evaluation of flexural strength and surface properties of prepolymerized CAD/CAM PMMA-based polymers used for digital 3D complete dentures. *Int. J.Comput. Dent.* **2018**, *21*, 31–40. [PubMed]
47. Atalaya, S.; Çakmakb, G.; Fonsecac, M.; Schimmel, M.; Yilmazcef, B. Effect of thermocycling on the surface properties of CAD-CAM denture base materials after different surface treatments. *J. Mech. Behav. Biomed. Mater.* **2021**, *121*, 104646. [CrossRef] [PubMed]
48. Liebermann, A.; Wimmer, T.; Schmidlin, P.R.; Scherer, H.; Loffler, P.; Roos, M.; Stawarczyk, B. Physicomechanical characterization of polyetheretherketone and current esthetic dental CAD/CAM polymers after aging in different storage media. *J. Prosthet. Dent.* **2016**, *115*, 321–328.e2. [CrossRef]

Article

Glass Fiber Reinforced Composite Orthodontic Retainer: In Vitro Effect of Tooth Brushing on the Surface Wear and Mechanical Properties

Maria Francesca Sfondrini [1], Pekka Kalevi Vallittu [2], Lippo Veli Juhana Lassila [2], Annalisa Viola [1], Paola Gandini [1] and Andrea Scribante [1,*]

1. Unit of Orthodontics and Paediatric Dentistry, Section of Dentistry, Department of Clinical, Surgical, Diagnostic and Paediatric Sciences, University of Pavia, 27100 Pavia, Italy; francesca.sfondrini@unipv.it (M.F.S.); annalisa.viola01@universitadipavia.it (A.V.); paola.gandini@unipv.it (P.G.)
2. Department of Biomaterial Science and Turku Clinical Biomaterials Centre (TCBC), Institute of Dentistry, University of Turku, 20100 Turku, Finland; pekval@utu.fi (P.K.V.); liplas@utu.fi (L.V.J.L.)
* Correspondence: andrea.scribante@unipv.it; Tel.: +39-382-516223; Fax: +39-382-516224

Received: 27 January 2020; Accepted: 21 February 2020; Published: 25 February 2020

Abstract: Fiber reinforced composites (FRCs) are metal free materials that have many applications in dentistry. In clinical orthodontics, they are used as retainers after active treatment in order to avoid relapse. However, although the modulus of the elasticity of FRCs is low, the rigidity of the material in the form of a relatively thick retainer with a surface cover of a flowable resin composite is known to have higher structural rigidity than stainless steel splints. The aim of the present study is to measure load and bending stress of stainless steel wires, as well as flowable resin composite covered and spot-bonded FRC retainer materials after tooth brushing. These materials were tested with a three point bending test for three different conditions: no brushing, 26 min of brushing, and 60 min of brushing. SEM images were taken before and after different times of tooth brushing. Results showed that stainless steel was not significantly affected by tooth brushing. On the other hand, a significant reduction of values at maximum load at fracture was reported for both FRC groups, and uncovered FRCs were most affected. Concerning maximum bending stress, no significant reduction by pretreatment conditions was reported for the materials tested. SEM images showed no evident wear for stainless steel. Flowable resin composite covered FRCs showed some signs of composite wear, whereas spot-bonded FRCs, i.e., without the surface cover of a flowable resin composite, showed signs of wear on the FRC and exposed glass fibers from the FRC's polymer matrix. Because of the significant changes of the reduction of maximum load values and the wear for spot-bonded FRCs, this technique needs further in vitro and in vivo tests before it can be performed routinely in clinical practice.

Keywords: FRC; bonding; technique; fiber; reinforced; composite; spot; mechanical; deflection; orthodontics; brushing; wear; retainer; splint; load

1. Introduction

During the last years, fiber reinforced composites (FRCs) have been proposed for many clinical applications because they are easy to customize and manipulate, and they showed high improvements in properties if compared to unreinforced resins [1,2].

FRCs, if compared with other materials, have a high strength/weight and stiffness/weight ratio [3]. The most common fibers used in dentistry are glass fibers because of their low extensibility, high tensile strength, and aesthetic and optical qualities [4,5]. In dentistry, these materials are used for various purposes: fixed dental prostheses, fillings and core-built ups, removable devices, root canal anchoring systems, periodontal and trauma splints, orthodontic frameworks, and retainers [6–13].

Retention is an important phase of orthodontic treatment, especially because without any type of retention there is a tendency for the teeth to relapse [14]. Fixed retainers are the most common retention systems as they have a number of advantages. First of all, they provide better aesthetics, additional patient cooperation is not needed, and they are suitable for lifelong retention [4]. In the last 10 years, multi-stranded wires were the most popular type of retention in regards to bonded fixed retainers. At the same, time resin fiberglass bondings were introduced as an aesthetic alternative and they are widely used. Nowadays, there are many different types of retainers, removable and fixed, and it is unclear which ones are the best and for how long they should be used.

To our knowledge, there are no studies in literature about the prevalence of different bonding techniques, and each clinician uses the one that is more suitable according to its experience. There are only studies about the survival of different splinting techniques [9,15]. An extensive study of the literature suggests that there are significant variations in the results describing the effectiveness, cost factors, survival times, oral hygiene status, and regimen of various orthodontic retention appliances [16].

FRC retainers are composed of glass fibers, thermoplastic polymer, and light-cured resin matrix for the reinforcement of the dental polymer. [17,18]. Continuous unidirectional FRCs which are used in dental applications have a flexural modulus of 17 GPa and are influenced by absorption of water and volume fraction of fibers in the composite [19,20]. When glass FRC is used as an orthodontic retainer, the cross-sectional diameter of the retainer increases and this increases shear stresses within the FRC, as well as structural rigidity of the retainer. Some studies demonstrated that, if compared with metallic wires, FRC splints present high deflection values, showing mean indicative stiffness about 30 and 40 N under deflections of 1 and 2 mm with the average span length and cross-sectional dimension of an orthodontic retainer [21,22]. These values are higher than those for metallic wires, which are thinner in cross-sectional diameter. A high stiffness is useful in prosthodontics but is less desirable for splints and retainers as it can be in contrast with physiological tooth movement increasing the ankylosis risk, even if this concern has been tested in a single in vivo study, using an animal model [23]. Other studies demonstrated that the rigidity of a FRC splint is magnified by the FRC application technique [24,25].

In fact, following the manufacturer's instructions, the composite covers the entire surface of the retainer. On the contrary, metallic splints are manufactured with a spot-bonding technique, thus allowing the wire to be covered with a composite only on the tooth surface, while the wire is left exposed in interproximal areas. The application of the spot bonding technique to FRCs implies a significant decrease in the rigidity of the framework if compared with a conventional full-bonded technique [24]. However, in this case, the fiber is exposed to the oral environment, thus leading to a higher wear risk, especially when the patient is eating or brushing their teeth. Wear is defined as the progressive loss of substance. It depends on the material type and geometry, on the interactions (stresses and forces), and on the environmental conditions (temperature, chemistry) [26].

Previous authors demonstrated that, increasing the number of brushing cycles, the abrasion of composite resins increased in a linear way [27]. There are no studies evaluating force levels of FRCs bonded with a spot technique simulating a different tooth brushing entity. Tooth brushing affects mechanical properties and wear of restorative materials. Authors evaluated its effects on a composite and we can suppose that FRC materials could probably also be affected, even if there is a lack of studies on this topic. The rationale of the present study was to test different FRC coverages for different variables.

The rationale of the present report is based on previous studies that demonstrated that FRC retainers have a higher rigidity if compared to metal splints and that this feature is less desirable. Previous studies also demonstrated that retainers made of FRCs without composite coverage have values of rigidity more similar to values of metal retainers, and this is a positive feature [21,22,24,25]. Finally brushing interferes with all restorative materials previously tested changing their mechanical properties [27]. Therefore, the data presented have the aim to test mechanical properties, maximum load and maximum bending stress, of covered and uncovered FRC, compare them to stainless steel, and evaluate eventual changing after tooth brushing. Additionally, their aim is to visually observe surfaces.

The purpose of this study was to evaluate mechanical properties (maximum bending stress and maximum load) and surface wear (SEM analysis) of metallic and FRC splints after various amounts of electrical tooth brushing, comparing the effects of tooth brushing on different kinds of materials. The null hypothesis of the present report was that no significant differences were reported in mechanical properties and wear among various materials tested.

2. Materials and Methods

In the present report, flat metallic splints (Straight 8 Lingual Retainer Wire 6' length. DB Orthodontics, Silsden, United Kingdom), FRCs (Everstick ORTHO, StickTech, Turku, Finland) with composite coverage, and FRCs (Everstick ORTHO, StickTech, Turku, Finland) without composite coverage were tested. The fiber reinforced composite that was tested in the present study was a unidirectional FRC reinforced with silanized-treated glass fibers. This fiber-reinforced retainer contained 1000 silanized glass fibers plunged in a monomer-polymer gel matrix. With these materials, 72 specimens were prepared, 24 for each group. All specimens were cut with scissors to a size of 20 mm and handled accordingly to the manufacturer's guidelines.

Metal Specimens Were Cut

Covered FRCs specimens were cut, covered with resin (Everstick Resin, GC America, Alsip, IL, USA), and subsequently light cured by hand using a halogen lamp (D-Light Pro, GC Europe, Leuven, Belgium) with a 1400 mW/cm^2 wave length range of 430–480 nm for 40 s. Subsequently, the specimens were covered with a flowable particulate filler composite by hand (G-aenial Universal Injectable A2, GC America, Alsip, IL, USA) and light cured with the same halogen unit for 40 s [28]. The light tip of the curing unit was kept in 3 mm distance from the materials during light curing. Non-covered FRC specimens were prepared with the same technique without final flow composite coverage. All specimens were than stored in an incubator at 37 degrees Celsius for 24 h.

For each group (metal, covered FRC, non-covered FRC) 8 specimens were not brushed, 8 were subjected to electronic brushing for 26 min, and 8 were subjected to electronic brushing for 60 min. Brushed specimens were fixed at their edges on a laboratory glass (Figure 1) and brushed with an electric toothbrush (Oral B PRO 670 with Oral B Crossaction brush heads, Procter & Gamble, Cincinnati, USA) using a 124 RDA toothpaste (MaxWhite-white crystals, Colgate-Palmolive, New York, NY, USA).

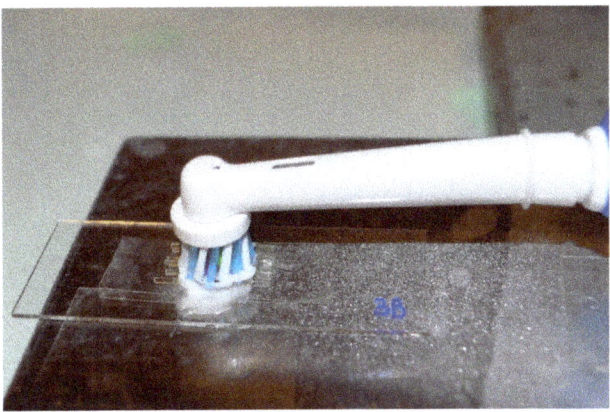

Figure 1. Brushing apparatus.

The brushing set up was performed following a previous study [29]. Our aim was to attain contact between brush heads and specimens, which were posed on laboratory glasses and fixed by their extremity. Toothpaste was put on toothbrushes and toothbrushes were activated by pressing the start button. The chronometer was used in order to respect the time point.

Retainer material grouping was divided as follows:

(1) Flat Metal—Not brushed;
(2) Flat Metal—26 min brushed;
(3) Flat Metal—60 min brushed;
(4) Covered FRC—Not brushed;
(5) Covered FRC—26 min brushed;
(6) Covered FRC—60 min brushed;
(7) Non-covered FRC—Not brushed;
(8) Non-covered FRC—26 min brushed;
(9) Non-covered FRC—60 min brushed.

The retainer materials were subsequently evaluated with a three-point bending test (Figure 2). The span length was 14 mm, and the crosshead speed was 1 mm/min [30]. A universal testing machine (Lloyd LRX; Lloyd Instruments, Fareham, UK) was used to apply the load on the middle of the specimens tested. The middle point of the machine was moved using a computer-controlled stepper motor, the force was recorded by electronic sensors, and the position of the middle point of the machine was associated with the passive position. The flexural strength values were recorded with the Nexygen MT software (Lloyd Instruments) [31].

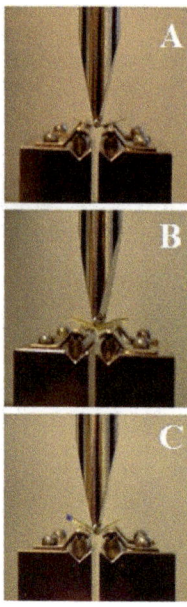

Figure 2. Three point bending test. (**A**) Metal wire, (**B**) fiber reinforced composite (FRC) full coverage, (**C**) FRC no coverage.

Maximum load (N) and maximum bending stress (MPa) were measured for each group [32]. Using a scanning electron microscope (JEOL 5500, JEOL Ltd., Tokyo, Japan), microphotographs for all the materials tested were taken with a magnification of 35×, 100×, and 250×. In order to use the scanning electron microscope, specimens were first submitted to sputter coating (BAL-TEC SCD050 Sputter Coater, Capovani Brothers Inc., New York, NY, USA).

Statistical analysis was performed with a computer software (R®version 3.1.3, R Development Core Team, R Foundation for Statistical Computing, Wien, Austria). For all groups, descriptive statistics, including mean, standard deviation, median, minimum, and maximum values, were calculated. The Kolmogorov–Smirnov test assessed normality of distributions. Analysis of variance (ANOVA) and post-hoc Tukey tests were used for inferential statistics. Significance was predetermined at $p < 0.05$ for all statistical tests.

3. Results

The descriptive statistics for the maximum load evaluation values are listed in Table 1. Significant differences among various groups ($p < 0.05$) were demonstrated. The post-hoc Tukey test showed that the lowest values ($p < 0.05$) were reported with stainless steel wires (Groups 1, 2, and 3). The highest forces ($p < 0.05$) were demonstrated in FRCs bonded with a conventional covered technique (Groups 4, 5, and 6). When the experimental non-covered technique was tested (Groups 7, 8, and 9), intermediate measures were reported (Figure 3), with significantly higher values than the metal groups ($p < 0.001$) and significantly lower values than the covered groups ($p < 0.001$).

Table 1. Descriptive statistics of the different groups at maximum load (N).

Group	Material	Coverage	Brushing	Mean	St Dev	Min	Mdn	Max	Significance *
1	SS	no	no	2.98	0.91	2.23	2.38	4.17	A
2	SS	no	26 min	2.91	0.21	2.40	2.99	3.00	A
3	SS	no	60 min	3.02	0.08	2.90	3.08	3.08	A
4	FRC	full	no	40.82	11.13	26.58	40.15	60.30	B
5	FRC	full	26 min	34.20	9.67	23.17	32.14	48.07	C
6	FRC	full	60 min	32.84	5.99	24.30	32.46	40.91	C
7	FRC	no	no	10.89	2.49	8.85	9.57	15.22	E
8	FRC	no	26 min	10.90	2.26	7.24	11.22	14.19	E
9	FRC	no	60 min	7.97	2.54	5.54	7.43	12.59	F

* Means with the same letters are not significantly different ($p > 0.05$).

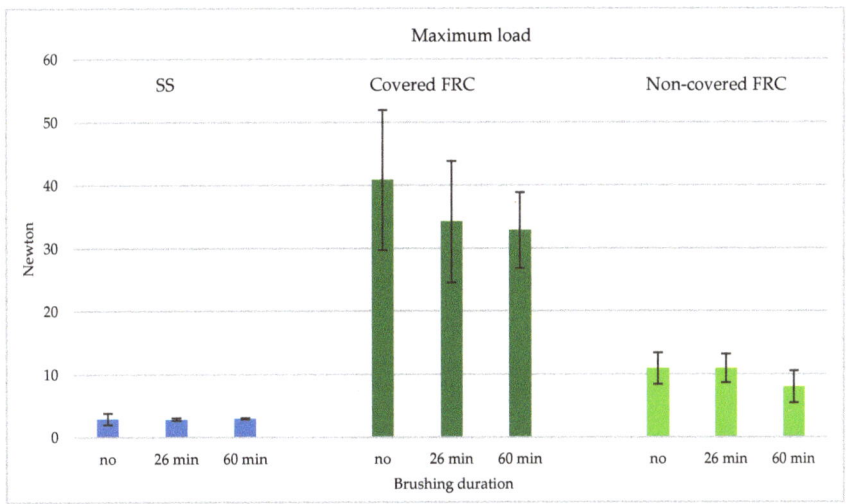

Figure 3. Graphical representation of maximum load values (Mean and SD) of the various conditions (metal, covered FRC, and non-covered FRC) after different brushing times (no brushing, 26 min of brushing, and 60 min of brushing).

No difference was reported ($p > 0.05$) between not-brushed and brushed stainless steel groups (Groups 1–3). Conventional covered FRC groups (Groups 4–6) showed decreased values after 26 min of brushing ($p < 0.05$), and no differences between 26 and 60 min of brushing ($p > 0.05$). On the other hand, non-covered FRCs (Groups 7–9) showed a reduction only after 60 min of brushing ($p < 0.05$).

Maximum bending stress results (Table 2) showed significant differences among the various groups tested ($p < 0.05$). Post-hoc analysis showed that the highest values ($p < 0.05$) were reported with stainless steel wires (Groups 1–3). Significantly lower values ($p < 0.05$) were reported for FRCs bonded with both conventional (Groups 4–6) and experimental (Groups 7–9) techniques (Figure 4) that showed no significant differences between them ($p > 0.05$). No significant difference was reported ($p > 0.05$) between not-brushed and brushed groups for all the three different conditions tested, even if there was a decrease of maximum bending stress values after tooth brushing in the FRCs groups.

Table 2. Descriptive statistics (MPa) of the maximum bending stress of the different groups.

Group	Material	Coverage	Brushing	Mean	St Dev	Min	Mdn	Max	Significance *
1	SS	no	no	748.61	21.14	712.33	750.36	777.77	A
2	SS	no	26 min	766.88	157.10	546.00	784.00	967.00	A
3	SS	no	60 min	764.26	139.16	578.72	759.55	957.00	A
4	FRC	full	no	409.35	89.70	275.07	392.21	571.71	B
5	FRC	full	26 min	377.65	151.78	219.73	315.24	626.15	B
6	FRC	full	60 min	348.25	82.53	225.39	343.33	442.79	B
7	FRC	no	no	381.03	139.23	205.57	337.30	612.25	B
8	FRC	no	26 min	363.61	143.04	208.41	318.67	588.38	B
9	FRC	no	60 min	263.32	85.64	181.18	245.61	454.61	B

* Means with the same letters are not significantly different ($p > 0.05$).

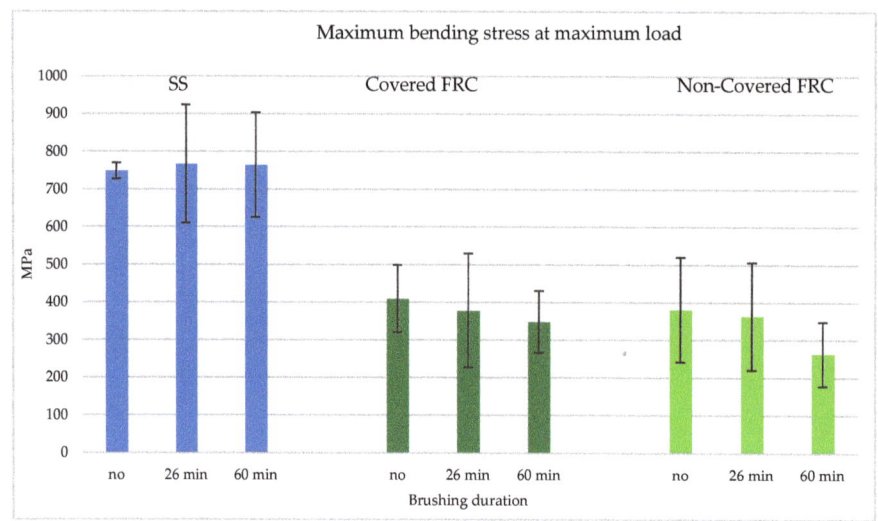

Figure 4. Graphical representation of maximum bending stress (Mean and SD) of the various conditions (metal, covered FRC, and non-covered FRC) after different brushing times (no brushing, 26 min of brushing, and 60 min of brushing).

4. Discussion

The null hypothesis of the present investigation was rejected: significant differences were reported among various groups. In this report, three different materials, which are used in orthodontic splints, were tested for different times of tooth brushing: stainless steel, FRC covered with a composite (used in the FRC covered conventional technique), and non-covered FRC (used in the experimental uncovered spot-bonding FRC technique) [24,25].

Rigidity (or bending stress) and maximum load are characteristics used to evaluate the longevity of retainers [21,22]. The load value in N with the predetermined magnitude of deflection was used as a descriptive value of the retainer's rigidity. The highest values of maximum load were reported for FRC covered specimens (conventional technique), and these results confirm previous studies [21,22,29] which demonstrated the higher rigidity of fully covered FRC frameworks if compared to the metal ones. Uncovered FRCs showed lower values, as reported in previous studies [24,25]. The results of the present report are in agreement with previous investigations, showing that maximum load of uncovered FRC presented intermediate values between the conventional full covered FRC technique and metal splints.

Nowadays, there is only one in vitro study about the tooth brushing effect on these materials [30]. The previous study was made on Frasaco models splinted with different techniques in order to simulate a canine-to-canine splint. Mechanical (load at 0.1 mm deflection and at maximum load) and surface properties were tested. An experimental FRC spot bonding technique and metal splint technique seemed to have similar mechanical properties.

However, tooth brushing is a daily routine for every patient, which is why further studies are needed before a clinical use of the experimental FRC technique. In the present report the same materials of the previous study were tested before and after different times of tooth brushing with a three point bending test (evaluating mechanical properties, such as maximum load and maximum bending stress) and with SEM microphotographs.

Plaque is a predisposing factor to caries and periodontal disease, and dental hygiene is extremely important for oral health [33] as a soiled acidic environment could damage enamel [34], dentin [35], restorative materials [36], and prosthodontic frameworks [37]. In our research, a rotating oscillating electrical device was used, as rotation oscillation powered brushes significantly reduce plaque. The mechanical tooth brushing movement can lead to surface wear [38].

In dentistry, there are different kinds of wear: attrition, abrasion, and chemical wear. Concerning abrasive wear, material is scraped off the surface and this variable represents an important mechanism into oral environment [39,40]. Occlusal wear only concerns contact surfaces while tooth brushing abrasion can affect any exposed surface [41]. The abrasive composites wear was the subject of many studies and it is influenced by different factors, such as the size, shape, and the amount of filler, the resin matrix, and the bonding between the two phases [26]. In addition, the abrasive surface has an important role: its hardness is directly linked with the abrasion amount [42].

In the present test, the toothbrush used for all specimens was the Oral B PRO 670 toothbrush, Oral B crossaction brush heads (Procter & Gamble, Cincinnati, OH, USA). It was used in combination with a toothpaste (MaxWhite-white crystals, Colgate-Palmolive, New York, NY, USA) with a RDA (Relative Dentin Abrasion) of 124. Our aim was to analyze how mechanical properties and surface wear of different materials change with the same abrasive. Further studies can be made by testing the abrasion of different brands of toothbrushes and toothpastes, evaluating if their different abrasive power will produce different effects in mechanical properties and surface wear of FRC splints.

As previous authors reported on in vitro brushing test [29], the present test was performed as a result of a simulated six-month long tooth brushing period, supposing that patients brush their teeth everyday with an average of twice a day during 6 months. An average brushing time of 2 min for each time was considered, two times per day. Next, the result was divided by 28, the expected number of teeth, in order to find the mean brushing time of each single tooth. The result was 26 min, thus specimens were experimentally brushed for this amount of time. Moreover, in the present report, the same times of tooth brushing were applied. Six months (26 min of simulated continuous brushing) and 14 months (60 min of simulated continuous brushing) were tested.

Additionally, the three point bending test was conducted after the specimens were stored in an incubator at 37 degrees Celsius for 24 h. It would be interesting to perform further tests after subjecting the glass reinforced composite into artificial saliva, in order to ascertain if oral lubrification could influence surface wear.

In the present report, mechanical properties were evaluated with a three point bending test, no cyclical loading test was performed. A bending test would relate more directly to orthodontic retention and relapse with a cyclical loading test, so further studies would be welcomed in order to test this additional variable.

Stainless steel is the most common material used for orthodontic retainers, and it is well accepted by patients [43–45]. Only one in vitro study [29] tested orthodontic stainless steel wires after tooth brushing, showing no significant differences in mechanical properties over time. No three point bending tests were performed after tooth brushing.

In this report stainless steel specimens were tested before tooth brushing, after 26 min, and after 60 min of tooth brushing: no significant difference was shown both for maximum load and for maximum bending stress. Conversely, a significant reduction of maximum load values was reported in a previous study [29], but the reduction of values after tooth brushing were due to the wear composite used to fix the wires. Therefore, tooth brushing does not damage metal wires.

Concerning FRC with coverage, after tooth brushing, a significant decrease of maximum load values was reported after 26 min. After 60 min of tooth brushing, no significant difference was reported. Instead, for the maximum bending stress at maximum load, there was no significant reduction of values after tooth brushing. A previous report tested maximum load of conventional FRCs after 26 min of brushing, showing no significant differences. The variability of the results is probably due to the different investigation methods (Frasaco models and three point bending tests).

In the present study, FRC without coverage after tooth brushing presented a significant decrease of maximum load values after 60 min, but no significant decrease between 0 and 26 min. No significant reduction of maximum bending stress was reported before and after tooth brushing.

Concerning the differences between metal and FRCs behavior, they seem to be ascribed to the intrinsic differences of the two materials tested and to their different responses into a three point bending test. Concerning maximum bending stress, stainless steel wires showed no differences after brushing. On the other hand, both FRC groups showed a decrease in mean values, but the decrease was not significant. Further tests should be conducted in the future analyzing FRC maximum bending stress behavior after longer brushing times in order to determine if the decrease will be significant with more brushing time.

Pictures of all specimens with a magnification of 35× (Figure 5), 100× (Figure 6), and 250× (Figure 7) were taken with a scanning electron microscope. SEM images provided a qualitative, not quantitative, evaluation, which is a limitation of the test. However, SEM images showed visual signs of wear. No visual signs of wear were reported on stainless steel specimens, both after twenty-six minutes and one hour of tooth brushing. Common signs of composite wear were reported on FRC covered with a composite, similar to those reported in other studies [29,46–48]. These scratches are more evident after one hour of tooth brushing than after 26 min of tooth brushing. Remarkable signs of wear were reported on FRC left uncovered, and higher after 60 min with disarranged and broken fibers. All dental materials are subject to wear. Wear can reduce the resistance of the material, change its mechanical and aesthetic properties, and lead to bacterial adhesion [46,47].

Results on FRCs without coverage after 26 min of tooth brushing are in agreement with a previous study in which microphotographs were taken with a magnification of 35× and 100× [29]. No studies have been conducted on the FRC surface after 1-h tooth brushing, thus no direct comparison can be made on wear sign results. Before a clinical use of this experimental spot-bonded FRC technique, further in vitro and in vivo tests are needed to evaluate other variables of fibers left uncovered, such as bonding efficiency, duration, and bacterial adhesion.

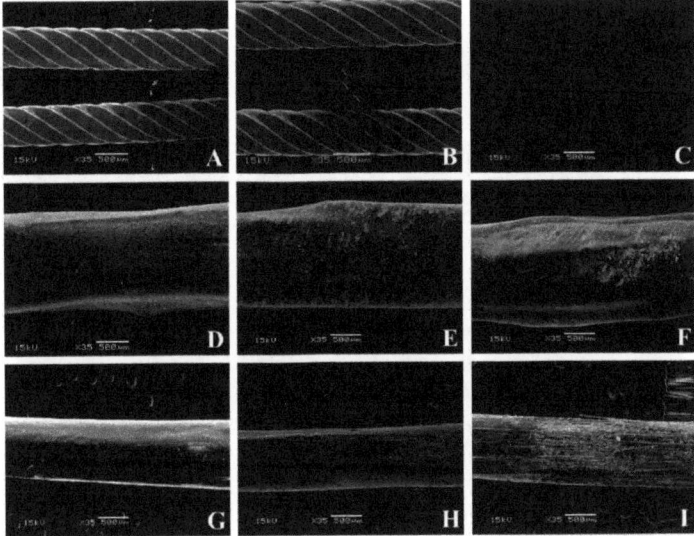

Figure 5. SEM pictures of various conditions tested at 35× magnification. (**A**) Flat metallic wire not brushed, (**B**) flat metallic wire brushed for 26 min, (**C**) flat metallic wire brushed for 60 min, (**D**) full-bonded FRC not brushed, (**E**) full-bonded FRC brushed for 26 min, (**F**) full-bonded FRC brushed for 60 min, (**G**) spot-bonded FRC not brushed, (**H**) full-bonded FRC brushed for 26 min, (**I**) full-bonded FRC brushed for 60 min.

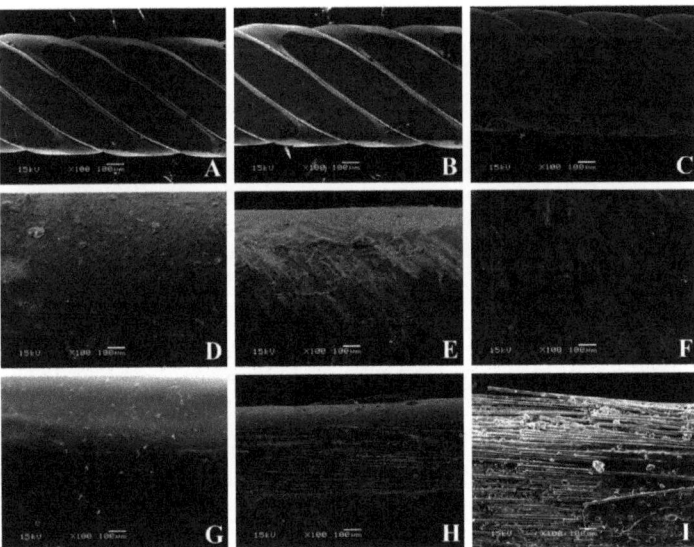

Figure 6. SEM pictures of various conditions tested at 100× magnification. (**A**) Flat metallic wire not brushed, (**B**) flat metallic wire brushed for 26 min, (**C**) flat metallic wire brushed for 60 min, (**D**) full-bonded FRC not brushed, (**E**) full-bonded FRC brushed for 26 min (**F**) full-bonded FRC brushed for 60 min, (**G**) spot-bonded FRC not brushed, (**H**) full-bonded FRC brushed for 26 min, (**I**) full-bonded FRC brushed for 60 min.

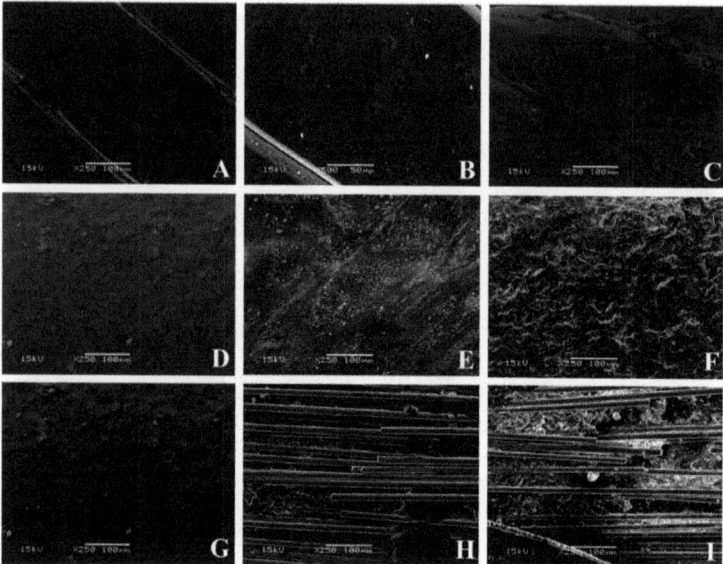

Figure 7. SEM pictures of various conditions tested at 250× magnification. (**A**) Flat metallic wire not brushed (**B**) flat metallic wire brushed for 26 min, (**C**) flat metallic wire brushed for 60 min, (**D**) full-bonded FRC not brushed, (**E**) full-bonded FRC brushed for 26 min, (**F**) full-bonded FRC brushed for 60 min, (**G**) spot-bonded FRC not brushed, (**H**) full-bonded FRC brushed for 26 min, (**I**) full-bonded FRC brushed for 60 min.

Limitations of the present study are related to the materials used. In fact, only some materials have been taken into consideration, even if other FRC materials are nowadays present on the market with different shapes, sizes, and diameters of fibers. Different geometries of materials tested could influence results. Moreover, our in vitro study cannot completely simulate real clinical conditions. In fact, even though many in vitro studies have been conducted, the main limitation of a FRC clinical use is that research is still lacking regarding long-term clinical performance [49]. As wear, delamination and fracture of FRC devices have been reported [48]. In order to confirm the results of the present investigation, which is a pilot study, randomized controlled clinical trials would be welcomed.

5. Conclusions

The present study demonstrated that after tooth brushing no mechanical differences were reported for metal specimens, while a reduction of maximum load values was reported for all FRC specimens. SEM images showed that after tooth brushing metal specimens were not visually affected, while signs of wear were reported for all FRC groups, especially for specimens not covered with a composite.

Author Contributions: Investigation, L.V.J.L. and A.V.; methodology, M.F.S., P.K.V., and L.V.J.L.; project administration and text revision, P.K.V. and M.F.S.; software, L.V.J.L., A.V., and A.S.; data curation, A.V., L.V.J.L., and A.S.; supervision, P.K.V. and P.G.; validation, L.V.J.L.; visualization, A.V. and A.S.; writing, A.V. and A.S.; language review, M.F.S; formal analysis, L.V.J.L.; conceptualization, M.F.S., P.K.V., and A.S; resource, P.K.V. and L.V.J.L. All authors have read and agreed to the published version of the manuscript.

Funding: This research received no external funding.

Acknowledgments: The authors would like to thank StickTech for providing the materials tested in the present research study and Roosa Prinssi for the excellent theoretical and technical assistance.

Conflicts of Interest: P.K.V. consults StickTech and GC in RD and education. Other authors declare no conflict of interest.

References

1. Goldberg, A.; Burstone, C. The use of continuous fiber reinforcement in dentistry. *Dent. Mater.* **1992**, *8*, 197–202. [CrossRef]
2. Vallittu, P.K.; Lassila, V.P. Reinforcement of acrylic resin denture base material with metal or fibre strengtheners. *J. Oral Rehabil.* **1992**, *19*, 225–230. [CrossRef] [PubMed]
3. Nayar, S.; Ganesh, R.; Santhosh, S. Fiber reinforced composites in prosthodontics—A systematic review. *J. Pharm. Bioallied Sci.* **2015**, *7*, S220–S222. [CrossRef] [PubMed]
4. Scribante, A.; Vallittu, P.K.; Özcan, M. Fiber-Reinforced Composites for Dental Applications. *BioMed Res. Int.* **2018**, *2018*, 1–2. [CrossRef]
5. Vallittu, P.K. Flexural properties of acrylic resin polymers reinforced with unidirectional and woven glass fibers. *J. Prosthet. Dent.* **1999**, *81*, 318–326. [CrossRef]
6. Vallittu, P.K.; Sevelius, C. Resin-bonded, glass fiber-reinforced composite fixed partial dentures: A clinical study. *J. Prosthet. Dent.* **2000**, *84*, 413–418. [CrossRef]
7. Dijken, J.W.V.; Sunnegårdh-Grönberg, K. Fiber-reinforced packable resin composites in Class II cavities. *J. Dent.* **2006**, *34*, 763–769. [CrossRef]
8. Lassila, L.V.; Vallittu, P.K.; Garoushi, S.K.; Tezvergil, A. Fiber-reinforced Composite Substructure: Load-bearing Capacity of an Onlay Restoration and Flexural Properties of the Material. *J. Contemp. Dent. Pr.* **2006**, *7*, 1–8. [CrossRef]
9. Scribante, A.; Sfondrini, M.F.; Broggini, S.; D'Allocco, M.; Gandini, P. Efficacy of Esthetic Retainers: Clinical Comparison between Multistranded Wires and Direct-Bond Glass Fiber-Reinforced Composite Splints. *Int. J. Dent.* **2011**, *2011*, 1–5. [CrossRef]
10. Fragkouli, M.; Tzoutzas, I.; Eliades, G. Bonding of Core Build-Up Composites with Glass Fiber-Reinforced Posts. *Dent. J.* **2019**, *7*, 105. [CrossRef]
11. Ohtonen, J.; Lassila, L.; Vallittu, P.K. Effect of monomer composition of polymer matrix on flexural properties of glass fibre-reinforced orthodontic arch wire. *Eur. J. Orthod.* **2011**, *35*, 110–114. [CrossRef] [PubMed]
12. Mosharraf, R.; Torkan, S. Fracture Resistance of Composite Fixed Partial Dentures Reinforced with Pre-impregnated and Non-impregnated Fibers. *J. Dent. Res. Dent. Clin. Dent. Prospect.* **2012**, *6*, 12–16.
13. Polacek, P.; Pavelka, V.; Ozcan, M. Adhesion of resin materials to S2-glass unidirectional and E-glass multidirectional fiber reinforced composites: Effect of polymerization sequence protocols. *J. Adhes. Dent.* **2013**, *15*.
14. Littlewood, S.J.; Millett, D.T.; Doubleday, B.; Bearn, D.R.; Worthington, H. Retention procedures for stabilising tooth position after treatment with orthodontic braces. *Cochrane Database Syst. Rev.* **2016**, CD002283. [CrossRef]
15. Årtun, J.; Spadafora, A.T.; Shapiro, P.A. A 3-year follow-up study of various types of orthodontic canine-to-canine retainers. *Eur. J. Orthod.* **1997**, *19*, 501–509. [CrossRef]
16. Padmos, J.A.; Fudalej, P.S.; Renkema, A.M. Epidemiologic study of orthodontic retention procedures. *Am. J. Orthod. Dentofac. Orthop.* **2018**, *153*, 496–504. [CrossRef]
17. Burstone, C.J.; Kuhlberg, A.J. Fiber-reinforced composite in orthodontics. *J. Clin. Orthod.* **2000**, *34*, 271–279.
18. Tanimoto, Y.; Inami, T.; Yamaguchi, M.; Nishiyama, N.; Kasai, K. Preparation, mechanical, and in vitro properties of glass fiber-reinforced polycarbonate composites for orthodontic application. *J. Biomed. Mater. Res. Part B Appl. Biomater.* **2014**, *103*, 743–750. [CrossRef]
19. Lassila, L.V.; Tanner, J.; Le Bell, A.-M.; Narva, K.; Vallittu, P.K. Flexural properties of fiber reinforced root canal posts. *Dent. Mater.* **2004**, *20*, 29–36. [CrossRef]
20. Pastila, P.; Lassila, L.V.; Jokinen, M.; Vuorinen, J.; Vallittu, P.K.; Mäntylä, T. Effect of short-term water storage on the elastic properties of some dental restorative materials—A resonant ultrasound spectroscopy study. *Dent. Mater.* **2007**, *23*, 878–884. [CrossRef]
21. Cacciafesta, V.; Sfondrini, M.F.; Lena, A.; Scribante, A.; Vallittu, P.K.; Lassila, L.V. Force levels of fiber-reinforced composites and orthodontic stainless steel wires: A 3-point bending test. *Am. J. Orthod. Dentofac. Orthop.* **2008**, *133*, 410–413. [CrossRef] [PubMed]
22. Alavi, S.; Mamavi, T. Evaluation of load-deflection properties of fiber-reinforced composites and its comparison with stainless steel wires. *Dent. Res. J.* **2014**, *11*, 234–239.

23. Oshagh, M.; Heidary, S.; Nazhvani, A.D.; Koohpeima, F.; Hosseinabadi, O.K. Evaluation of Histological Impacts of Three Types of Orthodontic Fixed Retainers on Periodontium of Rabbits. *J. Dent.* **2014**, *15*, 104–111.
24. Scribante, A.; Gandini, P.; Tessera, P.; Vallittu, P.K.; Lassila, L.; Sfondrini, M.F. Spot-Bonding and Full-Bonding Techniques for Fiber Reinforced Composite (FRC) and Metallic Retainers. *Int. J. Mol. Sci.* **2017**, *18*, 2096. [CrossRef]
25. Sfondrini, M.F.; Gandini, P.; Tessera, P.; Vallittu, P.K.; Lassila, L.; Scribante, A. Bending Properties of Fiber-Reinforced Composites Retainers Bonded with Spot-Composite Coverage. *BioMed Res. Int.* **2017**, *2017*, 1–6. [CrossRef] [PubMed]
26. Tsujimoto, A.; Barkmeier, W.W.; Fischer, N.; Nojiri, K.; Nagura, Y.; Takamizawa, T.; Latta, M.A.; Miazaki, M. Wear of resin composites: Current insights into underlying mechanisms, evaluation methods and influential factors. *Jpn. Dent. Sci. Rev.* **2017**, *54*, 76–87. [CrossRef]
27. Kyoizumi, H.; Yamada, J.; Suzuki, T.; Kanehira, M.; Finger, W.J.; Sasaki, K. Effects of Toothbrush Hardness on in vitro Wear and Roughness of Composite Resins. *J. Contemp. Dent. Pract.* **2013**, *14*, 1137–1144. [CrossRef]
28. Lassila, L.; Garoushi, S.; Vallittu, P.K.; Säilynoja, E. Mechanical properties of fiber reinforced restorative composite with two distinguished fiber length distribution. *J. Mech. Behav. Biomed. Mater.* **2016**, *60*, 331–338. [CrossRef]
29. Scribante, A.; Vallittu, P.; Lassila, L.; Viola, A.; Tessera, P.; Gandini, P.; Sfondrini, M.F. Effect of Long-Term Brushing on Deflection, Maximum Load, and Wear of Stainless Steel Wires and Conventional and Spot Bonded Fiber-Reinforced Composites. *Int. J. Mol. Sci.* **2019**, *20*, 6043. [CrossRef]
30. Scribante, A.; Massironi, S.; Pieraccini, G.; Vallittu, P.K.; Lassila, L.; Sfondrini, M.F.; Gandini, P. Effects of nanofillers on mechanical properties of fiber-reinforced composites polymerized with light-curing and additional postcuring. *J. Appl. Biomater. Funct. Mater.* **2015**, *13*. [CrossRef]
31. Scribante, A.; Bollardi, M.; Chiesa, M.; Poggio, C.; Colombo, M. Flexural Properties and Elastic Modulus of Different Esthetic Restorative Materials: Evaluation after Exposure to Acidic Drink. *BioMed Res. Int.* **2019**, *2019*, 1–8. [CrossRef] [PubMed]
32. Dahl, K.A.; Moritz, N.; Vallittu, P.K. Flexural and torsional properties of a glass fiber-reinforced composite diaphyseal bone model with multidirectional fiber orientation. *J. Mech. Behav. Biomed. Mater.* **2018**, *87*, 143–147. [CrossRef] [PubMed]
33. Deery, C.; Heanue, M.; Deacon, S.; Robinson, P.; Walmsley, D.; Worthington, H.; Shaw, W.; Glenny, A.-M. The effectiveness of manual versus powered toothbrushes for dental health: A systematic review. *J. Dent.* **2004**, *32*, 197–211. [CrossRef] [PubMed]
34. Soares, L.E.S.; Soares, A.L.S.; De Oliveira, R.; Nahórny, S. The effects of acid erosion and remineralization on enamel and three different dental materials: FT-Raman spectroscopy and scanning electron microscopy analysis. *Microsc. Res. Tech.* **2016**, *79*, 646–656. [CrossRef]
35. Caneppele, T.M.F.; Jeronymo, R.D.I.; Di Nicoló, R.; De Araújo, M.A.M.; Soares, L.E.S. In Vitro assessment of dentin erosion after immersion in acidic beverages: Surface profile analysis and energy-dispersive X-ray fluorescence spectrometry study. *Braz. Dent. J.* **2012**, *23*, 373–378. [CrossRef]
36. Poggio, C.; Dagna, A.; Chiesa, M.; Colombo, M.; Scribante, A. Surface roughness of flowable resin composites eroded by acidic and alcoholic drinks. *J. Conserv. Dent.* **2012**, *15*, 137–140. [CrossRef]
37. Colombo, M.; Poggio, C.; Lasagna, A.; Chiesa, M.; Scribante, A. Vickers Micro-Hardness of New Restorative CAD/CAM Dental Materials: Evaluation and Comparison after Exposure to Acidic Drink. *Materials* **2019**, *12*, 1246. [CrossRef]
38. Lefever, D.; Perakis, N.; Roig, M.; Krejci, I.; Ardu, S. The effect of toothbrushing on surface gloss of resin composites. *Am. J. Dent.* **2012**, *25*, 54–58.
39. Mair, L. Wear in dentistry—current terminology. *J. Dent.* **1992**, *20*, 140–144. [CrossRef]
40. Mair, L.H.; Stolarski, T.A.; Vowles, W.; Lloyd, C.H. Wear: Mechanisms, manifestations and measurement. Report of a workshop. *J. Dent.* **1996**, *24*, 141–148. [CrossRef]
41. Cavalcante, L.M.; Masouras, K.; Watts, D.; A Pimenta, L.; Silikas, N. Effect of nanofillers' size on surface properties after toothbrush abrasion. *Am. J. Dent.* **2009**, *22*, 60–64. [PubMed]
42. Manhart, J.; Kunzelmann, K.-H.; Chen, H.Y.; Hickel, R. Mechanical properties and wear behavior of light-cured packable composite resins. *Dent. Mater.* **2000**, *16*, 33–40. [CrossRef]

43. A Beckett, H.; Evans, R.D.; Gilmour, A.G. Permanent retention in orthodontic patients with reduced levels of bone support: A pin stabilised resin bonded splint. *Br. Dent. J.* **1992**, *173*, 272–274. [CrossRef] [PubMed]
44. Filippi, A.; Von Arx, T.; Lussi, A. Comfort and discomfort of dental trauma splints—A comparison of a new device (TTS) with three commonly used splinting techniques. *Dent. Traumatol.* **2002**, *18*, 275–280. [CrossRef] [PubMed]
45. Geramy, A.; Retrouvey, J.; Sobuti, F.; Salehi, H. Anterior Teeth Splinting After Orthodontic Treatment: 3D Analysis Using Finite Element Method. *J. Dent.* **2012**, *9*, 90–98.
46. Dalla-Vecchia, K.B.; Taborda, T.D.; Stona, D.; Pressi, H.; Júnior, L.H.B.; Rodrigues-Junior, S.A. Influence of polishing on surface roughness following toothbrushing wear of composite resins. *Gen. Dent.* **2017**, *65*, 68–74.
47. Dos Santos, J.-H.-A.; Silva, N.-M.-D.L.; Gomes, M.-G.-N.; Paschoal, M.-A.-B.; Gomes, I.-A. Whitening toothpastes effect on nanoparticle resin composite roughness after a brushing challenge: An in vitro study. *J. Clin. Exp. Dent.* **2019**, *11*, e334–e339.
48. Shimokawa, C.; Giannini, M.; André, C.; Sahadi, B.; Faraoni, J.; Palma-Dibb, R.; Soares, C.; Price, R. In Vitro Evaluation of Surface Properties and Wear Resistance of Conventional and Bulk-fill Resin-based Composites after Brushing with a Dentifrice. *Oper. Dent.* **2019**, *44*, 637–647. [CrossRef]
49. Scribante, A.; Vallittu, P.K.; Özcan, M.; Lassila, L.V.J.; Gandini, P.; Sfondrini, M.F. Travel beyond Clinical Uses of Fiber Reinforced Composites (FRCs) in Dentistry: A Review of Past Employments, Present Applications, and Future Perspectives. *BioMed Res. Int.* **2018**, *2018*, 1–8. [CrossRef]

© 2020 by the authors. Licensee MDPI, Basel, Switzerland. This article is an open access article distributed under the terms and conditions of the Creative Commons Attribution (CC BY) license (http://creativecommons.org/licenses/by/4.0/).

Article

The One-Year In Vivo Comparison of Lithium Disilicate and Zirconium Dioxide Inlays

Rini Behera [1], Lora Mishra [1], Darshan Devang Divakar [2], Abdulaziz A. Al-Kheraif [2], Naomi Ranjan Singh [1] and Monika Lukomska-Szymanska [3,*]

1. Department of Conservative Dentistry & Endodontics, Institute of Dental Sciences, Siksha 'O' Anusandhan, Bhubaneswar P.O. Box 751003, India; rinibehera@soa.ac.in (R.B.); loramishra@soa.ac.in (L.M.); naomiranjansingh@soa.ac.in (N.R.S.)
2. Dental Biomaterials Research Chair, Department of Health Department, College of Applied Medical Sciences, King Saud University, Riyadh P.O. Box 10219, Saudi Arabia; ddivakar@ksu.edu.sa (D.D.D.); aalkhuraif@ksu.edu.sa (A.A.A.-K.)
3. Department of General Dentistry, Medical University of Lodz, 251 Pomorska St, 92-213 Lodz, Poland
* Correspondence: monika.lukomska-szymanska@umed.lodz.pl

Citation: Behera, R.; Mishra, L.; Divakar, D.D.; Al-Kheraif, A.A.; Singh, N.R.; Lukomska-Szymanska, M. The One-Year In Vivo Comparison of Lithium Disilicate and Zirconium Dioxide Inlays. *Materials* **2021**, *14*, 3102. https://doi.org/10.3390/ma14113102

Academic Editor: Enrico Marchetti

Received: 27 April 2021
Accepted: 1 June 2021
Published: 5 June 2021

Publisher's Note: MDPI stays neutral with regard to jurisdictional claims in published maps and institutional affiliations.

Copyright: © 2021 by the authors. Licensee MDPI, Basel, Switzerland. This article is an open access article distributed under the terms and conditions of the Creative Commons Attribution (CC BY) license (https://creativecommons.org/licenses/by/4.0/).

Abstract: The objective of the present study was to evaluate the one-year clinical performance of lithium disilicate (LD) and zirconium dioxide (ZrO_2) class II inlay restorations. Thirty healthy individuals who met the inclusion criteria were enrolled for the study. The patients were randomly divided into two study groups ($n = 15$): LD (IPS e.max press) and ZrO_2 (Dentcare Zirconia). In the ZrO_2 group, the internal surfaces of the inlays were sandblasted and silanized with Monobond N (Ivoclar, Leichsteistein, Germany). In the LD group, the internal surfaces of the inlays were etched with 5% hydrofluoric acid. The ceramic inlays were cemented with self-cure resin cement (Multilink N). Clinical examinations were performed using modified United State Public Health Codes and Criteria (USPHS) after 2 weeks, 4 weeks, 6 months and 1 year. The one-year survival rate was evaluated. In total, one failure was observed in the ZrO_2 group. The survival probability after 1 year for the ZrO_2 inlays was 93%, and for the LD inlays was 100%, which was statistically insignificant. The differences between both groups for most USPHS criteria (except for colour match) were statistically insignificant. Within the imitations of the present study, the lithium disilicate- and zirconia dioxide-based inlays exhibited comparable clinical performances. However, the colour and translucency match was superior for the lithium disilicate restorations.

Keywords: CAD-CAM; inlay; lithium disilicate; zirconium dioxide

1. Introduction

The prevalence of dental caries is estimated by the WHO to be over 90% [1]. The extension of caries is the prime dominance factor in choice of reconstruction method. Currently, composite restorations, crowns, inlays or onlays are recommended to reconstruct extensive class II MOD cavities [2]. However, in these cases, the establishment of occlusal anatomy, proximal contact and the contour, finishing and polishing of indirect restorations are far superior to direct reconstructions [3].

Ceramic and zirconium dioxide-based reconstructions provide enhanced strength and aesthetics [3,4]. Both materials offer the opportunity to maintain the tooth structure while providing the mechanical benefits of modern adhesive technology. Lithium disilicate (LD) glass ceramic is excellent for highly aesthetic restorations providing good mechanical properties. LD ceramic, the strongest and the toughest of the glass-ceramics available, exhibits moderate flexural strength (360–440 MPa) [5] and fracture toughness (2.5–3 MPa m$^{1/2}$) [6], yet provides excellent translucency and shade matching properties [7,8].

On the other hand, zirconium oxide (ZrO_2) is largely used due to its favourable mechanical properties and good fracture resistance. The biocompatibility, optical properties and translucency of ZrO_2 make it an alternative to porcelain-fused-to-metal restorations [3]. Additionally, ZrO_2 is the strongest and most robust of all dental ceramics with a flexural strength of 800–1200 MPa and fracture toughness of 6–8 MPa $m^{1/2}$ [4]. Therefore, it meets the mechanical requirements for high-stress bearing posterior restoration. Unfortunately, the limited translucency and poor adhesion to tooth structure, due to its inert and non-polar nature, are major disadvantages [9,10].

The survival rate of ceramic restorations has been largely investigated [8,11–16]. However, due to a lack of clinical studies, there is a great need to evaluate LD and ZrO_2 inlays in in vivo studies. Therefore, the objective of the present study was to evaluate and compare the one-year clinical performance of LD and ZrO_2 inlay restorations. The null hypothesis was that there is no difference in the survival rate and quality between LD inlay and ZrO_2 inlay restorations.

2. Materials and Methods

2.1. Study Design

This research protocol and design was approved by the Institutional ethical committee (Ref. No/DMR/IMS.SH/SOA/180035). Thirty healthy individuals who met the inclusion criteria were enrolled for the study (Table 1) [16]. The patients were randomly divided into two groups ($n = 15$) with online software www.randomizer.org (first accessed on 28 May 2018) (Urbaniak, G. C., & Plous, S. 2013, Research Randomizer, Version 4.0, Computer software). The cavity distribution within the study groups is presented in Table 2. The distribution of the tooth and cavity type was not significant at $p < 0.05$.

Table 1. Inclusion and exclusion criteria of patients.

Inclusion Criteria	Exclusion Criteria
class II cavities in permanent teeth	severe systematic diseases and allergies
isthmus size of the treated cavities at least half of the intercuspal distance	severe salivary gland dysfunction
	severe periodontal problems
no clinical signs and symptoms of pulp and periapical pathology	poor plaque control
at least one neighbouring tooth in occlusion to antagonistic teeth	parafunctional habits like bruxism or clenching restricted mouth opening
good oral hygiene	history of orthodontic treatment
over 18 years old	preparations extending below the gingiva margin and close to the pulp
willing to participate in the study	initial defects, i.e., discoloured pits and fissures and caries restricted to enamel only

Table 2. The Class II mesio-occlusal (MO) and occluso-distal (DO) cavity distribution within study groups.

Tooth Type	Type of Class II Cavity	LD	ZrO_2
Premolars	MO	4	5
	OD	3	2
Molars	MO	4	4
	OD	4	4
-Total no of teeth	30	15	15

2.2. Tooth Preparation

After administering the local anaesthetic (Indoco, Warren Lignox with Adrenaline, Mumbai, India), class II cavities were prepared using a high-speed handpiece with inlay diamond points (Coltene Diatech Inlay & Crown preparation kit 11312, Altstätten, Switzerland) under a constant, copious water supply. The isthmus width was established at a minimum of 2.5 mm, the pulpal floor depth amounted up to at least 1.5–2.0 mm, the axial wall depth was up to 1.5 mm, the internal line angles were rounded and the divergence angle of the cavity was approximately 10°–15° with no bevel (Figure 1a). The enamel margins were refined using an enamel hatchet hand instrument (Hu-Friedy Mfg. Chicago, IL, USA).

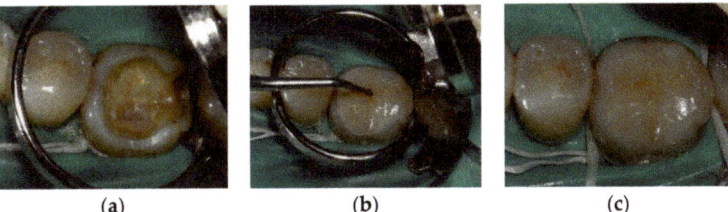

Figure 1. (a) The inlay preparation; lower molar, class IIOD cavity; (b) Try-in of the inlay; (c) Lithium disilicate inlay in situ.

2.3. Impressions

Gingival retraction was achieved with gel (Racegel, Septodont Saint Maur des Fosses, France) applied for 2 min.

2.3.1. Zirconium Oxide (ZrO$_2$) Group

In the zirconium oxide (ZrO$_2$) group, teeth were digitally scanned with an intraoral scanner (CEREC Omnicam scanner; Dentsply Sirona, Bensheim, Germany), followed by conversion to a 3-dimensional (3D) virtual model (CEREC AC software 4.3; Dentsply Sirona, Charlotte, NC, USA). An irreversible hydrocolloid impression (Alginate, Zelgan Plus, Dentsply, Gurgaon, India) for the antagonist arch was taken and disinfected with 0.5% sodium hypochlorite solution. Antagonist impression casts were immediately poured with dental stone type IV (Durone, Dentsply, Petropolis, RJ, Brazil). The impressions scans and antagonist cast were sent to the laboratory for fabrication of the inlays.

2.3.2. Lithium Disilicate (LD) Group

In the lithium disilicate (LD) glass ceramic group, full arch impressions (the two-step putty wash technique) with elastomeric putty impression material (Silagum-Putty, DMG, Germany) and light body impression material (Silagum-Light, DMG, Hamburg, Germany) using stock trays (GDC Dentulous Perforated Impression Trays, Hoshiarpur, Punjab, India) were taken. The impressions were disinfected for 10 min in glutaraldehyde (2%) solution and rinsed with water for 15 s. The antagonist arch impressions were taken and disinfected as described for the ZrO$_2$ group. The impression and antagonist cast were sent to the laboratory for fabrication of the inlays.

2.4. Shade Selection, Occlusion Registration and Temporalization

For both groups, shades were selected from the Classical Vita shade guide (VITA Zahnfabrik, Germany) [17]. Occlusion registration was performed using bite registration wax (Denar® Bite Registration Wax, Whip Mix Corp, Louisville, KY, USA). Patients were temporized with Orafil LC (Prevest, Brahmana, Jammu, India) until the delivery of the final inlay for one week.

2.5. Fabrication of Inlay

2.5.1. ZrO$_2$ Group

The master cast models were poured using type IV dental stone (Elite stone, Zhermack, Badia Polesine (RO), Italy). The inlay design and finish line marking were planned with CEREC AC 4.3 (Dentsply Sirona, Charlotte, NC, USA) software. The marginal discrepancy was set at 0.0 mm, and the margin thickness was at 0.2 mm. The simulated die spacer was programmed at 30 µm, starting 1.0 mm away from the margin [18]. This was followed by an assessment of the master cast model physically and virtually. The inlays were fabricated from monolithic zirconia (DentCare Zirconia, Weiland Zenostar, Ivoclar Vivadent, Pforzheim, Germany) in CORiTec 250i milling unit (imes-icore dental solutions, Eiterfeld, Germany).

The milled discs were manually separated from the zirconia blanks and sintered using Austromat µSiC furnace (Dekema, Freilassing, Germany) for 9 h at 1450 °C. Then, the inlays were glazed by applying Ivoclar glazing paste e-max (Ivoclar Vivadent, Liechtenstein, Germany), with the thickness ranging between 20 and 50 µm, and fired in a furnace (Ivoclar P310 furnace, Liechtenstein, Germany). The restorations were then mirror-finished with diamond-impregnated silicone instruments (Brasseler, Savannah, GA, USA) and polishing pastes (Perfect Polish, Henry Schein, Melville, NY, USA). Finally, the occlusion and proximal contacts were checked and adjusted on the master cast model using stereomicroscope $5\times$ magnification (Labomed CZM6, Labo America Inc., Houston, TX, USA).

2.5.2. LD Group

Master casts were poured using type IV stone gypsum (Elite stone, Zhermack, Badia Polesine (RO) Italy). The inlay wax patterns were fabricated and invested in a phosphate bonded investment, IPS PressVEST Speed (Ivoclar Vivadent, Schaan, Leichsteistein, Germany). The restorations were fabricated from lithium disilicate ingots (IPS e.max Press, Ivoclar Vivadent, Schaan, Leichsteistein, Germany) in a press furnace EP600 (Ivoclar Vivadent, Schaan, Leichsteistein, Germany) at 920 °C at 600 kPa pressure following the manufacturer's recommendations with the lost-wax technique (spacer of 60 mm).

Glazing (IPS e.max Ceram Glaze Liquid, Ivoclar Vivadent, Schaan, Leichsteistein, Germany) firing was performed in a P200 furnace (Ivoclar Vivadent, Schaan, Leichsteistein, Germany). The restorations were adjusted with water cooled diamond rotary instruments (Set 4562, Brasseler GmbH, Savannah, GA, Germany). The internal surface of the restorations was sandblasted with 50-mm aluminium oxide particles at a pressure of 6 Bar (Opiblast, Buffalo Dental Mfg., Inc. Syosset, NY, USA). An initial assessment of the inlays on the master model with stereomicroscope (Labomed CZM6, Labo America Inc., Fremont, CA, USA) at $5\times$ magnification was performed.

2.6. Clinical Try-In and Luting Procedure

The temporary restorations were removed using a probe. The inlays were carefully tried in under the split rubber dam isolation technique (Figure 1b). With the aid of Optra-Stick (Ivoclar, Vivadent, Lienchtenstein, Germany), the inlays were handled and securely positioned within the cavity, and the fit was evaluated. Next, the interproximal contacts and colour were examined. The tooth was cleaned with a slurry of ultrafine pumice and water and then air dried before luting.

In the ZrO_2 group, the internal surfaces of the inlays were sandblasted with aluminium oxide particles. The surface was then silanized with Monobond N (Ivoclar, Leichsteistein, Germany). While, in the LD group, the internal surfaces of the inlays were etched with 5% hydrofluoric acid (IPS Ceramic Kit, Ivoclar, Leichsteistein, Germany) for 20 s, cleaned with water and dried. Self-etch adhesive luting cement (Multilink N-system, Ivoclar, Leichsteistein, Germany) was used according to the manufacturer's recommendations. The restoration was then seated with slight pressure.

The excess resin cement was light cured (Mectron, Starlight P, Mectron Pvt Ltd., Karnataka, India) for 1–2 s for smooth excess removal (Figure 1c). Subsequently, additional light-curing for 20 s per surface was performed. Margins of luted restorations were refined using fine round tapered diamond burs (MANI Diamond Burs, CR series, Takanezawa factory, Shioya, Tochigi, Japan) and rubber points (Brasseler, Savannah, GA, USA) under water cooling. After removal of the rubber dam, the occlusal contacts were checked, and interferences were removed. Next, final finishing and polishing was performed.

The bitewing and intraoral periapical radiograph (IOPA) of the cemented inlays in both groups were taken to assess the immediate post-op marginal adaptation.

2.7. Evaluation

The overall survival probability of the restorations in the LD and ZrO_2 groups after 1 year was evaluated. Direct intraoral clinical examination was carried out by two calibrated examiners independent of the investigation (Cohen's Kappa 0.76). The double-blind evaluation was performed. The restorations were clinically observed under $20\times$ magnification (Seiler, Mitron Instrument Revelation, St. Louis, MO, USA).

The quality of the restorations was evaluated according to modified USPHS criteria (United State Public Health Codes and Criteria) (Table 3) [15]. Immediate occlusal evaluation was carried out after bonding. The minor adjustments then considered necessary were performed. The tightness of the interproximal contact was verified using metal strips of 50 μm (Shimstock-Folie, Coltene, Altstätten, Switzerland) placed between the inlay and the adjacent tooth. At 2 weeks, 4 weeks, 6 months and 1 year, follow up evaluations were performed [15]. If any difference was found between both examiners, a third calibrated examiner (Cohen's Kappa 0.76) established the final decision.

Table 3. The post-operative review assessment codes and criteria-USPHS criteria.

Assessment Criteria	Parameters
(1) Occlusal and interproximal contact	(A) Normal (B) Heavy (C) Light (D) Open
(2) Anatomic form	(A) Continuous with existing anatomy (B) Discontinuous with existing anatomy, but not sufficient to expose dentine/base exposed (C) Dentine/base exposed
(3) Marginal adaptation	(A) Closely adapted no evidence of a catch or crevice at any point (B) Visible evidence of a crevice. Fine probe will not penetrate (C) Visible evidence of a crevice. Fine probe will penetrate (D) Evidence of a positive step when probe drawn from tooth to restoration
(4) Surface roughness	(A) Smooth (B) Slightly pitted
(5) Colour Match	(A) Matches colour and translucency of adjacent tooth structure. (B) Mismatch in colour and translucency is within the acceptable range
(6) Sensitivity	(A) None (B) Mild but bearable (C) Uncomfortable (D) Very painful data
(7) Overall survival probability of restorations after one year	(A) In percentage

2.8. Statistical Analysis

Statistical analysis was performed using IBM SPSS statistics 24.0, SPSS (South Asia PVT LTD., www.spss.co.in, India, accessed on 27 December 2019). Comparison of the mean age by group was carried out following independent sample t-test. The categorical variable of gender was tabulated using a frequency procedure. The chi-square test was used to assess the association of groups, the association of anatomic deformity at follow-up visits with restorative materials in groups and the failure and the survival rate of restorations. A p value less than 0.05 was considered significant.

3. Results

The survival probability in the ZrO_2 group amounted up to 93%, while in the LD group, this was 100%. The difference between groups was statically insignificant (Table 4). One restoration debonded completely in the ZrO_2 group (class II MO, molar) just before the completion of one year of service. This restoration exhibited flaws (open/absent occlusal and proximal contacts, discontinuous with the existing anatomy, evidence of a positive step at margin, slightly pitted surface, and mild postoperative sensitivity) during all follow up-periods (2, 4 weeks and 6 months) (Tables 5–10, marked with *)

Table 4. The survival probability in the study groups.

Study Group	Survival Probability	
	No.	%
LD	15	100.0
ZrO$_2$	14	93.0
Total	29	96.0
Chi-square and p value	$\chi^2 = 5.9032; p = 0.522$	

In the ZrO$_2$ group, the mean patient age amounted up to 36.27 ± 9.48 years, while in the LD group, this was 36.93 ± 8.65 years, and there was an insignificant difference between these values (p = 0.842). In both groups, the male to female ratio of 60% and 40% was found to be absolute matching (p = 1.0000).

Occlusal evaluation was carried out after bonding. Any necessary adjustments were performed, and the majority were minor. The ZrO$_2$ group exhibited 80% normal occlusal and interproximal contact, while in LD group, this was 66.7% at all follow-up periods of 2 weeks, 4 weeks, 6 months and 1 year. However, 6.7% cases in ZrO$_2$ showed open contact at all follow-up visits after 2 weeks, 4 weeks, 6 months and 1 year. The difference between groups was statically insignificant at all follow-up visits (Table 5).

Table 5. Occlusal and proximal contact in study groups.

Follow-Up Periods	Occlusal and Proximal Contact	Group ($\chi^2 = 4.612, p = 0.242$)					
		LD		ZrO$_2$		Total	
		No.	%	No.	%	No.	%
2 weeks	Normal	10	66.7	12	80.0	22	73.3
4 weeks	Heavy	3	20.0	0	0	3	10.0
6 months	Light	2	13.3	2	13.3	4	13.3
1 year	Open/Absent	0	0	1 *	6.7	1	3.3

* One inlay was lost just before completion of the 1-year evaluation.

There was no significant difference between the anatomical form in both groups for the anatomy of inlays (Table 6). The ZrO$_2$ group remained continuous only in 73.3% cases, whereas 100% of the restorations in the LD group exhibited proper anatomic form.

Table 6. The anatomic form in the study groups.

Follow-Up Periods	Anatomic Form	Group ($\chi^2 = 4.615, p = 0.032$)					
		LD		ZrO$_2$		Total	
		No.	%	No	%	No.	%
2 weeks	Continuous with the existing anatomy	15	100.0	11	73.3	26	86.7
4 weeks 6 months 1 year	Discontinuous with the existing anatomy but not sufficient enough to expose dentin/base	0	0	4 *	26.7	4	13.3

* One inlay was lost just before completion of the 1-year evaluation.

In the ZrO$_2$ group, 73.3% of cases exhibited closely adapted margins at the 2- and 4-week follow-ups. This percentage decreased to 66.7% after 6 months to 1 year. Whereas, in the LD group, 80% of restorations were closely adapted. However, the difference between groups was statically insignificant at all follow-up visits (Table 7). In the ZrO_2 group, four restorations (26.7%) had a visible crevice; however, the sharp point of a probe (point diameter 0.5 mm, GDC Exs6XL, India) could not penetrate it. Moreover, one restoration had evidence of a step when the probe was drawn from the tooth for the restoration.

Table 7. The marginal adaptation in study groups.

Follow-Up Periods	Marginal Adaptation	Group (χ^2 = 1.043; p = 0.593)					
		LD		ZrO_2		Total	
		No.	%	No	%	No.	%
2 weeks 4 weeks	Closely adapted. No evidence of a catch or crevice at any point	12	80.0	11.0	73.3	26.0	86.7
	Visible evidence of a crevice. Fine probe will not penetrate	3	20.0	3.0	20.0	6.0	20.0
	Visible evidence of a crevice. Fine probe will penetrate	0	0	0	0	0	0
	Evidence of a positive step when probe drawn from tooth to restoration	0	0	1.0	6.7	1.0	3.3
6 months 1 year	Closely adapted. No evidence of a catch or crevice at any point	12	80.0	10.0	66.7	22.0	73.3
	Visible evidence of a crevice. Fine probe will not penetrate	3	20.0	4.0	26.7	7.0	23.3
	Visible evidence of a crevice. Fine probe will penetrate	0	0	0	0	0	0
	Evidence of a positive step when probe drawn from tooth to restoration	0	0	1.0 *	6.7	1	3.3

* One inlay was lost just before completion of the 1-year evaluation.

In the LD group, all restorations exhibited a smooth surface at all time intervals. In the ZrO_2 group, this feature was observed for 93.3% of cases after 2 weeks, 4 weeks and 6 months. However, after 1 year, this value decreased to 80%. In contrast, in the LD group, only three restorations (20%) exhibited visible evidence of a crevice, but a sharp pointed probe was not able to penetrate even after one-year of follow-up. The difference between groups was statically insignificant at all follow-up visits (Table 8).

Table 8. The surface roughness in the study groups.

Follow-Up Periods	Surface Roughness	Group (χ^2 =1.034; p = 0.309)					
		LD		ZrO_2		Total	
		No.	%	No	%	No.	%
2 weeks 4 weeks 6 months	Smooth	15	100	14	93.3	29	96.7
	Slightly pitted	0	0	1	6.7	1	3.3
	Deeply pitted	0	0	0	0	0	0
	Surface fractured	0	0	0	0	0	0
1 year	Smooth	15	100	12	80	27	90
	Slightly pitted	0	0	1 *	6.7	1	3.3
	Deeply pitted	0	0	0	0	0	0
	Surface fractured	0	0	2	13.3	2	6.7

* One inlay was lost just before completion of the 1-year evaluation.

In the LD group, all restorations exhibited proper colour and translucency match, while in the ZrO_2 group only 26.7% matched the colour of the tooth being restored. The difference between groups was statistically significant post-immediate placement of the restoration at all follow-up visits (Table 9).

Table 9. The colour match in the study groups.

Immediate Colour Match	Groups ($\chi^2 = 17.368, p = 0.000$)					
	LD		ZrO$_2$		Total	
	No.	%	No.	%	No.	%
Matches colour and translucency of adjacent tooth structure	15	100	4	26.7	19	63.3
Mismatch in colour and translucency	0	0	11	73.3	11	36.7

Post-cementation of three restorations (20%) in group ZrO$_2$ patients experienced mild, but bearable sensitivity at all follow-up time periods up to one year. However, the difference between groups was statically insignificant at all follow-up visits (Table 10).

Table 10. The occurrence of sensitivity in the study.

Follow-Up	Sensitivity	Groups ($\chi^2 = 3.33, p = 0.068$)					
		LD		ZrO$_2$		Total	
		No.	%	No.	%	No.	%
2 weeks	None	15	100	12	80	27	90
4 weeks	Mild but bearable	0	0	3 *	20	3	10
6 months	Uncomfortable	0	0	0	0	0	0
1 year	Very painful	0	0	0	0	0	0

* One inlay was lost just before completion of the 1-year evaluation.

4. Discussion

In this investigation, the null hypothesis was accepted. There was no difference between the clinical performance of the CAD-CAM zirconia dioxide and lithium disilicate inlays. It is worth emphasizing that this is the first clinical study on posterior indirect inlays comparing two different ceramic materials with a one-year follow-up.

It was evident that, due to the debonding of one restoration in the clinical scenario in group ZrO$_2$, this group showed a slightly lower survival rate than did the LD restorations (93% and 100%, respectively) although this finding was not statistically significant. The present results are in agreement with other studies that evaluated partial coverage restorations and crowns [19]. The high survival probability of the restorations in the LD group could be a result of micromechanical and chemical bonds to the etched silica [20–26]. Consequently, micromechanical interlocking between the rough surface of the restoration and resin-based cement is created, which enhances the bond strength [27].

In addition, chemical bonds can be increased by silanization of the restoration bonding surface. Silane forms strong siloxane linkages between the restoration and resin interface [4]. The silane agent used in the present study was Monobond N (Monobond N, Ivoclar Vivadent Schaan, Liechtenstein, Germany), which is composed of three different functional monomers, namely silane methacrylate, phosphoric methacrylate and sulphide methacrylate [4,28]. However, in the ZrO$_2$ group, one restoration (7%) debonded due to adhesive failure within two months. The remnants of the adhesive cement were located on the tooth surface. The adhesive procedure (Monobond N) of the CAD-CAM inlays did not result in a chemical bond to the ZrO$_2$ restoration [29].

The traditional silanization is not effective in the case of restorations lacking a glass phase [30]. On the contrary, it was proven that the addition of MDP (methacryloyloxydcyl dihydrogen phosphate) to silane or to primer enhanced the bond strength of resin materials to zirconium oxide-based restorations [31–34]. MDP is a monomer derived from the reaction of methacrylic acid with phosphoric acid or carboxylic acid. It creates chemical (P = O, OH = Zr) or ionic bonds with ZrO$_2$ [35].

Another possible reason for the debonding of two ZrO$_2$ restorations could be due to the poor adhesion of this cement system (Multilink N, Ivoclar Vivadent Schaan, Liechtenstein) to dentine. The resin system either led to partial demineralization of the dentine substrate or to incomplete polymerization of the adhesive and cement, resulting in premature degradation of the interface [27,29,33,35–41]. The survival rates for all-ceramic restorations were found to be over 90% after 10 years of service [42].

In the present study, the ZrO_2 (80%) and LD (66.7%) restorations exhibited and maintained normal occlusal and interproximal contacts at all follow-up periods up to one year. No statistical difference was observed. A similar outcome was seen in another study where there was no significant difference between LD and ZrO_2 full-coverage crowns regarding the marginal, axial and occlusal fit [43]. Open proximal contact can contribute to, for instance, the formation of periodontal pockets, gingival inflammation, or proximal caries [44]. This can occur due to imperfections in impressions (traditional or digital), during fabrication (firing or sintering) of the ceramics or through wear at the interproximal surface [45].

The anatomical form in both the ZrO_2 (73.3%) and LD (100%) groups remained continuous. The present study also evaluated the marginal adaptation of the restorations. Only 66.7% of restorations in the ZrO_2 group and 80% in the LD group exhibited close marginal adaptation with no evidence of a catch or crevice up to the one-year follow-up. These findings are supported by several studies [45–56]. The most probable reason for visible crevices (26.7% after 1 year) in the ZrO_2 group could be that ceramic veneering and layering on zirconia copings may result in an increased marginal gap compared with press techniques [56–58].

The marginal fit is one of the factors influencing possible restoration failure due to secondary caries and retention loss [59]. The marginal discrepancies can be observed due to the dissolution of luting cement, polymerization shrinkage of cement, occlusal load, type of finish line and margin placement (supra-gingival, sub-gingival or crestal gingival margin), salivary pH and brushing technique [60]. Moreover, the marginal gap can accumulate bacterial plaque and consequently result in carious lesions [57].

The clinical acceptable marginal discrepancy between prosthodontic restoration and the prepared tooth surface is approximately 50–120 μm [61–64]. However, minor marginal discrepancies in an indirect restoration may be compensated by the dual cure resin composite luting system [65]. The present study used conventional impressions in the LD group and digital ones in the ZrO_2 group according to the recommendations of other studies [42,66,67]. The internal fit of restorations was proven to be comparable for both impression techniques [67].

In the LD group, all cases presented a smooth surface up to one year. This finding is in consensus with similar clinical studies and laboratory studies that evaluated the surface smoothness of all ceramic restorations [2,12,15,16,42,48,66,68–72]. In the ZrO_2 group, 13.3% inlays exhibited surface fracture/chipping of the veneering ceramic after one year. These results are supported by similar studies that evaluated the clinical chipping of porcelain from zirconium dioxide substructures [19,73]. The crack formation and propagation occurs when the tensile strength within the ceramic exceeds the tensile strength of the veneering ceramic [68].

The tensile strength of the ceramic is the sum of the external and residual stresses. Without any load applied, residual stress persists, which can cause immediate or delayed ceramic cracks. On the contrary, external stress is formed within the structure by externally applied loads that occur during function and mastication [42,69]. Moreover, LD ceramic has an extended microcrystal structure (3–6 μm), which provides a strong bond with tooth structure after cementation [3–6,49]. This structure perfectly distributes forces due to the increased surface area of the crack and the interlocking microstructure of the ceramic. The crack propagation is described as an intragranular process and is characterised by a meandering line. Thus, the spread of a crack through this material is stopped by lithium disilicate crystals, providing a substantial increase in the bending strength and fracture toughness [70].

In the present study, all the LD group cases matched the colour and translucency, whereas in the ZrO$_2$ group, the matches amounted to up to 26.3% of cases. The clinical evaluation of the surface and colour of the LD crowns (Empress 2) after 14 years was found to be in the range of excellence [64,71,72,74–76]. The perfect aesthetics outcome of LD restorations were in accordance with several other studies [8,12,15,16,19,61,63,70,71]. The reasons for the high aesthetics of LD restorations are polyvalent ions in the glass that provide the desired colour, the even distribution of glass ceramics with leucite and lithium-disilicate-reinforced crystals in the single-phase equipment and the elimination of pigment defects in the microstructure [67].

Moreover, the similar light refraction index between glass ceramics with leucite and lithium-disilicate-reinforced crystals leads to high translucency [49]. However, in some cases, the complex optical characteristics of tooth colour makes it difficult to achieve a close shade match of an artificial restoration to the natural tooth structure [25]. On the contrary, ZrO$_2$ is white in colour and opaque. In the ZrO$_2$ group, all restorations were performed with single blocks of the same colour and opacity, which may have hampered the ability to mimic a natural appearance.

Therefore, this procedure does not always provide an optimal aesthetic integration, and consequently a veneering material should be applied [72]. In the present study, a glaze was applied to increase the gloss of the restorations, and tints were used to mimic the pits and fissures. The glaze resulted in a darker appearance of some restorations at the baseline recall, but the glaze was mostly lost after 1 year. A decrease in the translucency of some restorations was observed.

Additional reasons for a poor colour match could be the repeated firing of all ceramic zirconia cores and the thickness of the dentine porcelain [73,76]. Certain metal oxides are not colour stable after they are subjected to firing temperatures due to pigment breakdown of surface colorants [25,28,30]. Additionally, visual shade selection could contribute to the colour mismatch. However, several studies found no difference between visual and instrumental shade selection techniques [42,68–70,72].

In this study, all patients in the LD group and 80% in the ZrO$_2$ group did not report post-operative sensitivity at all follow-up periods. However, there was no significant difference in sensitivity between the two groups in the follow-up period. Postoperative sensitivity has been attributed to several factors, including trauma due to dentin preparation, dentin etching, bacterial penetration of the pulp, occlusal discrepancies, the extent of cavity preparation, type of bonding, luting procedure and polymerization shrinkage [76,77]. A relatively low post-operative sensitivity rate was observed. A possible reason could be the mild-etching potential of the self-etch adhesive luting cement (Multilink N), which did not cause over-etching and created a uniform hybrid layer. These findings are in agreement with several studies showing a low or lack of post-operative sensitivity for restorations luted using a self-etch mode [19,72–75,78,79].

There are several in vitro and clinical studies comparing fixed prosthesis, including ceramic restorations, using different parameters [27,28,30,34,36–41,45,55,61,80]. Several clinical studies used USPHS criteria for tooth-coloured restorations in posterior teeth [15,66,78]. Therefore, this method was used to assess zirconia dioxide and lithium disilicate inlay restorations in the present study.

A low number of restorations was investigated in this study, and thus evaluations on larger study groups are needed. Moreover, only two ceramic materials and one adhesive agent and cement were used. Similar studies embracing more materials should be conducted in the future. There is a need to prolong the follow-up period to investigate both techniques in long-term studies. Additionally, the investigation was performed at one university, and thus more multicentre studies should be carried out; private dental offices should be also included to provide a wider perspective.

5. Conclusions

Within the imitations of the present study, the lithium disilicate- and zirconia dioxide-based inlays exhibited comparable clinical performance. However, the colour and translucency match was superior for the lithium disilicate restorations.

Author Contributions: For Conceptualization, R.B. and L.M.; methodology, R.B., L.M. and N.R.S.; software, D.D.D.; validation, R.B., L.M. and M.L.-S.; formal analysis, N.R.S. and A.A.A.-K.; investigation, R.B., L.M. and N.R.S.; resources, L.M.; data curation, M.L.-S.; writing—original draft preparation, R.B. and L.M.; writing—review and editing, R.B., L.M. and M.L.-S.; visualization, N.R.S.; supervision, M.L.-S.; project administration, L.M.; funding acquisition, D.D.D. and A.A.A.-K. All authors have read and agreed to the published version of the manuscript.

Funding: The authors are grateful to the deanship of Scientific Research, King Saud University for funding through Vice Deanship of Scientific Research Chairs.

Institutional Review Board Statement: The study was conducted according to the guidelines of the Declaration of Helsinki, and approved by the Ethics Committee of Siksha 'O' Anusandhan University (Protocol: Ref. No/DMR/ims.SH/SOA/180035 approved on 25 May 2018).

Informed Consent Statement: Informed consent was obtained from all subjects involved in the study.

Data Availability Statement: The data presented in this study are available on request from the corresponding author.

Acknowledgments: The authors are grateful to Swadheena Patro, Sonia Aggarwal and Amit jena for providing resources and clinical guidance during the inception of this study.

Conflicts of Interest: The authors declare no conflict of interest. The funders had no role in the design of the study; in the collection, analyses, or interpretation of data; in the writing of the manuscript, or in the decision to publish the results.

References

1. Petersen, P.E. The World Oral Health Report 2003: Continuous improvement of oral health in the 21st century—The approach of the WHO Global Oral Health Programme. *Community Dent. Oral Epidemiol.* **2003**, *31*, 3–24. [CrossRef] [PubMed]
2. Bartlett, D.; Ricketts, D. Inlays, Onlays and Veneers. In *Advanced Operative Dentistry*; Ricketts, D., Bartlett, D., Eds.; Churchill Livingstone: Edinburgh, UK, 2011; pp. 151–162, ISBN 9780702031267.
3. Zarone, F.; Di Mauro, M.I.; Ausiello, P.; Ruggiero, G.; Sorrentino, R. Current status on lithium disilicate and zirconia: A narrative review. *BMC Oral Health* **2019**, *19*, 1–14. [CrossRef] [PubMed]
4. Al-Amleh, B.; Lyons, K.; Swain, M. Clinical trials in zirconia: A systematic review. *J. Oral Rehabil.* **2010**, *37*, 641–652. [CrossRef] [PubMed]
5. Albakry, M.; Guazzato, M.; Swain, M.V. Biaxial flexural strength, elastic moduli, and x-ray diffraction characterization of three pressable all-ceramic materials. *J. Prosthet. Dent.* **2003**, *89*, 374–380. [CrossRef]
6. Guazzato, M.; Albakry, M.; Ringer, S.P.; Swain, M.V. Strength, fracture toughness and microstructure of a selection of all-ceramic materials. Part I. Pressable and alumina glass-infiltrated ceramics. *Dent. Mater.* **2004**, *20*, 441–448. [CrossRef] [PubMed]
7. Heffernan, M.J.; Aquilino, S.A.; Diaz-Arnold, A.M.; Haselton, D.R.; Stanford, C.M.; Vargas, M.A. Relative translucency of six all-ceramic systems. Part I: Core materials. *J. Prosthet. Dent.* **2002**, *88*, 4–9. [CrossRef]
8. Heffernan, J.M.; Aquilino, A.S.; Diaz-Arnold, M.A.; Haselton, R.D.; Stanford, M.C.; Vargas, A.M. Relative translucency of six all-ceramic systems. Part II: Core and veneer materials. *J. Prosthet. Dent.* **2002**, *88*, 10–15. [CrossRef] [PubMed]
9. Suputtamongkol, K.; Tulapornchai, C.; Mamani, J.; Kamchatphai, W.; Thongpun, N. Effect of the shades of background substructures on the overall color of zirconia-based all-ceramic crowns. *J. Adv. Prosthodont.* **2013**, *5*, 319–325. [CrossRef]
10. Mopkar, M.; Aras, M.A.; Chitre, V.; Mysore, A.; Coutinho, I.; Rajagopal, P. Factors affecting shade of all ceramic restorations. A literature review. *J. Dent. Appl.* **2018**, *5*, 417–424.
11. Chritchlow, S. Ceramic materials have similar short term survival rates to other materials on posterior teeth. *Evid. Based. Dent.* **2012**, *13*, 49. [CrossRef]
12. Beier, U.S.; Kapferer, I.; Dumfahrt, H. Clinical long-term evaluation and failure characteristics of 1,335 all-ceramic restorations. *Int. J. Prosthodont.* **2012**, *25*, 25.
13. Stoll, R.; Cappel, I.; Jablonski-Momeni, A.; Pieper, K.; Stachniss, V. Survival of inlays and partial crowns made of IPS empress after a 10-year observation period and in relation to various treatment parameters. *Oper. Dent.* **2007**, *32*, 556–563. [CrossRef] [PubMed]
14. Fernandes, N.A.; Vally, Z.I.; Sykes, L.M. The longevity of restorations—A literature review. *South Afr. Dent. J.* **2015**, *70*, 410–413.
15. Qualtrough, A.J.; Wilson, N.H. A 3-year clinical evaluation of a porcelain inlay system. *J. Dent.* **1996**, *24*, 317–323. [CrossRef]

16. Fabianelli, A.; Goracci, C.; Bertelli, E.; Davidson, C.L.; Ferrari, M. A clinical trial of Empress II porcelain inlays luted to vital teeth with a dual-curing adhesive system and a self-curing resin cement. *J. Adhes. Dent.* **2006**, *8*, 427–431. [PubMed]
17. Nakhaei, M.; Ghanbarzadeh, J.; Alavi, S.; Amirinejad, S.; Rajatihaghi, H. The influence of dental shade guides and experience on the accuracy of shade matching. *J. Contemp. Dent. Pr.* **2016**, *17*, 22–26. [CrossRef] [PubMed]
18. Homsy, F.R.; Özcan, M.; Khoury, M.; Majzoub, Z.A. Marginal and internal fit of pressed lithium disilicate inlays fabricated with milling, 3D printing, and conventional technologies. *J. Prosthet. Dent.* **2018**, *119*, 783–790. [CrossRef]
19. Guess, P.C.; Selz, C.F.; Steinhart, Y.-N.; Stampf, S.; Strub, J.R. Prospective clinical split-mouth study of pressed and CAD/CAM all-ceramic partial-coverage restorations: 7-year results. *Int. J. Prosthodont.* **2013**, *26*, 21–25. [CrossRef] [PubMed]
20. Alshiddi, I.F.; Richards, L.C. A comparison of conventional visual and spectrophotometric shade taking by trained and untrained dental students. *Aust. Dent. J.* **2015**, *60*, 176–181. [CrossRef]
21. Liberato, W.F.; Barreto, I.C.; Costa, P.P.; de Almeida, C.C.; Pimentel, W.; Tiossi, R. A comparison between visual, intraoral scanner, and spectrophotometer shade matching: A clinical study. *J. Prosthet. Dent.* **2019**, *121*, 271–275. [CrossRef]
22. Lapinska, B.; Rogowski, J.; Nowak, J.; Nissan, J.; Sokolowski, J.; Lukomska-Szymanska, M. Effect of surface cleaning regimen on glass ceramic bond strength. *Molecules* **2019**, *24*, 389. [CrossRef] [PubMed]
23. Łapińska, B. Changes in dental ceramic surface structure and their influence on the bond strength to composite material (Zmiany struktury powierzchni ceramik dentystycznych oraz ich wpływ na wytrzymałość połączenia z materiałem kompozytowym). *Przemysł Chem.* **2017**, *1*, 124–128. [CrossRef]
24. Łapińska, B. Lithium silicate ceramic surface properties after surface treatment (Właściwości ceramiki litowo-silikatowej po obróbce jej powierzchni). *Przemysł Chem.* **2017**, *1*, 145–149. [CrossRef]
25. Łapińska, B.; Sokołowski, J.; Klimek, L.; Łukomska-Szymańska, M. Ocena zmian struktury i składu chemicznego ceramiki dwukrzemianu litu trawionej kwasem fluorowodorowym po zanieczyszczeniu śliną i zastosowaniu różnych metod oczyszczania powierzchni. (Surface Structure and Chemical Composition of Hydrofluoric Acid-Etched Lithium Disilicate Ceramic After Application of Different Cleaning Methods of Saliva Contamination Removal). *Dent. Med Probl.* **2015**, *52*, 71–77.
26. Succaria, F.; Morgano, S.M. Prescribing a dental ceramic material: Zirconia vs lithium-disilicate. *Saudi Dent. J.* **2011**, *23*, 165–166. [CrossRef] [PubMed]
27. Tanış, M.Ç.; Akay, C.; Karakış, D. Resin cementation of zirconia ceramics with different bonding agents. *Biotechnol. Biotechnol. Equip.* **2015**, *29*, 363–367. [CrossRef]
28. Stawarczyk, B.; Teuss, S.; Eichberger, M.; Roos, M.; Keul, C. Retention Strength of PMMA/UDMA-Based Crowns Bonded to Dentin: Impact of Different Coupling Agents for Pretreatment. *Materials* **2015**, *8*, 7486–7497. [CrossRef] [PubMed]
29. Chai, J.; Chu, F.C.S.; Chow, T.W. Effect of surface treatment on shear bond strength of zirconia to human dentin. *J. Prosthodont.* **2011**, *20*, 173–179. [CrossRef] [PubMed]
30. Wat, P.; Cheung, G.S. Incidence of post-operative sensitivity following indirect porcelain onlay restorations: Preliminary results. *Asian J. Aesthetic Dent.* **1995**, *3*, 3–7.
31. Christensen, G.J. Why use resin cements? *J. Am. Dent. Assoc.* **2010**, *141*, 204–206. [CrossRef]
32. Costa, T.; Rezende, M.; Sakamoto, A.; Bittencourt, B.; Dalzochio, P.; Loguercio, A.D.; Reis, A. Influence of adhesive type and placement technique on postoperative sensitivity in posterior composite restorations. *Oper. Dent.* **2017**, *42*, 143–154. [CrossRef] [PubMed]
33. Hiraishi, N.; Breschi, L.; Prati, C.; Ferrari, M.; Tagami, J.; King, N. Technique sensitivity associated with air-drying of HEMA-free, single-bottle, one-step self-etch adhesives. *Dent. Mater.* **2007**, *23*, 498–505. [CrossRef] [PubMed]
34. Moura, D.M.D.; Januário, A.B.D.N.; de Araújo, A.M.M.; Piva, A.M.D.O.D.; Özcan, M.; Bottino, M.A.; Souza, R.O.A. Effect of primer-cement systems with different functional phosphate monomers on the adhesion of zirconia to dentin. *J. Mech. Behav. Biomed. Mater.* **2018**, *88*, 69–77. [CrossRef]
35. Roman-Rodriguez, J.; Roig-Vanaclocha, A.; Fons, A.; Granell-Ruiz, M.; Sola-Ruiz, M.; Amigó, V.; Busquets-Mataix, D.; Vicente-Escuder, A. In vitro experimental study of bonding between aluminium oxide ceramics and resin cements. *Med. Oral Patol. Oral. Cir. Bucal.* **2009**, *15*, e95–e100. [CrossRef]
36. Nagaoka, N.; Yoshihara, K.; Feitosa, V.P.; Tamada, Y.; Irie, M.; Yoshida, Y.; Van Meerbeek, B.; Hayakawa, S. Chemical interaction mechanism of 10-MDP with zirconia. *Sci. Rep.* **2017**, *7*, srep45563. [CrossRef]
37. Perdigão, J.; Loguercio, A.D. Universal or multi-mode adhesives: Why and how? *J. Adhes. Dent.* **2014**, *16*, 193–194. [CrossRef]
38. Mounajjed, R.; Layton, D.; Azar, B. The marginal fit of E.max Press and E.max CAD lithium disilicate restorations: A critical review. *Dent. Mater. J.* **2016**, *35*, 835–844. [CrossRef]
39. Prasad, P.; Gaur, A.; Kumar, V.; Chauhan, M. To Evaluate and compare postcementation sensitivity under class II composite inlays with three different luting cements: An in vivo study. *J. Int. Oral Health* **2017**, *9*, 165–173. [CrossRef]
40. Tanaka, R.; Fujishima, A.; Shibata, Y.; Manabe, A.; Miyazaki, T. Cooperation of phosphate monomer and silica modification on zirconia. *J. Dent. Res.* **2008**, *87*, 666–670. [CrossRef] [PubMed]
41. Al Hamad, K.Q.; Al Quran, F.A.; AlJalam, S.A.; Baba, N.Z. Comparison of the accuracy of fit of metal, zirconia, and lithium disilicate crowns made from different manufacturing techniques. *J. Prosthodont.* **2018**, *28*, 497–503. [CrossRef]
42. Almalki, A.D.; Al-Rafee, M.A. Evaluation of presence of proximal contacts on recently inserted posterior crowns in different health sectors in Riyadh City, Saudi Arabia. *J. Fam. Med. Prim. Care* **2019**, *8*, 3549–3553. [CrossRef] [PubMed]

43. Kohorst, P.; Junghanns, J.; Dittmer, M.P.; Borchers, L.; Stiesch, M. Different CAD/CAM-processing routes for zirconia restorations: Influence on fitting accuracy. *Clin. Oral Investig.* **2010**, *15*, 527–536. [CrossRef]
44. Wittneben, J.; Gavric, J.; Belser, U.; Bornstein, M.; Joda, T.; Chappuis, V.; Sailer, I.; Brägger, U. Esthetic and clinical performance of implant-supported all-ceramic crowns made with prefabricated or CAD/CAM Zirconia abutments: A Randomized, multicenter clinical trial. *J. Dent. Res.* **2017**, *96*, 163–170. [CrossRef]
45. Zarone, F.; Di Mauro, M.I.; Spagnuolo, G.; Gherlone, E.; Sorrentino, R. Fourteen-year evaluation of posterior zirconia-based three-unit fixed dental prostheses. *J. Dent.* **2020**, *101*, 103419. [CrossRef]
46. Abou-Steit, S.; Elguindy, J.; Zaki, A. Evaluation of patient satisfaction and shade matching of Vita Suprinity versus lithium disilicate (E-max) ceramic crowns in the esthetic zone: A randomized controlled clinical trial. *F1000Research* **2019**, *8*, 371. [CrossRef]
47. Brandt, S.; Winter, A.; Lauer, H.-C.; Kollmar, F.; Portscher-Kim, S.-J.; Romanos, G.E. IPS e.max for all-ceramic restorations: Clinical survival and success rates of full-coverage crowns and fixed partial dentures. *Materials* **2019**, *12*, 462. [CrossRef] [PubMed]
48. Hamza, T.A.; Sherif, R.M. fracture resistance of monolithic glass-ceramics versus bilayered zirconia-based restorations. *J. Prosthodont.* **2017**, *28*, e259–e264. [CrossRef]
49. Rosentritt, M.; Schumann, F.; Krifka, S.; Preis, V. Influence of zirconia and lithium disilicate tooth- or implant-supported crowns on wear of antagonistic and adjacent teeth. *J. Adv. Prosthodont.* **2020**, *12*, 1–8. [CrossRef]
50. Aladağ, A.; Oğuz, D.; Çömlekoğlu, M.E.; Akan, E. In vivo wear determination of novel CAD/CAM ceramic crowns by using 3D alignment. *J. Adv. Prosthodont.* **2019**, *11*, 120–127. [CrossRef] [PubMed]
51. Vargas, S.P.; Neves, A.C.C.; Vitti, R.; Amaral, M.; Henrique, M.N.; Silva-Concílio, L.R. Influence of different ceramic systems on marginal misfit. *Eur. J. Prosthodont. Restor. Dent.* **2017**, *25*, 127–130.
52. Saridag, S.; Sevimay, M.; Pekkan, G. Fracture resistance of teeth restored with all-ceramic inlays and onlays: An in vitro study. *Oper. Dent.* **2013**, *38*, 626–634. [CrossRef]
53. Seidel, A.; Belli, R.; Breidebach, N.; Wichmann, M.; Matta, R.E. The occlusal wear of ceramic fixed dental prostheses: 3-Year results in a randomized controlled clinical trial with split-mouth design. *J. Dent.* **2020**, *103*, 103500. [CrossRef] [PubMed]
54. De Angelis, F.; D'Arcangelo, C.; Mališková, N.; Vanini, L.; Vadini, M. Wear properties of different additive restorative materials used for onlay/overlay posterior restorations. *Oper. Dent.* **2020**, *45*, E156–E166. [CrossRef] [PubMed]
55. Ahmed, W.M.; Shariati, B.; Gazzaz, A.Z.; Sayed, M.E.; Carvalho, R.M. Fit of tooth-supported zirconia single crowns—A systematic review of the literature. *Clin. Exp. Dent. Res.* **2020**, *6*, 700–716. [CrossRef]
56. El-Dessouky, R.; Salama, M.; Shakal, M.; Korsel, A. Marginal adaptation of CAD/CAM zirconia-based crown during fabrication steps. *Tanta Dent. J.* **2015**, *12*, 81–88. [CrossRef]
57. Sailer, I.; Makarov, N.A.; Thoma, D.S.; Zwahlen, M.; Pjetursson, B.E. All-ceramic or metal-ceramic tooth-supported fixed dental prostheses (FDPs)? A systematic review of the survival and complication rates. Part I: Single crowns (SCs). *Dent. Mater.* **2015**, *31*, 603–623. [CrossRef] [PubMed]
58. Abad-Coronel, C.; Naranjo, B.; Valdiviezo, P. Adhesive Systems used in indirect restorations cementation: Review of the literature. *Dent. J.* **2019**, *7*, 71. [CrossRef]
59. Guess, P.C.; A Zavanelli, R.; A Silva, N.R.F.; A Bonfante, E.; Coelho, P.G.; Thompson, V.P. Monolithic CAD/CAM lithium disilicate versus veneered Y-TZP crowns: Comparison of failure modes and reliability after fatigue. *Int. J. Prosthodont.* **2010**, *23*, 434–442. [PubMed]
60. Luciano, M.; Francesca, Z.; Michela, S.; Tommaso, M.; Massimo, A. Lithium disilicate posterior overlays: Clinical and biomechanical features. *Clin. Oral Investig.* **2019**, *24*, 841–848. [CrossRef]
61. Toman, M.; Toksavul, S. Clinical evaluation of 121 lithium disilicate allceramic crowns up to 9 Years. *Quintessence Int.* **2015**, *46*, 189–197.
62. Esquivel-Upshaw, J.F. Four-year clinical performance of a lithium disilicate-based core ceramic for posterior fixed partial dentures. *Int. J. Prosthodont.* **2008**, *21*, 155–160. [PubMed]
63. Owitayakul, D.; Lertrid, W.; Anatamana, C.; Pittayachawan, P. The Comparison of the marginal gaps of zirconia framework luted with different types of phosphate based-resin cements. *M. Dent. J.* **2015**, *35*, 237–251.
64. Miura, S.; Kasahara, S.; Kudo, M.; Okuyama, Y.; Izumida, A.; Yoda, M.; Egusa, H.; Sasaki, K. Clinical Chipping of Zirconia All-Ceramic Restorations. In *Interface Oral Health Science*; Springer Science and Business Media LLC.: Tokyo, Japan, 2015; pp. 317–323.
65. Della Bona, A.; Kelly, J.R. The clinical success of all-ceramic restorations. *J. Am. Dent. Assoc.* **2008**, *139*, 8–13. [CrossRef]
66. Anusavice, K.J.; Kakar, K.; Ferree, N. Which mechanical and physical testing methods are relevant for predicting the clinical performance of ceramic-based dental prostheses? *Clin. Oral Implant. Res.* **2007**, *18*, 218–231. [CrossRef] [PubMed]
67. Shenoy, A.; Shenoy, N. Dental ceramics: An update. *J. Conserv. Dent.* **2010**, *13*, 195–203. [CrossRef] [PubMed]
68. Teichmann, M.; Göckler, F.; Rückbeil, M.; Weber, V.; Edelhoff, D.; Wolfart, S. Periodontal outcome and additional clinical quality criteria of lithium-disilicate restorations (Empress 2) after 14 years. *Clin. Oral Investig.* **2019**, *23*, 2153–2164. [CrossRef] [PubMed]
69. Santos, M.; Mondelli, R.; Navarro, M.F.L.; Francischone, C.; Rubo, J.; Santos, G. Clinical evaluation of ceramic inlays and onlays fabricated with two systems: Five-year follow-up. *Oper. Dent.* **2013**, *38*, 3–11. [CrossRef]
70. Vichi, A.; Louca, C.; Corciolani, G.; Ferrari, M. Color related to ceramic and zirconia restorations: A review. *Dent. Mater.* **2011**, *27*, 97–108. [CrossRef]

71. Habib, S.R.; Al Shiddi, I.F. Comparison of shade of ceramic with three different zirconia substructures using spectrophotometer. *J. Contemp. Dent. Pr.* **2015**, *16*, 135–140. [CrossRef]
72. Kimmich, M.; Stappert, C.F. Intraoral treatment of veneering porcelain chipping of fixed dental restorations. *J. Am. Dent. Assoc.* **2013**, *144*, 31–44. [CrossRef]
73. Ayash, G.M.; Osman, E.; Segaan, L.G.; Rayyan, M.M. Visual Versus Instrumental Shade Selection Techniques. *Egypt. Dent. J.* **2011**, *61*, 6.
74. Demir, N.; Ozturk, A.N.; Malkoc, M.A. Evaluation of the marginal fit of full ceramic crowns by the microcomputed tomography (micro-CT) technique. *Eur. J. Dent.* **2014**, *8*, 437–444. [CrossRef] [PubMed]
75. Anadioti, E.; Aquilino, S.A.; Gratton, D.G.; Holloway, J.A.; Denry, I.L.; Thomas, G.W.; Qian, F. Internal fit of pressed and computer-aided design/computer aided manufacturing ceramic crowns made from digital and conventional impressions. *J Prosthet. Dent.* **2015**, *113*, 304–309. [CrossRef] [PubMed]
76. Son, H.-J.; Kim, W.-C.; Jun, S.-H.; Kim, Y.-S.; Ju, S.-W.; Ahn, J.-S. Influence of dentin porcelain thickness on layered all-ceramic restoration color. *J. Dent.* **2010**, *38*, e71–e77. [CrossRef]
77. Judeh, A.; Al-Wahadni, A. A comparison between conventional visual and spectrophotometric methods for shade selection. *Quintessence Int.* **2009**, *40*, 69–79.
78. Patankar, A.H.; Miyajiwala, J.S.; Kheur, M.G.; Lakha, T.A. Comparison of photographic and conventional methods for tooth shade selection: A clinical evaluation. *J. Indian Prosthodont. Soc.* **2017**, *17*, 273–281. [CrossRef]
79. Kim, J.-H.; Chae, S.-Y.; Lee, Y.; Han, G.-J.; Cho, B.-H. Effects of multipurpose, universal adhesives on resin bonding to zirconia ceramic. *Oper. Dent.* **2015**, *40*, 55–62. [CrossRef] [PubMed]
80. Rodolpho, P.A.D.R.; Cenci, M.; Donassollo, T.A.; Loguércio, A.D.; Demarco, F.F. A clinical evaluation of posterior composite restorations: 17-year findings. *J. Dent.* **2006**, *34*, 427–435. [CrossRef]

Article

The Influence of Aging in Solvents on Dental Cements Hardness and Diametral Tensile Strength

Agata Szczesio-Wlodarczyk [1,*], Karolina Rams [2], Karolina Kopacz [3], Jerzy Sokolowski [2] and Kinga Bociong [1]

1. University Laboratory of Materials Research, Medical University of Lodz, ul Pomorska 251, 92-213 Lodz, Poland
2. Department of General Dentistry, Medical University of Lodz, ul, Pomorska 251, 92-213 Lodz, Poland
3. "DynamoLab" Academic Laboratory of Movement and Human Physical Performance, Medical University of Lodz, ul. Pomorska 251, 92-216 Lodz, Poland
* Correspondence: agata.szczesio@umed.lodz.pl; Tel.: +48-42-272-57-66

Received: 28 June 2019; Accepted: 31 July 2019; Published: 2 August 2019

Abstract: Prosthetic materials must exhibit adequate resistance to the oral environment. The aim of this paper was to study the resistance of selected cements used for cementing restorations (Breeze—composite, Adhesor Carbofine—zinc-polycarboxylate and IHDENT–Giz type II—glass-ionomer) against ethanol, soda and green tea solutions. The highest values of hardness and DTS (diametral tensile strength) were obtained by composite cement (HV = 15–31, DTS = 34–45 MPa). Ethanol solution had the greatest impact on the hardness value of composite cement, and soda solution on zinc-polycarboxylate cement. No significant differences were noted in the DTS values of composite cements after immersion in solvents; however, the DTS value of zinc-polycarboxylate cement increased after prolonged immersion time in ethanol and the DTS of glass-ionomer cement (IHDENT Giz type II) clearly decreased after submersion in soda solutions. Variation in pH across the range of 6 (tea) to 9 (soda solution) had a low impact on the properties of dental cements. Extended exposure to solvents appears to worsen the properties of cements.

Keywords: dental cements; ethanol; water; resistance; DTS; Vickers hardness

1. Introduction

Dental cements have a wide range of applications in modern dentistry. They are used for luting, fixation and cementation, i.e., luting inlays crowns, bridges, veneers on the prepared tooth. They protect pulp from heat ("thermal insulation") and from chemical irritation (liners and bases), they also stimulate secondary dentin formation and act as temporary filling material.

The rehabilitation of the stomatognathic system is a very important aspect for patients with partial edentulism. Prosthetic restorations used during treatment often restore lost function of chewing, improve aesthetics and speech [1]. Dental cements constitute an important element during prosthetic treatment. These materials are designed to bond the restoration to the natural teeth of the patient or metallic core. Cements affect the retention of a restoration and protect the exposed dentin against many mechanical, chemical, thermal and bacterial factors. Success of the treatment depends on the proper selection of material used to bond the restoration. As there is currently no ideal cement that can meet all the requirements in terms of mechanical and biological properties, it is important to make an appropriate individual selection for each patient based on the properties of the materials [2–4].

Nowadays, prosthetic restoration can be performed using composite, zinc-polycarboxylate, glass-ionomer, glass-ionomer reinforced with resin, zinc-phosphate and oxide-zinc-eugenol cements and compomers [5]. Of these, composite cements show the best mechanical properties: Their diametral tensile strength (DTS) ranges from 30 to 60 MPa (zinc phosphate cements approximately 10 MPa, polycarboxylate cements approximately 10 MPa, glass-ionomer cements greater than 15 MPa) and compressive strength from 140 to 200 MPa (zinc phosphate cements approximately 50 MPa, polycarboxylate cements greater than 60 MPa, glass-ionomer cements approximately 100 MPa) [6,7]. These materials consist of an organic matrix and powdered ceramics, e.g., aluminum-boron-bar glass or silanized silica. The filler constitutes from 30% to 80% of the volume of the material, and the size of their particles is from 0.04 to 5.0 μm. The organic phase (matrix) consists of resins such as bisphenol A glycol dimethacrylate (bis-GMA), urethane dimethacrylate (UDMA), hydroxyethyl methacrylate (HEMA), 4-methacryloxyethyl trimellitic anhydride (4-META). Most resin cements contain bis-GMA. Some materials have replaced bis-GMA partially or entirely with bis-EMA resin (ethoxylated bisphenol-A dimethacrylate) or UDMA to reduce their viscosity, eliminating the need for large quantities of diluents such as TEGDMA (triethylene glycol dimethacrylate) and DEGMA (di(ethylene glycol) dimethacrylate) monomers [4,8,9].

Zinc-polycarboxylate cements have been in use for over 30 years and are considered to be the safest cements due to their high biocompatibility. They are available in powder and liquid forms. The powder consists of zinc and magnesium oxide previously subjected to sintering and grinding; it also sometimes contains tin fluoride, which influences the setting time. The liquid form is usually based on a 40% aqueous solution of polyacrylic acid; however, cements can also be made with distilled water, the powder contains oxides particles coated with polyacrylic acid (15–18%). Mixing the powder with the liquid induces formation of complexes called chelates. The reactions occur between metal ions (zinc, calcium) and carboxylic groups derived from polyacrylic acid (liquid) [2,10,11].

Zinc-polycarboxylate cements not only offer very favorable biocompatible properties, but they also are characterized by good adhesion to dentine. Unfortunately, such cements also demonstrate poor mechanical properties, such as low compressive strength (67–91 MPa) and high solubility; in addition, due to their plasticity after curing, they are also not viable for cementation of restorations exposed to high short-circuit forces or those with large spans. Zinc-polycarboxylate cements also show low resistance to erosion in an acidic environment, which is a contraindication to use in patients struggling with reflux disease or consuming large amounts of acidic or carbonated drinks [12].

Glass-ionomer cements (GIC) are also available as powders and liquids. The powder consists of calcium-aluminum-fluorosilicate glass. The fluid is an aqueous solution of polyacrylic acid, maleic acid, tartaric acid and itaconic acid [13,14]. The glass-ionomer cements are cured through acid-base reactions comprising a three-step process of dissolving, gelling and hardening. Reactions occur between polyacids (polyanions) and glass (metal cations and fluoride anions) which leads to the formation of a gel matrix. The main advantage of glass ionomer cements is their ability to release fluoride ions, which has a remineralizing effect and prevents the development of caries. They bind physically and chemically with enamel and dentin [13]. Among other properties, they are also distinguished by adequate tensile and compression strength: The compressive strength of glass-ionomer cements is above 130 MPa and this value is sufficient to resist the masticatory forces in the posterior teeth [15]. They also have a similar coefficient of thermal expansion to dentine [15,16]. Glass-ionomer cements are characterized by high sensitivity to both moisture and dehydration. In all binding steps, they have high solubility in water, which can lead to modification of their mechanical properties. However, such materials are also characterized by low initial pH, which can lead to pulp hypersensitivity [3,17]. The above-mentioned cements seem to be most often used in today's dental prosthetics, hence their resistance to selected solvents are significant.

Prosthetic cements must be resistant to wide variations in the oral fluid environment, which is influenced by a range of factors including food, drinks and smoking. The aim of the work is to examine the resistance of selected cements used for cementing prosthetic restorations (composite, zinc-polycarboxylate and glass-ionomer) to solvents (75% ethanol, soda solution, green tea) and variations in pH = 5–9.

The following research hypothesis is stated at work: Solvents with different pH (across the range of 6 (tea) to 9 (soda solution)) has an impact on the properties of dental cements.

2. Materials and Methods

Three different cements were selected for the tests: Breeze, Adhesor Carbofine and IHADENTA Giz typ II. Selected cements are a representative example of luting cements (resin, glass-ionomer and zinc-polycarboxylate). Table 1 contains information on tested materials.

Table 1. The composition of tested materials.

Manufacturer	Material	Type	Composition	Preparation and Curing
Jeneric Pentron (Orange, CA, USA)	Breeze	Self-adhesive resin cement	Bisphenol A glycol dimethacrylate (bis-GMA), urethane dimethacrylate (UDMA), triethylene glycol dimethacrylate (TEGDMA), hydroxyethyl methacrylate (HEMA), 4-methacryloxyethyl trimellitic anhydride (4-MET), silanized barium glass, silica, BiOCl, curing system	Mixing: Self-mixing syringe, curing: Polymerized using a 3M ESPE EliparTM S10 diode lamp on the top and bottom surface of the sample as recommended by the manufacturer for 20 s
SpofaDental (Jičín, Czech Republic)	Adhesor Carbofine	Zinc-polycarboxylate cement	Powder: Zinc oxide, magnesium oxide, aluminum oxide, boric acid, liquid: Acrylic acid, maleic anhydride, distilled water	Mixing: 1:1 (power:liquid) mixing ratio recommended by manufacturer; curing: 7 min self-curing in plastic zip bag
Ihde Dental AG (Gommiswald, Switzerland)	Glass-ionomer cement (Ihdent® GIZ® fil Typ II)	Glass ionomer cement	Aluminum-fluoride-silicate powder, iron oxide, polyacrylic acid	Mixing: 1.8–2.2 g of powder per 1 g of liquid (mixing ratio recommended by manufacturer); curing: 5–8 min self-curing in plastic zip bag

Fifty samples of each cement were made according to the manufacturer's instructions (Table 1). All samples were shaped as cylinders with a diameter of 6 mm and a height of 3 mm. The material was cured according to manufacturer's instructions (Table 1) and the samples were placed in a container with distilled water for a period of seven days.

After seven days, control tests (hardness and DTS) were conducted. Five samples were put in separate containers that contained various substances:

- Solution of baking soda (2.5%);
- Solution of water and ethanol (75%);
- Green tea.

Most foods have a pH close to neutral or acidic. In order to widen the pH range, a 2.5% solution of baking soda was included. Baking soda (sodium bicarbonate) produces OH^- ions during hydrolysis, which can accelerate the hydrolysis of dental materials.

Samples were immersed in the solutions for 1, 7 and 30 days. The pH values of the solutions were measured using litmus paper and a pH meter (CPI-505, Elmetron, Zabrze, Polska). The obtained results are presented in Table 2.

Table 2. The results obtained during the pH measurement of the solutions.

Solution	pH (Litmus Paper)	pH (pH-Meter)
Green tea	5–6	6
Ethanol:water (75%)	7	8
Soda solution (2.5%)	9	9

The hardness of Breeze and Adhesor Carbofine cements were measured using the Vickers method. The Zwick ZHV2-m hardness tester (Zwick-Roell, Ulm, Germany) was used for the tests. The applied load was 1000 g and the penetration time was 10 s. Eleven measurements were performed on three out of five samples from container for each material at specific time intervals (1, 7 and 30 days).

For testing the diametral tensile strength (DTS), a Zwick Roell Z020 universal strength machine (Germany) was used. The traverse speed was 2 mm/min. The measurement was made on five samples from each material at specific time intervals (1, 7 and 30 days) making a total of 15 samples. The DTS values were calculated using the Formula (1):

$$DTS = \frac{2F}{\pi dh} (\text{MPa}) \qquad (1)$$

F—Compressive force, which caused the destruction of the sample (N)
d—Diameter of the sample (mm)
h—Height of the sample (mm).

For statistical calculations, Statistica v. 13.1 (Statsoft, Kraków, Poland) was used. The normality of the distribution of data was confirmed using the Shapiro—Wilk test; depending on the result, either parametric (F-test) or non-parametric (Kruskal—Wallis test) tests were used for statistical analysis (alpha = 0.05). Equality of variance was tested with Levene's test.

3. Results and Discussion

The composite cements, *viz.* Breeze and Adhesor Carbofine, were found to have a higher hardness than the zinc-polycarboxylate cement (Figures 1 and 2). DTS results indicate that Breeze has the highest strength (Figures 3–5). Cements modified with resin matrix demonstrate better properties than traditional cements and tend to be more popular [18]. The research hypothesis was not rejected. But it should be emphaticized that immersion in solutions of pH 6 (tea)–9 (soda solution) has little impact on the properties of the dental cements. The impact depends on the material and solution composition.

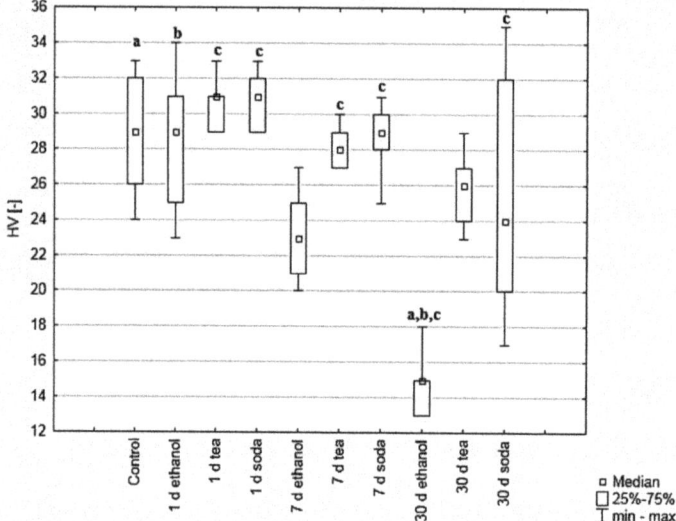

Figure 1. Box and whiskers plot of Vickers hardness of samples. Values were obtained for samples made of Breeze cement treated with ethanol (75%), green tea and baking soda (2.5%) solutions after one, seven or 30 days and samples immersed for seven days in distilled water as control. Statistically significant differences were detected between: (a) Control group vs. 30 days in ethanol solution, (b) 1 day in ethanol vs. 30 days in ethanol solution, (c) 30 days in ethanol solution vs. 1 day in tea solution, 7 day in tea solution, 1 day in soda solution, 7 day in soda solution, 30 days in soda solution.

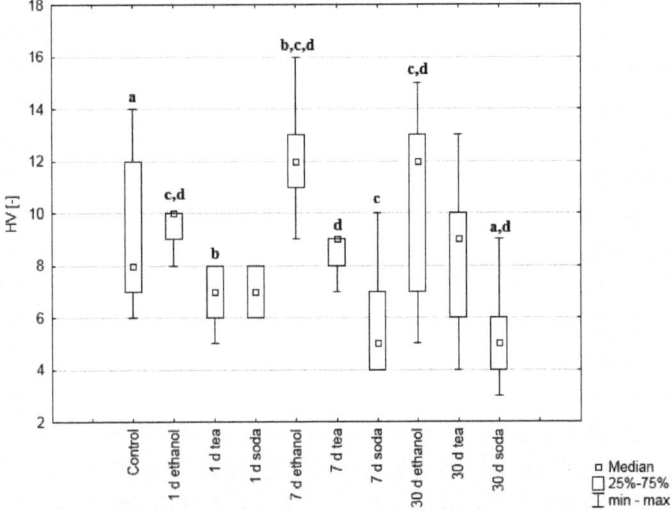

Figure 2. Box and whiskers plot of Vickers hardness. Values were obtained for samples made of Adhesor Carbonfine cement treated with ethanol (75%), green tea and baking soda (2.5%) solutions after one, seven or 30 days and samples immersed for seven days in distilled water as control. Statistically significant differences were detected between: (a) Control group vs. 30 days in soda solution, (b) 1 day in tea solution vs. 7 days in ethanol solution, (c) 7 days in soda solution vs. 1 day in ethanol solution, 7 days in ethanol solution, 30 days in ethanol solution, (d) 30 days in soda solution vs. 1 day in ethanol solution, 7 days in ethanol solution, 30 days in ethanol solution, 7 days in tea solution.

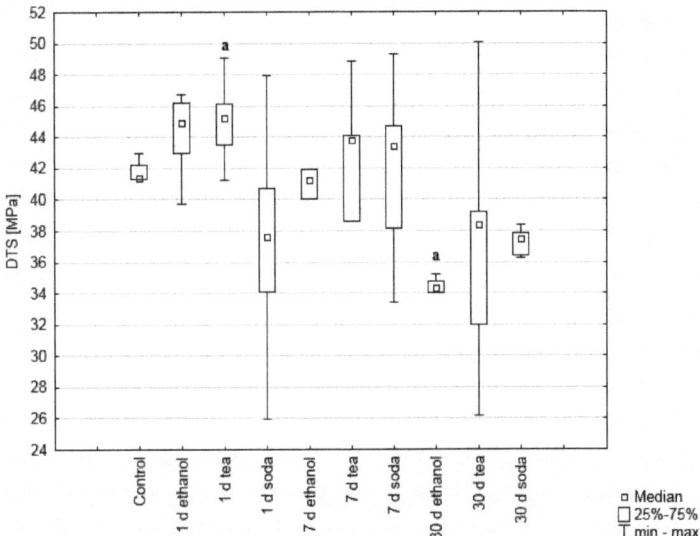

Figure 3. Box and whiskers plot of diametral tensile strength (DTS). Values were obtained for samples made of Breeze cement treated with ethanol (75%), green tea and baking soda (2.5%) solutions after one, seven or 30 days and samples immersed for seven days in distilled water as control. Statistically significant differences were detected between: (a) 1 day in tea solution vs. 30 days in ethanol solution.

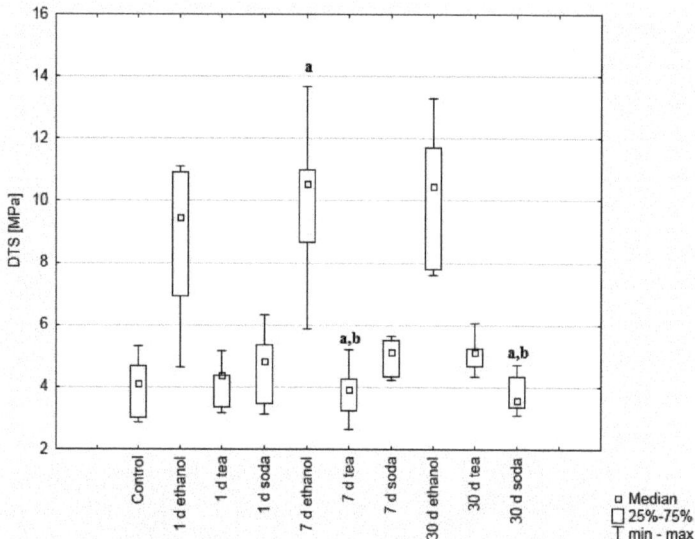

Figure 4. Box and whiskers plot of diametral tensile strength (DTS). Values were obtained for samples made of Adhesor Carbonfine cement treated with ethanol (75%), green tea and baking soda (2.5%) solutions after one, seven or 30 days and samples immersed for seven days in distilled water as control. Statistically significant differences were detected between: (a) 7 days in ethanol solution vs. 30 days in soda solution, 7 days in tea solution, (b) 30 days in tea solution vs. 30 days in soda solution, 7 days in tea solution.

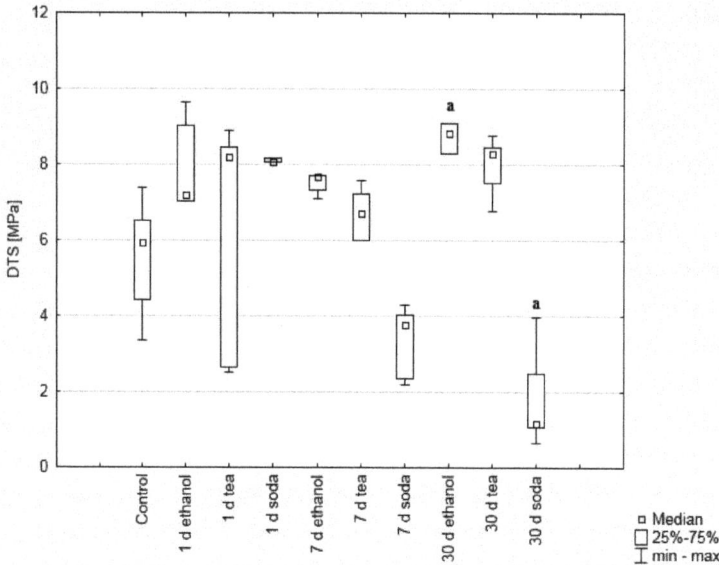

Figure 5. Box and whiskers plot of diametral tensile strength (DTS). Values were obtained for samples made of IHDENT Giz typ II cement treated with ethanol (75%), green tea and baking soda (2.5%) solutions after one, seven or 30 days and samples immersed for seven days in distilled water as control. Significant statistical differences were detected between: (a) 30 days in ethanol solution vs. 30 days in soda solution.

The hardness of the composite cement (Breeze) (Figure 1) stored in ethanol solution was found to decrease with time. This tendency is also seen for soda and tea solutions, but no statistically significant difference was found between the two. Significant differences were found between:

- Control group vs. 30 days in ethanol solution ($p = 0.0006$);
- 1 day in ethanol vs. 30 days in ethanol solution ($p = 0.0008$);
- 30 days in ethanol solution vs. 1 day in tea solution ($p = 0.0000$), 7 day in tea solution ($p = 0.0017$), 1 day in soda solution ($p = 0.0000$), 7 day in soda solution ($p = 0.0006$), 30 days in soda solution ($p = 0.0459$).

Although no studies have so far examined the effect of aging in various solvents on the properties of resin cements, Breeze has a similar structure and composition to that of a dental composite with a resin matrix. In contrast to conventional dental resin, cements are most often characterized by the addition of a chemical catalyst, which allows for the dual polymerization of the material. In addition, fluoride can be added as an anticaries agent, and is used to be competitive with GICs. To simplify the cementation procedure, self-adhesive resin cements consisting of adhesive monomers were designed [19]. The fracture toughness values of various dental composites are known to decrease after aging in ethanol solution for six months [20]. Ageing of dental composites in ethanol solution resulted in the elution of residual, unreacted monomers, filler/matrix interfacial failure, and reduction of the mechanical properties [20–28]. This can be explained by the structure and properties of the resin matrix. Some dental resins, such as Bis-GMA and HEMA, absorb a significant proportion of ethanol and water molecules thanks to their hydrophilic properties. Ethanol is considered to be good solvent of dental composites: Ethanol and the dimethacrylate resins used in these materials have similar Hoy's solubility parameters (26.1 and 19.2–23.6 $(J/cm^3)^{\frac{1}{2}}$ respectively) [29,30]. Hence, when immersed in ethanol, Breeze swells, resulting in lower moduli of elasticity and strength [31].

The hardness of zinc-polycarboxylate cement (Adhesor Carbofine) was found to decrease with duration of immersion in soda solution (Figure 2). No similar relationship was noticed for the soda or tea solutions.

Statistical analysis of hardness results allowed to show significant statistical differences between:

- Control group vs. 30 days in soda solution ($p = 0.0163$),
- 1 day in tea solution vs. 7 days in ethanol solution ($p = 0.0021$),
- 7 days in soda solution vs. 1 day in ethanol solution ($p = 0.0167$), 7 days in ethanol solution ($p = 0.0000$), 30 days in ethanol solution ($p = 0.0108$),
- 30 days in soda solution vs. 1 day in ethanol solution ($p = 0.0022$), 7 days in ethanol solution ($p = 0.0000$), 30 days in ethanol solution ($p = 0.0013$), 7 day in tea solution ($p = 0.0336$).

Adhesor Cabofine belongs to the group of zinc-polycarboxylate cements whose curing reaction consists of the formation of a complex binding between metal ions (zinc, calcium) and carboxylic groups derived from polyacrylic acid. The hardness of zinc-polycarboxylate cements has been found to increase after 35 days in distilled water [32]. This has been attributed to the presence of a solid polycarboxylic phase around the oxides (e.g., zinc oxide) responsible for curing. In addition, glass-ionomer cements and zinc-polycarboxylate cements have been found to be the most soluble of various tested cements [33]. As a result of solubility and sorption, margin integrity tends to reduce, resulting in improved surface properties and aesthetics. Water sorption adversely affects bending strength and hardness [33].

Dental cements were found demonstrate greater solubility with duration of immersion in the test solutions, at all tested pH values (pH 3, 7 and 9) [34]. However, for the Adhesor Carbofine cement samples in the present study, no great differences in solubility were observed for samples treated with tea and ethanol solution. Interestingly, a clear downward trend in the value of HV hardness was observed when the cement was immersed in soda solution. These findings can be explained by the fact that the cement contains zinc, magnesium and aluminum oxides, which are higher in the electrochemical series than sodium. Sodium, as a more active element, displaces the less active metal from the salt compound.

Few studies examine the strength properties of prosthetic cements; nevertheless, the most common methods used are the three-point bending and compression tests. The three-point bending test is very important to assess the suitability of materials. It is recognized by the International Organization for Standardization (ISO) as a valid strength test of composite materials [35]. However, it should be noted that sample preparation is difficult and may result in heterogeneous results being obtained for the degree of polymerization. Additionally, real dental restorations are several times smaller and this method is "clinically" unfounded. Furthermore, the tensile strength values of dental materials have greater clinical value than compressive strength, because many clinical failures are due to tensile forces [36]. Composite cements have been found to demonstrate higher DTS values (44 MPa) than zinc phosphate cements [37], with similar results being obtained in other studies (e.g., Li and White [6]., Kim et al. [38]). Sokolowski et al. [7] obtained similar result for a composite cement, i.e., Breeze.

Although it is difficult to compare our findings with those of previous studies, assuming that resin cements consist of similar components as dental composites, it would be reasonable to assume that the two sets of materials have similar strength properties. The DTS values for composite materials for fillings are usually in the range of 30–55 MPa [39–41]. Although some variation was observed between the mean DTS values of the Breeze cement, it is within the limits of measurement error (Figure 3). Significant differences in hardness were found between:

- 1 day in tea solution vs. 30 days in ethanol solution ($p = 0.0321$).

The mean DTS value of Adhesor Carbofine cement increased with duration of immersion in ethanol solution (Figure 4). The DTS values of the samples immersed in the tea and soda solutions are very similar and no differences were found. Significant differences in hardness were found between:

- 7 days in ethanol solution vs. 30 days in soda solution ($p = 0.0439$), 7 days in tea solution (0.0439).
- 30 days in tea solution vs. 30 days in soda solution ($p = 0.0348$), 7 days in tea solution (0.0348).

No information is available in the literature regarding the DTS of zinc-polycarboxylic cement (Adhesor Carbofine). However, literature findings indicate that the compressive strength of this material and of other tested zinc-polycarboxylate cements increased after 30 days in water. The reaction between the metal ions with the polyacrylic acid reaches stabilization after one day; therefore, some changes of the mechanical properties can be deduced after this time. This observation may be explained by the presence of the unreacted phase (solid polycarboxylate phase around the zinc oxide or other metal oxides), which can further react, causing hardening over time [32]. Our findings indicate that the DTS values of the samples contained in the ethanol solution increase according to time of immersion. The setting time of cement is believed to be stabilized and extended by the addition of alcohol to the fluid of zinc-polycarboxylic cements.

The highest DTS values of glass ionomer cement (IHDENT Giz type II) were observed for samples treated with ethanol solution after seven or 30 days. The lowest recorded values were for samples immersed in soda solution (Figure 5). Significant differences in hardness were found between:

- 30 days in ethanol solution vs. 30 days in soda solution ($p = 0.0093$).

It should be emphasized that the mechanical strength of glass ionomer cements depends on the water balance during the curing process. This process is based on acid-base reaction in which an acid reacts with the salts present in the powder; the process causes the release of metal ions and the formation and precipitation of polyolefin salts [42]. The initial curing is followed by a maturation process, which takes place more slowly. It was observed to be disturbed by the presence of soda solution [43]. As with the Adhesor Carbofine cement, the presence of sodium can cause changes in ionic reactions by displacing less active metals from the salt compound. Large changes were observed in the appearance of samples, which delaminated and crumbled after exposure to soda (Figure 6).

The DTS results for this material are characterized by low values, in other studies values for GIC are higher (>20 MPa), however, in these studies GIC for temporary fillings was tested [44]. It is know that the highest strength were obtained by the restorative glass-ionomers in comparison to luting type [45].

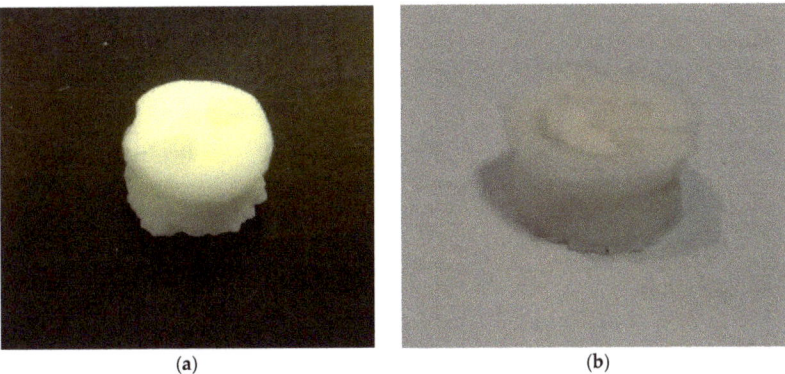

Figure 6. Samples made of IHDENT Giz type II ionomer glass cement removed from the soda solution after (**a**) 7 days and (**b**) 30 days.

4. Conclusions

1. The immersion in solvents (tea, ethanol and baking soda solutions) have influence on the diametral tensile strength and the hardness of analyzed cements. The impact depends on the material and solution composition.

2. Of the tested cements, composite cement obtained the highest hardness and diametral tensile strength values.

3. Ethanol affects studied composite cement and causes its diametral tensile strength and hardness to deteriorate in time.

4. Baking soda solution affects the properties of zinc-polycarboxylate. The prolongation of aging time results in a significant reduction in the hardness of the cement.

5. Baking soda solution influences glass-ionomer properties. Glass-ionomer samples immersed in a baking soda solution after seven and 30 days showed changes in appearance and DTS value.

Author Contributions: Conceptualization, K.B.; data curation, A.S.-W. and K.K.; funding acquisition, J.S.; investigation, A.S.-W. and K.R.; methodology, A.S.-W., K.K. and K.B.; supervision, J.S. and K.B.; writing—original draft, K.R. and K.K.; writing—review and editing, A.S.-W. and K.B.

Funding: This research received no external funding.

Conflicts of Interest: The authors declare no conflicts of interest. The founding sponsors had no role in the design of the study; in the collection, analyses, or interpretation of data; in the writing of the manuscript, or in the decision to publish the results.

References

1. Macura-Karbownik, A.; Chladek, G.; Żmudzki, J.; Kasperski, J. Chewing Efficiency and Occlusal Forces in PMMA, Acetal and Polyamide Removable Partial Denture Wearers. *Acta Bioeng. Biomech.* **2016**, *18*, 127–134. [CrossRef]
2. Diaz-Arnold, A.M.; Vargas, M.A.; Haselton, D.R. Current Status of Luting Agents for Fixed Prosthodontics. *J. Prosthet. Dent.* **1999**, *81*, 135–141. [CrossRef]
3. Hill, E.E.; Lott, J. A Clinically Focused Discussion of Luting Materials. *Aust. Dent. J.* **2011**, *56* (Suppl. S1), 67–76. [CrossRef] [PubMed]

4. Marzec-Gawron, M.; Michalska, S.; Dejak, B. Właściwości Współczesnych Cementów Kompozytowych Oraz Ich Mechanizm Wiązania Do Szkliwa i Zębiny to Enamel and Dentin. *Protet. Stomatol.* **2012**, *3*, 173–180. [CrossRef]
5. Craig, R.G. *Materiały Stomatologiczne*; Edra Urban & Partner: Wrocław, Poland, 2006.
6. Li, Z.C.; White, S.N. Mechanical Properties of Dental Luting Cements. *J. Prosthet. Dent.* **1999**, *81*, 597–609. [CrossRef]
7. Sokołowski, G.; Szczesio-Włodarczyk, A.; Konieczny, B.; Bociong, K.; Sokołowski, J. Ocena Porównawcza Właściwości Mechanicznych Cementów Żywicznych, Samoadhezyjnych i Samotrawiących. *Protet* **2018**, *68*, 415–424. [CrossRef]
8. Christensen, G.J. Why Use Resin Cements? *J. Am. Dent. Assoc.* **2010**, *141*, 204–206. [CrossRef] [PubMed]
9. Pawłowska, E.; Loba, K.; Błasiak, J.; Szczepańska, J. Właściwości i Ryzyko Stosowania Metakrylanu Bisfenolu A i Dimetakrylanu Uretanu – Podstawowych Monomerów Kompozytów Stomatologicznych. *Dent. Med. Probl.* **2009**, *46*, 477–485.
10. Nicholson, J.W.; Wasson, E.A. A Study of the Structure Dental Cements of Zinc Polycarboxylate. *J. Mater. Sci. Mater. Med.* **1993**, *4*, 32–35. [CrossRef]
11. Majewski, S.; Pryliński, M. *Materiały i Technologie Współczesnej Protetyki Stomatologicznej*; Czelej: Lublin, Poland, 2011.
12. Wagner, L.; Bączkowski, B. *Wprowadzenie do Ćwiczeń Przedklinicznych z Materiałoznawstwa: Materiały Stosowane w Protetyce. Skrypt dla Studentów*; Oficyna Wydawnicza WUM: Warszawa, Poland, 2010.
13. Bowen, R.L.; Marjenhoff, W.A. Dental Composites/Glass Ionomers: The Materials. *Adv. Dent. Res.* **1992**, *6*, 44–49. [CrossRef] [PubMed]
14. Kupka, T. Szkło-Jonomery—Przesłość, Teraźniejszość i Przyszłość Dentystyki Odtwórczej. *Mag. Stomatol.* **2014**, *4*, 28–32.
15. Pereira, L.C.; Nunes, M.C.P.; Dibb, R.G.P.; Powers, J.M.; Roulet, J.F.; Navarro, M.F.D.L. Mechanical Properties and Bond Strength of Glass-Ionomer Cements. *J. Adhes. Dent.* **2002**, *4*, 73–80. [PubMed]
16. Lohbauer, U. Dental Glass Ionomer Cements as Permanent Filling Materials?—Properties, Limitations and Future Trends. *Materials* **2010**, *3*, 76–96. [CrossRef]
17. Szczyrek, P.; Zadroga, K.; Mierzwińska-nastalska, E. Cementowanie Uzupełnień Pełnoceramicznych—Przegląd Piśmiennictwa. Część I. *Protet. Stomatol.* **2008**, *58*, 279–283.
18. Sokolowski, G.; Szczesio, A.; Bociong, K.; Kaluzinska, K.; Lapinska, B.; Sokolowski, J.; Domarecka, M.; Lukomska-Szymanska, M. Dental Resin Cements—The Influence of Water Sorption on Contraction Stress Changes and Hydroscopic Expansion. *Materials* **2018**, *11*, 973. [CrossRef] [PubMed]
19. Özcan, M. Luting Cements for Dental Applications. In *Non-Metallic Biomaterials for Tooth Repair and Replacement*; Woodhead Publishing: Cambridge, UK, 2013; pp. 375–394.
20. Ferracane, J.L.; Berge, H.X. Fracture Toughness of Experimental Dental Composites Aged in Ethanol. *J. Dent. Res.* **1995**, *74*, 1418–1423. [CrossRef] [PubMed]
21. Bauer, H.; Ilie, N. Effects of Aging and Irradiation Time on the Properties of a Highly Translucent Resin-Based Composite. *Dent. Mater. J.* **2013**, *32*, 592–599. [CrossRef]
22. Sideridou, I.D.; Karabela, M.M.; Bikiaris, D.N. Aging Studies of Light Cured Dimethacrylate-Based Dental Resins and a Resin Composite in Water or Ethanol/Water. *Dent. Mater.* **2007**, *23*, 1142–1149. [CrossRef]
23. Lee, S.Y.; Greener, E.H.; Menis, D.L. Detection of Leached Moieties from Dental Composites in Fluid Simulating Food and Saliva. *Dent. Mater.* **1995**, *11*, 348–353. [CrossRef]
24. Fonseca, A.S.Q.S.; Gerhardt, K.M.F.; Pereira, G.D.S.; Sinhoreti, M.A.C.; Schneider, L.F.J. Do New Matrix Formulations Improve Resin Composite Resistance to Degradation Processes? *Braz. Oral Res.* **2017**, *27*, 410–416. [CrossRef]
25. Ferracane, J.L.; Marker, V.A. Solvent Degradation and Reduced Fracture Toughness in Aged Composites. *J. Dent. Res.* **1992**, *71*, 13–19. [CrossRef]
26. McKinney, J.E.; Wu, W. Chemical Softening and Wear of Dental Composites. *J. Dent. Res.* **1985**, *64*, 1326–1331. [CrossRef]
27. Drummond, J.L.; Andronova, K.; Al-Turki, L.I.; Slaughter, L.D. Leaching and Mechanical Properties Characterization of Dental Composites. *J. Biomed. Mater. Res. Part B Appl. Biomater.* **2004**, *71*, 172–180. [CrossRef]

28. Al Badr, R.M.; Hassan, H.A. Effect of Immersion in Different Media on the Mechanical Properties of Dental Composite Resins. *Int. J. Appl. Dent. Sci.* **2017**, *3*, 81–88.
29. Malacarne, J.; Carvalho, R.M.; de Goes, M.F.; Svizero, N.; Pashley, D.H.; Tay, F.R.; Yiu, C.K.; Carrilho, M.R.D.O. Water Sorption/Solubility of Dental Adhesive Resins. *Dent. Mater.* **2006**, *22*, 973–980. [CrossRef]
30. Pashley, D.H.; Tay, F.R.; Carvalho, R.M.; Rueggeberg, F.A.; Agee, K.A.; Carrilho, M.; Donnelly, A.; García-Godoy, F. From Dry Bonding to Water-Wet Bonding to Ethanol-Wet Bonding. A Review of the Interactions between Dentin Matrix and Solvated Resins Using a Macromodel of the Hybrid Layer. *Am. J. Dent.* **2007**, *20*, 7–20.
31. Gavranović-Glamoč, A.; Ajanović, M.; Korać, S.; Zukić, S.; Strujić-Porović, S.; Kamber-Ćesir, A.; Kazazić, L.; Berhamović, E. Evaluation of the Water Sorption of Luting Cements in Different Solutions. *Acta Med. Acad.* **2017**, *46*, 124–132. [CrossRef]
32. Afife Binnaz Hazar, Y.; Karaaslan, A. Effect of Water Storage on the Mechanical Properties of Zinc Polycarboxylate Cements. *Dig. J. Nanomater. Biostruct.* **2007**, *2*, 243–252.
33. Keyf, F.; Tuna, S.H.; Şen, M.; Safrany, A. Water Sorption and Solubility of Different Luting and Restorative Dental Cements. *Turk. J. Med. Sci.* **2007**, *37*, 47–55.
34. Yanikoglu, N.; Duymus, Z. Evaluation of the Solubility of Dental Cements in Artificial Saliva of Different PH Values. *Dent. Mater. J.* **2007**, *26*, 62–67. [CrossRef]
35. PN-EN ISO 4049:2003. Available online: http://sklep.pkn.pl/pn-en-iso-4049-2003p.html (accessed on 1 August 2019).
36. Podlewska, M.; Nowak, J.; Półtorak, K.; Sokołowski, J.; Łukomska-Szymańska, M. Metody Badania Parametrów Wytrzymałości Mechanicznych Materiałów Kompozytowych. *E Dentico* **2015**, *5*, 92–98.
37. Fonseca, R.G.; Gomes, J.; Adabo, G.L. Influence of Activation Modes on Diametral Tensile Strength of Dual-Curing Resin Cements. *Braz. Oral Res.* **2005**, *19*, 267–271. [CrossRef]
38. Kim, A.R.; Jeon, Y.C.; Jeong, C.M.; Yun, M.J.; Choi, J.W.; Kwon, Y.H.; Huh, J.B. Effect of Activation Modes on the Compressive Strength, Diametral Tensile Strength and Microhardness of Dual-Cured Self-Adhesive Resin Cements. *Dent. Mater. J.* **2016**, *35*, 298–308. [CrossRef]
39. Stencel, R.; Kasperski, J.; Pakiela, W.; Mertas, A.; Bobela, E.; Barszczewska-Rybarek, I.; Chladek, G. Properties of Experimental Dental Composites Containing Antibacterial Silver-Releasing Filler. *Materials* **2018**, *11*, 1031. [CrossRef]
40. Alves, P.B.; Brandt, W.C.; Neves, A.C.C.; Cunha, L.G.; Silva-Concilio, L.R. Mechanical Properties of Direct and Indirect Composites after Storage for 24 Hours and 10 Months. *Eur. J. Dent.* **2013**, *7*, 117–122.
41. Zandinejad, A.A.; Atai, M.; Pahlevan, A. The Effect of Ceramic and Porous Fillers on the Mechanical Properties of Experimental Dental Composites. *Dent. Mater.* **2006**, *22*, 382–387. [CrossRef]
42. Crisp, S.; Wilson, A.D. Reactions in Glass Ionomer Cements I. Decomposition of the Powder. *J. Dent. Res.* **1974**, *53*, 1408–1413. [CrossRef]
43. Sidhu, S.; Nicholson, J. A Review of Glass-Ionomer Cements for Clinical Dentistry. *J. Funct. Biomater.* **2016**, *7*, 16. [CrossRef]
44. Xie, D.; Brantley, W.A.; Culbertson, B.M.; Wang, G. Mechanical Properties and Microstructures of Glass-Ionomer Cements. *Dent. Mater.* **2000**, *16*, 129–138. [CrossRef]
45. Cattani-Lorente, M.A.; Godin, C.; Meyer, J.M. Early Strength of Glass Ionomer Cements. *Dent. Mater.* **1993**, *9*, 57–62. [CrossRef]

© 2019 by the authors. Licensee MDPI, Basel, Switzerland. This article is an open access article distributed under the terms and conditions of the Creative Commons Attribution (CC BY) license (http://creativecommons.org/licenses/by/4.0/).

Article

Cytotoxic Potential of Denture Adhesives on Human Fibroblasts—In Vitro Study

Ewa Sobolewska [1], Piotr Makowiecki [2], Justyna Drozdowska [3], Ireneusz Dziuba [4,5], Alicja Nowicka [6], Marzena Wyganowska-Świątkowska [7], Joanna Janiszewska-Olszowska [3] and Katarzyna Grocholewicz [3,*]

1. Department of Dental Prosthetics, Pomeranian Medical University, 70-111 Szczecin, Poland; ewa.sobolewska@pum.edu.pl
2. Department of Radiology, Pomeranian Medical University, 70-111 Szczecin, Poland; piotr.makowiecki@pum.edu.pl
3. Department of Interdisciplinary Dentistry, Pomeranian Medical University, 70-111 Szczecin, Poland; justyna.drozdowska@pum.edu.pl (J.D.); jjo@pum.edu.pl (J.J.-O.)
4. Faculty of Medicine, University of Technology, 40-555 Katowice, Poland; mmid@wp.pl
5. Faculty of Medicine, Collegium Medicum, Cardinal Stefan Wyszyński University in Warsaw, 01-815 Warsaw, Poland
6. Department of Conservative Dentistry and Endodontics, Pomeranian Medical University, 70-111 Szczecin, Poland; alicja.nowicka@pum.edu.pl
7. Department of Oral Surgery and Periodontology, Poznan University of Medical Sciences, 60-812 Poznan, Poland; wyganowska@ump.edu.pl
* Correspondence: katarzyna.grocholewicz@pum.edu.pl; Tel.: +48-91-4661690

Abstract: (1) In recent years, there has been a significant increase in the availability of denture adhesives for stabilizing removable dentures. The aim of the present study was to assess the cytotoxicity of three denture adhesives on human fibroblasts. (2) Methods: Three denture adhesives were analyzed. Fibroblast cultures were established for the study and control groups in order to assess the incidence of necrosis and to evaluate the microscopic intracellular alterations induced. Following incubation with (study groups) or without adhesives (control group), trypan blue dye exclusion assay was used to determine the number of viable and/or dead cells. Microscopic specimens were stained with haematoxylin and eosin, scanned, digitally processed and then analyzed by a histopathologist. (3) Results: All three denture adhesives analyzed demonstrated various toxic effects in vitro on human fibroblast: quantitative evaluation—45.87–61.13% reduction of cell viability ($p = 0.0001$) and slight to moderate cytotoxicity in qualitative evaluation. (4) Conclusions: Denture adhesive creams demonstrated a toxic effect on human fibroblasts in vitro in quantitative and qualitative evaluation. In vivo observations are needed to find out if denture adhesives present a cytotoxic effect in patients.

Keywords: dentures; denture adhesives; human fibroblasts; cytotoxicity

1. Introduction

Epidemiological data indicate a continuous increase in the number of edentulous patients. It has been attributed to elongation of global average life expectancy [1–3]. Prosthetic rehabilitation of edentulous patients is difficult and requires knowledge and experience, both from dentists and dental technicians. Despite considerable advances in the field of prosthodontics, conventional complete dentures are still the most popular prosthetic restorations in edentulous patients [4]. Significant bone resorption following teeth extractions deteriorates the clinical conditions for satisfactory denture retention and stability; retention and stability clearly decrease after several years [5,6]. Efforts have been made to develop a material for dental prostheses with the best functional properties [7]. Retention of dentures can be improved by using denture adhesives or relining dentures. Properly used denture adhesives can improve the retention and stability of prosthetic restorations and prevent food residue accumulation under the denture [6,8–12].

In recent years, there has been a significant increase in the availability of adhesives for stabilizing removable dentures. The study of Okazaki et al. showed that 19% of denture wearers use denture adhesives [13]. Most denture adhesives contain non-toxic polymers of carboxymethyl cellulose [14]. All creams that improve the stability of dentures also contain swelling agents, such as karaya gum, Arabic gum, tragacanth gum, gelatin, pectin, methylcellulose, hydroxyethylcellulose, synthetic polyethylene polymers and others. Another group of ingredients are antibacterial and antifungal agents: sodium borate, hexachlorophene and polyhydroxybenzoate [15,16]. Adhesives are thus compound products; their use exerts not only a local effect on the oral mucosa, but also may influence the general health [17–19]. Ingredients of adhesives (e.g., formaldehyde) may produce allergenic and cytotoxic effects [20–22]. Another negative feature of denture adhesives is their low pH (5.5 on average), which is capable of dissolving enamel hydroxyapatites in the remaining dentition [23]. Denture adhesives are often used for an extended time period, which causes excessive pressure on the denture base and consequently its progressive wear. This may be a potential factor causing pathologies of the soft tissues [24]. In the leaflets for adhesive creams, manufacturers recommend that they be applied pointwise by squeezing out strips a few millimeters long from the tube. However, patients usually do not follow these recommendations and use too much of these materials. Considering all these problems associated with the use of denture adhesives, especially of formaldehyde content, there is a justified need for testing their cytotoxicity, irrespective of the data provided by their manufacturers.

Fibroblasts, the main group of connective tissue cells, are a heterogeneous group of cells, which, despite numerous similarities in structure and function, are characterized by significant differentiation depending on the anatomical location of the connective tissue, but those in the face and oral cavity are derived from the neural crest. There are also differences in fibroblasts isolated from healthy tissue and granulation tissue [25–31]. An important feature depending on the source of fibroblasts used in experimental studies is the rate of proliferation. Tooth pulp as an immature gelatinous tissue is rich in fibroblasts capable of rapid multiplication.

The aim of the present study was to evaluate the cytotoxicity of three denture adhesives on human fibroblasts and to compare the effect of the analyzed products.

2. Materials and Methods

2.1. Harvesting Fibroblasts

Fibroblasts were harvested from the pulp of 15 healthy (non-pathologically damaged) teeth extracted for orthodontic indications. All the patients involved were informed about the research project and signed an informed consent form according to guidelines from the Declaration of Helsinki. The study was approved by the Bioethics Committee of Pomeranian Medical University in Szczecin (Decision Reference No. KB-0012/05/13). Immediately after tooth extraction (up to 10 min) the pulp chamber was opened using a ball-shaped diamond drill in an air turbine head with water cooling. The pulp was removed using sterile root canal broaches and immediately suspended in Roswell Park Memorial Institute (RPMI) 1640 Medium (Sigma-Aldrich, St. Louis, MO, USA) supplemented with 20% fetal bovine serum (FBS; Sigma-Aldrich).

2.2. Fibroblast Cultures

The extracted dental pulp was homogenized and the fibroblast cultures were established in tissue culture flasks (Sarstedt Inc., Newton, MA, USA). Cells were cultured in RPMI 1640 Medium supplemented with 20% FBS (Biological Industries, Beit-Haemek, Izrael) in an incubator under standard conditions (48 h, 37 °C, CO_2 5%, relative humidity 99.6%). Fibroblast cultures for the study and control groups were prepared in the Laboratory of Cell and Tissue Culture, Department of Genetics and Pathomorphology, Pomeranian Medical University in Szczecin. The culture of fibroblasts from tooth no. 1 presented abnormal growth of cells, probably caused by incorrect handling of biological material (pulp) before placing it in the transport medium. Thus, the number of cultures was 14, and each culture was supplemented with tested denture adhesives.

2.3. Quantitative Evaluation

Three denture adhesives, commercially available in Poland, were tested. Their manufactures and compositions are presented in Table 1. It is visible that two of the adhesives tested (COREGA Extra Strong and PROTEFIX) do not contain zinc salts opposite to the other one (BLEND-A-DENT Plus). The composition of COREGA Extra Strong and PROTEFIX is very similar but not identical.

Table 1. Denture adhesives tested.

Denture Adhesive	Manufacturer	Composition
COREGA Extra Strong	GlaxoSmithKline, Consumer Healthcare SA. Stafford Miller (Ireland) Limited, Clochreane, Youghal Road, Dungarvan, Co. Waterford, Ireland	calcium/sodium PVM/MA copolymer, petrolatum, cellulose gum (carboxymethyl cellulose), paraffinum liquidum, propylparaben, aroma, CI 45430 (erythrosine)
PROTEFIX	Queisser Pharma GmbH&Co. KG, Schleswiger Straße, Flensburg, Germany	calcium/sodium PVM/MA copolymer, carboxymethyl Cellulose, paraffinum, petrolatum, silicon dioxide, menthol, azorubine, methyl benzoate
BLEND-A-DENT Plus	Procter & Gamble GmbH, Sulzbacher Straße, Schwalbach am Taunus, Germany	calcium/zinc PVM/MA copolymer, paraffinum liquidum, petrolatum, cellulose gum (carboxymethyl cellulose), silica, CI 15985 (Yellow 6), menthyl lactate, aroma, CI 45410 (phloxin B), sodium saccharin, limonene, cinnamal, eugenol

The assay was conducted according to the following procedure: 0.5 mL of each tested adhesive was placed in a Petri dish with 3 mL RPMI 1640 Medium supplemented with 20% FBS to obtain a solution. The Petri dishes were then placed in an incubator and kept for 5 days under standard culture conditions. After 5 days the solution was transferred to 96-well tissue culture plates (Sarstedt Inc., Newton, MA, USA). Each denture adhesive was placed into 3 wells (study groups), and one well was filled with a pure medium (as a negative control) to be used as the control group (K). Cultures of fibroblasts were established in media prepared this way by placing about 100,000 cells from the first passage. Culture plates were moved to the incubator set at standard parameters and incubated for 72 h. After this time, trypan blue dye exclusion assay was used to determine the number of viable and/or dead cells. Trypan blue is a ~960 Daltons molecule, which is cell membrane impermeable and therefore only enters cells with compromised membranes. Upon entry into the cell, trypan blue binds to intracellular proteins thereby rendering the cells a bluish color. The trypan blue exclusion assay allows for a direct identification and enumeration of live (unstained) and dead (blue) cells in a given population. For that the cell culture was stained with 0.4% Trypan Blue solution (Sigma-Aldrich, St. Louis, MI, USA). Then, viable and necrotic fibroblasts were counted using an Axiovert 25 inverted transmitted light microscope (Carl-Zeiss, Jena, Germany) and a glass hemocytometer. Trypan blue was added to an Eppendorf tube with 100 μL of cells 400 μL 0.4% (final concentration 0.32%). Using a pipette, 100 μL of trypan blue-treated cell suspension was applied to the hemocytometer. Viable (unstained) and necrotic (blue stained) cells were counted in all 16 squares under the microscope with a 100× magnification. Cell counting was performed 3 times for each well. Counting was carried out by the same person, unfamiliar with the tested materials. The results from all wells for a given adhesive were summed up and averaged. For the control culture, counting of viable and necrotic cells was carried out in the same way, using a glass hemocytometer, but the cells were taken from three different places of the well. The results were also summed up and averaged.

In order to assess the incidence of necrosis after in vitro cell culture, an AI (apoptotic index) according to Prieto was used [32]. It is calculated by dividing the percentage of apoptotic cells by the total percentage of cells in the sample. In the present study the index was modified by using it to calculate the percentage of necrotic cells.

The results were subjected to statistical analysis. Statistical analysis was performed using STATA 11 software. All continuous variables were verified for distribution normality using a Shapiro-Wilk test. Statistical significance of differences between two groups was analyzed using a Mann-Whitney test. To investigate the relationship between two variables a chi^2 Pearson test and Spearman's rank correlation test were used. The level of statistical significance was set at α = 0.05. The risk of cell necrosis was expressed as an odds ratio (OR) at 95% confidence interval (CI). Differences were considered significant if the level of significance was α = 0.05.

2.4. Qualitative Evaluation

In parallel, fibroblasts from dental pulp were cultured in order to assess the microscopic changes induced in the cells and to prepare microscope slides of cells damaged by the tested adhesives. Microscope slides were placed on Petri dishes with fibroblasts from the first passage cultured in a mixture of RPMI 1640 Medium, 20% FBS and different denture adhesives. These were the study groups. The same procedure was followed to establish the control group (K), which was a fibroblast culture in pure RPMI 1640 Medium. The cultures were placed in an incubator and kept for 72 h under standard conditions. After incubation the fibroblasts attached to the slides were stained with haematoxylin (Haematoxylin, Fluka, Switzerland) and eosin (Eosin Yellowish, Loba Chemie, Mumbai, India) in a standard procedure (HE). Prepared microscopic slides were assessed using a light standard laboratory microscope (Olympus BX 43, Olympus Corporation, Tokyo, Japan) with magnification of 100× and 200×. Then, the slides were scanned using an Aperio CS2 pathology scanner (Leica Microsystems, Wetzlar, Germany) to take a photograph at a magnification of 100× and 200×. The histopathologist did now know the materials were being assessed.

2.5. Determination of Cytotoxicity

The cytotoxic effect was evaluated quantitatively and qualitatively according to INTERNATIONAL STANDARD ISO 10993-5:2009(E) [33]. According to this standard, reduction of cell viability by more than 30% is considered a cytotoxic effect. Qualitative morphological grading of cytotoxicity is based on assessing of general morphology, vacuolization, detachment, cell lysis and membrane integrity and expressed on a five-point scale.

Comparison of the necrotic effect of the adhesives on fibroblasts made it possible to divide the creams into three classes and identify products which induced the lowest (CLASS 1), moderate (CLASS 2) or the highest (CLASS 3) number of necrotic cells. CLASS 1 included all cases of the tested cream in which the number of necrotic cells was lower than that of both samples of the other two materials. CLASS 3 included all the cases of the tested cream in which the number of necrotic cells was higher than that of the samples of other materials. If the number of necrotic cells in the sample with the tested material was smaller than in the sample with the second material and at the same time higher than in the sample with the third material, it was classified as CLASS 2.

3. Results

3.1. Qantitative Evaluation of Cytotoxic Effect

Table 2 presents descriptive statistics for the value of necrotic fibroblasts in the study and control groups expressed in %. For all tested materials, a significantly higher percentage of necrotic cells was found compared to the control cultures (p < 0.0001). The highest percentage of necrotic cells was observed in culture supplemented with COREGA Extra

Strong. Although COREGA Extra Strong and PROTEFIX have a similar composition, their necrotic effect on pulp fibroblast is different. Quantitative evaluation showed a reduction of cell viability from 45.87% to 61.13%, which means that all tested materials induce a cytotoxic effect on fibroblasts. In control groups the reduction of viability was 4.56–6.16%.

Table 2. Descriptive statistic for the value of necrotic fibroblasts and differences between study and control groups analyzed using Mann-Whitney test.

Group	Necrotic Cells %							p
	Mean	SD	Min.	Max.	Q25	Median	Q75	
PROTEFIX	52.70	7.89	40.44	64.18	45.58	54.21	60.65	<0.0001
K	5.10	2.65	2.59	10.68	3.00	4.41	5.83	
COREGA Extra Strong	61.13	4.02	54.99	69.30	58.89	60.11	63.39	<0.0001
K	6.16	2.82	3.06	10.58	4.00	4.95	9.44	
BLENDA-A-DENT Plus	45.87	5.58	36.44	56.76	42.19	45.07	48.82	<0.0001
K	4.56	1.69	2.42	7.80	3.58	4.00	5.90	

The percentage of necrotic cells caused by tested adhesives was different. All differences were statistically significant; the levels of differences are presented in Table 3.

Table 3. Significance levels of differences between percentages of necrotic cells in Mann-Whitney test.

Compared Adhesives	p
PROTEFIX vs. COREGA Extra Strong	0.0058
PROTEFIX vs. BLENDA-A-DENT Plus	0.0274
COREGA Extra Strong vs. BLENDA-A-DENT Plus	<0.0001

The modified apoptotic index for BLEND-A DENT Plus was 45.87, for PROTEFIX it was 52.70 and for COREGA Extra Strong it was 61.13.

The risk of detecting necrotic cells for all tested adhesives are presented in Table 4. In each case we assessed the risk of detecting necrotic cells in the study group for each dental adhesive compared to the control group. Results were expressed as the odds ratio (OR) with a 95% confidence interval (95% CI) at significance level p. The analysis revealed a higher risk for OR > 0, lower risk for OR < 0 and no risk for OR = 0. With regard to the control group the highest risk of detecting necrotic cells was for COREGA Extra Strong and the lowest for BLEND-A-DENT Plus.

Table 4. Risk of detecting of necrotic cells in study groups versus control groups.

Necrotic Cells	OR	95% CI		p
BLEND-A-DENT Plus vs. K	17.19	17.12	17.27	<0.0001
PROTEFIX vs. K	19.44	19.35	19.52	<0.0001
COREGA Extra Strong vs. K	23.16	23.07	23.26	<0.0001

OR (odds ratio)—relative risk; 95% CI—95% confidence interval; p—significance level.

Table 5 presents a comparison of odds ratio for detecting necrotic cells between adhesives. The risk of detecting necrotic cells was 1.74 times higher for COREGA Extra Strong than for BLEND-A-DENT Plus and 1.38 times higher than for PROTEFIX. Comparing PROTEFIX and BLEND-A-DENT Plus, the risk was 1.26 times higher for the first adhesive.

Table 5. Risk of detecting of necrotic cells for different adhesives.

Necrotic Cells	OR	95% CI		p
COREGA Extra Strong vs. BLEND-A-DENT Plus	1.74	1.73	1.75	<0.0001
COREGA Extra Strong vs. PROTEFIX	1.38	1.38	1.39	<0.0001
PROTEFIX vs. BLEND-A-DENT Plus	1.26	1.25	1.26	<0.0001

The classification of adhesives tested is presented in Table 6. For BLEND-A-DENT Plus in 11 cases the number of necrotic cells was lower than for PROTEFIX and COREGA Extra Strong, and only in 1 case the number of necrotic cells was higher than in PROTEFIX and COREGA Extra Strong. For PROTEFIX, in 3 cases the number of necrotic cells was lower than in COREGA Extra Strong and BLENDA-A-DENT Plus, and in 3 cases the number of necrotic cells was higher than for both the other adhesives. For COREGA Extra Strong, there was no case in which the number of necrotic cells was lower than in PROTEFIX and BLEND-A-DENT Plus, and in 10 cases the number of necrotic cells was higher than for both the other adhesives. CLASS 2 means that the tested adhesive compared with the one product induced more necrotic cells and compared to the second, less. In this classification BLEND-A-DENT has the highest number of cases in CLASS 1, which means the lowest cytotoxic effect, and COREGA Extra Strong has the highest number of cases in CLASS 3, which means the highest cytotoxic effect.

Table 6. Classification of denture adhesives BLEND-A-DENT Plus, PROTEFIX and COREGA Extra Strong for their cytotoxic effect.

Adhesive	Number of Classified Cases			
	CLASS 1	CLASS 2	CLASS 3	Total
BLEND-A-DENT Plus	11 78.57%	2 14.29%	1 7.14%	14
PROTEFIX	3 21.43%	8 57.14%	3 21.43%	14
COREGA Extra Strong	0 0.00%	4 28.57%	10 71.43%	14
Total	14	14	14	

Table 7 presents the values of the chi^2 Pearson test and Spearman's rank correlation test for (r) compared pairs of adhesives.

Table 7. Statistics for comparisons between adhesives.

Adhesive	chi^2	df	p	r	t	p
BLEND-A-DENT Plus vs. COREGA Extra Strong	19.03	2	<0.0001	−0.81	6.939	<0.0001
PROTEFIX vs. COREGA Extra Strong	8.10	2	0.0174	0.54	3.244	0.0032
BLEND-A-DENT Plus vs. PROTEFIX	9.17	2	0.0102	−0.53	3.226	0.0038

3.2. Qualitative Evaluation of Cytotoxic Effect

Analysis of the histopathologic image indicated a small number of degenerative changes in fibroblasts cultured with BLEND-A-DENT Plus. Observation of fibroblasts cultured with COREGA Extra Strong showed the highest diversity of damage and a higher severity of cell damage. In fibroblasts cultured with PROTEFIX signs of cell damage were moderate. The histopathologic images of control cells culture and cells cultured with the tested materials are presented in Figures 1–8.

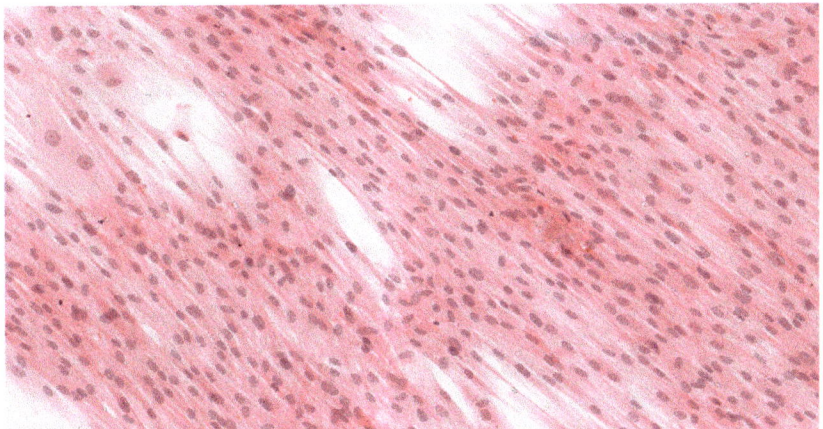

Figure 1. Image of control culture (K); 100× magnification.

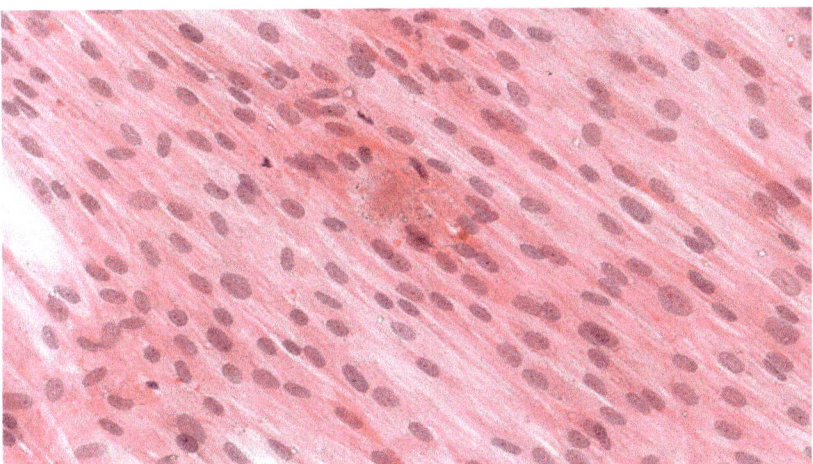

Figure 2. Image of control culture (K); 200× magnification.

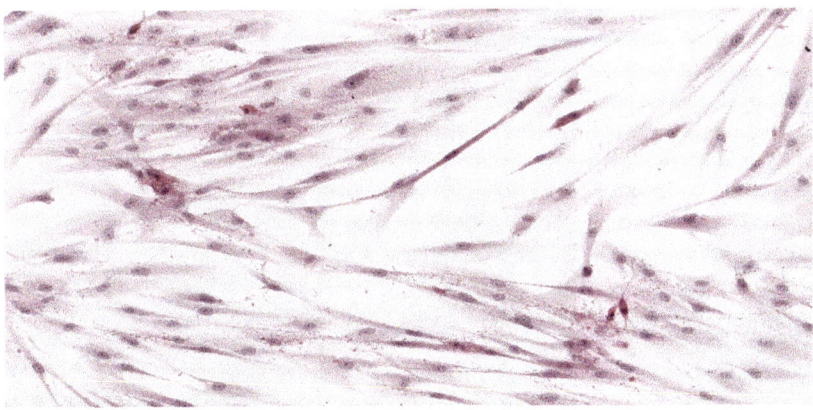

Figure 3. Image of cells cultured on medium with BLEND-A-DENT Plus; 100× magnification.

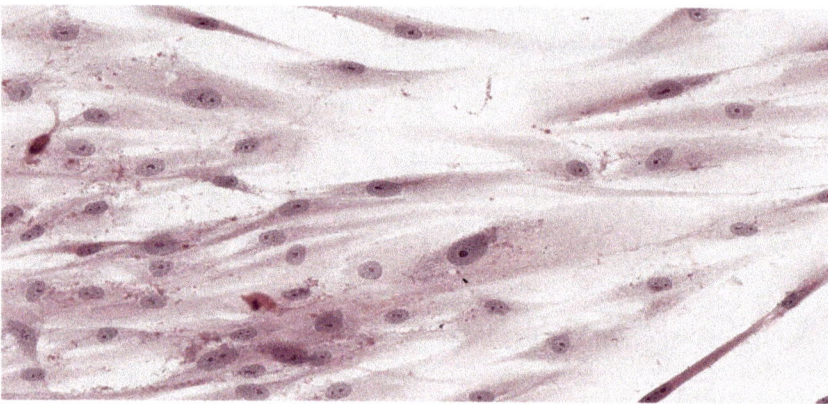

Figure 4. Image of cells cultured on medium with BLEND-A-DENT Plus; 200× magnification.

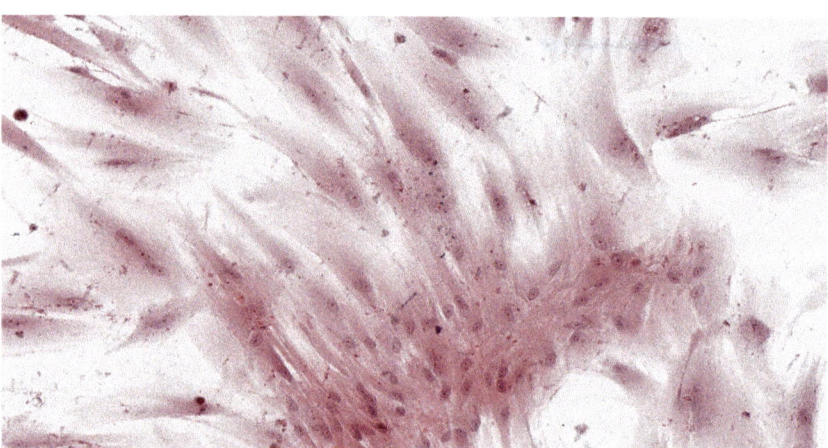

Figure 5. Image of cells cultured on medium with PROTEFIX; 100× magnification.

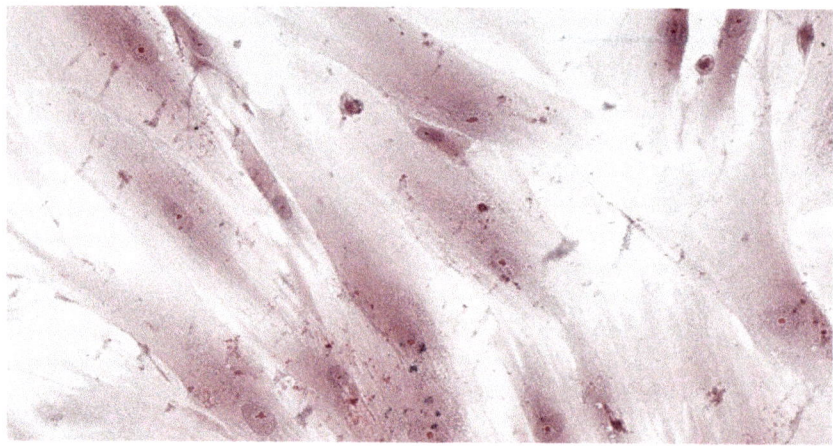

Figure 6. Image of cells cultured on medium with PROTEFIX; 200× magnification.

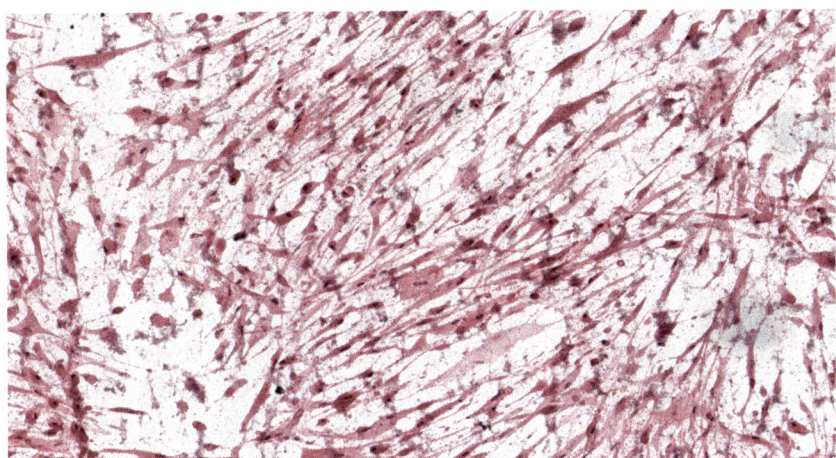

Figure 7. Image of cells cultured on medium with COREGA Extra Strong; 100× magnification.

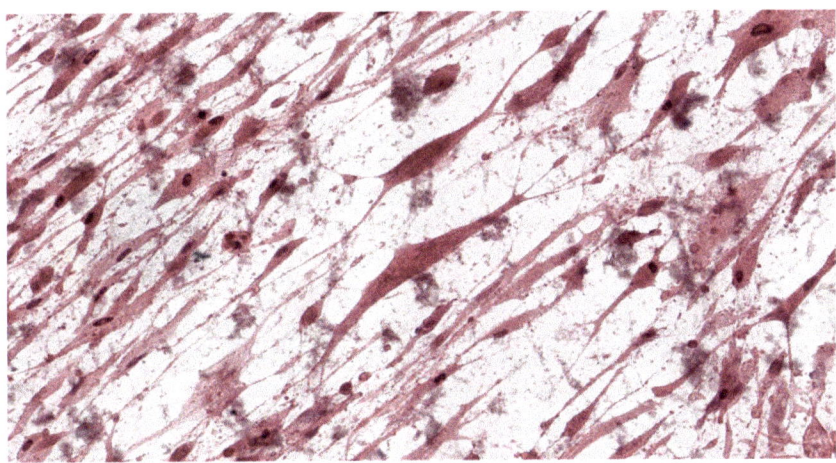

Figure 8. Image of cells cultured on medium with COREGA Extra Strong; 200× magnification.

Figures 1 and 2 show the histopathologic images of control cells culture (K) at 100× and 200× magnifications. It is a homogeneous population of proliferating spindle-shaped fibroblasts with tapering ends of the cells; there is no cell lysis and no reduction of cell growth. Oval nuclei can be in the central part of the cell with distinct ruby nucleoli. Intense cytoplasmic staining indicates active protein synthesis. Numerous visible shape changes occurred during mitosis. This image represents grade 0 (no reactivity) in qualitative morphological grading of cytotoxicity according to INTERNATIONAL STANDARD ISO 10993-5:2009(E).

Figures 3 and 4 show the histopathologic images of cells cultured with BLEND-A-DENT. No more than 20% of cells show changes in morphology. Spindle-shaped cells have obvious morphological features of damage. The pale cytoplasm is weakly stained, the cells lose their spindle shape, and the cell margins are blurred. Fibroblasts have different morphology, some with nuclei clearly displaced to one of the ends of the cell. Damaged fibroblasts are malformed and show different cytoplasm eosinophilicity. The number of cells is markedly reduced compared to the control culture. Cellular debris (fragments of disintegrated cells) is seen in the background of the image. This image corresponds to grade 1 (slight reactivity) of qualitative morphological grading of cytotoxicity.

Figures 5 and 6 show the histopathologic images of cells cultured with PROTEFIX. The changes in morphology are visible in 30% of cells, which do not have a typical spindle shape, and the cell margins are uneven and jagged. Nuclei are absent in some cells, others have pale nuclei without nuclear membrane (cariolysis), which reflects leakage of their contents into the cytoplasm. Cellular debris (fragments of disintegrated cells) is seen in the background of the image. These features indicate necrosis of fibroblasts. This means grade 2 (mild reactivity) cytotoxicity.

Figures 7 and 8 show the histopathologic images of cells cultured with COREGA Extra Strong. Fibroblasts demonstrate morphological features of acute damage. All cells are markedly malformed due to loss of cell membrane. There is a lack of integrity between cells. Nuclei are absent in most of the damaged fibroblast, others present with a disintegrating nucleus. Cytoplasm is excessively eosinophilic. Cellular debris (fragments of disintegrated cells) is seen in the background of the image. The changes are observed in more than 70% of cells, therefore, it can be concluded grade 3 (moderate reactivity) cytotoxicity exists.

4. Discussion

The presented study analyzed the biocompatibility of three denture adhesives. The cytotoxicity of the adhesives was assessed in an assay with fibroblasts extracted from mature permanent human teeth, a model reflecting the effect of denture adhesives on fibroblasts from oral tissues. Mesenchymal-derived connective tissues including heart, lung, gastrointestinal tract and muscle contain fibroblasts that fulfill specialized functions [25–30]. Differences in gene expression have been demonstrated between dermal and nondermal fibroblasts, and fibroblasts derived from different anatomical sites have differing developmental origins, including the neural crest, lateral plate mesoderm and dermatomyotome [31]. Some studies on fibroblasts from different anatomical sites found marked topographic differences in expression of genes related to growth and differentiation, ECM production, cell migration, lipid metabolism and various genodermatoses, which are molecularly regulated [27,34], but the reaction on toxic materials is similar, regardless of the place of origin. There are a lot of studies that have evaluated the effect of dental materials not having contact with gingiva on gingival fibroblasts [35–39]. Thus, an assumption was made that all fibroblasts from oral tissues follow the same metabolic traits, and for the experiment, dental pulp fibroblasts were used.

After a predefined culture time the rates of viable and necrotic cells were estimated. For all assays using cultured cells as a model system, it is valuable to know how many live and dead cells are present during or after the end of the experiment. Commonly used direct methods of estimating dead cells take advantage of the loss of membrane integrity and the ability of indicator molecules to partition into a compartment, which is not achievable if the cell membrane is intact. The selective staining of dead cells with trypan blue and microscopic examination is one of the most frequently used routine methods to determine the cell number and percent viability in a population of cells. Viable cells have a clear cytoplasm, whereas dead cells have a blue cytoplasm. All tested adhesives demonstrated a significantly higher amount/percentage of necrotic fibroblasts compared to controls, which testifies to their cytotoxic effect. The adhesives differed regarding their cytotoxic potential. The weakest negative effect was found for BLEND-A-DENT Plus, and the strongest for COREGA Extra Strong. PROTEFIX demonstrated a moderately toxic effect on cell cultures.

There is a limited number of reports on the cytotoxicity of denture adhesives. Papers published concern COREGA Extra Strong and PROTEFIX [18–20]. However, we found no studies investigating the effects of BLEND-A-DENT Plus. Results reported by other researchers seem to be consistent with those presented in our paper, despite the use of different types of tests evaluating cell viability. Depending on the method used, the toxicity of the tested adhesives was defined as mild to moderate. Ekstrand et al. [40] reported that in addition to the lysis of cultured cells, samples showed microbial growth despite the addition of antibiotics to growth media, indicating microbial contamination of denture adhesives. Other researchers [18] reported that denture adhesives, including PROTEFIX, showed significantly stronger cytotoxicity compared to the controls in the MTT assay (colorimetric assay for assessing cell metabolic activity) and in the flow cytometric apoptosis assay. Yamada et al. studied the cytotoxicity of six denture adhesives in direct and indirect human epidermal keratinocyte cells and human oral fibroblasts cultures [41]. They observed the cytotoxicity of all tested materials in both cell culture systems and suggested patients should be careful regarding overuse of denture adhesives in terms of amount and duration.

On the other hand, Al et al. [42] found no cytotoxic effect of PROTEFIX on murine fibroblasts in the MTT assay. The inconsistency of the results may be attributed to different species (human and murine) used in the abovementioned studies. Similarly, de Gomes et al. [22] also used MTT assay and cultures of L929 fibroblasts on agar gels containing denture adhesives, including COREGA, and demonstrated its low cytotoxicity. Chen et al. [21] defined the cytotoxic effect of PROTEFIX as mild or moderate, depending on the used culture medium. López-García et al. evaluated the viability of gingival fibroblasts in the presence of six different denture adhesives using MTT assay [43]. Two of them were equivalent to products evaluated in the present study. Poligrip Flavour Free (GlaxoSmithKline, Consumer Healthcare SA. Stafford-Miller Ireland Ltd., Waterford, Ireland) is an equivalent of COREGA Extra Strong, and Fixodent Pro Plus Duo Protection (Procter & Gamble Portugal S.A., Qta da fonte, Ed. Álvares Cabral, 2774-527, Paço de Arcos, Portugal) is an equivalent of BLEND-A-DENT Plus. They found that denture adhesive containing zinc (Fixodent Pro Plus Duo Protection) could be responsible for the decrease of cell viability and aberrant cell morphology as well as induction of apoptosis and cell death. Our study provided contrary results; the necrosis induced by zinc containing BLEND-A-DENT Plus was lower than that induced by zinc-free PROTEFIX and COREGA Extra Strong. The differences between our observations and those made by López-García et al. seem interesting, but require further research, since other components in denture adhesives might be responsible for cell apoptosis. After all, zinc has been used for a very long time as a therapeutic agent in skin and wound care. Rembe at al. showed relevant pro-proliferative, antimicrobial and tendential anti-apoptotic properties of zinc derivatives in an in vitro study [44].

Results obtained from laboratory cultures and viability evaluation of cells are supported by findings from microscopic analysis of morphological changes. Pathomorphological assessment suggests a lower degree of damage to the morphology of fibroblasts in samples with BLEND-A-DENT Plus—grade 1 cytotoxicity with slight reactivity—and the highest in samples with COREGA Extra Strong—grade 3 cytotoxicity with moderate reactivity. The authors found no publications describing the results of similar studies.

This study demonstrated differences in the cytotoxic effect of three denture adhesives on fibroblasts. This may be caused by potentially toxic ingredients. Researchers have attributed this effect to different ingredients [20,21]: formaldehyde is associated with cytotoxic and allergenic effects, whereas karaya gum reduces pH below the critical value for enamel [23]. A similar potential has also been reported for antibacterial and antifungal compounds of adhesive creams [15,16]. It is difficult to identify any specific factor responsible for the adverse effects reported because detailed information regarding the composition and concentration of individual ingredients of adhesives is rarely provided by manufacturers.

The composition of three analyzed denture adhesives is similar but not identical. The most important difference refers to the preservatives. Perhaps the different cytotoxic effect on pulp fibroblasts may be due to the content of different preservatives. Research has shown that propylparaben exerts a cytotoxic effect on human fibroblasts in vitro [45]. It serves as an antifungal and an antimicrobial agent. Corega Extra Strong containing propylparaben demonstrated in this study the strongest toxicity. Protefix contains methyl benzoate, a substance that kills or slows the growth of microorganisms, including bacteria, viruses, fungi and protozoans. Methyl benzoate seems to be less cytotoxic than propylparaben, but the authors did not find any relevant comparative study. In an in vitro study Bunch et al. found that methyl benzoate made cells less viable, but they grew well compared to the control [46]. Thus, the cytotoxic effect was considered as minimal. The manufacturer of BLEND-A-DENT Plus does not provide any preservative, and this adhesive demonstrated the lowest cytotoxic effect compared to the other two tested materials. Perhaps the cause of the cytotoxicity is not the zinc content, but the preservatives. This requires clarification in further research.

In 2010 the European Union Scientific Committee on Consumer Safety stated that the use of butylparaben and propylparaben as preservatives in finished cosmetic products may be considered safe to the consumer, as long as the sum of their individual concentrations does not exceed 0.19% [47].

It is clear that many other materials or drugs may have an effect on the oral mucosa, either directly or indirectly through biofilm formation [48]. Further research in the field of the cytotoxic effects of various dental materials could be focused on stem cells, which can be isolated from oral tissues and contribute to their regeneration [49]. Another important issue for future research could be the effects of lasers used in dentistry on oral cells, since laser therapy has gained an important role in contemporary dental therapy [50,51].

Possible limitations of the present study may be associated with its in vitro design, duration and concentration. In vitro studies carried on various cell types (human epidermal keratinocyte cells, human oral fibroblasts cultures, gingival fibroblasts) have shown the cytotoxic effect of adhesive creams, as shown by the results of this study. It can be suspected that the use of denture adhesives may cause cellular damage in human fibroblasts in vivo resulting in adverse health effects. The manufacturers' recommendations regarding the amount of the product used are intended to prevent exceeding the permissible doses of any ingredients. However, the observations show that patients use too much of denture adhesives and for an extended time period, which may have undesirable effects.

Thus, dentists should advise patients not to overuse denture adhesives, both in terms of product quantity applied and using time. We also suggest that the use of these products should be limited only to cases where the denture does not show proper retention and only in exceptional situations. After all, there is a need for in vivo studies in this field.

5. Conclusions

All the three adhesive creams analyzed, PROTEFIX, COREGA Extra Strong and BLEND-A-DENT Plus, demonstrated slight to moderate toxic effects on human fibroblasts in in vitro quantitative and qualitative evaluation. The strongest toxicity was demonstrated by COREGA Extra Strong and the weakest by BLEND-A-DENT Plus. In vivo observations are needed to find out if denture adhesives cause a cytotoxic effect in patients.

Author Contributions: Conceptualization, E.S. and K.G.; methodology, E.S. and A.N.; formal analysis, J.J.-O. and M.W.-Ś.; investigation, I.D.; resources, J.D. and I.D.; data curation, P.M. and K.G.; writing—original draft preparation, P.M., K.G. and J.J.-O.; writing—review and editing, K.G. and E.S. and M.W.-Ś.; visualization, J.D. and I.D.; supervision, E.S. and A.N.; project administration, E.S. All authors have read and agreed to the published version of the manuscript.

Funding: This research received no external funding.

Institutional Review Board Statement: The study was conducted according to the guidelines of the Declaration of Helsinki and approved by the Bioethics Committee of the Pomeranian Medical University in Szczecin (Resolution No. KB-0012/05/13).

Informed Consent Statement: Informed consent was obtained from all subjects involved in the study.

Acknowledgments: The authors would like to thank Elżbieta Kowalska for preparation of the cell cultures and microscopic specimens, as well as counting of the necrotic and viable cells.

Conflicts of Interest: The authors declare no conflict of interest.

References

1. Assuncao, W.G.; Barao, V.A.; Delben, J.A.; Gomes, E.A.; Tabata, L.F. A comparison of patient satisfaction between treatment with conventional complete dentures and overdentures in the elderly: A literature review. *Gerodontology* **2010**, *27*, 154–162. [CrossRef] [PubMed]
2. Douglass, C.W.; Shih, A.; Ostry, L. Will there be a need for complete dentures in the United States in 2020? *J. Prosthet. Dent.* **2008**, *87*, 5–8. [CrossRef] [PubMed]
3. Thomason, J.M.; Lund, J.P.; Chehade, A.; Feine, J.S. Patient satisfaction with mandibular implant overdentures and conventional dentures 6 months after delivery. *Int. J. Prosthodont.* **2003**, *16*, 467–473. [CrossRef] [PubMed]
4. Anastassiadou, V.; Heath, M.R. The effect of denture quality attributes on satisfaction and eating difficulties. *Gerodontology* **2006**, *23*, 23–32. [CrossRef] [PubMed]
5. Tallgren, A. The continuing reduction of the residual alveolar ridges in complete denture wearers: A mixed-longitudinal study covering 25 years. *J. Prosthet. Dent.* **1972**, *27*, 120–132. [CrossRef]
6. Nishi, Y.; Nomura, T.; Murakami, M.; Kawai, Y.; Nishimura, M.; Kondo, H.; Ito, Y.; Tsuboi, A.; Hong, G.; Kimoto, S.; et al. Effect of denture adhesives on oral moisture: A multicenter randomized controlled trial. *J. Prosthodont. Res.* **2020**, *64*, 281–288. [CrossRef]
7. Gawdzinska, K.; Paszkiewicz, S.; Piesowicz, E.; Bryll, K.; Irska, I.; Lapis, A.; Sobolewska, E.; Kochmanska, A.; Slaczka, W. Preparation and Characterization of Hybrid Nanocomposites for Dental Applications. *Appl. Sci.* **2019**, *9*, 1381. [CrossRef]
8. Uysal, H.; Altay, O.T.; Alparslan, N.; Bilge, A. Comparison of four different denture cushion adhesives—A subjective study. *J. Oral Rehabil.* **1998**, *25*, 209–213. [CrossRef]
9. Bo, T.M.; Hama, Y.; Akiba, N.; Minakuchi, S. Utilization of denture adhesives and the factors associated with its use: A cross-sectional survey. *BMC Oral Health* **2020**, *20*, 194. [CrossRef]
10. Ito, Y.; Hong, G.; Tsuboi, A.; Kawai, Y.; Kondo, H.; Nomura, T.; Kimoto, S.; Gunji, A.; Suzuki, A.; Ohwada, G.; et al. Multivariate analysis reveals oral health-related quality of life of complete denture wearers with denture adhesives: A multicenter randomized controlled trial. *J. Prosthodont. Res.* **2021**, *65*, 353–359. [CrossRef]
11. Zhao, K.; Tian, T.; Zhu, W.J.; Yu, S.H. Preparation and lab evaluation of a new denture adhesive. *J. Wuhan Univ. Technol. Mater. Sci. Ed.* **2011**, *26*, 1036–1040. [CrossRef]
12. Munoz, C.A.; Gendreau, L.; Shanga, G.; Magnuszewski, T.; Fernandez, P.; Durocher, J. A clinical study to evaluate denture adhesive use in well-fitting dentures. *J. Prosthodont.* **2012**, *21*, 123–129. [CrossRef] [PubMed]
13. Okazaki, Y.; Abe, Y.; Dainobu, K.; Iwaguro, S.; Kato, R.; Tsuga, K. A web-based survey of denture adhesive use among denture wearers 40 years of age and older. *J. Oral Sci.* **2020**, *63*, 98–100. [CrossRef] [PubMed]
14. Zhao, K.; Cheng, X.R.; Chao, Y.L.; Li, Z.A.; Han, G.L. Laboratory evaluation of a new denture adhesive. *Dent. Mater.* **2004**, *20*, 419–424. [CrossRef] [PubMed]
15. Murata, H.; Hong, G.; Yamakado, C.; Kurogi, T.; Kano, H.; Hamada, T. Dynamic viscoelastic properties, water absorption, and solubility of home reliners. *Dent. Mater. J.* **2010**, *29*, 554–561. [CrossRef] [PubMed]
16. Darwish, M.; Nassani, M.Z. Evaluation of the effect of denture adhesives on surface roughness of two chemically different denture base resins. *Eur. J. Dent.* **2016**, *10*, 321–326. [CrossRef]
17. Nations, S.P.; Boyer, P.J.; Love, L.A.; Burritt, M.F.; Butz, J.A.; Wolfe, G.I.; Hynan, L.S.; Reisch, J.; Trivedi, J.R. Denture cream: An unusual source of excess zinc, leading to hypocupremia and neurologic disease. *Neurology* **2008**, *71*, 639–643. [CrossRef]
18. Wernke, M.; Wurzel, K.A. Zinc-Containing denture adhesives, toxicity and causal inference. *Clin. Toxicol.* **2013**, *51*, 641.
19. Hedera, P.; Peltier, A.; Fink, J.K.; Wilcock, S.; London, Z.; Brewer, G.J. Myelopolyneuropathy and pancytopenia due to copper deficiency and high zinc levels of unknown origin II: The denture cream is a primary source of excessive zinc. *Neurotoxicology* **2009**, *30*, 996–999. [CrossRef]
20. Lee, Y.; Ahn, J.S.; Yi, Y.A.; Chung, S.H.; Yoo, Y.J.; Ju, S.W.; Hwang, J.Y.; Seo, D.G. Cytotoxicity of four denture adhesives on human gingival fibroblast cells. *Acta Odontol. Scand.* **2015**, *73*, 87–92. [CrossRef]
21. Chen, F.; Wu, T.; Cheng, X. Cytotoxic effects of denture adhesives on primary human oral keratinocytes, fibroblasts and permanent L929 cell lines. *Gerodontology* **2014**, *31*, 4–10. [CrossRef] [PubMed]
22. de Gomes, P.S.; Figueiral, M.H.; Fernandes, M.H.; Scully, C. Cytotoxicity of denture adhesives. *Clin. Oral Investig.* **2011**, *15*, 885–893. [CrossRef] [PubMed]
23. Love, W.B.; Biswas, S. Denture adhesives–pH and buffering capacity. *J. Prosthet. Dent.* **1991**, *66*, 356–360. [CrossRef]
24. Dahl, J.E. Potential of dental adhesives to induce mucosal irritation evaluated by the HET-CAM method. *Acta Odontol. Scand.* **2007**, *65*, 275–283. [CrossRef] [PubMed]
25. Shah, M.; Patel, A.; Patel, S.; Surani, J. Fibroblast heterogeneity and its implications. *Monali. S NJIRM* **2016**, *7*, 92–94.
26. Klewin-Steinböck, S.; Adamski, Z.; Wyganowska-Świątkowska, M. Potential usefulness of enamel matrix derivative in skin and mucosal injury treatment. *Adv. Dermatol. Allergol.* **2021**, *3*, 351–358. [CrossRef]
27. Hua, S.; Bartold, P.M.; Gulati, K.; Moran, C.S.; Ivanovski, S.; Han, P. Periodontal and dental pulp cell-derived small extracellular vesicles: A review of the current status. *Nanomaterials* **2021**, *11*, 1858. [CrossRef]
28. Kaufman, G.; Skrtic, D. Spatial development of gingival fibroblasts and dental pulp cells: Effect of extracellular matrix. *Tissue Cell* **2017**, *49*, 401–409. [CrossRef]
29. le Lièvre, C.S.; le Douarin, N.M. Mesenchymal derivatives of the neural crest: Analysis of chimaeric quail and chick embryos. *J. Embryol. Exp. Morphol.* **1975**, *34*, 125–154. [CrossRef]

30. Sriram, G.; Bigliardi, P.L.; Bigliardi-Qi, M. Fibroblast heterogeneity and its implications for engineering organotypic skin models in vitro. *Eur. J. Cell Biol.* **2015**, *94*, 483–512. [CrossRef]
31. Lynch, M.D.; Watt, F.M. Fibroblast heterogeneity: Implications for human disease. *J. Clin. Investig.* **2018**, *128*, 26–35. [CrossRef]
32. Prieto, A.; Díaz, D.; Barcenilla, H.; García-Suárez, J.; Reyes, E.; Monserrat, J.; San Antonio, E.; Melero, D.; de la Hera, A.; Orfao, A.; et al. Apoptotic rate: A new indicator for the quantification of the incidence of apoptosis in cell cultures. *Cytometry* **2002**, *48*, 185–193. [CrossRef] [PubMed]
33. International Standard. *ISO 10993-5:2009(E)*; Biological Evaluation of Medical Devices—Part 5: Tests for In Vitro Cytotoxicity. Available online: https://www.iso.org/standard/36406.html(accessed on 1 November 2021).
34. Chang, H.Y.; Chi, J.-T.; Dudoit, S.; Bondre, C.; van de Rijn, M.; Botstein, D.; Brown, P.O. Diversity, topographic differentiation, and positional memory in human fibroblasts. *Proc. Natl. Acad. Sci. USA* **2002**, *99*, 12877–12882. [CrossRef] [PubMed]
35. Nirwana, I.; Munadziroh, E.; Yogiartono, R.M.; Thiyagu, C.; Ying, C.S.; Dinaryanti, A. Cytotoxicity and proliferation evaluation on fibroblast after combining calcium hydroxide and ellagic acid. *J. Adv. Pharm. Technol. Res.* **2021**, *12*, 27–31. [CrossRef] [PubMed]
36. Javidi, M.; Dastmalchi, P.; Zarei, M.; Rad, M.S.; Ghorbani, A. In vitro cytotoxicity of a new nano root canal sealer on human gingival fibroblasts. *Iran Endod. J.* **2017**, *12*, 220–225. [PubMed]
37. Jose, J.; Palanivelu, A.; Subbaiyan, H. Cytotoxicity evaluation of calcium hypochlorite and other commonly used root canal irrigants against human gingival fibroblast cells: An in vitro evaluation. *Dent. Med. Probl.* **2021**, *58*, 31–37. [CrossRef] [PubMed]
38. Teixeira, A.B.V.; Moreira, N.C.S.; Takahashi, C.S.; Schiavon, M.A.; Alves, O.L.; Reis, A.C. Cytotoxic and genotoxic effects in human gingival fibroblast and ions release of endodontic sealers incorporated with nanostructured silver vanadate. *J. Biomed. Mater. Res. B Appl. Biomater.* **2021**, *109*, 1380–1388. [CrossRef] [PubMed]
39. Narvaez-Flores, J.J.; Vilar-Pineda, G.; Acosta-Torres, L.S.; Garcia-Contreras, R. Cytotoxic and anti-inflammatory effects of chitosan and hemostatic gelatin in oral cell culture. *Acta Odontol. Latinoam.* **2021**, *99*, 98–103. [CrossRef]
40. Ekstrand, E.; Hensten-Pettersen, A.; Kullmann, A. Denture adhesives: Cytotoxicity, microbial contamination, and formaldehyde content. *J. Prosthet. Dent.* **1993**, *69*, 314–317. [CrossRef]
41. Yamada, M.; Takase, K.; Suehiro, F.; Nishimura, M.; Murata, H. Effects of denture adhesives and mouth moisturizers to human oral fibroblast and human keratinocyte cells using direct and indirect cell culture systems. *Dent. Mater. J.* **2020**, *39*, 571–576. [CrossRef]
42. Al, R.H.; Morisbak, E.; Polyzois, G.L. Irritation and cytotoxic potential of denture adhesive. *Gerodontology* **2005**, *22*, 177–183. [CrossRef] [PubMed]
43. López-García, S.; Pecci-Lloret, M.P.; García-Bernal, D.; Guerrero-Gironés, J.; Pecci-Lloret, M.R.; Rodríguez-Lozano, F.J. Are Denture Adhesives Safe for Oral Cells? *J. Prosthodont.* **2021**, *30*, 65–70. [CrossRef] [PubMed]
44. Rembe, J.D.; Boehm, J.K.; Fromm-Dornieden, C.; Hauer, N.; Stuermer, E.K. Comprehensive analysis of zinc derivatives pro-proliferative, anti-apoptotic and antimicrobial effect on human fibroblasts and keratinocytes in a simulated, nutrient-deficient environment in vitro. *Int. J. Mol. Cell Med.* **2020**, *9*, 165–178. [PubMed]
45. de Carvalho, C.M.; Menezes, P.F.C.; Letenski, G.C.; Praes, C.E.O.; Feferman, I.H.S.; Lorencini, M. In vitro induction of apoptosis, necrosis and genotoxicity by cosmetic preservatives: Application of flow cytometry as a complementary analysis by NRU. *Int. J. Cosmet. Sci.* **2012**, *34*, 176–182. [CrossRef] [PubMed]
46. Bunch, H.; Park, J.; Choe, H.; Mostafiz, M.M.; Kim, J.-E.; Lee, K.-Y. Evaluating cytotoxicity of methyl benzoate in vitro. *Heliyon* **2020**, *6*, e03251. [CrossRef]
47. Directorate-General for Consumer Safety, European Union. Scientific Committee on Consumer Safety Opinion on Parabens COLIPA n° P82. 2011. Available online: https://ec.europa.eu/health/scientific_committees/consumer_safety/docs/sccs_o_041.pdf (accessed on 1 November 2021).
48. Cazzaniga, G.; Ottobelli, M.; Ionescu, A.C.; Paolone, G.; Gherlone, E.; Jack, L.; Ferracane, J.L.; Brambilla, E. In vitro biofilm formation on resin-based composites after different finishing and polishing procedures. *J. Dent.* **2017**, *67*, 43–52. [CrossRef]
49. Capparè, P.; Tetè, G.; Sberna, M.T.; Panina-Bordignon, P. The emerging role of stem cells in regenerative dentistry. *Curr. Gene. Ther.* **2020**, *20*, 259–268. [CrossRef]
50. Lucchese, A.; Matarese, G.; Ghislanzoni, L.H.; Gastaldi, G.; Manuelli, M.; Gherlone, E. Efficacy and effects of palifermin for the treatment of oral mucositis in patients affected by acute lymphoblastic leukemia. *Leuk. Lymphoma* **2016**, *57*, 820–827. [CrossRef]
51. Jedliński, M.; Romeo, U.; del Vecchio, A.; Palaia, G.; Galluccio, G. Comparison of the effects of photobiomodulation with different lasers on orthodontic movement and reduction of the treatment time with fixed appliances in novel scientific reports: A systematic review with meta-analysis. *Photobiomodul. Photomed. Laser. Surg.* **2020**, *38*, 455–465. [CrossRef]

Article

Degradation of Polylactide and Polycaprolactone as a Result of Biofilm Formation Assessed under Experimental Conditions Simulating the Oral Cavity Environment

Dawid Łysik [1,*], Piotr Deptuła [2], Sylwia Chmielewska [2], Robert Bucki [2] and Joanna Mystkowska [1]

[1] Institute of Biomedical Engineering, Bialystok University of Technology, 15-351 Bialystok, Poland
[2] Department of Microbiological and Nanobiomedical Engineering, Medical University of Bialystok, 15-222 Bialystok, Poland
* Correspondence: d.lysik@pb.edu.pl

Abstract: Polylactide (PLA) and polycaprolactone (PCL) are biodegradable and bioabsorbable thermoplastic polymers considered as promising materials for oral applications. However, any abiotic surface used, especially in areas naturally colonized by microorganisms, provides a favorable interface for microbial growth and biofilm development. In this study, we investigated the biofilm formation of *C. krusei* and *S. mutans* on the surface of PLA and PCL immersed in the artificial saliva. Using microscopic (AFM, CLSM) observations and spectrometric measurements, we assessed the mass and topography of biofilm that developed on PLA and PCL surfaces. Incubated up to 56 days in specially prepared saliva and microorganisms medium, solid polymer samples were examined for surface properties (wettability, roughness, elastic modulus of the surface layer), structure (molecular weight, crystallinity), and mechanical properties (hardness, tensile strength). It has been shown that biofilm, especially *S. mutans*, promotes polymer degradation. Our findings indicate the need for additional antimicrobial strategies for the effective oral applications of PLA and PCL.

Keywords: biomaterial; polymer; microbial degradation; biofilm; artificial saliva; mucins

1. Introduction

Polylactide (PLA) and polycaprolactone (PCL) are commonly used biodegradable and bioabsorbable aliphatic polyesters in a variety of biomedical applications [1–4]. Recently, PLA and PCL have been considered as promising materials in oral and pharyngeal appliances, including for 3D printed dentures [5,6], drug-enriched denture coatings [7], pharyngeal and laryngeal implants [8,9], and mucosal drug carriers [10]. Specific conditions for biofilm formation on the abiotic surfaces in the semi-open oral environment can create susceptibility to infection and reduce the reliability of polymeric medical devices. This study looks into the processes occurring on the polymer–biofilm interface.

In vivo, the degradation of PLA and PCL is dominated by non-enzymatic mechanisms, mainly hydrolysis, during which polymer absorbs water and their chains break down into smaller pieces over time [11]. This process leads to polymer erosion and alters their structure and properties, including the deterioration of mechanical strength [12]. PLA and PCL degradation kinetics depend on the polymer and environmental factors. In general, PLA degrades faster than PCL, and in the blends of these polymers, the proportions can regulate the degradation time [13]. Due to the chirality of the lactide molecule in two optically active forms (L-lactide and D-lactide), PLA can polymerize in various forms, such as poly (L-lactide) (PLLA), poly (D-lactide) (PDLA), poly (D, L-lactide) (PDLLA), and mesopolylactide. Of these, PLLA and PDLLA are mainly used for biomedical applications. Depending on the D-isomer content of PLA, fully crystalline PLLA degrades slower than amorphous PDLLA [14]. Large samples of these polymers (with dimensions of a few millimeters) decompose faster than small samples (films or microparticles) due to the autocatalytic hydrolysis inside the polymer [15]. The higher the initial degree of crystallinity and the crystallite size, the faster the degradation progresses [16]. Other material-related factors, including fillers and porosity, have been extensively discussed in the literature [17–19].

The three main environmental-related degradation factors of the oral cavity are temperature, pH, and the diversity of the microflora. Lyu et al. [19,20] discuss the influence of the first two on the degradation rate. Our previous in vitro studies on PLA and PCL degradation in model oral cavity conditions [20] showed a significant increase in the degradation rate of these polymers at 42 °C compared to 37 °C and no impact of pH level. We also considered the aspects of saliva and its influence on the degradation rate, pointing out its significant impact on biofilm formation [21–23].

The problem of material degradation in the oral cavity microbiological environment is rarely addressed, although there are many indications that the presence of biofilm may contribute to faster biomaterials degradation [24,25]. In some cases, the hydrolytic degradation of PLA and PCL can be enzymatically catalyzed with the participation of microorganisms. For PLA, these include actinomycetes, such as *Amycolatopsis* [26], *Saccharothrix* [27], *Kibdelosporangium* [28], *Actinomadura* [29], *Laceyella* [30], and *Pseudonocardia* [31] which are involved in degradation mediated by proteases; bacteria, such as *Bacillus* [32], *Geobacillus* [33], *Paenibacillus* [34], *Stenotrophomonas* [35], *Pseudomonas* [36], and the types of enzymes they produce—lipases and esterases; and fungi, such as cutinase-producing *Aspergillus* [37] and *Cryptococcus* [38], protease-producing *Tritirachium* [39], and *Trichoderma* [40], for which the type of enzyme affecting degradation has not been identified. The enzymatic degradation of PCL is involved in bacteria (mainly *Pseudomonas* and *Lactobacillus* [41–45]) and fungi (mainly *Aspergillus*, *Candida*, *Mucor*, *Rhizopus*, and *Thermomyces*) lipases [41,46–50].

We designed a set of experiments in which polylactide and polycaprolactone were incubated in the specially prepared artificial saliva in the presence of common oral microorganisms, such as *Streptococcus mutans* and *Candida krusei*. We assessed the biofilm development on these polymers' surfaces and its influence on material properties.

2. Materials and Methods

2.1. Polymer Samples

PLA (3001D, content of D isomers ~ 1.6%, M_n ~ 90,000 g/mol, NatureWorks, Minnetonka, MN, USA) and PCL granules (M_n ~ 45,000 g/mol, Sigma-Aldrich, Burlington, MA, USA) were dried for approximately 3 h at 45 °C. Polymer specimens were made by injection molding on a BS60 device (Borche, Kanton, P.R.C.). After that, molded specimens with dimensions of 30 mm × 5 mm × 4 mm, were pre-conditioned in the air at room temperature for 24 h.

2.2. Artificial Saliva

Artificial saliva (AS) was prepared based on a Phosphate-buffered saline (PBS) in which porcine gastric mucin (M1778, Sigma-Aldrich, Saint Louis, MO, USA) at a concentration of 10 g/L and xanthan gum (G1253, Sigma-Aldrich, Saint Louis, MO, USA) at a concentration of 4 g/L were dissolved.

2.3. Incubation Conditions

The prepared polymer samples were incubated in the 6-well polystyrene plates in the mixture of AS and the cell culture medium (Lysogeny broth and brain heart infusion) (1:1 volume). The research used two microbial strains: the fungus *Candida krusei* ATCC 6258 and the bacteria *Streptococcus mutans* ATCC 35668 (Biomaxima, Poland). Incubation was carried out at 37 °C for 56 days. The medium was changed every 48 h (the biofilm from the surface of the samples was not removed, the samples were only rinsed in PBS). Control samples were incubated in the medium. Polymer tests were carried out after 24 and 72 h and after 28 and 56 days of incubation.

2.4. Biofilm Characterization on the Surface of the Polymer

The crystal violet (CV) method was used to determine the biofilm mass on the surface of polymers. In the first step, the polymers in the plates were washed with PBS to remove planktonic bacteria and 0.1% CV solution made in water was added to each well. After 15 minutes of incubation at room temperature (RT), the dye was rinsed with deionized water, and then 70% ethanol solution was added to dissolve the CV. Next, a microplate reader (Varioskan LUX, Thermo Fisher Scientific, Waltham, MA, USA) measured the absorbance at a wavelength $\lambda = 570$ nm. In each measurement cycle, 3 samples were tested ($n = 3$). The obtained test results represent mean values ± SD. Additionally, biofilm morphology was characterized by scanning laser confocal microscopy (CLSM). The samples were observed immediately after being removed from the incubation containers and rinsed with water. A laser scans of the surface topography were performed on the area of 128×128 micrometers with an accuracy of 0.01 micrometers on the z-axis.

2.5. AFM Biofilm Topography and Surface Layer Stiffness

Using atomic force microscopy, imaging of the biofilm structure and measurements of the surface layer stiffness was performed. Measurement needles with an elasticity constant of 0.37 N/m (AppNano NITRA-TALL-V-G) were used in the research. Both to obtain topographic images and measurements of the surface layer stiffness modulus, the Quantitative Imaging mode was used, based on measurements of probe insertion force curves (at a constant speed of 300 μm/s) at each measurement point. Topographic maps covered an area of 25 μm × 25 μm with a resolution of 128×128 pixels. The stiffness modulus maps covered an area of 10 μm × 10 μm with a resolution of 8×8 pixels. The Hertz–Sneddon model was used to calculate the elastic modulus E from the force curves.

2.6. Contact Angle

An Ossila Contact Angle Goniometer (Ossila, Sheffield, UK) equipped with a digital camera and a leveling table was used to measure the surface contact angle. On the dry surface of the polymers, with the aid of a micropipette, 25 μL of deionized water was placed and the drop image on the surface was recorded for 5 s. The image was analyzed with the included software with a drop edge detection function. In each measurement cycle, 5 polymer samples ($n = 5$) were tested. The obtained test results represent mean values ± standard deviation (SD). Statistically significant differences were determined at $p < 0.05$ (Student's t-test).

2.7. Polymer Mass Loss

The weight loss of the incubated PLA and PCL was determined by a gravimetric method using a laboratory balance (Mettler Toledo, Columbus, OH, USA) with an accuracy of 0.01 mg. After removing from the containers, the samples were rinsed several times in water and ethanol and dried in a moisture analyzer at 37 °C for 24 h, and then the weight of the dry sample (m_d) was measured. The weight loss was determined by Formula (1), in which m_0 is the initial mass of the sample.

$$Mass\ loss = \frac{m_0 - m_d}{m_0} \times 100\% \qquad (1)$$

2.8. Hardness and Tensile Strength

The hardness of PLA and PCL was determined using a Shore durometer (type D) (Zwick Roell, Ulm, Germany) following the ASTM D2240 standard. Each time, 5 polymer samples were tested, and 6 measurements were made for each of them. The tensile strength R_m was determined using the Zwick/Roell Z010 testing machine (Zwick Roell, Ulm, Germany) using the ISO 527 standard. The highest force obtained during the static tensile test (F_m) related to the initial cross-section of the specimen (A0) was used for the calculations.

Five polymer samples were tested. The results presented in the paper represent mean values ± SD.

2.9. Molecular Weight of Polymers Surface Layer

The average viscosity molecular weight $M\eta$ of polymers (assumed in the work as the molecular weight) was estimated based on measurements of the intrinsic viscosity $[\eta]$ of PLA or PCL solutions (with a concentration of 0.001 g/mL) in chloroform using a capillary viscometer with an automatic viscosity measurement system (iVisc Capillary Viscometer, Lauda-Brinkmann LP, Delran, NJ, USA; Ubbelohde capillary constant $c = 0.003$ mm^2/s^2). The solutions were prepared by dissolving precisely cut fragments of polymer samples (50 ± 1 µm thick) (using a rotating microtome (Struers, Ballerup, Denmark)) in chloroform. The molecular weight was determined based on the Mark–Houwink Equation (2), in which K and a are constants determined for the appropriate polymer-solvent system at a given temperature (for PLA–chloroform at 25 °C: $K = 6.06 \times 10^{-2}$ g/cm^3, $a = 0.64$ [51]; for PCL–chloroform at 30 °C: $K = 1.298 \times 10^{-2}$ g/cm^3, $a = 0.828$ [52]).

$$[\eta] = KM_\eta^a \tag{2}$$

In each measurement cycle, 3 polymer solutions were tested, with each viscosity test being repeated 3 times. The results represent mean values ± SD ($n = 9$).

2.10. Differential Scanning Calorimetry Measurements

Polymer samples were tested by Differential Scanning Calorimetry using a DSC Discovery device (TA Instruments, New Castle, DE, USA) in a heat–cool–heat mode in the range from 0 to 200 °C for PLA and from −80 to 80 °C for PCL with 5°/min heating/cooling rate. The results analyzed in the study concern the second course of heating the sample (the first heating was carried out to remove the thermal history of the sample). As in the case of the molecular weight tests, the measurements were carried out on precisely cut fragments of polylactide and polycaprolactone samples (50 ± 1 µm thick). In each measurement cycle, 5 polymer samples ($n = 5$) were tested. On this basis, the TRIOS program obtained several data, such as glass transition temperature, cold crystallization temperature, cold crystallization enthalpy ΔH_{cc}, melting point T_m, and melting enthalpy ΔH_m. Crystallinity X_c was determined based on Equation (3):

$$X_c = \frac{\Delta H_m - \Delta H_{cc}}{\Delta H_m^{100}} \tag{3}$$

where ΔH_m^{100} is the melting enthalpy of 100% crystalline PLA/PCL (93.7 J/g for PLA and 135.3 J/g for PCL).

The obtained results of thermal properties tests show mean values ± SD. Statistically significant differences were determined at $p < 0.05$ (Student's t-test).

3. Results

The collected data describing the development of *C. krusei*, *S. mutans*, and *C. krusei* + *S. mutans* biofilms on the PLA and PCL surfaces are summarized in Figure 1. Figure 1a shows the AFM surface topography of biofilm on PLA after 72 h incubation. At this stage of biofilm development, individual cells can be distinguished—large and oval *C. krusei*, small and round *S. mutans*, and the structure of a mixed biofilm, where *C. krusei* cells are surrounded by *S. mutans* cells. Figure 1b shows the CLSM surface topography of *C. krusei* + *S. mutans* biofilm on PLA after 72 h and 56 days of incubation. The clusters of fungal cells surrounded by bacterial cells evolved into a more homogeneous structure. Figure 1c shows how the coverage of the polymer surface by the biofilm increased for the first 72 hours. After that time, *C. krusei* biofilm covered less than 20% of the surface, *S. mutans* nearly 40%, and *C. krusei* + *S. mutans* nearly 90%. Figure 1d shows the biofilm mass development (linearly correlated with the optical density of the CV dyed biofilm) after 24 and 72 h on the surface of PLA and PCL. The results of spectrophotometric studies are consistent with the CLSM and indicate that mixed biofilm *C. krusei* + *S. mutans* develops the most. After 56 days of incubation, the highest biofilm mass was observed for *S. mutans* (Figure 1e).

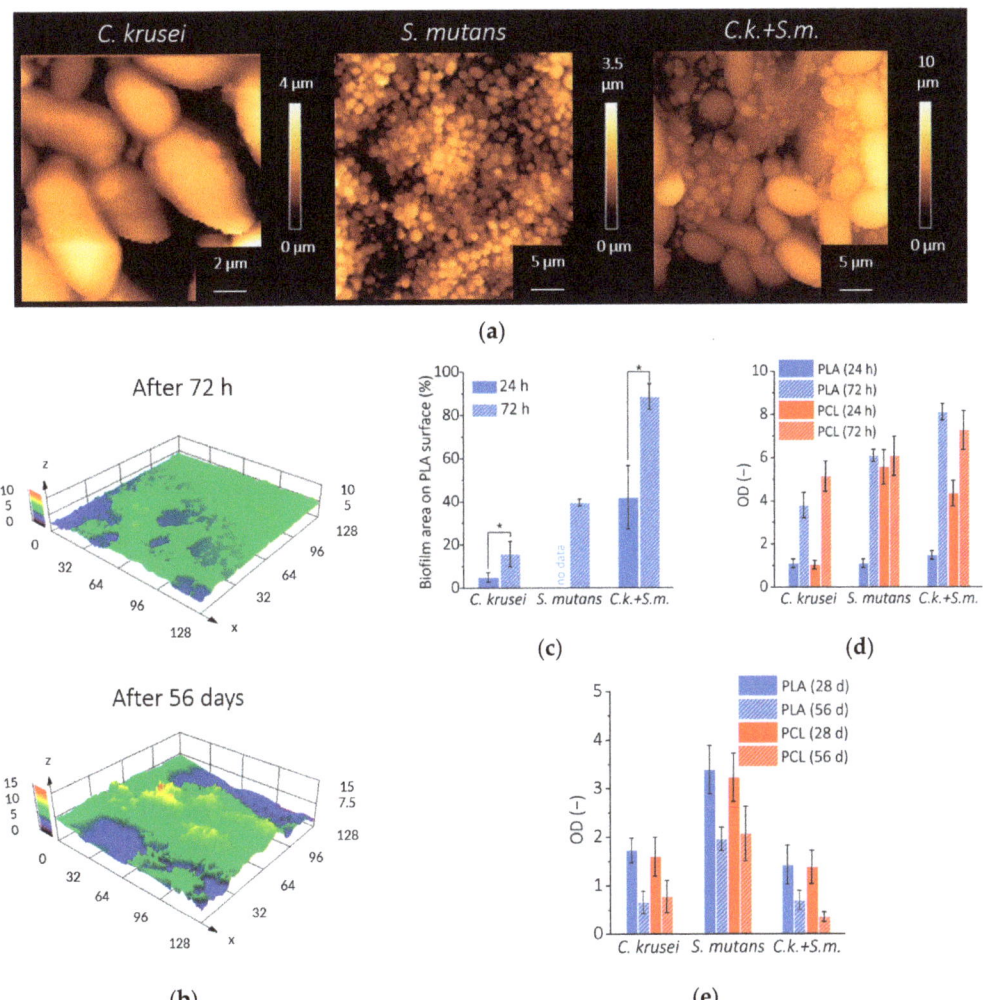

Figure 1. Biofilm (*C. krusei*, *S. mutans*, or *C. krusei* + *S. mutans*) on the surface of the tested polymers: (**a**) AFM topography of individual types of biofilm on the surface of PLA after 72 h; (**b**) CLSM topography of the *C. krusei* + *S. mutans* biofilm on the PLA surface after 72 h and 56 days; (**c**) biofilm development on the PLA surface during the first 72 h (based on CLSM); (**d**) biofilm development on the surface of PLA or PCL during the first 72 h (based on the optical density of the biofilm mass stained with CV); (**e**) biofilm development on the surface of PLA or PCL for 56 days (as measured by the optical density of the biofilm mass stained with CV). Results mean ± SD (*n* = 3). (*) $p < 0.05$.

We believe that the biofilm on the surface of PLA or PCL may contribute to the surface layer degradation that is associated with a reduction in the physical properties of a polymer caused by changes in its chemical structure. Several tests were performed to verify this hypothesis.

After removing the biofilm from the tested polymers, no visual changes were noticed between the samples incubated in artificial saliva with and without the microorganisms. However, we observe some changes in the properties of the surface layer. The results of the contact angle measurements presented in Figure 2 show that the incubation of polymers for 56 days significantly increases the hydrophilic properties of the surfaces of both PLA and PCL. In the AS, the contact angle of PLA decreased from 55° to 42°. Biofilm has intensified this phenomenon and the contact angle after 56 days dropped to 38° in *C. krusei*, to 30° in *S. mutans*, and 35° in *C. krusei* + *S. mutans*. A similar trend was observed for PCL, in which the contact angle after incubation in AS decreased from 59° to 50°, while in the presence of *C. krusei* microorganisms it dropped to 47°, in *S. mutans* to 37°, and in *C. krusei* + *S. mutans* up to 40°.

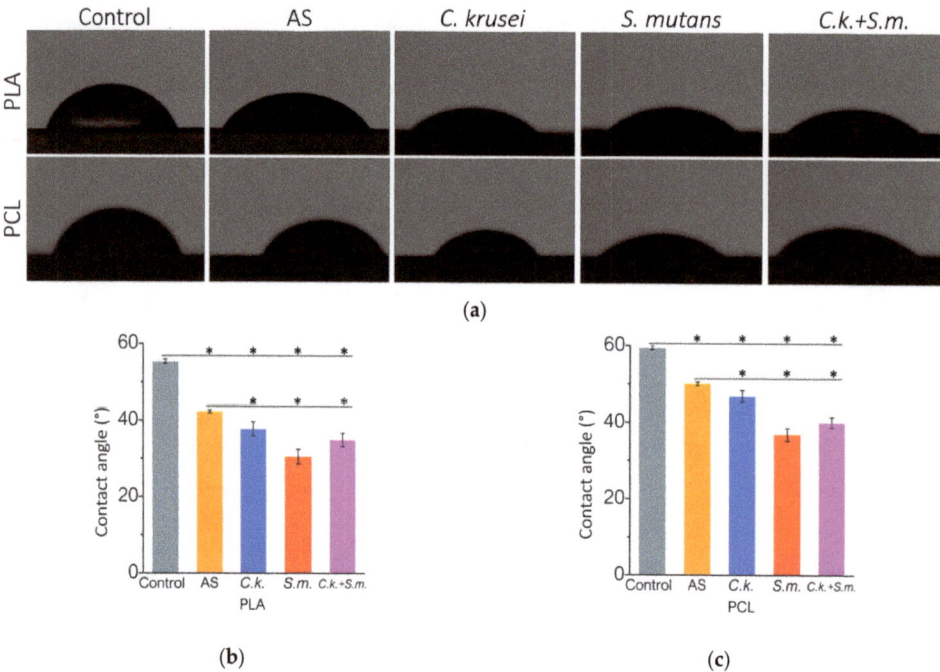

Figure 2. Wettability of PLA and PCL surfaces after incubation in the presence of artificial saliva and microorganisms: (**a**) water droplets on the surface of the tested polymers; (**b**) contact angle at the surface of the PLA; (**c**) contact angle at the surface of the PCL. Results mean ± SD (n = 3). (*) $p < 0.05$.

The topography and mechanical properties of the surface layer also changed after incubation in a biological environment (Figure 3). PLA roughness measurements showed a slight increase in the R_a parameter (the arithmetic average of profile height deviations from the mean line) for samples in the *C. krusei* and *C. krusei* + *S. mutans* environment as compared to non-incubated samples (Control) (Figure 3a). The profilometric analysis of the PCL surface showed a significant decrease in the R_a value after incubation in AS and the presence of microorganisms (Figure 3b). The results of mechanical properties of the PLA surface layer show that as a result of hydrolytic degradation in AS, the elastic modulus of the surface layer decreased by nearly 45% (E = 514 MPa) compared to non-incubated PLA (E = 937 MPa). The interaction of PLA with the biological environment significantly ($p < 0.05$) changed the elastic modulus of the polymer surface layer (Figure 3c). In the presence of *C. krusei* biofilm, the elastic modulus of PLA decreased by 83% (E = 88 MPa) in samples incubated in AS and by over 90% in non-incubated samples. In the presence of *S. mutans* biofilm (E = 45 MPa) and a mixed biofilm of *C. krusei* + *S. mutans* (E = 42 MPa), the elastic modulus decreased by more than 91% compared to AS and by more than 95% compared to non-incubated samples. In the case of PCL (Figure 3d), statistically significant ($p < 0.05$) differences in the elastic modulus were observed after contact with *S. mutans* biofilm ($E = 195$ MPa), where the E modulus decreased by 63% compared to PCL incubated in AS ($E = 531$ MPa). It is worth emphasizing that in the mixed biofilm environment there was no decrease in the elastic modulus of the PCL surface layer as was the case with PLA.

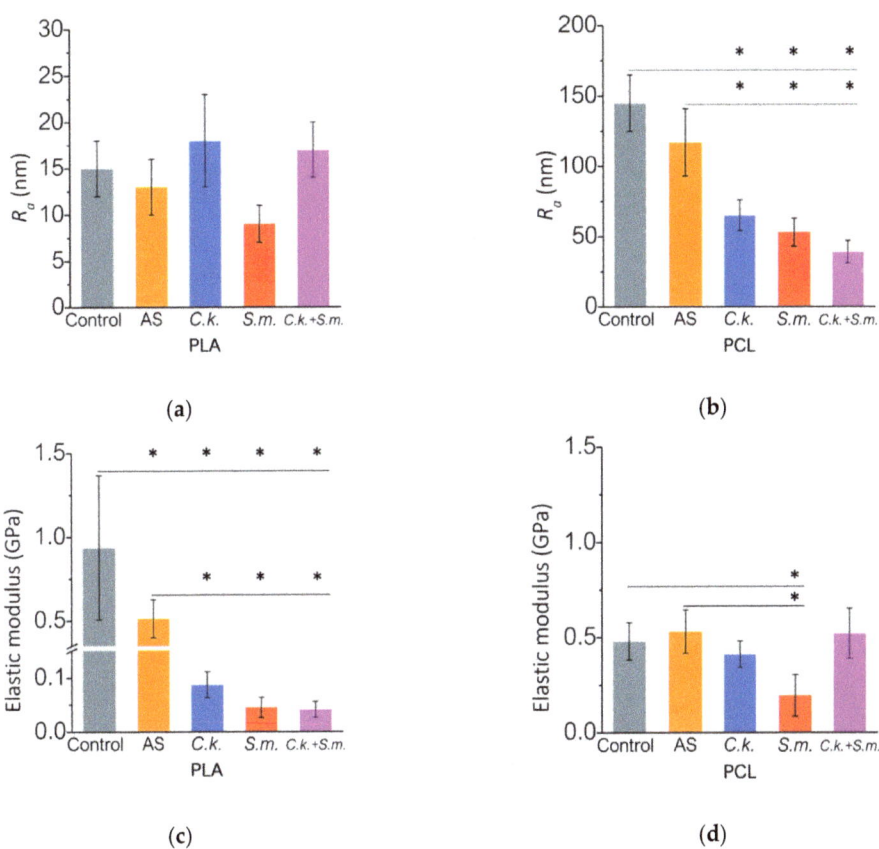

Figure 3. Surface properties of PLA and PCL after incubation: (**a**,**b**) roughness before and after incubation expressed by the R_a parameter for PLA and PCL, respectively; (**c**,**d**) surface elasticity modulus before and after incubation for PLA and PCL, respectively. Results mean ± SD ($n = 3$). (*) $p < 0.05$.

In Figure 4a,b we show the mass loss during incubation of PLA and PCL, respectively. For PLA in the AS, after 56 days the mass loss was 0.61% but is much higher in the biofilm—in the presence of *C. krusei* the mass loss after 56 days was 0.81%, and in the presence of *S. mutans* bacteria 0.95%, and the presence of mixed biofilm 0.83%. For PCL in AS, after 56 days of incubation, the mass loss was about 0.5% and no significant changes were observed in the biofilm environment.

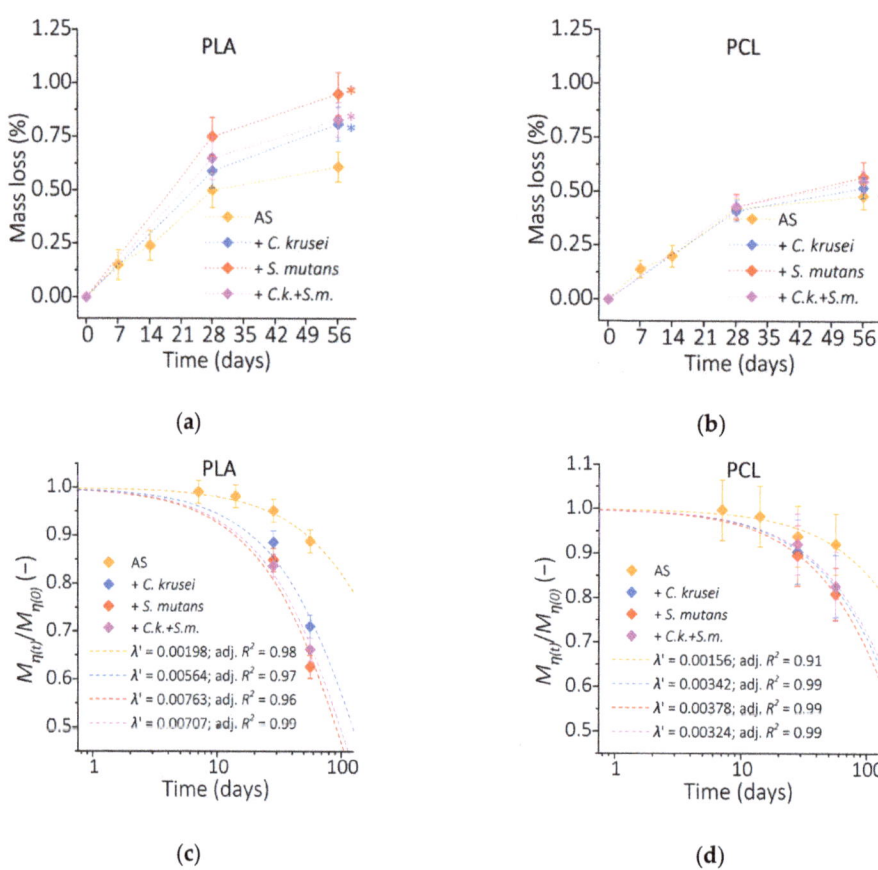

Figure 4. (**a**,**b**) Mass loss of PLA and PCL, respectively; (**c**,**d**) the ratio of the molecular weight after time t to the initial molecular weight of PLA and PCL, respectively. (*) $p < 0.05$.

Polymer erosion is caused by the reduction of polymer molecular weight. In this study, we used the viscosity molecular weight M_η (Figure 4c,d). The initial value of $M_{\eta(0)}$ at time "0" for PLA was 166,000 g/mol, while for PCL was 58,000 g/mol (for comparison, the values determined by the manufacturer using chromatographic methods are $Mn \sim 90{,}000$ g/mol for PLA and $Mn \sim 45{,}000$ g/mol for PCL). Changes in the molecular weight of the polymers during incubation are shown as the ratio of the molecular weight over time $M_{\eta(t)}$ to the initial molecular weight $M_{\eta(0)}$. The ratio $M_{\eta(t)}/M_{\eta(0)}$ decreases exponentially over time for PLA and PCL in all tested conditions, which was illustrated by fitting the exponential function $M_{\eta(t)}/M_{\eta(0)} = e^{-\lambda' t}$ with the parameter λ' (expressed in units of 1/day) as the degradation rate (Figure 4c,d) (discussed in more detail in [12]). For PLA samples incubated in the AS, the value of λ' was 1.98×10^{-3}. In the *C. krusei* biofilm, it was 5.64×10^{-3}; in *S. mutans*, 7.63×10^{-3}; and in the mixed biofilm, 7.07×10^{-3}. For PCL in AS, the parameter λ' was 1.56×10^{-3}; in *C. krusei*, 3.42×10^{-3}; in *S. mutans*, 3.78×10^{-3}; and w *C. krusei* + *S. mutans*, 3.24×10^{-3}. It can be clearly stated that the presence of microorganisms on the surface of PLA during incubation in the artificial saliva significantly increases the degradation rate of the surface layer, with the highest degradation rate observed in the presence of *S. mutans*. The presence of microorganisms also increases the rate of PCL degradation.

The results of differential scanning calorimetry (DSC) analyses of the tested PLA and PCL polymers are presented in Figure 5. Figure 5a,b shows the DSC curves of the second heating for PLA and PCL, respectively. For the PLA control (not incubated), the glass transition temperature was ~55 °C, the cold crystallization temperature was 101 °C, the cold crystallization enthalpy was ~31 J/g, and the melting point was 168 °C, while the fusion enthalpy was ~43 J/g. After 56 days in AS, the glass transition temperature decreased to ~54 °C, the cold crystallization temperature decreased to ~91 °C, the cold crystallization enthalpy decreased to ~12 J/g, and the melting point to ~165 °C, while the melting enthalpy remained at a similar level ~43 J/g. The presence of microorganisms changed the cold crystallization enthalpy, while other properties, including the fusion enthalpy, did not change significantly. Thus, the influence of environmental conditions on the properties of polymers is most evident in the change of the crystallinity, which for PLA is the difference between the fusion enthalpy and the cold crystallization enthalpy related to the fusion enthalpy of fully crystalline PLA equal to 93.7 J/g (Figure 5c). The degree of crystallinity of non-incubated PLA was 12%, while after 56 days of incubation in AS it increased almost threefold (to ~33%). The presence of microorganisms in the artificial saliva during the incubation of PLA samples significantly ($p < 0.05$) increases their crystallinity to ~59% for *C. krusei*, ~56% for *S. mutans*, and ~57.5% for the mixed biofilm. From the DSC curves of the second PCL heating (Figure 5b), two parameters were determined—the melting point and the melting enthalpy. For the control, the melting temperature was ~56 °C and melting enthalpy ~56 J/g. Incubation in the environment of artificial saliva and microorganisms has little effect on the value of the melting enthalpy, so the changes in crystallinity are not significant (Figure 5d).

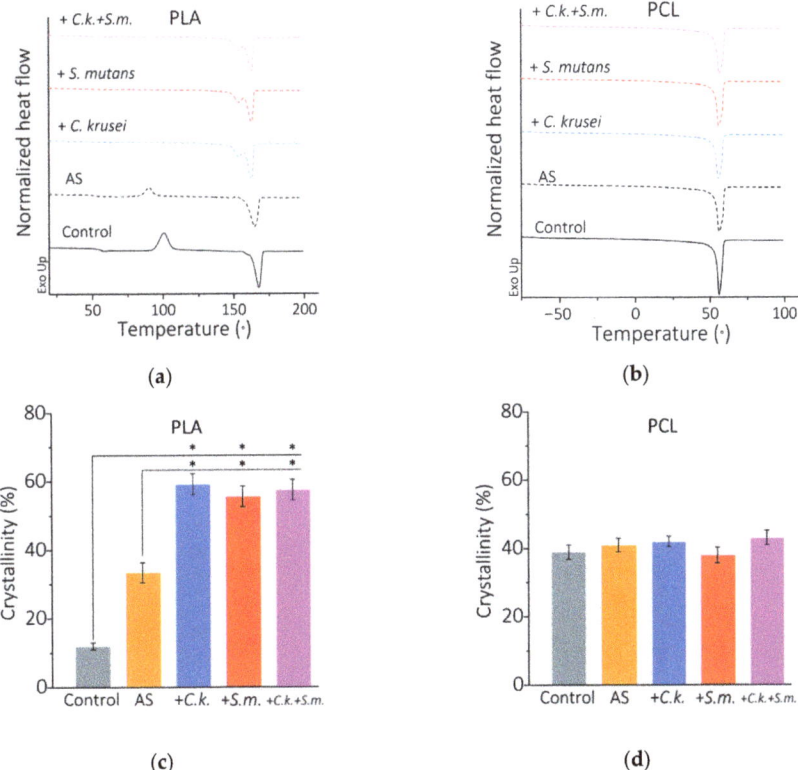

Figure 5. DSC of PLA and PCL after 56 days of incubation: (**a**) DSC curves (second heating) for PLA; (**b**) DSC (second heat) curves for PCL; (**c**) crystallinity for PLA; (**d**) crystallinity X_c for PCL. Results mean ± SD ($n = 5$). (*) $p < 0.05$.

The biofilm degradation of PLA and PCL, resulting in the erosion, decrease in molecular weight, and change of crystallinity, may cause a deterioration of polymer mechanical properties. Figure 6a shows the changes in PLA hardness during incubation. The initial PLA hardness was 68 ShD, and an increase in hardness of 71 ShD was observed during incubation in AS. After 56 days in *C. krusei*, the hardness did not change, in *S. mutans* and mixed biofilm, it dropped slightly to 67 ShD. Figure 6b shows the changes in PCL hardness during incubation. The initial hardness of PCL was 51 ShD, similar to PLA, and incubation in AS resulted in a slight increase in hardness (to 53 ShD). During incubation in the biofilm, an increase in hardness was observed after 28 days and a return to the initial value. Figure 6c shows the changes in PLA tensile strength during incubation. Incubation in AS resulted in a drop in tensile strength from 65 to about 63 MPa. The presence of microorganisms negatively affected the tensile strength, which fell below 62 MPa. Figure 6d shows the changes in the tensile strength of PCL during incubation. In AS, an initial increase in strength of about 1 MPa from 18.5 MPa was observed, followed by a decrease to 19 MPa. The presence of microorganisms did not change the strength of PCL. Overall, no significant reduction in the strength of the tested polymers was observed.

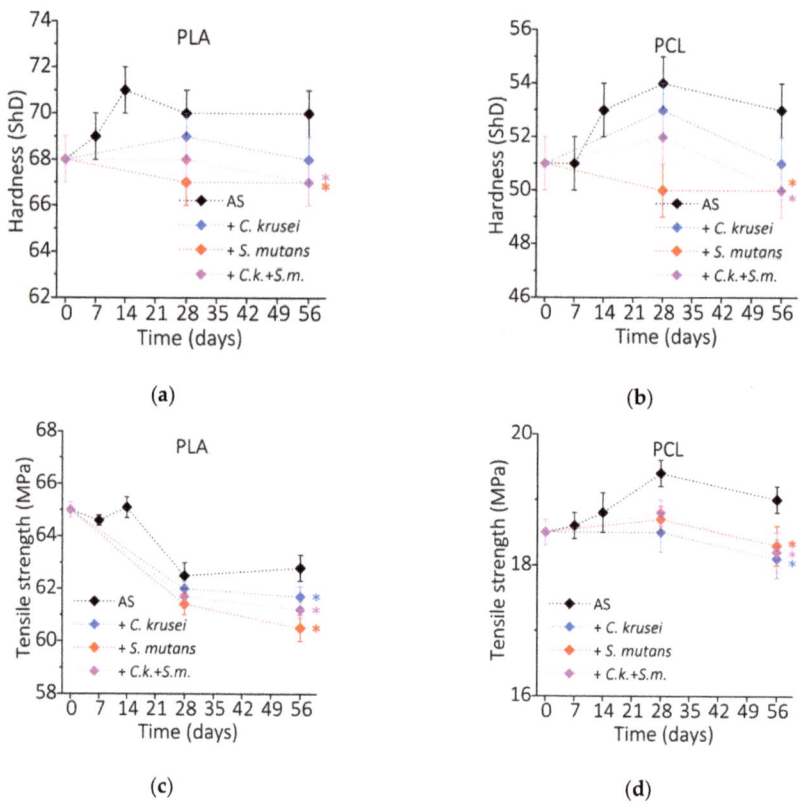

Figure 6. Mechanical properties of PLA and PCL during incubation in artificial saliva and microorganisms: (**a**,**b**) Shore hardness; (**c**,**d**) tensile strength. (*) $p < 0.05$.

4. Discussion

All surfaces in the oral cavity, including implanted biomaterials, are covered with the salivary pellicle—a layer of a viscoelastic gel composed of saliva components, such as mucins MUC5B and MUC7, proline-rich proteins (gPRP, aPRP, bPRP), amylase, cystatin, statherin, histatin, and immunoglobulins (IgA, IgM). The thickness of the acquired pellicle ranges from 300 nm to even over 1000 nm [23,53]. This layer acts as an interface between the microbiological environment and tissues or biomaterials surfaces. Our previous studies [20] have shown that the incubation of polymers in artificial saliva based on porcine gastric mucins results in the adsorption of a layer with similar physicochemical properties (height, elastic modulus, pH response) to the natural acquired pellicle. This better reflects the physiological conditions in the oral cavity and allows a more accurate assessment of the processes on the biomaterial surface, especially the biofilm development and material degradation.

The processes of microbial adhesion to the acquired pellicle, the co-adhesion of microbes to microbes on the surface, and co-aggregation in the suspension are responsible for the biofilm formation in the oral cavity [54]. Intercellular interactions are specific in the oral environment and most species of microorganisms closely coexist with at least one other species. This study assessed the development of *C. krusei* and *S. mutans* biofilms as a model of bacterial and fungal strains that co-exist in the oral cavity. Both microorganisms cause oral diseases, and the development of their biofilms is a common challenge for dentists. *S. mutans* is the predominant caries pathogenic species and its main virulence characteristics are organic acid production and biofilm formation [55]. *C. krusei* belongs to the *Candida*, and together with *C. albicans*, *C. glabrata* and *C. tropicalis* are responsible for the development of oral candidiasis. It is commonly found in saliva and areas such as periodontal pockets, root canals, mucosal surfaces, and enamel as well as orthodontic appliances and dentures. *Candida* spp. actively participate in cariogenic biofilms through synergistic interaction with *S. mutans*. Their ability to secrete a large amount of extracellular matrix, combined with a large surface area of hyphal networks, supports the growth of pathogenic mixed communities. In addition, *C. krusei* has been implicated in periodontal disease, endodontic infections, and denture stomatitis [56].

In the initial stages, the mixed biofilm of both studied microorganisms developed better than single species biofilms, which indicates the possible cooperation of *C. krusei* and *S. mutans*. After 56 days of incubation, *S. mutans* biofilm developed the most. Streptococci, in particular *S. mutans*, belong to the group of early colonizers in the oral cavity and attach to the surface by recognizing molecule receptors, such as statherin, proline-rich proteins, salivary α-amylase, agglutinin, and mucins. Bamford et al. [57] showed that streptococci can increase the production of hyphae in *Candida*, which increases their adhesion to the surface and increases the survival rate of the biofilm. In later stages, the better development of *S. mutans* may be associated with the aciduricity of mature *S. mutans* biofilms, which, by secreting large amounts of organic acids, lower the pH and, due to evolutionary pressure, outcompete other microorganisms in the oral cavity [58].

The presence of mucins change the conditions for the development of such a biofilm—they reduce the virulence of pathogenic *Candida*, responsible for many types of infections [59]. The mechanism of mucin interaction takes place at the expression level of genes responsible for the development of morphological features, enabling attachment to the surface of host cells and the surface of biomaterials. Mucins also inhibit the growth of hyphae that penetrate the colonized surfaces and promote the development of biofilm [60]. The relationship of the saliva, and mucin in particular, with the microbial environment seems to be very close. The use of a saliva substitute in the biomimetics of the oral cavity, in particular the biofilm development, is a new approach to biomaterial degradation studies.

Biofilm significantly changes the surface properties of PLA and PCL, which has far-reaching consequences. The observed reduction of hydrophobicity as a result of degradation may have an impact on the further colonization of microorganisms. The general biofilm formation trends differ depending on the site (the application of the biomaterial) in the oral cavity [23]. In the subgingival areas, less biofilm is formed on hydrophobic than hydrophilic surfaces. However, such a tendency is not observed in hydrophobic supragingival areas, where the biofilm can be detached more easily by shear forces. It should be borne in mind that the presence of saliva molecules may change the surface properties of these biomaterials. Saliva molecules adsorb better to hydrophobic surfaces [61]; therefore, due to the degradation of polymeric biomaterials, the layers of mucins and other molecules may be less stable and more susceptible to shear forces. On the other hand, glycosylated mucin chain fragments may change the surface character to a more hydrophilic. However, these processes have not been fully understood so far.

In this study, the increase in the roughness of PLA samples incubated in the presence of *C. krusei* and a decrease in the presence of *S. mutans* bacteria was observed. The presence of microorganisms on the surface of PCL reduced the surface roughness. The surface topography of biomaterials is one of the key factors in microbial adhesion and biofilm development. The increase in surface roughness associated with the larger contact surface increases microbial adhesion, especially in the initial stage of biofilm development, and reduces its susceptibility to oral shear forces [62,63]. However, it should be kept in mind that the surface topography may change due to the adsorption of saliva molecules. In addition, the saliva molecules and exopolysaccharides (formed in situ by the enzymatic reactions induced by microorganisms) can locally create zones of increased adhesion and increase the virulence of bacteria, such as *S. mutans* [64].

We have also shown that the presence of microorganisms, such as *C. krusei* or *S. mutans*, reduces the elastic modulus of the surface layer both PLA and PCL. Polylactide is more susceptible to microbial degradation, in particular against *S. mutans*. The deterioration of the substrate's mechanical properties at the nanoscale has mechanobiological consequences for further biofilm development. Studies have shown that the adhesion of microorganisms, such as *E. coli* or *Lactococcus lactis*, decreases with increasing substrate stiffness [65]. In addition, a decrease in substrate stiffness may increase the size of the microorganisms and their antibiotic susceptibility [66]. Thus, the degradation of PLA and PCL in the oral cavity may be associated with the susceptibility to biofilm development. Hypothetically, biofilm, especially fungal biofilm, may cause it to grow into the partially degraded polymer surface and cause chronic oral infections.

During 56 days of incubation, the mass loss of PLA and PCL did not exceed 1%. A greater mass loss and greater susceptibility to microbial degradation were observed for PLA. The cause of PLA erosion in the presence of *S. mutans* may be esterases and acid metabolic products, which may contribute to the catalysis of the PLA hydrolysis reaction. The erosion of PLA in the presence of esterases and other enzymes, such as lipase or proteinase K, has been investigated previously [67]. For PCL, however, no effect of these microorganisms/enzymes on weight loss was observed. Typically, the process of microbial degradation of PLA or PCL follows a similar route for most enzymes. In the first phase of microbial degradation of aliphatic polyesters, microorganisms secrete depolymerases, the production of which requires stimulation by enzymatic inducers, such as elastin, gelatine, certain proteins, or amino acids. Subsequently, the depolymerases interact with the ester bonds, resulting in the breakdown of the polymer chains into smaller segments of oligo- and monomers, which then penetrate through the cell membranes of the microorganisms into their interior, where they are broken down to carbon dioxide, water, or methane by intracellular enzymes [68].

It has been shown that the ratio of molecular weight during degradation to initial molecular weight decreases exponentially for both PLA and PCL. The shortening of polymer chains is the basic effect of degradation and affects all changes in the structure and properties of these materials. We have shown that the degradation rate of PCL is lower than that of PLA and biofilm contributes to an increase in degradation rate. These changes were much greater in PLA than in PCL. The presence of *C. krusei* biofilm nearly 3-fold increased the λ' value for PLA and more than 2-fold for PCL. A similar effect was observed for PCL in *S. mutans* and mixed biofilm. The presence of *S. mutans* on the surface of PLA, both in the form of a single-species biofilm and along with *C. krusei*, turned out to be a significant catalyst for the degradation processes of the surface layer, increasing the λ' parameter over 3.5 times. So far, the processes of microbial degradation of PLA or PCL in the presence of *Candida* or *Streptococcus* has not been observed.

DSC studies show that the degradation of PLA and PCL leads to changes in the crystal structure, and the presence of biofilm on the surface accelerates these processes. This is influenced by two effects, the first one concerns the faster hydrolysis processes in amorphous rather than crystalline regions, and the second is the reorganization of the structure due to hydrolysis. In the presented studies, due to the low mass loss, the second mechanism has greater importance. Hydrolysis shortening the polymer chains facilitates the recrystallization of amorphous regions, resulting in the formation of regions with an ordered structure. When the molecular weight of the polymer drops below the solubility limit and the polymer mass begins to drop rapidly, the spaces between the chains increases, the ability to recrystallize decreases, and the first mechanism becomes more important. Degradation kinetics significantly depends on the relative amount of the amorphous phase. In subsequent degradation steps, the degree of crystallinity may decrease as the recrystallization process ceases.

Over the observed period of 56 days, for both PLA and PCL, no decrease in hardness or tensile strength was noted. This means that both the water absorption and the structural changes due to the decrease in molecular weight and the erosion of the polymer did not affect these basic properties during this period.

5. Conclusions

This work concerned the interaction processes at the polymer (PLA, PCL)–biofilm interface under simulated oral conditions. The key observation of the conducted research is the phenomenon of faster decomposition of the surface layer of polymers in the presence of microorganisms, which may lead to further unfavorable consequences, especially the increased colonization of microorganisms and the development of biofilm. We have summarized the processes of biofilm interaction with the surface of biodegradable polymers in Figure 7.

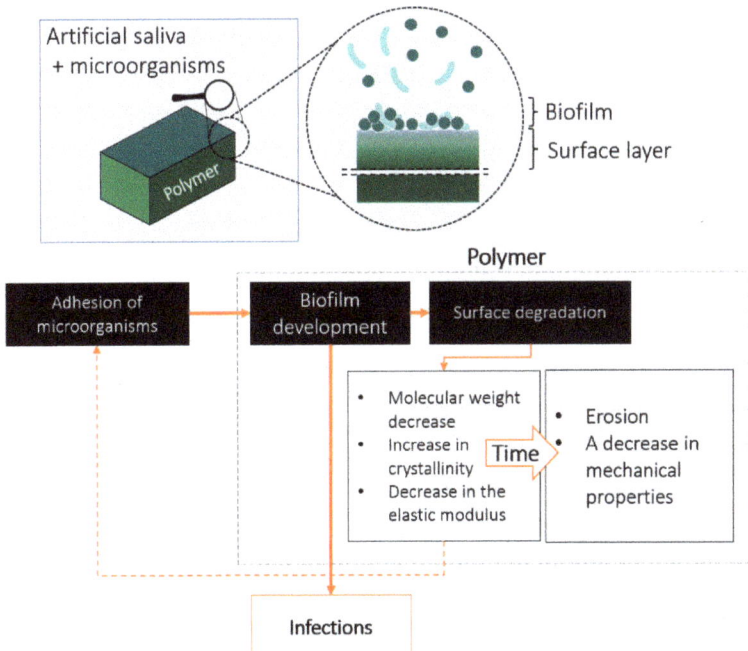

Figure 7. Diagram of the processes taking place at the polymer–biofilm interface in the oral cavity environment on the example of the conducted studies.

Author Contributions: Conceptualization, D.Ł. and J.M.; methodology, D.Ł. and J.M.; investigation, D.Ł., P.D. and S.C.; resources, D.Ł., J.M. and R.B.; data curation, D.Ł. and P.D.; writing—original draft preparation, D.Ł.; writing—review and editing, R.B. and J.M.; visualization, D.Ł.; supervision, J.M. and R.B.; project administration, D.Ł.; funding acquisition, J.M. and R.B. All authors have read and agreed to the published version of the manuscript.

Funding: This scientific work was realized within the context of work No. WZ/WM-IIB/2/2020 and financed by research funds of the Ministry of Education and Science, Poland.

Institutional Review Board Statement: Not applicable.

Informed Consent Statement: Not applicable.

Data Availability Statement: The data presented in this study are available on request from the corresponding author.

Conflicts of Interest: The authors declare no conflict of interest.

References

1. Narayanan, G.; Vernekar, V.N.; Kuyinu, E.; Laurencin, C.T. Poly (lactic acid)-based biomaterials for orthopaedic regenerative engineering. *Adv. Drug Deliv. Rev.* **2016**, *107*, 247–276. [CrossRef] [PubMed]
2. Woodruff, M.A.; Hutmacher, D.W. The return of a forgotten polymer—Polycaprolactone in the 21st century. *Prog. Polym. Sci.* **2010**, *35*, 1217–1256. [CrossRef]
3. Santoro, M.; Shah, S.R.; Walker, J.L.; Mikos, A.G. Poly(lactic acid) nanofibrous scaffolds for tissue engineering. *Adv. Drug Deliv. Rev.* **2016**, *107*, 206–212. [CrossRef]
4. James, R.; Manoukian, O.S.; Kumbar, S.G. Poly(lactic acid) for delivery of bioactive macromolecules. *Adv. Drug Deliv. Rev.* **2016**, *107*, 277–288. [CrossRef] [PubMed]
5. Goyanes, A.; Det-Amornrat, U.; Wang, J.; Basit, A.W.; Gaisford, S. 3D scanning and 3D printing as innovative technologies for fabricating personalized topical drug delivery systems. *J. Control. Release* **2016**, *234*, 41–48. [CrossRef] [PubMed]
6. Norman, J.; Madurawe, R.D.; Moore, C.M.V.; Khan, M.A.; Khairuzzaman, A. A new chapter in pharmaceutical manufacturing: 3D-printed drug products. *Adv. Drug Deliv. Rev.* **2017**, *108*, 39–50. [CrossRef]
7. Abid, Z.; Mosgaard, M.D.; Manfroni, G.; Petersen, R.S.; Nielsen, L.H.; Müllertz, A.; Boisen, A.; Keller, S.S. Investigation of Mucoadhesion and Degradation of PCL and PLGA Microcontainers for Oral Drug Delivery. *Polymers* **2019**, *11*, 1828. [CrossRef]
8. Kang, Y.; Wang, C.; Qiao, Y.; Gu, J.; Zhang, H.; Peijs, T.; Kong, J.; Zhang, G.; Shi, X. Tissue-Engineered Trachea Consisting of Electrospun Patterned sc-PLA/GO-g-IL Fibrous Membranes with Antibacterial Property and 3D-Printed Skeletons with Elasticity. *Biomacromolecules* **2019**, *20*, 1765–1776. [CrossRef] [PubMed]
9. Gao, M.; Zhang, H.; Dong, W.; Bai, J.; Gao, B.; Xia, D.; Feng, B.; Chen, M.; He, X.; Yin, M.; et al. Tissue-engineered trachea from a 3D-printed scaffold enhances whole-segment tracheal repair. *Sci. Rep.* **2017**, *7*, 1–12. [CrossRef] [PubMed]
10. Ensign, L.M.; Schneider, C.; Suk, J.S.; Cone, R.; Hanes, J. Mucus Penetrating Nanoparticles: Biophysical Tool and Method of Drug and Gene Delivery. *Adv. Mater.* **2012**, *24*, 3887–3894. [CrossRef]
11. Pitt, G.G.; Gratzl, M.M.; Kimmel, G.L.; Surles, J.; Sohindler, A. Aliphatic polyesters II. The degradation of poly (DL-lactide), poly (ε-caprolactone), and their copolymers in vivo. *Biomaterials* **1981**, *2*, 215–220. [CrossRef]
12. Laycock, B.; Nikolić, M.; Colwell, J.M.; Gauthier, E.; Halley, P.; Bottle, S.; George, G. Lifetime prediction of biodegradable polymers. *Prog. Polym. Sci.* **2017**, *71*, 144–189. [CrossRef]
13. Middleton, J.C.; Tipton, A.J. Synthetic biodegradable polymers as orthopedic devices. *Biomaterials* **2000**, *21*, 2335–2346. [CrossRef]
14. Tsuji, H. Poly(lactide) Stereocomplexes: Formation, Structure, Properties, Degradation, and Applications. *Macromol. Biosci.* **2005**, *5*, 569–597. [CrossRef]
15. Grizzi, I.; Garreau, H.; Li, S.; Vert, M. Hydrolytic degradation of devices based on poly(dl-lactic acid) size-dependence. *Biomaterials* **1995**, *16*, 305–311. [CrossRef]
16. Tsuji, H.; Ikada, Y. Properties and morphology of poly(l-lactide) 4. Effects of structural parameters on long-term hydrolysis of poly(l-lactide) in phosphate-buffered solution. *Polym. Degrad. Stab.* **2000**, *67*, 179–189. [CrossRef]
17. Elsawy, M.A.; Kim, K.-H.; Park, J.-W.; Deep, A. Hydrolytic degradation of polylactic acid (PLA) and its composites. *Renew. Sustain. Energy Rev.* **2017**, *79*, 1346–1352. [CrossRef]
18. Bartnikowski, M.; Dargaville, T.R.; Ivanovski, S.; Hutmacher, D.W. Degradation mechanisms of polycaprolactone in the context of chemistry, geometry and environment. *Prog. Polym. Sci.* **2019**, *96*, 1–20. [CrossRef]
19. Oksiuta, Z.; Jalbrzykowski, M.; Mystkowska, J.; Romanczuk, E.; Osiecki, T. Mechanical and Thermal Properties of Polylactide (PLA) Composites Modified with Mg, Fe, and Polyethylene (PE) Additives. *Polymers* **2020**, *12*, 2939. [CrossRef] [PubMed]
20. Łysik, D.; Mystkowska, J.; Markiewicz, G.; Deptuła, P.; Bucki, R. The Influence of Mucin-Based Artificial Saliva on Properties of Polycaprolactone and Polylactide. *Polymers* **2019**, *11*, 1880. [CrossRef]
21. Marsh, P.D.; Do, T.; Beighton, D.; Devine, D.A. Influence of saliva on the oral microbiota. *Periodontology 2000* **2015**, *70*, 80–92. [CrossRef] [PubMed]

22. Werlang, C.; Cárcarmo-Oyarce, G.; Ribbeck, K. Engineering mucus to study and influence the microbiome. *Nat. Rev. Mater.* **2019**, *4*, 134–145. [CrossRef]
23. Gibbins, H.L.; Yakubov, G.E.; Proctor, G.B.; Wilson, S.; Carpenter, G.H. What interactions drive the salivary mucosal pellicle formation? *Colloids Surf. B Biointerfaces* **2014**, *120*, 184–192. [CrossRef]
24. Mystkowska, J.; Niemirowicz-Laskowska, K.; Łysik, D.; Tokajuk, G.; Dąbrowski, J.R.; Bucki, R. The Role of Oral Cavity Biofilm on Metallic Biomaterial Surface Destruction–Corrosion and Friction Aspects. *Int. J. Mol. Sci.* **2018**, *19*, 743. [CrossRef] [PubMed]
25. Bettencourt, A.F.; Neves, C.B.; de Almeida, M.S.; Pinheiro, L.M.; e Oliveira, S.A.; Lopes, L.P.; Castro, M.F. Biodegradation of acrylic based resins: A review. *Dent. Mater.* **2010**, *26*, e171–e180. [CrossRef] [PubMed]
26. Pranamuda, H.; Tsuchii, A.; Tokiwa, Y. Poly (L-Lactide)-Degrading Enzyme Produced by Amycolatopsis Sp. *Macromol. Biosci.* **2001**, *1*, 25–29. [CrossRef]
27. Jarerat, A.; Tokiwa, Y. Poly(L-lactide) degradation by Saccharothrix waywayandensis. *Biotechnol. Lett.* **2003**, *25*, 401–404. [CrossRef]
28. Jarerat, A.; Tokiwa, Y.; Tanaka, H. Poly(l-lactide) degradation by *Kibdelosporangium aridum*. *Biotechnol. Lett.* **2003**, *25*, 2035–2038. [CrossRef]
29. Sukkhum, S.; Tokuyama, S.; Kitpreechavanich, V. Development of fermentation process for PLA-degrading enzyme production by a new thermophilic Actinomadura sp. T16-1. *Biotechnol. Bioprocess Eng.* **2009**, *14*, 302–306. [CrossRef]
30. Hanphakphoom, S.; Maneewong, N.; Sukkhum, S.; Tokuyama, S.; Kitpreechavanich, V. Characterization of poly(L-lactide)-degrading enzyme produced by thermophilic filamentous bacteria Laceyella sacchari LP175. *J. Gen. Appl. Microbiol.* **2014**, *60*, 13–22. [CrossRef]
31. Konkit, M.; Jarerat, A.; Khanongnuch, C.; Lumyong, S.; Pathom-Aree, W. Poly(Lactide) Degradation by Pseudonocardia Alni AS4.1531t. *Chiang Mai J. Sci.* **2012**, *39*, 128–132.
32. Tomita, K.; Kuroki, Y.; Nagai, K. Isolation of thermophiles degrading poly(l-lactic acid). *J. Biosci. Bioeng.* **1999**, *87*, 752–755. [CrossRef]
33. Tomita, K.; Nakajima, T.; Kikuchi, Y.; Miwa, N. Degradation of poly(l-lactic acid) by a newly isolated thermophile. *Polym. Degrad. Stab.* **2004**, *84*, 433–438. [CrossRef]
34. Akutsu-Shigeno, Y.; Teeraphatpornchai, T.; Teamtisong, K.; Nomura, N.; Uchiyama, H.; Nakahara, T.; Nakajima-Kambe, T. Cloning and Sequencing of a Poly(dl -Lactic Acid) Depolymerase Gene from *Paenibacillus amylolyticus* Strain TB-13 and Its Functional Expression in *Escherichia coli*. *Appl. Environ. Microbiol.* **2003**, *69*, 2498–2504. [CrossRef]
35. Jeon, H.J.; Kim, M.N. Biodegradation of poly(l-lactide) (PLA) exposed to UV irradiation by a mesophilic bacterium. *Int. Biodeterior. Biodegradation* **2013**, *85*, 289–293. [CrossRef]
36. Liang, T.-W.; Jen, S.-N.; Nguyen, A.D.; Wang, S.-L. Application of Chitinous Materials in Production and Purification of a Poly(l-lactic acid) Depolymerase from Pseudomonas tamsuii TKU015. *Polymers* **2016**, *8*, 98. [CrossRef]
37. Maeda, H.; Yamagata, Y.; Abe, K.; Hasegawa, F.; Machida, M.; Ishioka, R.; Gomi, K.; Nakajima, T. Purification and characterization of a biodegradable plastic-degrading enzyme from Aspergillus oryzae. *Appl. Microbiol. Biotechnol.* **2005**, *67*, 778–788. [CrossRef]
38. Masaki, K.; Kamini, N.R.; Ikeda, H.; Iefuji, H. Cutinase-Like Enzyme from the Yeast *Cryptococcus* sp. Strain S-2 Hydrolyzes Polylactic Acid and Other Biodegradable Plastics. *Appl. Environ. Microbiol.* **2005**, *71*, 7548–7550. [CrossRef]
39. Jarerat, A.; Tokiwa, Y. Degradation of Poly(L-Lactide) by a Fungus. *Macromol. Biosci.* **2001**, *1*, 136–140. [CrossRef]
40. Lipsa, R.; Tudorachi, N.; Darie-Nita, R.N.; Oprică, L.; Vasile, C.; Chiriac, A. Biodegradation of poly(lactic acid) and some of its based systems with Trichoderma viride. *Int. J. Biol. Macromol.* **2016**, *88*, 515–526. [CrossRef]
41. Gan, Z.; Liang, Q.; Zhang, J.; Jing, X. Enzymatic degradation of poly(ε-caprolactone) film in phosphate buffer solution containing lipases. *Polym. Degrad. Stab.* **1997**, *56*, 209–213. [CrossRef]
42. Chen, D.R.; Bei, J.Z.; Wang, S.G. Polycaprolactone microparticles and their biodegradation. *Polym. Degrad. Stab.* **2000**, *67*, 455–459. [CrossRef]
43. Ashton, J.; Mertz, J.; Harper, J.; Slepian, M.; Mills, J.; McGrath, D.; Geest, J.V. Polymeric endoaortic paving: Mechanical, thermoforming, and degradation properties of polycaprolactone/polyurethane blends for cardiovascular applications. *Acta Biomater.* **2011**, *7*, 287–294. [CrossRef]
44. Castilla-Cortázar, I.; Más-Estellés, J.; Meseguer-Dueñas, J.M.; Ivirico, J.E.; Marí, B.; Vidaurre, A. Hydrolytic and enzymatic degradation of a poly(ε-caprolactone) network. *Polym. Degrad. Stab.* **2012**, *97*, 1241–1248. [CrossRef]
45. Khan, I.; Dutta, J.R.; Ganesan, R. Lactobacillus sps. lipase mediated poly (ε-caprolactone) degradation. *Int. J. Biol. Macromol.* **2017**, *95*, 126–131. [CrossRef] [PubMed]
46. Sivalingam, G.; Chattopadhyay, S.; Madras, G. Enzymatic degradation of poly (ε-caprolactone), poly (vinyl acetate) and their blends by lipases. *Chem. Eng. Sci.* **2003**, *58*, 2911–2919. [CrossRef]
47. Sivalingam, G.; Chattopadhyay, S.; Madras, G. Solvent effects on the lipase catalyzed biodegradation of poly (ε-caprolactone) in solution. *Polym. Degrad. Stab.* **2003**, *79*, 413–418. [CrossRef]
48. Ebata, H.; Toshima, K.; Matsumura, S. Lipase-Catalyzed Transformation of Poly(ε-caprolactone) into Cyclic Dicaprolactone. *Biomacromolecules* **2000**, *1*, 511–514. [CrossRef]
49. Yang, L.; Li, J.; Jin, Y.; Li, M.; Gu, Z. In vitro enzymatic degradation of the cross-linked poly(ε-caprolactone) implants. *Polym. Degrad. Stab.* **2015**, *112*, 10–19. [CrossRef]

50. Sivalingam, G.; Vijayalakshmi, S.P.; Madras, G. Enzymatic and Thermal Degradation of Poly(ε-caprolactone), Poly(d,l-lactide), and Their Blends. *Ind. Eng. Chem. Res.* **2004**, *43*, 7702–7709. [CrossRef]
51. Rak, J.; Ford, J.L.; Rostron, C.; Walters, V. The preparation and characterization of poly(D,L-lactic acid) for use as a biodegradable drug carrier. *Pharm. Acta Helv.* **1985**, *60*, 162–169. [PubMed]
52. Kouparitsas, I.K.; Mele, E.; Ronca, S. Synthesis and Electrospinning of Polycaprolactone from an Aluminium-Based Catalyst: Influence of the Ancillary Ligand and Initiators on Catalytic Efficiency and Fibre Structure. *Polymers* **2019**, *11*, 677. [CrossRef] [PubMed]
53. Amaechi, B.; Higham, S.; Edgar, W.; Milosevic, A. Thickness of acquired salivary pellicle as a determinant of the sites of dental erosion. *J. Dent. Res.* **1999**, *78*, 1821–1828. [CrossRef]
54. Bowen, W.H.; Burne, R.A.; Wu, H.; Koo, H. Oral Biofilms: Pathogens, Matrix, and Polymicrobial Interactions in Microenvironments. *Trends Microbiol.* **2018**, *26*, 229–242. [CrossRef] [PubMed]
55. Scharnow, A.M.; Solinski, A.E.; Wuest, W.M. Targeting *S. mutans* biofilms: A perspective on preventing dental caries. *MedChemComm* **2019**, *10*, 1057–1067. [CrossRef]
56. O'Donnell, L.E.; Millhouse, E.; Sherry, L.; Kean, R.; Malcolm, J.; Nile, C.J.; Ramage, G. Polymicrobial *Candida* biofilms: Friends and foe in the oral cavity. *FEMS Yeast Res.* **2015**, *15*, fov077. [CrossRef]
57. Bamford, C.V.; D'Mello, A.; Nobbs, A.H.; Dutton, L.C.; Vickerman, M.M.; Jenkinson, H.F. *Streptococcus gordonii* Modulates *Candida albicans* Biofilm Formation through Intergeneric Communication. *Infect. Immun.* **2009**, *77*, 3696–3704. [CrossRef]
58. Matsui, R.; Cvitkovitch, D. Acid tolerance mechanisms utilized by *Streptococcus mutans*. *Futur. Microbiol.* **2010**, *5*, 403–417. [CrossRef]
59. Kavanaugh, N.L.; Zhang, A.Q.; Nobile, C.; Johnson, A.D.; Ribbeck, K. Mucins Suppress Virulence Traits of Candida albicans. *mBio* **2014**, *5*, e01911. [CrossRef]
60. Arevalo, A.V.; Nobile, C.J. Interactions of microorganisms with host mucins: A focus on *Candida albicans*. *FEMS Microbiol. Rev.* **2020**, *44*, 645–654. [CrossRef]
61. Bansil, R.; Turner, B.S. The biology of mucus: Composition, synthesis and organization. *Adv. Drug Deliv. Rev.* **2018**, *124*, 3–15. [CrossRef] [PubMed]
62. Anselme, K.; Davidson, P.; Popa, A.M.; Giazzon, M.; Liley, M.; Ploux, L. The interaction of cells and bacteria with surfaces structured at the nanometre scale. *Acta Biomater.* **2010**, *6*, 3824–3846. [CrossRef] [PubMed]
63. Hao, Y.; Huang, X.; Zhou, X.; Li, M.; Ren, B.; Peng, X.; Cheng, L. Influence of Dental Prosthesis and Restorative Materials Interface on Oral Biofilms. *Int. J. Mol. Sci.* **2018**, *19*, 3157. [CrossRef] [PubMed]
64. Bowen, W.H.; Koo, H. Biology of Streptococcus mutans-Derived Glucosyltransferases: Role in Extracellular Matrix Formation of Cariogenic Biofilms. *Caries Res.* **2011**, *45*, 69–86. [CrossRef]
65. Saha, N.; Monge, C.; Dulong, V.; Picart, C.; Glinel, K. Influence of Polyelectrolyte Film Stiffness on Bacterial Growth. *Biomacromolecules* **2013**, *14*, 520–528. [CrossRef] [PubMed]
66. Song, F.; Ren, D. Stiffness of Cross-Linked Poly(Dimethylsiloxane) Affects Bacterial Adhesion and Antibiotic Susceptibility of Attached Cells. *Langmuir* **2014**, *30*, 10354–10362. [CrossRef]
67. Stepczyńska, M.; Rytlewski, P. Enzymatic degradation of flax-fibers reinforced polylactide. *Int. Biodeterior. Biodegradation* **2018**, *126*, 160–166. [CrossRef]
68. Qi, X.; Ren, Y.; Wang, X. New advances in the biodegradation of Poly(lactic) acid. *Int. Biodeterior. Biodegradation* **2017**, *117*, 215–223. [CrossRef]

Influence of ZrO$_2$ Addition on Structural and Biological Activity of Phosphate Glasses for Bone Regeneration

M. Mohan Babu [1], P. Syam Prasad [1,*], P. Venkateswara Rao [2], S. Hima Bindu [1], A. Prasad [1], N. Veeraiah [3] and Mutlu Özcan [4]

1. Department of Physics, National Institute of Technology Warangal, Warangal 506004, India; mmbabu771@gmail.com (M.M.B.); h.bindu05@gmail.com (S.H.B.); prasadbabunitw@gmail.com (A.P.)
2. Department of Physics, The University of the West Indies, Mona Campus, Kignston 7, Jamaica; pvrao54@gmail.com
3. Department of Physics, Acharya Nagarjuna University, Nagarjuna Nagar, Guntur, AP 522510, India; profnvr@gmail.com
4. Center for Dental and Oral Medicine, Division of Dental Biomaterials, Clinic for Reconstructive Dentistry, University of Zurich, 8032 Zurich, Switzerland; mutluozcan@hotmail.com
* Correspondence: syamprasad@nitw.ac.in

Received: 4 August 2020; Accepted: 8 September 2020; Published: 12 September 2020

Abstract: Zirconium doped calcium phosphate-based bioglasses are the most prominent bioactive materials for bone and dental repair and regeneration implants. In the present study, a 8ZnO–22Na$_2$O–(24 − x)CaO–46P$_2$O$_5$–xZrO$_2$ ($0.1 \leq x \leq 0.7$, all are in mol%) bioglass system was synthesized by the conventional melt-quenching process at 1100 °C. The glass-forming ability and thermal stability of the glasses were determined by measuring the glass transition temperature (T_g), crystallization temperature (T_c), and melting temperature (T_m), using differential thermal analysis (DTA). The biological activity of the prepared samples was identified by analyzing X-ray diffraction (XRD), Fourier transform infrared spectroscopy (FTIR) and scanning electron microscopy-energy dispersive spectra (SEM-EDS), before and after immersion in simulated body fluid (SBF) for various intervals of 0, 1 and 5 days, along with the magnitude of pH and the degradation of glasses also evaluated. The obtained results revealed that the glass-forming ability and thermal stability of glasses increased with the increase in zirconia mol%. The XRD, FTIR, and SEM-EDS data confirmed a thin hydroxyapatite (HAp) layer over the sample surface after incubation in SBF for 1 and 5 days. Furthermore, the development of layer found to be increased with the increase of incubation time. The degradation of the glasses in SBF increased with incubation time and decreased gradually with the increase content of ZrO$_2$ mol% in the host glass matrix. A sudden rise in initial pH values of residual SBF for 1 day owing to ion leaching and increase of Ca^{2+} and PO$_4^{3-}$ ions and then decreased. These findings confirmed the suitability of choosing material for bone-related applications.

Keywords: P$_2$O$_5$-bioglass; zirconia; melt-quenching; SBF; hydroxyapatite; in vitro bioactivity

1. Introduction

Bioactive glasses are the widely used surface reactive inorganic biomaterials in engineering, essentially for the repair and regeneration of damaged soft and hard bone tissues [1–4]. 45S5 bioglass is the well-known and widely used bioactive glass developed by Professor Larry Hench and his co-workers in 1969 [5]; comprised of inorganic oxides (viz., SiO$_2$, Na$_2$O, CaO, and P$_2$O$_5$) in a specific molar ratios and has exhibited thriving biological properties such as in vitro bioactivity, osteostimulative and osteoconductive properties. These special qualities made the bioglass a biocompatible and bioresorbable material in comparison to the natural bone. Various types of melt-derived bioactive glasses (SiO_2, B_2O_3, P_2O_5 etc.) have been developed and tested because of their capability to interact with living bone tissues [3,6]. Out of all these glasses, phosphate-based glasses have many advantages, due to solubility, high biocompatibility and low melting temperature etc., which made them suitable to be used as biomaterials [7]. The strong glass former P_2O_5 contributes in the glass network, as PO_4 structural units [6,8,9] are due to covalent bonding by the bridging oxygens. This bioglass material upon immersion in the simulated body fluid (SBF) solution shows a state where the ionic and pH conditions perfectly simulate with human blood plasma [7]. The ability to form rich calcium phosphate hydroxyapatite (HAp) layer on the surface of the bioglass samples when it comes into contact with the SBF confirms the in vitro bioactivity. The main disadvantage of the phosphate-based glasses is their poor mechanical strength, which limits the applications in implant development related to hard tissue replacement. This can be resolved by incorporating suitable transition metal ions, such as TiO_2, MgO, ZnO, CuO, Fe_2O_3, etc., to phosphate glass network in appropriate amounts. The first generations of biomaterials are directly related to the bone and tissue engineering by the implantation of ZrO_2 and TiO_2-based materials [9–11]. The first research paper on zirconia was published by Helmer and Driskel in 1969 and mentioned that ZrO_2 can be used as a biomaterial [12]. ZrO_2 is one of the common trace elements present in the human body, and thus its inclusion in the bioglass could be exploited for stimulating bioactivity. Both the biological activity and the mechanical properties can improve with the incorporation of ZrO_2 to the phosphate glass network. The phosphate glass structure is mainly strengthened by forming of Zr–O–P covalent bonds due to the entering of ZrO_6 octahedra structural units of zirconia [13,14]. Most of the available research reports related to zirconia mixed silica based bioactive glasses revealed the gradual decrease in their biological activity with the increase of ZrO_2 concentration. However, considerably limited work is available on ZrO_2 contain phosphate based bioglasses and glass ceramics [15]. Zirconia mixed porous calcium titanium phosphate glass ceramic system was fabricated by V. K. Marghussian et al. and studied the effect of zirconia on chemical durability and mechanical strength, which are observed to be improved [15]. V. Rajendran et al., synthesized P_2O_5–Na_2O–CaO–ZrO_2 glasses with the addition of ZrO_2 up to 1.0 mol% by replacing the Na_2O and observed that considerable high bioactivity along with improved mechanical strength at 0.75 mol% of ZrO_2 out of other concentrations [16]. Caiyun Zheng et al. prepared $60CaO$–$30P_2O_5$–$3TiO_2$–$xZrO_2$–$(7-x)Na_2O$ (x = 0, 1, 3) glass ceramics and studied the influence of ZrO_2 on mechanical and bioactive properties, and found that the toughening of the system with 1 mol% of ZrO_2 added has no adverse effect on the bioactivity [17]. Contrary to earlier reports, we have considered that silica free zinc calcium phosphate glasses mixed with small quantities of ZrO_2 might improve the structural and bioactivity suitable for tissue engineering implant applications. Moreover, the result of ZrO_2 on structural and biological properties of the P_2O_5 glasses is not yet revealed completely. Therefore, ZrO_2 doped bioactive glass system was prepared and analyzed to probe some light on structural and biological properties, by performing some experiments, such as XRD, FTIR, and SEM-EDS, etc., pre- and post-soaked in SBF, and monitoring the degradation and pH variation of the bioactive materials suitable for bone regeneration applications.

In this present study, we have developed the novel ZnO–Na_2O–CaO–P_2O_5 bioglass system by doping less than 1 mol% of ZrO_2, and explored the properties (thermal and structural) that stimulate the deposition of HAp layer over the glass surface.

2. Materials and Methods

2.1. Preparation of Bioglass Materials

The bioglass composition is given in Table 1. The ZrO_2 mixed calcium phosphate glasses were synthesized by taking high purity (99.9%) P_2O_5, ZnO, CaO, Na_2O and ZrO_2 chemical compounds from Sigma-Aldrich (St. Louis, MO, USA) by the melt-quenching method. The details of the glass composition chosen for the present study and their corresponding codes are given in Table 1.

The appropriate proportions of the chemicals (20 g batches) were homogeneously mixed in an agate mortar and melted in platinum crucibles at the temperature 1000 °C for 2 h. During melting, in order to ensure homogeneity, the melt was stirred for every half hour. The obtained melt was casted into pre-heated graphite molds and then shifted the glasses in an annealing chamber maintained at 250 °C; the annealing was carried out for 2 h and subsequently cooled at the rate of 1 °C/min to the room temperature, to make the samples free from the internal cracks, residual stress etc. The obtained samples were ground and well-polished to the final dimension of 1.5 cm × 1.5 cm × 0.2 cm.

Table 1. The Nominal bioactive glass (mol%) composition.

Glass Code	ZnO	Na$_2$O	CaO	P$_2$O$_5$	ZrO$_2$
Z.1	8.0	22.0	23.9	46.0	0.1
Z.3	8.0	22.0	23.7	46.0	0.3
Z.5	8.0	22.0	23.5	46.0	0.5
Z.7	8.0	22.0	23.3	46.0	0.7

2.2. Thermal Analysis

Thermal characterization was carried out on the glass powder by heating from room temperature to 1100 °C under Argon atmosphere, with a heating rate of 10 °C/min, using a standard NETZ5CH-STA 2500 (NETZSCH, Selb, Germany) Regulus thermal analysis system. A 20 mg of glass powder in an alumina pan is used for heating, with identical alumina pan as reference material. The obtained DTA patterns are used to identify the T_g, T_c, T_m, and other thermal parameters, such as thermal stability (ΔT) and Hruby's criterion (K_H). The difference between T_c and T_g representing the thermal stability (ΔT) and the ration between $T_c - T_g$ and $T_m - T_c$ gives Hruby's criterion (K_H) of the glass system [18]:

$$\Delta T = Tc - Tg \tag{1}$$

$$K_H = \frac{Tc - Tg}{Tm - Tc} \tag{2}$$

2.3. Bioactivity Assessment

The SBF is arranged in a polyethylene vessel by mixing appropriate quantities of analar grade reagents of NaCl, KCl, NaHCO$_3$, MgCl$_2$, 6H$_2$O, CaCl$_2$ and KH$_2$PO$_4$ (99.95%, Sigma-Aldrich) to distilled water with continuous stirring. The solution is buffered to pH 7.4 by adding Tris-buffer and hydrochloric acid is kept at 278 K for 48 h to trace the presence of any precipitates, the process recommended by Kokubo and Takadama [7]. After confirming the lack of precipitate, the obtained solution was used for in vitro studies. It was ensured that the ratios of concentration (mM) of different ions in the prepared solution were like those of human blood plasma (Table 2). In vitro bioactivity studies were performed (so as to achieve HAp layer formation on the surface of the samples) by immersing each sample (0.10g of glass powder) separately in 50 mL of SBF at 37 °C [19]. The weight loss measurements and the pH of the solution were performed after different incubation periods (viz. 0, 1 and 5 days).

Table 2. The concentration of various ions in the simulated body fluid (SBF) solution.

Ion Type	Na$^+$	K$^+$	Mg^{2+}	Ca^{2+}	Cl$^-$	HCO$_3^-$	HPO$_4^{2-}$	SO$_4^-$
Concentration (mM)	142.0	5.0	1.5	2.5	148.8	4.2	1.0	0.5
Human blood plasma	142.0	5.0	1.5	2.5	103.0	27.0	1.0	0.5

2.4. Powder XRD

The structural phases of the glasses were analyzed by PANalytical X'pert Powder (Malvern Panalytical Ltd., Malvern, UK), using Cu-Kα as a radiation (λ = 1.540598 Å) source. The XRD patterns were recorded diffraction angle range 2θ = 10–80°. Using the International Center for Diffraction Data cards (ISO 9001:2015 certified by DEKRA) were preferred to identify the crystal phases corresponding to each diffraction peak observed in the XRD pattern.

2.5. Fourier Transform Infrared Spectroscopy Analysis

The FTIR (model: S100, PerkinElmer, Shelton, CT, USA) transmittance spectra of the samples were recorded (for understanding the internal structural variations of the glass network) in the wavenumber range of 4000–400 cm^{-1}. The mixture of 1 mg of glass powder and 300 mg of Potassium bromide powders was used to make pellets prepared under vacuum pressure.

2.6. SEM-EDS Micrographs

The surface morphology and microstructure of the bioglass samples were analyzed with the help of Scanning electron microscopy of VEGA 3 LMU, TESCAN (Brno, Czech Republic), pre- and post-immersion in SBF. The elemental analysis of the glass samples was estimated by Energy dispersive X-ray analyzer connected to SEM. In order to get clear images, the samples should be electroconductive; for this, the samples were coated with a thin gold layer.

2.7. Degradation Behavior

The dissolution behavior of the glasses immersed in SBF (pH 7.4 at 37.5 °C) was measured by a weight loss process. Initially, the powder samples were weighed pre immersion and then immersed for different days (0, 1 and 5) in SBF solution. Next, the glasses were removed from the solution and dried at 80 °C, and again measure the weight of each sample to determine the weight loss by the following:

$$\text{Weight loss} = \frac{W_o - W_t}{W_o} \times 100\% \tag{3}$$

where W_o is pre-immersion weight and W_t is post-immersion weight.

2.8. pH Evaluation

The pH values of the residual SBF were measured (so as to have the information on dissolution behavior of the bioactive glass) pre-and post-immersion of the samples, containing different contents of ZnO for different intervals of time by a pH meter (ORION pH 7000). The pH meter (Thermo Scientific, Beverly, MA, USA) pre-calibrated to 4.01, 7.00, and 9.20 was used for these measurements.

3. Results and Discussion

3.1. Thermal Properties

The traces of differential thermal analyses (DTA) are shown in Figures 1 and 2, and the corresponding temperature values T_g, T_c, and T_m of ZrO_2 doped glass samples are tabulated in Table 3. The values of glass transition temperature (T_g) from 262.21 (± 1.11) °C to 301.15 (± 1.25) °C, crystallization temperature (T_c) from 335.67 °C to 386.32 °C, and melting temperature (T_m) from 672.67 °C to 697.49 °C increased with ZrO_2 mol% (from 0.1 to 0.7). The increase in T_g values with an increase of zirconia is due to an increase in the average crosslink density through non-bridging oxygen ions (NBO) and the number of bonds per unit volume. In addition, an increase in T_g can also be due to the increasing aggregation effect of ZrO_2 on the glass network and slow mobility of large Zr^{4+} ions, which lead to more rigidity of the glass network. The exothermic T_c and endothermic T_m peaks are also increased gradually with the addition of ZrO_2 (of Zr^{4+} ions (0.72 A °)) [15]. Furthermore, it is found to raise the viscosity of glass with a gradual increase of zirconia content due to the decrement of NBO's and/or high ionic field strength [1,3,11].

In the current ZrO_2 doped glasses, the glass transition temperature (T_g) and stability (ΔT (°C)) values increased from 262.21 ± 1.11 to 301.15 ± 1.25 and from 73.46 ± 0.45 °C to 85.16 ± 0.35 °C respectively, whereas the Hruby criterion (K_H) increased from 0.22 to 0.27 with the content of ZrO_2, which undoubtedly designates the high stability and the good glass-forming tendency of as-prepared glasses. The obtained results confirm the structural modification and thermal stability of the zirconia incorporated bioactive glasses.

Table 3. Thermal properties of the ZrO_2 containing bioglasses.

Sample Code	T_g (°C)	T_c (°C)	T_m (°C)	ΔT (°C)	K_H
Z.1	262.21 (± 1.11)	335.67	672.67	73.46 (± 0.45)	0.22
Z.3	271.10 (± 1.30)	349.21	686.92	78.10 (± 0.38)	0.23
Z.5	280.40 (± 1.24)	362.87	691.84	82.47 (± 0.43)	0.25
Z.7	301.15 (± 1.25)	386.32	697.49	85.16 (± 0.35)	0.27

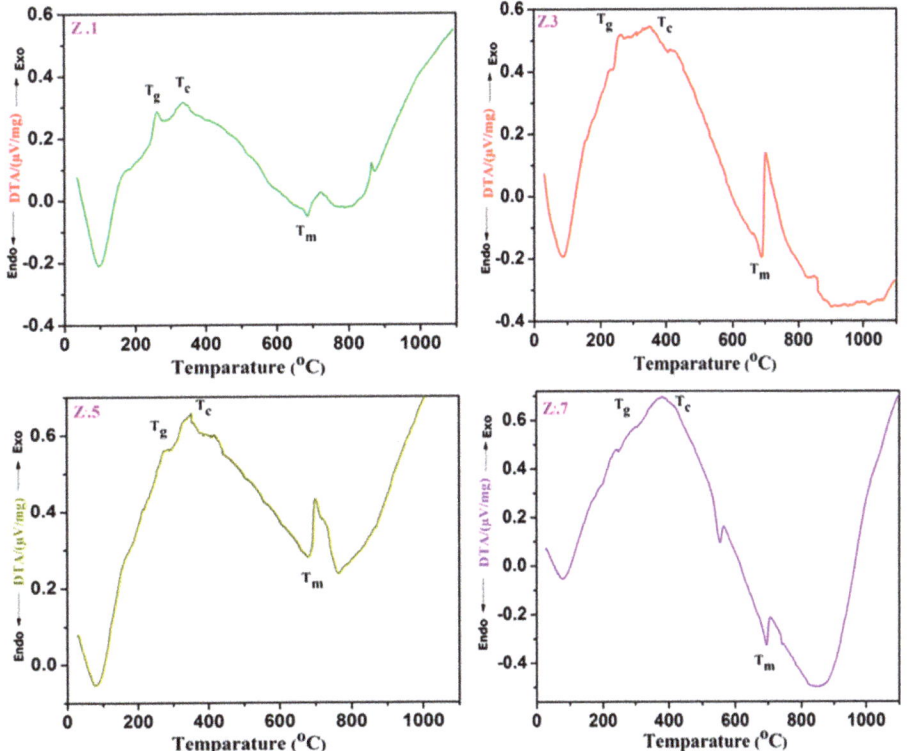

Figure 1. DTA analysis curves of ZrO_2 doped bioglass samples Z.1, Z.3, Z.5 and Z.7.

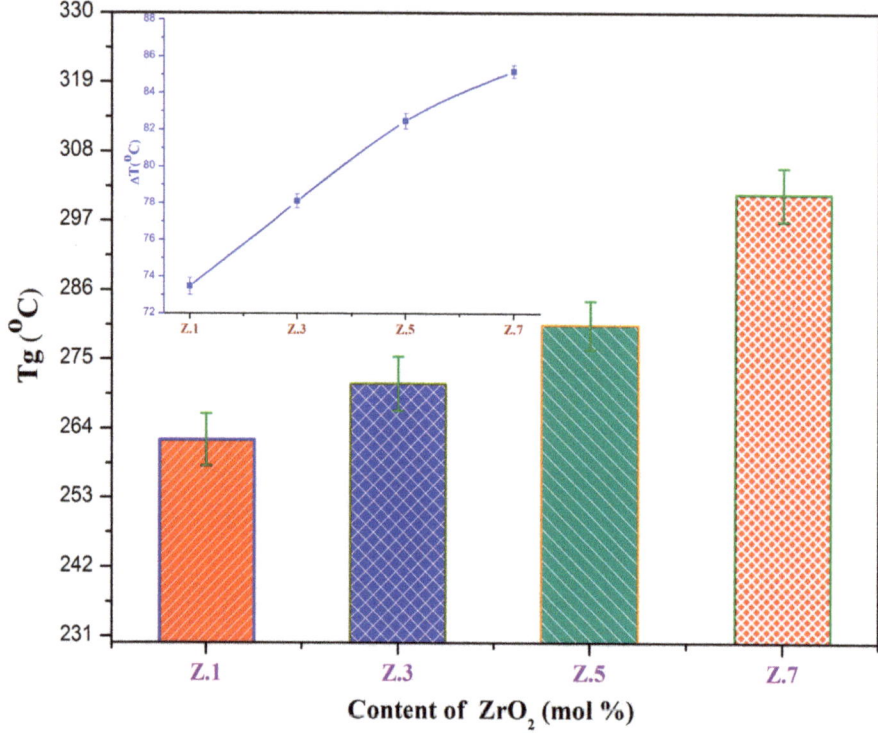

Figure 2. Variation of glass transition (T_g) as a function of ZrO_2 (mol%) content (inset shows the thermal stability vs ZrO_2 concentration).

3.2. XRD Analysis

Figure 3 illustrates the X-ray diffractograms of ZrO_2 mixed phosphate glasses immersed in SBF in 0, 1 and 5 days of intervals recorded at the diffraction angle (2θ) in between 10–80°. Before immersion, the samples do not show any sharp crystalline peaks, which indicate the non-crystalline nature of the samples [8] shown in Figure 3a. After immersion, the same samples exhibited prominent crystalline peaks; this indicates the formation of hydroxyapatite layer (HAp: $Ca_{10}(PO_4)_6(OH)_2$) on the surface of the glass samples, due to ion leaching from glass to SBF and vice versa. The intensity reflections at (1 0 0), (0 2 1), (2 0 0), (0 0 2), (2 1 1), (2 0 3), (5 0 0), and (2 1 5), in accordance with (h k l) values, are represented formation of the hydroxyapatite layer on the glass surface samples. These intensity reflections of HAp from XRD patterns were indexed using JCPDS card No: 72-1243. After soaking for 1 day (Figure 3b), the presence of intense diffraction peaks at 31.74° (2 1 1) and 45.32° (2 0 3) in the XRD diffractograms indicates the growth of crystalline calcium phosphate hydroxide [18,19]. With the increasing immersion time (1 day to 5 days), additional peaks are appearing laterally with the noticeable intense peaks existing during 1 day of incubation. This is predicted due to penetration of Ca^{2+} ions into PO_4^{3-} glass network leading to the formation of a crystalline HAp layer on glass surface [3,10]. Moreover, it is clearly noticed that the peak intensities are increased with the incubation time and decreased with the concentration of ZrO_2, due to the dissolution kinetics of the glass in SBF solution. This deposition of thin HAp layer over the glasses soaked in physiological fluid can be directly correlated to their capacity to generate effective chemical bonds with natural bone tissues [20,21]. The attained results from XRD on bioactivity of as prepared bioglasses are also confirmed further by FTIR and SEM studies.

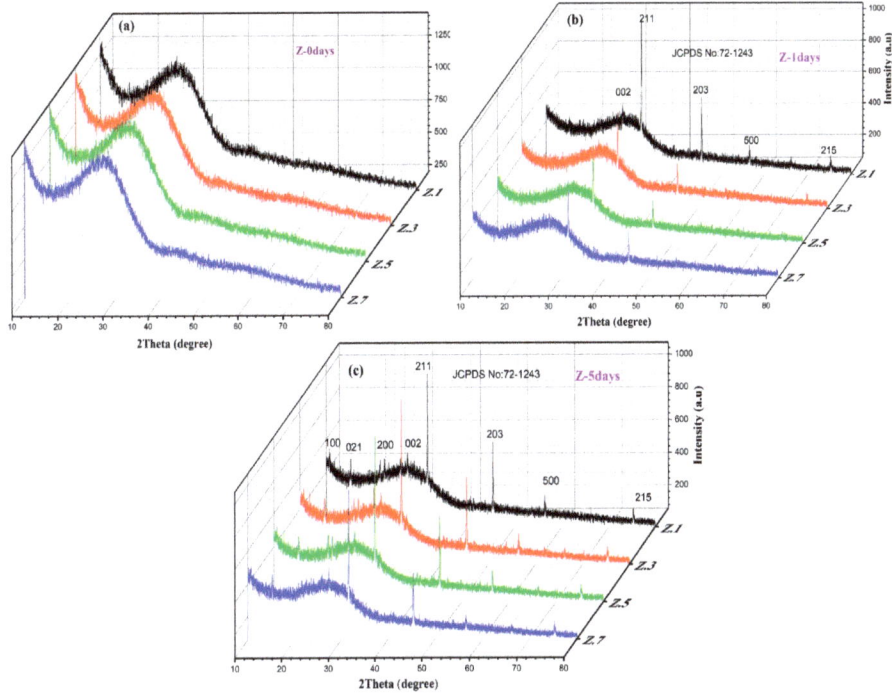

Figure 3. X-ray diffraction patterns of the bioglass samples after immersion in SBF (**a**) 0 day, (**b**) 1 day and (**c**) 5 days.

3.3. FTIR Spectroscopic Analysis

Figure 4 demonstrates the infrared transmission spectrum of zirconia doped phosphate samples pre and post immersion in SBF, and the assignment of various bands presented in Table 4. The FTIR spectra of before soaked samples (Figure 4a) are shown in the characteristic phosphate bands located at 506 cm^{-1}, 752 cm^{-1}, 978 cm^{-1}, 1642 cm^{-1}, 2382 cm^{-1} and 3464 cm^{-1}. The band at around 506 cm^{-1} is attributed to the bending vibrations (PO_4^{3-}) of O–P–O [15,16,22] The absorption bands at 752 cm^{-1} and 978 cm^{-1} are due to the (P–O–P) symmetric stretching vibrations of phosphate group [10,23]. The peak observed at 1642 cm^{-1} is assigned due to the stretching vibrations of P–O–H groups (water molecule) [9]. A minor peak at around 2382 cm^{-1} is due to CO_3^{2-} and HCO_3^- groups [23,24] and a wide intense band at 3464 cm^{-1} is ascribed to the symmetric stretching of O–H groups [9].

After immersion of 1 day and 5 days (Figure 4b,c) in SBF, the spectra showed the presence of additional bands (phosphate structural group) 557 cm^{-1}, 736–738 cm^{-1}, 916–918 cm^{-1}, 1126–1143 cm^{-1}, 1263 cm^{-1}, 1652 cm^{-1}, 1543 cm^{-1}, 2998 cm^{-1}, 3198 cm^{-1}, 3414 cm^{-1}, 3553 cm^{-1}, along with the bands before incubation in SBF. After immersion, the band at 555-557 cm^{-1} represents the HAp typical bond (PO_3^4) of a phosphate group and due to the hydroxyapatite crystallization [25]. The medium bands 736–738 cm^{-1} (P–O–P) can be assigned to the pyrophosphate (P_2O_7)$^{4-}$ group and the bands at around 916–918 cm^{-1} are corresponding to the P–O–P stretching vibrations [26]. The strong bands appearing at 1126–1143 cm^{-1} are attributed to the PO_2 symmetric stretching vibration, which is related to the calcium phosphate surface layer. The lower band at 1263 cm^{-1} is assigned to P=O stretching vibrations/anti-symmetrical vibrations of PO_2^- groups. The 1414 cm^{-1} band is –OH hydroxyl carbonate group, while the band noticed at 1543 cm^{-1} band is the deformation of –OH groups [22,25,27].

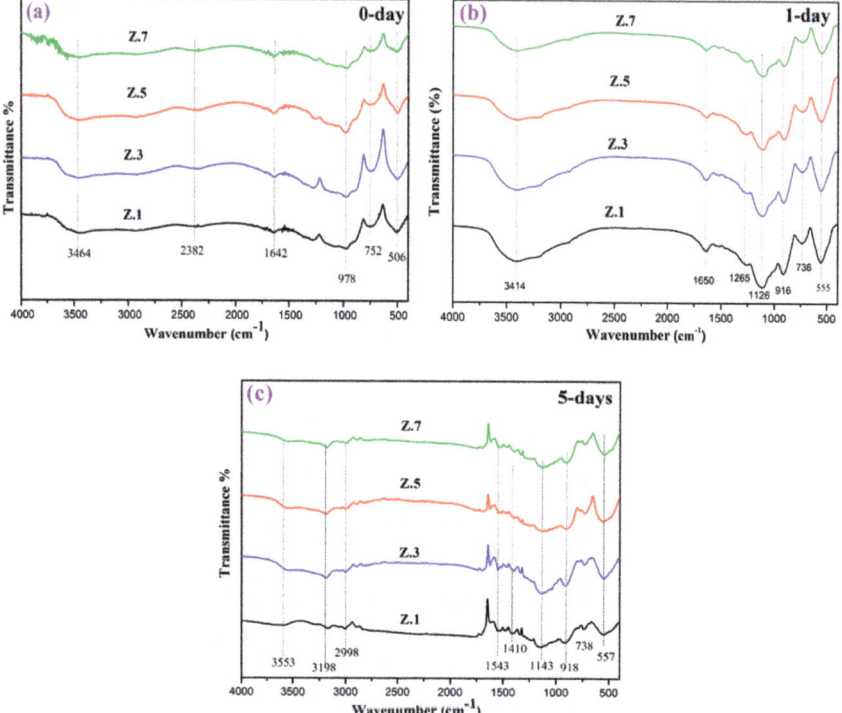

Figure 4. FTIR absorption spectra from the glass surface: (**a**) 0 day before and after (**b**) 1 day, (**c**) 5 days' immersion in several days in SBF.

Table 4. Assignments of various bands from FTIR spectra of the bioglasses.

Wavenumber (cm^{-1})			Assignments	References
0 day	1 day	5 days		
506	555	557	~506 PO_4^{3-} O–P–O bending vibrations/P–O amorphous	[15,16,22]
			~555–557 HAp (PO_3^4)	[25]
752	736	738	~752 P–O–P symmetric stretching	[10,23]
			~736–738 P–O–P pyrophosphate $(P_2O_7)^{4-}$ group	[26]
978	916	918	~978cm^{-1} P–O–P stretching vibrations	[10,23]
			~916–918 P–O–P stretching vibrations	[26]
-	1126	1143	PO_2 symmetric stretching vibration	[22,25,27]
-	1265	-	PO_2^- asymmetric group /P=O stretching vibration	[22,25,27]
-	-	1410	–OH, hydroxyl carbonate group	[22,25,27]
-	-	1543	–OH groups	[22,25,27]
1642	1650	-	~1642–1650 cm^{-1} stretching vibrations of P-O-H group	[9,28]
2382	-	-	P–O–H group /CO_3^{2-} and HCO_3^- groups	[23,24]
-	2998	-	C–H stretching vibrations	[22,29,30]
-	-	3198	C–H	[22,29,30]
3464	3414	3553	~3414–3464 H–O–H bond /CO_3^{2-} and HCO^{3-}	[9,25,29]
			~3553 O–H symmetric stretching	[25,29]

The bands appearing between 1642–1650 cm^{-1} are due to stretching vibrations of P–O–H group [28] and the bands around 3414–3464 cm^{-1} and 3553 cm^{-1} are ascribed to the O–H symmetric stretching [25,29]. The band at 2382 cm^{-1} is assigned to the O–H stretching vibrations of hydrogen-bonded H–O–H groups on the surfaces of the sample. Very small peaks could also be observed at 2998 cm^{-1} and 3198 cm^{-1} due to stretching vibrations in C–H groups [22,29,30]. The entrance of PO_3^{2-}, CO_3^{2-} and OH groups present in the FTIR spectra of the glasses confirmed the growth of the thin crystalline HCA layer. It was also observed that the intensities of the absorption peaks increased with an increase of soaking time, as well as the zirconia inclusion in the glass network [15,31]. Moreover, obtain results correlated with the XRD results. Furthermore, the formation of the HAp layer was confirmed by the SEM-EDS results.

3.4. SEM-EDS Analysis

Figure 5 displays the surface morphology of bioglasses pre- and post-immersion in SBF by SEM-EDS analysis. The micrographs reveal the precipitation over the samples after incubation, indicating the formation of the hydroxyapatite (HAp) layer. Morphologies of the Z.5% (Figure 5a) glass sample before incubation (0-day) in SBF visualize the plane surface by SEM and existence of minimal elements P, Ca, Zn, Na, Zr and O of the glass by EDS, clearly indicating their amorphous nature and lack of any precipitation formation of the samples [6]. After 1 day and 5 days of immersion in SBF, the bioglass surface shows changes in surface morphology (cotton-like structures) and creates the appearance of additional elements (P, Ca, Zn, Na, Zr, O, Cl, K, Mg) besides the authentic glass compositional elements, indicating the deposition of apatite layer and therefore, it can be concluded that the prepared bioglass samples are bioactive. After 1 day, as observed in Figure 5b, the SEM images exhibited small concentrations of precipitation, as the crystalline nature of the apatite layer is identified from XRD analysis. SEM images also have been taken for samples after 5 days of immersion, where the apatite layer on the glass sample surfaces are noticed to be fully covered, and the dense HAp layer (and also from EDS Ca-P ratios data) has been developed (~39 microns thick) [32]. In addition, there is an increment in the Ca and P intensity peaks from EDS after immersion in SBF. The obtained Ca/P ratios of the synthesized glasses are changing from 1.45–1.68 and are very near to human bone Ca/P ratios [12,33]. This is mainly because of Ca^{2+}, Na^+ and phosphate ions releasing from glass surface to SBF solution and transfer of ions from solution to the glass surface. Moreover, it leads to the easy super saturation of Ca^{2+} and PO_4^{3-} ions on the surface of glasses and is favorable for HAp to nucleate and grow [7]. These obtained results are in good agreement with the above mentioned XRD and FTIR data.

3.5. pH Measurement and Weight Loss Studies

From Figure 6, it can be noticed that the pH values are increased for after immersion in the SBF solution for 1 day suddenly, while after, there is a gradual reduction of the pH values for five days of immersion time. The initial raise of pH values of the residual SBF (during the immersion from 0 to three days) is due to the release of alkali/alkaline ions (viz., Na^+, or Ca^{2+}) and even the migration of zirconium species into the SBF that causes the increase of the basicity of SBF. The detected pH reduction in the SBF with the rise in the soaking period is owing to the formation of more concentrated phosphoric acid in SBF and might be the transfer of alkaline Ca^{2+} ions from SBF to the surface of the sample (to form HAp layer) [34]. Moreover, it is observed that the pH of the residual SBF decreases slightly with the increase in content of ZrO_2 and obeys the same trend as that of degradation. Figure 7 illustrates the weight loss of glass samples upon SBF treatment for different time intervals (1 day and 5 days). With an increase in the immersion time, the gradual degradation of glass samples also increased. Therefore, the accurate dissolution rate of Z.1 to Z.7 glass samples' increments is based on the Ca^{2+} and PO_4^{3-} ions from samples, due to their dissolution in the SBF solution. As we can see, the weight loss of each bioglass increased along with the immersion time, but decreases slightly by adding of ZrO_2 up to 0.7 mol% in the phosphate glass matrix and resemble earlier reported literature [35,36]. The slight decrease in the rate of degradation along with increasing ZrO_2 content is observed, which is mainly because of highly established cross-linked dense structures and a reduction of the degradation of the glass. This result implies a strengthening network with the addition of ZrO$_2$ to phosphate glasses [37,38]. These considerable changes occurred in pH, and the weight loss of as-developed bioglasses is necessary for the development of bone-like apatite.

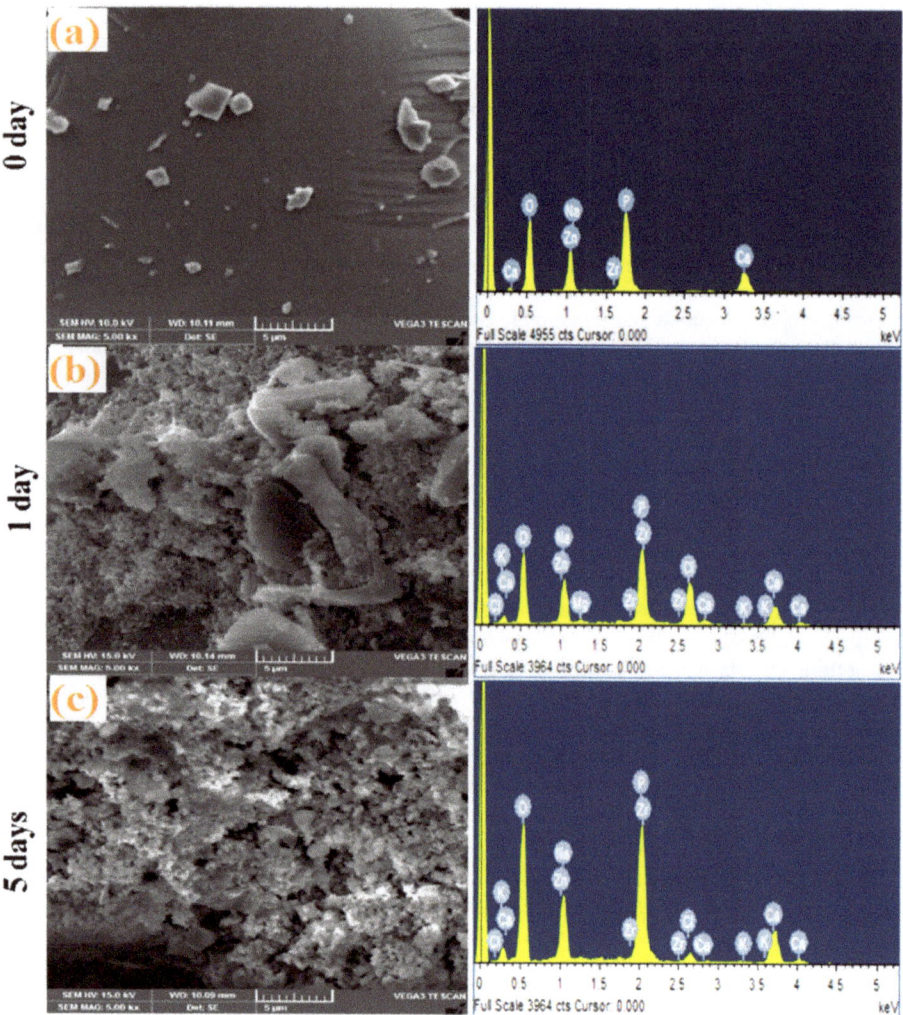

Figure 5. SEM micrograph and EDS analysis of the bioglass sample Z.5 magnifications: for (**a**) 0 day before and (**b**) 1 day, (**c**) 5 days after immersion in SBF solution.

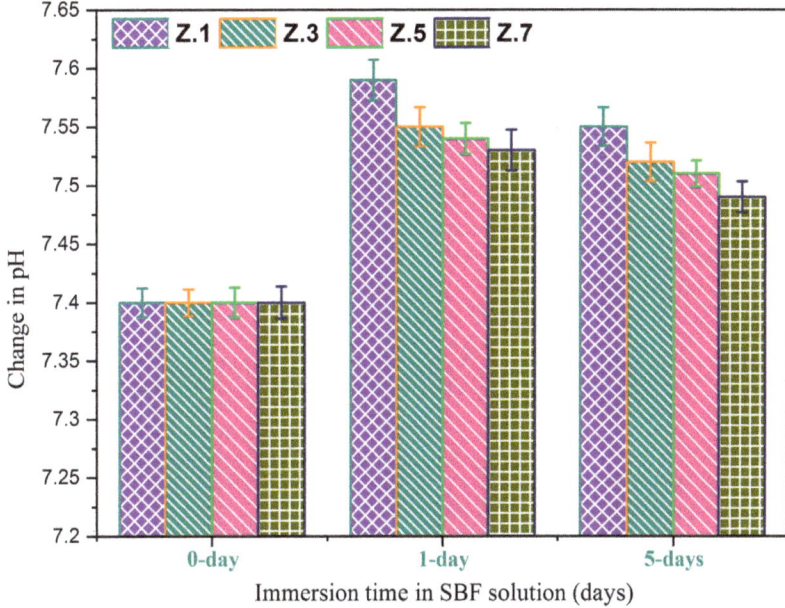

Figure 6. pH variation for all the bioglass samples after immersion in SBF solution.

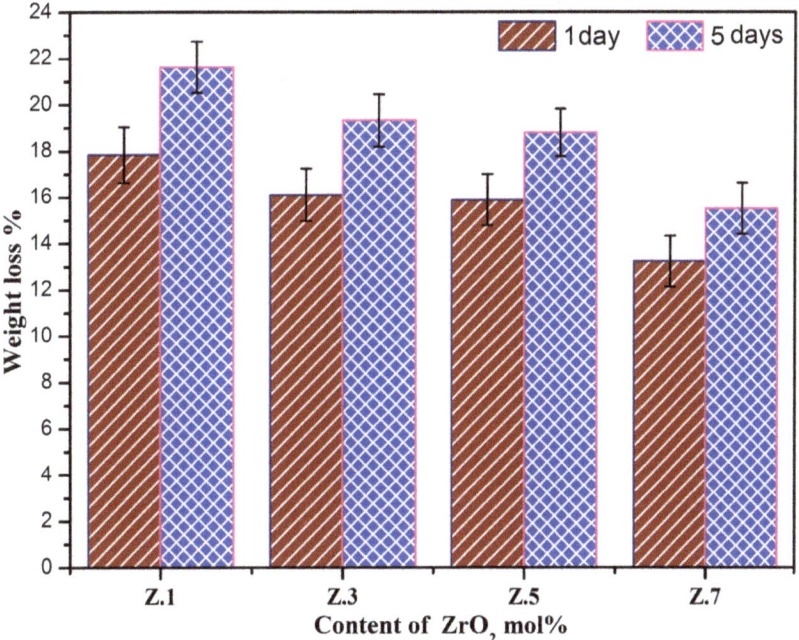

Figure 7. Weight loss of ZrO_2 doped glasses after immersion in SBF for 1 day and 5 days.

4. Conclusions

ZrO$_2$ doped calcium phosphate bioactive glasses were successfully synthesized through the melt-quenching technique. Thermal parameters such as glass stability (ΔT) and Hruby criterion (K_H) values increase with the content of zirconia, which describes the high stability and the good glass-forming tendency of as-prepared glasses. The formation of the hydroxyapatite layer was confirmed by structural studies by means of XRD, FTIR, and SEM. The ratio of Ca and P from EDS is around 1.67, which is almost equal to bone composition and the ability to produce bone-like apatite structures on the surface. The changes that occurred in pH and weight loss of bioglasses with immersion time and zirconia content are desirable for the formation of bone-like apatite. In vitro studies revealed that the ZrO$_2$ incorporated phosphate glasses exhibit high bioactivity relevant for bone tissue engineering applications.

Author Contributions: Conceptualization, P.S.P., M.M.B. and P.V.R.; methodology, P.S.P., M.M.B.; software, M.M.B., S.H.B. and A.P.; validation, P.S.P., P.V.R. and N.V.; formal analysis, M.M.B.; investigation, P.S.P., N.V.; resources, P.S.P.; data curation, P.S.P., M.M.B.; writing–original draft preparation, P.S.P., M.M.B. and S.H.B.; writing–review and editing, P.S.P., M.Ö. and P.V.R.; visualization, P.S.P., and M.Ö.; supervision, P.S.P., P.V.R. and N.V.; project administration, P.S.P. and M.Ö.; funding acquisition, M.Ö. All authors have read and agreed to the published version of the manuscript.

Funding: This research received no external funding.

Acknowledgments: One of the authors thanks CSIR-UGC JRF, New Delhi-India for providing financial assistance and NIT Warangal for providing the necessary facilities via the Centre for Automation and Instrumentation, and Centre for Advanced Materials in the field of research work.

Conflicts of Interest: The authors declare no conflict of interest.

References

1. Hench, L.L.; Splinter, R.J.; Allen, W.C.; Greenlee, T.K. Bonding mechanisms at the interface of ceramic prosthetic materials. *J. Biomed. Mater. Res.* **1971**, *5*, 117–141. [CrossRef]
2. Hench, L.L. Bioceramics: From Concept to Clinic. *J. Am. Ceram. Soc.* **1991**, *74*, 1487–1510. [CrossRef]
3. Kokubo, T.; Kushitani, H.; Sakka, S.; Kitsugi, T.; Yamamuro, T. Solutions able to reproduce in vivo surface-structure changes in bioactive glass-ceramic A-W3. *J. Biomed. Mater. Res.* **1990**, *24*, 721–734. [CrossRef] [PubMed]
4. Bellucci, D.; Bianchi, M.; Graziani, G.; Gambardella, A.; Berni, M.; Russo, A.; Cannillo, V. Pulsed Electron Deposition of nanostructured bioactive glass coatings for biomedical applications. *Ceram. Int.* **2017**, *43*, 15862–15867. [CrossRef]
5. Shirtliff, V.J.; Hench, L.L. Bioactive materials for tissue engineering, regeneration and repair. *J. Mater. Sci.* **2003**, *38*, 4697–4707. [CrossRef]
6. Hench, L.L.; Xynos, I.D.; Polak, J.M.; Hench, L.L.; Xynos, I.D.; Polak, J.M. Bioactive glasses for in situ tissue regeneration. *J. Biomater. Sci.* **2012**, *15*, 543–562. [CrossRef] [PubMed]
7. Kokubo, T.; Takadama, H. How useful is SBF in predicting in vivo bone bioactivity? *Biomaterials* **2006**, *27*, 2907–2915. [CrossRef]
8. Kokubo, T.; Ohtsuki, C.; Sakka, S.; Yamamuro, T. Chemical reaction of bioactive glass and glass-ceramics with a simulated body fluid. *Mater. Med.* **1992**, *3*, 79–83. [CrossRef]
9. Samudrala, R.; Reddy, G.V.N.; Manavathi, B.; Azeem, P.A. Synthesis, characterization and cytocompatibility of ZrO$_2$ doped borosilicate bioglasses. *J. Non Cryst. Solids* **2016**, *447*, 150–155. [CrossRef]
10. Boi, M.; Bianchi, M.; Gambardella, A.; Liscio, F.; Kaciulis, S.; Visani, A.; Barbalinardo, M.; Valle, F.; Iafisco, M.; Lungaro, L.; et al. Tough Adhes. Nanostructured Calcium Phosphate Thin Film. Depos. By Pulsed Plasma Depos. Method. *Rsc Adv.* **2015**, *5*, 78561–78571. [CrossRef]
11. Mollazadeh, S.; Eftekhari Yekta, B.; Javadpour, J.; Yusefi, A.; Jafarzadeh, T.S. The role of TiO$_2$, ZrO$_2$, BaO and SiO$_2$ on the mechanical properties and crystallization behavior of fluorapatite-mullite glass-ceramics. *J. Non Cryst. Solids* **2013**, *361*, 70–77. [CrossRef]
12. Gautam, C.; Joyner, J.; Gautam, A.; Rao, J.; Vajtai, R. Zirconia based dental ceramics: Structure, mechanical properties, biocompatibility and applications. *Dalt. Trans.* **2016**, *45*, 19194–19215. [CrossRef] [PubMed]

13. Mondal, D.; So-Ra, S.; Lee, B.T. Fabrication and characterization of ZrO_2-CaO-P_2O_5-Na_2O-SiO_2 bioactive glass ceramics. *J. Mater. Sci.* **2013**, *48*, 1863–1872. [CrossRef]
14. Lu, X.; Deng, L.; Du, J. Effect of ZrO_2 on the structure and properties of soda-lime silicate glasses from molecular dynamics simulations. *J. Non Cryst. Solids* **2018**, *491*, 141–150. [CrossRef]
15. Kord, M.; Marghussian, V.K.; Eftekhari-yekta, B.; Bahrami, A. Effect of ZrO_2 addition on crystallization behaviour, porosity and chemical-mechanical properties of a CaO-TiO_2-P_2O_5 microporous glass ceramic. *Mater. Res. Bull.* **2009**, *44*, 1670–1675. [CrossRef]
16. Rajkumar, G.; Aravindan, S.; Rajendran, V. Structural analysis of zirconia-doped calcium phosphate glasses. *J. Non Cryst. Solids* **2010**, *356*, 1432–1438. [CrossRef]
17. Zheng, C.Y.; Li, S.J.; Hao, Y.L.; Yang, R. Effect of ZrO_2 on Mechanical and Biological Properties of Calcium Phosphate-Based Glass-Ceramics for Biomedical Applications. *Key Eng. Mater.* **2008**, *368*, 1429–1432. [CrossRef]
18. Sergi, R.; Bellucci, D.; Cannillo, V. A Comprehensive Review of Bioactive Glass Coatings: State of the Art, Challenges and Future Perspectives. *Coatings* **2020**, *10*, 757. [CrossRef]
19. Goel, A.; Rajagopal, R.R.; Ferreira, J.M.F. Influence of strontium on structure, sintering and biodegradation behaviour of CaO–MgO–SrO–SiO_2–P_2O_5–CaF_2 glasses. *Acta Biomater.* **2011**, *7*, 4071–4080. [CrossRef]
20. Wheeler, D.L.; Eschbach, E.J.; Hoellrich, R.G.; Lmontfort, T.M.J.; Chamberland, L. Assessment of Resorbable Bioactive Material for Grafting of Critical-size Cancellous Defects. *J. Orthop. Res.* **2000**, *18*, 140–148. [CrossRef]
21. Sene, F.F.; Martinelli, J.R.; Gomes, L. Synthesis and characterization of niobium phosphate glasses containing barium and potassium. *Proc. J. Non-Cryst. Solids* **2004**, *348*, 30–37. [CrossRef]
22. Terra, J.; Dourado, E.R.; Eon, J.G.; Ellis, D.E.; Gonzalez, G.; Rossi, A.M. The structure of strontium-doped hydroxyapatite: An experimental and theoretical study. *Phys. Chem. Chem. Phys.* **2009**, *11*, 568–577. [CrossRef] [PubMed]
23. Cai, S.; Zhang, W.J.; Xu, G.H.; Li, J.Y.; Wang, D.M.; Jiang, W. Microstructural characteristics and crystallization of CaO-P_2O_5-Na_2O-ZnO glass ceramics prepared by sol-gel method. *J. Non Cryst. Solids* **2009**, *355*, 273–279. [CrossRef]
24. Brauer, D.S.; Karpukhina, N.; O'Donnell, M.D.; Law, R.V.; Hill, R.G. Fluoride-containing bioactive glasses: Effect of glass design and structure on degradation, pH and apatite formation in simulated body fluid. *Acta Biomater.* **2010**, *6*, 3275–3282. [CrossRef] [PubMed]
25. Babu, M.M.; Prasad, P.S.; Venkateswara Rao, P.; Govindan, N.P.; Singh, R.K.; Kim, H.-W.; Veeraiah, N. Titanium incorporated Zinc-Phosphate bioactive glasses for bone tissue repair and regeneration: Impact of Ti^{4+} on physico-mechanical and in vitro bioactivity. *Ceram. Int.* **2019**, *45*, 23715–23727. [CrossRef]
26. Li, H.C.; Wang, D.G.; Hu, J.H.; Chen, C.Z. Crystallization, mechanical properties and in vitro bioactivity of sol-gel derived Na_2O-CaO-SiO_2-P_2O_5 glass-ceramics by partial substitution of CaF_2 for CaO. *J. Sol-Gel Sci. Technol.* **2013**, *67*, 56–65. [CrossRef]
27. Little Flower, G.; Sahaya Baskaran, G.; Srinivasa Reddy, M.; Veeraiah, N. The structural investigations of PbO-P_2O_5-Sb_2O_3 glasses with MoO_3 as additive by means of dielectric, spectroscopic and magnetic studies. *Phys. B Condens. Matter* **2007**, *393*, 61–72. [CrossRef]
28. Babu, M.M.; Venkateswara Rao, P.; Veeraiah, N.; Prasad, P.S. Effect of Al^{3+} ions substitution in novel zinc phosphate glasses on formation of HAp layer for bone graft applications. *Colloids Surf. B Biointerfaces* **2020**, *185*, 110591. [CrossRef]
29. Kalita, H.; Prashanth Kumar, B.N.; Konar, S.; Tantubay, S.; Kr. Mahto, M.; Mandal, M.; Pathak, A. Sonochemically synthesized biocompatible zirconium phosphate nanoparticles for pH sensitive drug delivery application. *Mater. Sci. Eng. C* **2016**, *60*, 84–91. [CrossRef]
30. Lucacel Ciceo, R.; Trandafir, D.L.; Radu, T.; Ponta, O.; Simon, V. Synthesis, characterisation and in vitro evaluation of sol-gel derived SiO_2-P_2O_5-CaO-B_2O_3 bioactive system. *Ceram. Int.* **2014**, *40*, 9517–9524. [CrossRef]
31. Abo-Naf, S.M.; Khalil, E.S.M.; El-Sayed, E.S.M.; Zayed, H.A.; Youness, R.A. In vitro bioactivity evaluation, mechanical properties and microstructural characterization of Na_2O-CaO-B_2O_3-P_2O_5 glasses. *Spectrochim. Acta Part A Mol. Biomol. Spectrosc.* **2015**, *144*, 88–98. [CrossRef] [PubMed]

32. Zhang, Y.; Mizuno, M.; Yanagisawa, M.; Takadama, H. Bioactive behaviors of porous apatite- and wollastonite-containing glass-ceramic in two kinds of simulated body fluid. *J. Mater. Res.* **2003**, *18*, 433–441. [CrossRef]
33. Saldaña, L.; Méndez-Vilas, A.; Jiang, L.; Multigner, M.; González-Carrasco, J.L.; Pérez-Prado, M.T.; González-Martín, M.L.; Munuera, L.; Vilaboa, N. In vitro biocompatibility of an ultrafine grained zirconium. *Biomaterials* **2007**, *28*, 4343–4354. [CrossRef]
34. Balamurugan, A.; Balossier, G.; Kannan, S.; Michel, J.; Rebelo, A.H.S.; Ferreira, J.M.F. Development and in vitro characterization of sol-gel derived CaO-P_2O_5-SiO_2-ZnO bioglass. *Acta Biomater.* **2007**, *3*, 255–262. [CrossRef] [PubMed]
35. Vallet-Regí, M.; Romero, A.M.; Ragel, C.V.; LeGeros, R.Z. XRD, SEM-EDS, and FTIR studies of in vitro growth of an apatite-like layer on sol-gel glasses. *J. Biomed. Mater. Res.* **1999**, *44*, 416–421. [CrossRef]
36. Yin, P.; Yuan, J.W.; Liu, L.H.; Xiao, T.; Lei, T. Effect of ZrO_2 on the bioactivity properties of gel-derived CaO-P_2O_5-SiO_2-SrO glasses. *Ceram. Int.* **2017**, *43*, 9691–9698. [CrossRef]
37. Chen, X.; Meng, Y.; Li, Y.; Zhao, N. Investigation on bio-mineralization of melt and sol-gel derived bioactive glasses. *Appl. Surf. Sci.* **2008**, *255*, 562–564. [CrossRef]
38. Krishnamacharyulu, N.; Mohini, G.J.; Baskaran, G.S.; Kumar, V.R.; Veeraiah, N. Effect of ZrO_2 on the bioactive properties of B_2O_3–SiO_2–P_2O_5–Na_2O–CaO glass system. *J. Non Cryst. Solids* **2016**, *452*, 23–29. [CrossRef]

 © 2020 by the authors. Licensee MDPI, Basel, Switzerland. This article is an open access article distributed under the terms and conditions of the Creative Commons Attribution (CC BY) license (http://creativecommons.org/licenses/by/4.0/).

MDPI
St. Alban-Anlage 66
4052 Basel
Switzerland
Tel. +41 61 683 77 34
Fax +41 61 302 89 18
www.mdpi.com

Materials Editorial Office
E-mail: materials@mdpi.com
www.mdpi.com/journal/materials